PROCEEDINGS
SEVENTH INTERNATIONAL CONGRESS
INTERNATIONAL ASSOCIATION OF ENGINEERING GEOLOGY
VOLUME 1

COMPTES-RENDUS
SEPTIEME CONGRES INTERNATIONAL
ASSOCIATION INTERNATIONALE DE GEOLOGIE DE L'INGENIEUR
VOLUME 1

Comptes-rendus Septième Congrès International Association Internationale de Géologie de l'Ingénieur

5–9 SEPTEMBRE 1994 / LISBOA / PORTUGAL

Rédacteurs
R.OLIVEIRA, L.F.RODRIGUES, A.G.COELHO & A.P.CUNHA
LNEC, Lisboa, Portugal

VOLUME 1

Thème 1 *Développements récents dans les domaines de l'étude des sites et de la cartographie géotechnique*

A.A.BALKEMA / ROTTERDAM / BROOKFIELD / 1994

Proceedings
Seventh International Congress
International Association
of Engineering Geology

5–9 SEPTEMBER 1994 / LISBOA / PORTUGAL

Editors
R.OLIVEIRA, L. F. RODRIGUES, A.G.COELHO & A.P.CUNHA
LNEC, Lisboa, Portugal

VOLUME 1
Theme 1 Developments in site investigation and in engineering geological mapping

A.A. BALKEMA / ROTTERDAM / BROOKFIELD / 1994

INTERNATIONAL ASSOCIATION OF ENGINEERING GEOLOGY
ASSOCIATION INTERNATIONALE DE GEOLOGIE DE L'INGENIEUR

The publication of these proceedings has been partially funded by
La publication des comptes-rendus a été partiellement supportée par

Junta Nacional de Investigação Cientifica e Tecnológica (JNICT), Portugal
Fundação Calouste Gulbenkian, Portugal

The texts of the various papers in this volume were set individually by typists under the supervision of each of the authors concerned.

Les textes des divers articles dans ce volume ont été dactylographiés sous la supervision de chacun des auteurs concernés.

Complete set of six volumes / Collection complète de six volumes: ISBN 90 5410 503 8
Volume 1: ISBN 90 5410 504 6
Volume 2: ISBN 90 5410 505 4
Volume 3: ISBN 90 5410 506 2
Volume 4: ISBN 90 5410 507 0
Volume 5: ISBN 90 5410 508 9
Volume 6: ISBN 90 5410 509 7

Published by:
© 1994 A.A. Balkema, Postbus 1675, 3000 BR Rotterdam, Netherlands (Fax: +31.10.413.5947)
Distributed in the USA & Canada by: A.A. Balkema Publishers, Old Post Road, Brookfield, VT 05036, USA
(Fax: +1.802.276.3837)
Printed in the Netherlands

Publié par:
© 1994 A.A. Balkema, Postbus 1675, 3000 BR Rotterdam, Pays-Bas
Distribué aux USA & Canada par: A.A. Balkema Publishers, Old Post Road, Brookfield, VT 05036, USA
Imprimé aux Pays-Bas

SCHEME OF THE WORK
SCHÉMA DE L'OUVRAGE

INTERNATIONAL ASSOCIATION OF ENGINEERING GEOLOGY
ASSOCIATION INTERNATIONALE DE GÉOLOGIE DE L'INGÉNIEUR

EXECUTIVE COMMITTEE/COMITÉ EXÉCUTIF

President/Président	*Prof. Ricardo Oliveira*	Portugal
Secretary General/Secrétaire Général	*Dr Louis Primel*	France
Treasurer/Trésorier	*Dr Philippe Masure*	France
Vice-Presidents/Vice-Présidents:		
Africa/Afrique	*Dr Ben Dhia*	Tunisia
North America/Amérique du Nord	*Mr L. Espinosa-Graham*	Mexico
South America/Amérique du Sud	*Mr Guido Guidicini*	Brazil
Asia/Asie	*Prof. Keiji Kojima*	Japan
Australasia/Australasie	*Mr John Braybrooke*	Australia
Eastern Europe/Europe de l'Est	*Dr Kiril Anguelov*	Bulgaria
Western Europe/Europe de l'Ouest	*Prof. Paul Marinos*	Greece
Past President/Ancien Président	*Dr Owen White*	Canada

CONGRESS ORGANIZING COMMITTEE/COMITÉ D'ORGANISATION DU CONGRÈS

Chairman/Président	*Ricardo Oliveira,*	LNEC and UNL, National Civil Engineering Laboratory and New University of Lisbon
Vice-Chairman/Vice-Président	*L. Fialho Rodrigues*	LNEC, National Civil Engineering Laboratory
Secretary General/Secrétaire Général	*A. Gomes Coelho*	LNEC, National Civil Engineering Laboratory
Treasurer/Trésorier	*P. Sêco e Pinto*	LNEC, National Civil Engineering Laboratory
Members/Membres:	*Manuel Abrunhosa*	FC/UP, Faculty of Sciences/University of Oporto
	Manuel Barroso	LNEC, National Civil Engineering Laboratory
	J.A. Rodrigues de Carvalho	FCT/UNL, Faculty of Sciences and Technology/New University of Lisbon
	Rui Correia	LNEC, National Civil Engineering Laboratory
	A. Pinto da Cunha	LNEC, National Civil Engineering Laboratory
	J. Moura Esteves	LNEC, National Civil Engineering Laboratory
	M. Matos Fernandes	FE/UP, Faculty of Engineering/University of Oporto

M. Quinta Ferreira	FCT/UC, Faculty of Sciences and Technology/University of Coimbra
C. Dinis da Gama	IST/UTL, Technical University of Lisbon
Nuno Grossman	LNEC, National Civil Engineering Laboratory
Fernando Ladeira	UA, University of Aveiro
J.A. Azeredo Leme	JAE, Road Directorate
A. Janela Leitão	INAG, Institute of Water
Celso Lima	EDP, Electricity of Portugal
E. Maranha das Neves	LNEC and UNL, National Civil Engineering Laboratory and New University of Lisbon
Manuel de Oliveira	FC/UL, Faculty of Sciences/University of Lisbon
A. Costa Pereira	CÊGÊ Lda
J. Delgado Rodrigues	LNEC, National Civil Engineering Laboratory
Henrique Silva	LNEC, National Civil Engineering Laboratory

Foreword

At the 6th International Congress of the IAEG which took place in Amsterdam in 1990, the Council of the Association accepted the formal invitation of the Sociedade Portuguesa de Geotecnia – The Portuguese National Group of IAEG – to host the 7th International Congress in Lisboa, Portugal. The date was set for 5th to 9th September 1994.

The proposed motto for the 7th Congress was 'Turning the Century with Engineering Geology' and the subjects were chosen to highlight the importance and the wide range of interests and the scientific and technic activity of Engineering Geology.

Congresses are the appropriate occasion to gather some of the most significant and challenging topics in order to create a forum for engineering geologists coming from different countries and working in different areas to meet and discuss their points of view.

In fact, at the end of this century increasing environmental problems and development of large cities with the construction of large engineering works at surface and underground, and with mining activities, will require very accurate engineering geological assessment of the ground. Developing countries must experience a fast economic growth and improvement of the quality of life of their population, but this should be done in a sustainable manner which also demands a good standard of engineering geology practice. Restrictions on the exploitation of geological formations for construction materials will also increase, this calling for innovation and ingenuity which will also require new technological developments.

Information technologies are changing the world and certainly also the world of Engineering Geology. At the end of the century many new tools will be currently available to better understand and analyse the ground and the associated phenomena. Better teaching and training of Engineering Geology will be required in order to cope with all these problems.

The themes and workshops of this Congress were selected by the Organizing Committee, in consultation with the Executive Committee of the IAEG, aiming at a comprehensive exchange of knowledge and experiences. The six themes and two workshops are the following:

1. Developments in site investigation and in engineering geological mapping;
2. Engineering geology and natural hazards;
3. Engineering geology and environmental protection;
4. Construction materials;
5. Case histories in surface workings;
6. Case histories in underground workings.

Workshop A – Information technologies applied to engineering geology;
Workshop B – Teaching and training in engineering geology.
                    Professional practice and registration.

The appropriate selection of the themes called a large number of engineering geologists all over the world and more than 700 papers were received by the Organizing Committee spread all over the themes and topics of the Congress.

The screening of the papers was made by a special committee of the organization. A number of these papers could not be accepted for publication in the proceedings which include the significant number of 641 papers.

For each theme a recognized speaker was invited to present a lecture on a relevant topic of the theme and for each workshop a specialist was invited to coordinate the subject and to present his views on the advances of the matter. All these special contributions are included in the Congress volumes which will be distributed to the participants before the beginning of the Congress.

Ricardo Oliveira
Chairman of the Organizing Committee

Avant-propos

Au 6ème Congrès International de l'AIGI, organisé à Amsterdam en 1990, le Conseil de l'Association a accepté l'invitation formelle de la Sociedade Portuguesa de Geotecnia – Le Groupe National Portugais de l'AIGI – pour organiser le 7ème Congrès International à Lisboa, Portugal. La date retenue a été la semaine du 5 au 9 Septembre 1994.

La devise du Congrès 'Au seuil du 21 ème siècle avec la Géologie de l'Ingénieur' et les thèmes ont été choisis de façon à démontrer l'importance et la diversité des activités scientifiques et techniques de la Géologie de l'Ingénieur.

Les Congrès Internationaux sont les occasions appropriées pour discuter ensemble des thèmes de pointe en appelant l'attention d'un forum d'ingénieurs géologues venant de plusieurs pays et travaillant dans des domaines assez différents et les faire présenter et discuter leurs points de vue.

En effet, à la fin de ce siècle, les problèmes de l'environnement et le développement des grandes villes associés à la construction de grands travaux en surface et en souterrain et à l'activité minière demanderont une évaluation précise des sites.

Les pays en développement ont besoin d'établir une économie plus solide et d'améliorer la qualité de vie de leurs citoyens et, en même temps, assurer un développement durable. Cela demandera aussi des prestations de bon niveau dans la pratique de la Géologie de l'Ingénieur.

Des restrictions dans l'exploitation des formations géologiques utilisécs comme matériaux de construction seront aussi croissantes, situation qui obligera à faire appel à l'innovation et à l'ingéniosité, ce qui exige le développement de nouvelles méthodes d'étude et nouvelles technologies. L'informatique est en train de changer le monde où nous vivons. À la fin du siècle beaucoup de nouvelles technologies informatiques seront à la disposition de la Géologie de l'Ingénieur et permettront une connaissance plus détaillée des propriétés des terrains et de leur stabilité.

Il faut un enseignement et une formation de qualité en Géologie de l'Ingénieur afin d'obtenir des résultats adaptés à la solution de toutes ces questions.

Les thèmes et les ateliers de ce Congrès ont été choisis par le Comité d'Organisation en consultation avec le Comité Exécutif de l'AIGI dans le but d'assurer un échange approfondi de points de vue et d'expériences.

Les six thèmes et les deux ateliers sont les suivants:
1. Développements récents dans les domaines de l'étude des sites et de la cartographie géotechnique.
2. La Géologie de l'Ingénieur et les risques naturels.
3. La Géologie de l'Ingénieur et la protection de l'environnement.
4. Matériaux de construction.
5. Études de cas dans les travaux de surface.
6. Études de cas dans les travaux souterrains.

Atelier A: L'informatique appliquée à la Géologie de l'Ingénieur.

Atelier B: Enseignement et formation en Géologie de l'Ingénieur. Pratique professionnelle et qualification.

Le choix approprié des sujets a attiré l'attention d'un très grand nombre de scientifiques et d'ingénieurs de tout le monde et, en conséquence, plus de 700 articles ont étés reçus par le Comité d'Organisation relatifs à tous les thèmes du Congrès. La sélection des articles a été faite par un comité de sélection du Congrès qui a rejeté environ 10% des articles reçus.

Pour chaque thème l'organisation a invité un conférencier renommé à faire une conférence sur un des sujets du thème respectif et pour chaque atelier l'organisation a invité un spécialiste pour coordonner le sujet et pour présenter ses points de vue sur les développements récents et à venir. Toutes cettes conférences spéciales sont publiées dans les volumes des compte-rendus qui seront distribués à tous les participants avant le Congrès.

Ricardo Oliveira
Président du Comité d'Organisation

Author index (volumes 1,2,3,4,5,6)
Index des auteurs (volumes 1,2,3,4,5,6)

VOLUME 2

XXI

XXVIII

VOLUME 5

4 Construction materials
Matériaux de construction

VOLUME 6

9 *Supplement*
Supplément

VOLUME I

Keynote lecture: Some aspects of modern engineering geophysics
Conférence spéciale: Quelques applications récentes de la géophysique de l'ingénieur

Lucien Halleux
University of Liège, LGIH & G-TEC, Spa, Belgium

ABSTRACT : The technical part of the keynote lecture deals mainly with engineering applications of recent developments in borehole geophysics : borehole radar (tomography, dipole reflection, directional reflection) and seismic tomography. Logging and some surface methods (Time Domain EM, shallow seismics, GPR) are also briefly commented. The problem of quality in engineering geophysics is analysed in the second part of the paper.

RESUME : La partie technique de l'exposé décrit certaines applications récentes de la géophysique de l'ingénieur en forage : radar en forage (tomographie, réflexion dipôle, réflexion directionnelle) et tomographie sismique. Les diagraphies et quelques méthodes de surface (EM en domaine temporel, sismique superficielle et radar) sont présentées brièvement. La deuxième partie de l'exposé analyse le problème de la qualité en géophysique de l'ingénieur.

1 INTRODUCTION

Engineering geophysics, together with the closely related domains of environmental and ground water geophysics, is emerging as a separate branch within the earth sciences, as demonstrated by new specialised journals (e. g. Applied Geophysics), Societies (e.g. Environmental and Engineering Geophysics Society, EEGS), or sections of larger geophysical Societies (e.g. Near Surface Geophysical Section of the SEG). This trend is justified considering that traditional geologists or geophysisicists are no longer able to cope with the very wide range of methods and problems related to shallow geophysics, unless they have extensive training or experience in this field.

The importance of maintaining close ties with both geophysicists and engineering geologists is however evident :
  - understanding the possibilities and limitations of each method is only possible with a good theoretical knowledge of geophysics ;
  - choosing the right method or combination of methods, and integrating geophysics with other exploration techniques, requires good engineering geological skills .

As a consequence of his multidisciplinary activities, the engineering geophysicist is often criticised as lacking theoretical knowledge or competence by the specialists of each discipline. In order to infirm such critical comments, top ranking priority should be given to the quality of engineering geophysical work, without succumbing to bureaucratic quality control. These problems will be addressed in the second part of the lecture.

The first part of the lecture deals with recent technical developments. Main emphasis will be given to borehole geophysics while surface methods will only be briefly commented.

2. BOREHOLE GEOPHYSICS

Logging techniques have been used for over half a century in oil exploration, and for twenty years at least in engineering geology (Monjoie 1975) and are thus classical borehole geophysics. They provide very

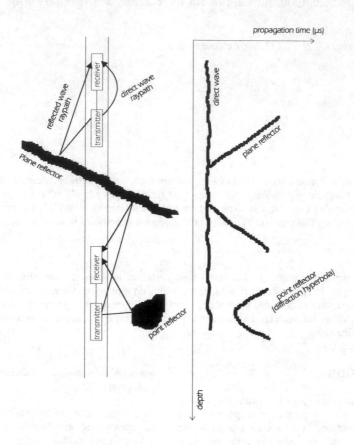

Borehole

Radar section

Figure 1 : principle of borehole radar reflection survey (after ABEM)

accurate, high resolution informations about the immediate surroundings of the boreholes, but have no significant penetration beyond the borehole wall.

As such, they should be considered as geophysical measurements, rather than geophysical exploration techniques.

Engineering geologists, unlike many geotechnical engineers, have always been aware that the reconstruction of a realistic subsurface model from pin pointed borehole data is a difficult, and dangerous task. Borehole geophysical methods able to explore the rockmass itself, with an investigation radius of several tens of meters may therefore be considered as major breakthroughs in exploration techniques.

Some of these methods will be reviewed and illustrated with case histories.

2.1 *Logging techniques*

As mentioned earlier, logging has become a classical technique and will be reviewed very briefly.

In oil geophysics, logging is usually carried out from mud filled holes, allowing direct contact or coupling between the probes and the borehole wall. Therefore, a very wide range of techniques could be developed, based on electrical, acoustic, nuclear and other measurements. In engineering geology, most holes are drilled without mud, above or below the water table, in soils or weathered bedrock which have to be supported with a casing. If a PVC casing is used, following logs are usually run :

- natural radioactivity or gamma ray, giving a qualitative measure of clay content ;
- gamma gamma giving an excellent in situ measure of density, and neutron giving a measure of water content. It must be stressed that the quantitative interpretation of these logs is less accurate or even impossible in a PVC cased hole because the probes cannot be put in direct contact with the borehole wall. In addition the PVC influences the neutron reading.
- sonic (under the water table only), giving the sonic velocity, very similar to seismic velocity, an excellent indicator of compacity ;
- electromagnetic induction logging, giving the electrical conductivity.

Logging techniques are characterised by very good vertical resolution and low penetration. The variety of geophysical parameters which may be acquired adds significantly to the quality of the engineering geological or geotechnical descriptions of the boreholes, provided they are interpreted correctly. Logging costs are normally low compared to drilling costs : depending upon the tools or combination of tools which are run, costs usually range from 3 $ to 15 $ per meter and log.

In addition to the valuable geophysical information, logging often contributes to the quality control of the drilling itself : discrepancies between core description and logs may often be traced back to inverted or wrong positions of the cores in the boxes.

Logging may even play the key role in the engineering geological interpretation of boreholes. A typical example is the detection of old mining works : in the Liège area, coal mining at shallow depth started around the 16th century, but mining plans are only available since the early 19th century. Intersection of such old works is a major problem for tunnel construction in the area, due to both stability problems and water inflow hazard. The absence of core recovery may indicate the presence of old works, but may also be due to other geological factors and poor drilling quality like bad coring of a gouge filled fracture or of a coal layer. Combined gamma ray and gamma gamma logging provides clear evidence for the interpretation of such zones.

2.2 *Borehole radar reflection techniques*

Dipole transmitter and receiver antennas are lowered in the same borehole and displaced stepwise along the measured interval as shown on figure 1. For each position, an electromagnetic impulse (usual central central frequencies are 22 Mhz or 60 Mhz) is generated by the transmitter antenna. Discontinuities in the rock or soil around the borehole, like contacts between layers, voids, fractures, reflect part of the impulse back to the receiver. The resulting signals are displayed in a way similar to reflection seismics, giving an image of the surrounding rockmass. Like in seismics, point reflectors are characterized by a hyperbolic diffraction pattern.
The investigation range mainly depends upon the electrical resistivity. It may be estimated as follows :

- a few meters or even less in clayey or silty

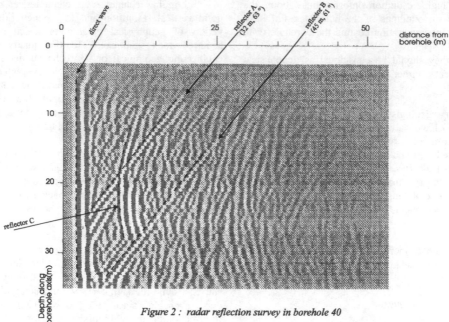

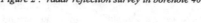

Figure 2 : radar reflection survey in borehole 40

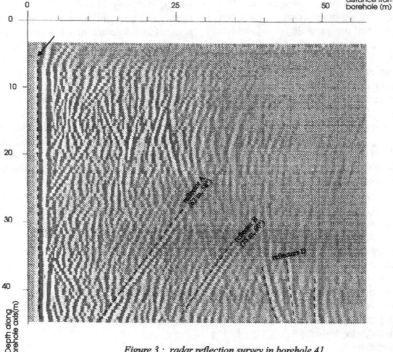

Figure 3 : radar reflection survey in borehole 41

ground : in most cases the method cannot be applied under such circumstances ;

    - 10 to 40 meters in normally fractured rock ;
    - 40 to 150 meters in massive rock ;
    - up to 300 meters and more in some exceptional situations (pure rocksalt, very massive limestone or granite, ice, ...).

In the reflection mode, borehole radar essentially gives a high resolution image of the rockmass, showing the geometry of plane structures which may, or not, intersect the borehole (contact between layers, thin marker beds, fractures), or showing the presence of local features around the borehole (voids, lenses, ...).

If normal dipole antennas are used, the image is not oriented, which means that the direction to a given reflector remains unknown, unless other geological data is available (e.g. strike of the stratification, ...). The most recent development of borehole radar includes directional receiver antennas which record reflections from all directions, but at the same time allow the determination of the azimut to each reflector. In this way, a full 3 D image of the rockmass around the borehole, within the investigation radius, may be reconstructed.

The method has already been used successfully in a wide range of engineering geological applications : investigation of rocksalt layers prior to dissolution mining for underground gas storage at depths up to 1600 m (Clerckx et al. 1994), investigation of underground repositories at intermediate depth (Olsson et al., 1992), structural exploration for tunnelling projects, ...

A selected example from the HST railway link between Paris and Cologne will be briefly presented. The project includes the 6 km long Soumagne tunnel, close to Liège (Belgium). In the investigated area, the tunnel crosscuts carboniferous limestone, with frequent karstic voids and faults. A borehole radar survey has been planned in order to detect such voids or faults around exploration boreholes, and to help correlating between boreholes.

The survey was carried out in the reflection mode, using 60 Mhz antennas. Figures 2 and 3 show the raw data for boreholes 40 and 41 with indication of the most prominent reflectors.

In borehole 40, two thin clay layers (stratigraphic marker beds) were intersected at a depths of 32 and 45 meters. The radar survey clearly shows them (reflectors A and B, fig. 2) . The intersection angle with the borehole can easily be determined (63° and 61°). A prominent point reflector, characterized by a typical hyperbolic reflection (C) is easily observed. Depth is 22 m, distance from borehole is 10 m. It is most probably due to a karstic cavity.

In borehole 41, the clay layers were not intersected. However the radar survey shows them (A and B, figure 3), intersecting the borehole axis at a depth of 62 and 75 m, i.e. below the drilled interval. The reflectors are observed out to a distance of at least 35 m, showing that the geological structure is regular (no faults or folds in the vicinity of the borehole). Reflectors D are most probably due to a major karstic cavity located 50 m away from the borehole.

2.3 *Radar Tomography*

The principle of crosshole radar tomography is illustrated on figure 4.

For crosshole surveys, two boreholes are required, one on each side of the investigated area. A transmitter antenna is lowered in one borehole, a receiver antenna in the other one. The transmitter radiates a very short EM impulse with a central frequency of 22 or 60 MHz. For each possible position of the transmitter and the receiver, the signal at the receiver is recorded. Each signal corresponds to one raypath between the boreholes. Along each raypath the propagation time and the amplitude are determined as shown on figure 5.

The area between the boreholes is discretized as cells or pixels. Using all the propagation times, it is possible to compute the radar velocity distribution between the boreholes. This procedure is called the tomographic inversion. Several methods have been developed. The distribution of radar attenuation is determined in a similar way, using the amplitude information. The results are usually displayed as gray scale or colour sections representing the velocity or attenuation.

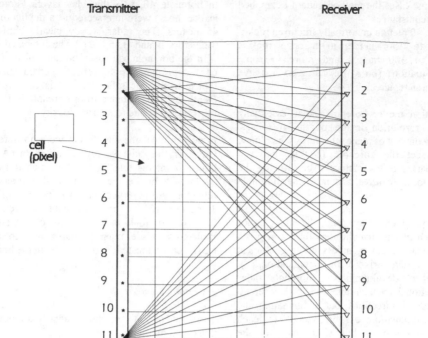

Figure 4 : principle of borehole radar tomography

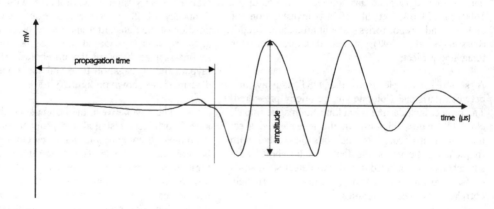

Figure 5 : radar signal

It must be stressed that there are several limitations inherent to the method itself, and to the interpretation technique :
- the resolution of the method is a complex function of wavelength, shot and receiver point spacing, distance between boreholes. Usually the resolution is of metric order. It is important to take this order of magnitude into account when comparing or correlating tomographic results with geotechnical data ;
- a layer parallel to the boreholes cannot be identified properly and will result in typical artefacts ;
- the interpretation assumes a bidimensionnal structure. Variations along the third dimension are not taken into account although they may influence the results ;
- the inversion procedure may be based upon a straigh or curved ray hypothesis.

Borehole radar tomography and reflection are totally distinct methods. The main differences in the resulting images are :

- the reflection mode gives an accurate description of the geometry, while characteristics can only be estimated ;
- in the tomographic mode two independent characteristics (radar wave velocity and attenuation) are determined, but the geometry is less accurate due to lower resolution.

The effective resistivity may be calculated from the radar attenuation. It is lower than the usual DC resistivity, but may be interpreted in a similar way for engineering geological purpose. It is mainly an indicator of clay content and of water mineralization.

Radar velocity (v) is directly related to the relative permittivity (ε_r) or dielectrical constant : $v=300/\sqrt{\varepsilon_r}$ (m/µs). It is a much less usual parameter for engineering geologists, although it is probably one of the most interesting geophysical characteristics for geotechnical applications. The permittivity is a measure of the electrical polarisability of the medium. For engineering geological interpretation, three constitutents must be considered : air in the non saturated zone ($\varepsilon_r=1$), minerals (mainly clay minerals, silica and calcite with $\varepsilon_r = 5$ to 9), and water ($\varepsilon_r=81$). The high value of ε_r for water is due to the dipole nature of the water molecule. It is thus obvious that water content is the most important factor controlling the permittity. Conversely, the water content may be estimated or even computed from the permittivity, whence its importance for engineering geological applications.

Although the method offers good possibilities for engineering applications, only few examples are documented in the litterature, for underground repositories (Ollson et al. 1992), quarry investigations (Siggins, 1992) or foundation studies (Corin et al., 1994).

A combined example of application of seismic and radar tomography will be presented in the next chapter.

2.4 *Seismic tomography*

From a geometrical point of view, radar and seismic tomography are nearly identical. For crosshole surveys, a seismic source is lowered in one borehole, seismic receivers in another one. For each source and receiver position, a seismic wave is generated and recorded, in order to achieve a good coverage of the investigated area between the boreholes. The propagation times are inverted to produce a section showing the seismic velocity.

Obviously, the physics of wave propagation in radar and seismic tomography are totally different. In seismics, either compressional (P) or shear (S) waves are used. S waves give valuable informations on dynamic soil characteristics but are much more inconvenient and expensive to use, because a mechanical coupling between the source or the receiver and the rockmass is necessary, which requires cementation of the casing and clamping of the devices. For P waves, below the water table, coupling is possible through the water and the PVC casing. This allows to use powerful sources (explosive, sparker, air gun) and simple receivers (hydrophones), without clamping, resulting in very fast measuring procedures. A promising technique is to use P wave systems and try to identify converted S wave, thus combining the advantages of both methods.

Radar and seismic tomography were used for foundation studies along the HST railway link Paris-Cologne. The example described concerns the 2 km long Arbre viaduct (western part of Belgium) which will be founded on heavily karstified limestone. Scope of the tomographic survey was to localize karstic voids under the future piers and to determine

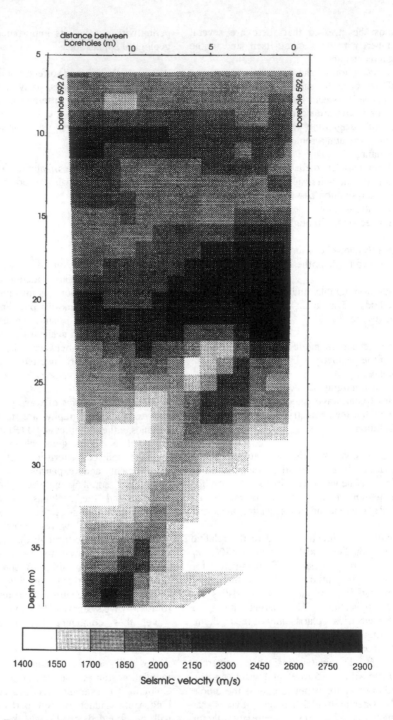

Figure 7 : seismic velocity tomogram between boreholes 592 A and 592 B

distance between boreholes (m)

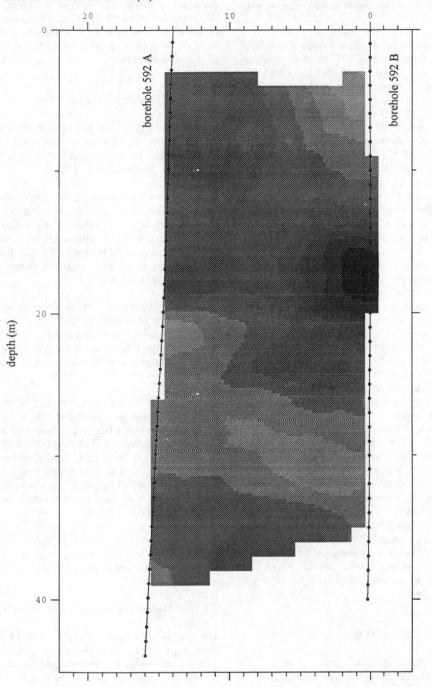

Figure 8 : radar velocity tomogram between boreholes 592 A and 592 B

whether they were filled or not. For each pier, two holes, about 40 m deep, were drilled, one on each side of the pier.

Figure 7 shows the result of the seismic tomography below the future pier 592. Spacing between boreholes is 14 m. The water table is located 5 m below surface. Seismic velocity usually ranges from 1800 to 2500 m/s, indicating fractured limestone. Velocities up to 2900 m/s are observed locally (around 20 m, 592B), showing more massive limestone. In the lower part of the borehole, from 25 m to 35 m (592 B), the velocities are distinctly lower. This area corresponds to a karstic cavity. The lowest velocities are around 1400 m/s, which may indicate that the cavity is either "void" (i.e. filled with water), or filled with loose material like silty sand. This ambiguity had to be solved, in order to plan appropriate grouting techniques.

Figure 8 shows the result of the radar tomography carried out from the same boreholes. The smoother aspect of the radar tomogram is due to different processing techniques. The smoothing results in a more "pleasant" aspect, but adds nothing to the results themselves. The geometry shown by both methods is fairly similar. Absolute similarity should not be expected since the methods are very different. As mentioned previously, radar velocity is a function of permittivity, itself mainly dependent upon volumetric water content (i.e. porosity under the water table). From 5 to about 20 m, the radar velocities range from 105 to 85 m/μs which corresponds to a computed porosity from 0 to 6 %, indicating fractured limestone. In the karstic cavity, from 20 to 32 m (592 A) and from 25 to 35 m (592 B), the radar velocity is about 75 to 65 m/μs corresponding to a porosity of 10 to 20 %. The effective resistivity, deduced from the radar attenuation tomogram (not shown) is around 60-80 Ω .m. These values indicate that the cavity is filled with silty sand, probably containing some blocks of limestone. An empty cavity would have been characterized by a radar velocity of 30 m/μs (i.e. velocity of water).

3. SURFACE GEOPHYSICS

Recent developments in surface engineering geophysics mainly concern new applications of techniques which have been used during several decades in other branches of geophysics, rather than the developments of truly new methods. This situation is partly due to the evolution of the equipment which has brought those techniques down to an affordable price for low budget engineering surveys. Only some of these methods will be briefly commented hereafter.

3.1 *Ground Penetrating Radar (GPR)*

The principle and applications of GPR are familiar to many engineering geologists and will not be discussed here. Descriptions of the technique may be found in e.g. Forkmann et al. (1989).

Radar exploration has been used in salt mines since the sixties (Unterberger,1978). The first commercial GPR systems became available in the seventies and have since been used extensively in engineering geological practice. GPR may be described as giving extremely good results in an extremely limited number of applications.

The main limitation of GPR is obviously related to the presence of clayey or silty conductive soils, resulting in complete attenuation of radar signals. This difficulty is inherent to the soil itself, and is unlikely to be overcome by technical evolution of the equipment.

The technique remains however very attractive for engineering geology whenever high resolution images of the subsurface are required in a not to conductive environment. The range is usually limited to a few meters in soils, up to several tens of meters in rocks. The method is unsensitive to vibrations, electrical or electromagnetic background, which makes it very well suited in an urban, quarry or mining environment.

Modern equipments are PC based and very lightweight. The commercialy available antennas have frequencies ranging from 10 MHz to 2.5 GHz.

3.2 *Time Domain Electromagnetics (TDEM)*

The method has been used for over 30 years in mining geophysics. It is based upon the decay of eddy currents induced in the ground by interrupting a current circulating in a loop at the surface. The eddy currents produce a secondary magnetic field detected by a receiver coil. Curves similar to resistivity sounding curves are obtained and inverted to give the vertical succession of resistivities.

Compared to conventional resistivity sounding TDEM is much faster. An additional advantage is the small size of the surface array, compared with investigation depth. This results in a better horizontal resolution than resistivity. Like most EM methods, TDEM is sensitive to the presence of electrical interferences, cables, fences, ...

3.3 Shallow reflection seismics

The most widely and nearly exclusively used method in oil geophysics was made available for shallow exploration due to the technical evolution of the equipment and to the computing power of modern PC's.

Excellent results have been obtained for many engineering applications (Steeples 1990). Good results may be obtained in a variety of situations, one of the most favourable targets being the detection of the bed rock at a depth exceeding 20 m. For shallower depth, refraction seismics is usually preferred.

4. PRESENT TRENDS AND PERSPECTIVES

The R & D efforts related to engineering geophysics show promising evolutions in at least four important fields.
   -Geophysical equipment is rapidly improving : dynamic range, acquisition rate, storage and processing capacity keep increasing, while cost decreases.
   -Processing software is being developped to allow more complex and realistic ground models to be obtained ;
   - Complex geophysical characteristics of the subsurface are being better understood. The best example probably come from the electrical characteristics. The frequency dependance of the real and imaginary parts of the conductivity may be interpreted in terms of pore and fluid characteristics (Börner et al. 1993) ;
   - New methods or combinations of methods are being applied. Nuclear Magnetic Resonance (NMR) is an interesting example for hydrogeology and engineering geology since it is directly sensitive

to the presence of water in the subsurface (Goldman et al. 1994).

5. QUALITY CONTROL IN ENGINEERING GEOPHYSICS

The once very fashionable concept of quality control (QC) becomes more and more associated with bureaucratic procedures like endless forms whose ultimate goal seems to transform the easiest technical task into an insurmountable administrative difficulty.

It is interesting to quote following recent comment about a proposed ASTM standard on refraction seismics : " ...the standard is quite general (...) . Basically it says to do the survey properly. It's real effect will be as an aid to mitigating bureaucracy. Now you can say to the "quality control" representative that you are conducting the survey according to the ASTM standard. He will be happy and you can do your job." (in : Newsletter of the NSG section of SEG, January 1994).

Even if such negative appreciations of QC are sometimes justified, the problem of quality in engineering geophysics is of crucial importance. This is easily demonstrated by the very bad reputation engineering geophysics still has among some potential clients. Although it is easy to justify such an attitude by invoking the client's ignorance, it is far more realistic to admit that it is frequently due to previous negative experiences with geophysics.

One of the most basic problems is probably the lack of professionalism of many engineering geophysicists due to the absence of adequate training, to restricted experience and to the wide range of techniques concerned. The "ideal" engineering geophysicist should be able to deal competently with all stages of a complete survey :
   - planning of the campaign and quoting, which requires proper understanding of the engineering problems to be solved and of the available geophysical techniques ;
   - realisation of the survey which requires technical expertise ;
   - processing of the results which requires familiarity with the theoretical aspects of each method, and a cautious use of processing softwares. Fully automatic processing softwares used by inexperienced personnel are probably the greatest

single threat to quality ;
    - engineering geological interpretation of the geophysical data and reporting.

Poor quality of any stage will irremediably affect the quality of the final result, and damage the reputation of geophysics.

In oil geophysics, each stage is handled by a specialist, or a group of specialists. This approach is impossible in engineering geophysics due to the small size of most surveys and limited budget. Accordingly, the engineering geophysicist, no matter whether he was originally trained as geologist, geophysicist or engineer, must acquire the relevant knowledge and experience in the other fields. He must of course be aware of the possibilities and limitations of each method, but also have enough restraint to admit his own limitations.

Considering the importance of shallow geophysics in many fields today, it is also time to revise academic training. Some universities already propose specialised degrees. This trend should be encouraged, but the basics of shallow geophysics should also be taught to civil engineers, architects, geologists, archaeologists, ... In this way, future users would have a better ability to integrate geophysics in preliminary surveys. The quality of the geophysical part of call for tenders might also be improved.

Codes of practice or books dealing with both technical and contractual aspects of shallow geophysics in a particular field (e.g. Vogelsang, 1993) can also play a helpful role, especially for the clients.

The client also bears part of the responsibility for the final quality. Choosing systematically the lowest bidder does not encourage highest quality, while organising an unbureaucratic, efficient quality control of field work and interpretation, together with some flexibility, always has very positive effects.

6. CONCLUSIONS

New engineering geophysical methods and new insight into the meaning of geophysical parameters, together with the evolution of equipment, computers and software are continuously broadening the range of applications and improving the accuracy of the results. In particular, borehole geophysics is being revolutionised by techniques investigating the rockmass within a radius of several tens of meters, thus adding a truly geological dimension to the pin pointed borehole information. A similar situation exists in surface geophysics.

Academic training alone is not sufficient to keep pace with the rapid evolution of engineering geophysics, so a major effort is required from the practitioner to continuously update his knowledge and experience, in order to guarantee a high quality service, essential for the reputation of our activity.

ACKNOWLEDGEMENTS

We thank TUCRAIL SA, the engineering company in charge of the HST project in Belgium for allowing us to use data from the Soumagne tunnel and the viaduct of Arbre.

REFERENCES

Börner F., Gruhne M., Schön J. 1993. Contamination indications derived from electrical properties in the low frequency range. *Geophysical Prospecting*, 41:83-98

Clerckx B. & Halleux L. 1994. Borehole radar surveys on the Epe salt cavern field. *Proc. Solution Mining Research Institute*, 1994 meeting, Hannover, to be published.

Corin L., Couchard I., Dethy B., Halleux L., Monjoie A., Richter T., Wauters JP. Radar tomography applied to Foundation Engineering, *Proc. conf. on Modern Geophysics in Engineering Geology*, Special Paper of the Geological Society, London, to be published.

Forkmann B. & Petzold H., 1989. Prinzip und Anwendung des Gesteinsradars zu Erkundung des Nahbereiches, *Freiberger Forschungshefte* C 432, Leipzig; Verlag für Grundstoffindustrie.

Goldman M., Rabinovitch B., Gilad D., Gev I., Schirov M. 1994. Application of the integrated NMR-TDEM method in groundwater exploration in Israel, Journal of applied geophysics, 31:27-52.

Monjoie A., 1975. Hydrogéologie du massif du Gran Sasso,*Coll. publ. Fac. des Sciences Appliquées*, Univ. de Liège, vol. 53; 1-60.

Ollson O., Falk L., Forslund O., Lundmark L., Sandberg E., 1992. Borehole radar applied to the characterization of hydraulically conductive fracture zones in crystalline rock. *Geophysical Prospecting*, vol 40;109-142.

Siggins A.F., 1992. Limitations of shallow cross-hole radar investigations, *Geological Survey of Finland, special paper* 16;307-315 .

Steeples D.W. & Mileer R. 1990. Seismic reflection methods applied to engineering, environmental and groundwater problems, *Soc. Explor. Geophys., Geotechnical and environmental Geophysics*, vol. 1;1-30

Vogelsang D. 1993. Geophysik an Altlasten. Berlin;Springer Verlag.

1 Developments in site investigation and in engineering geological mapping
  Développements récents dans les domaines de l'étude des sites
  de la cartographie géotechnique

Remote sensing and fluvial geomorphology for riverine engineering
Télédétection et géomorphologie fluviale pour le génie fluvial

S.M. Ramasamy & M.A. Paul
School of Earth Sciences, Bharathidasan University, Tiruchirapalli, India

ABSTRACT: In the site selection process of dams and reservoirs and also bridges and culverts, the understanding of fluvial geomorphology is very important. The remotely sensed photographs captured by the Polar orbiting satellites provide excellent information in such fluvial geomorphic mapping. This paper discusses the importance of such fluvial geomorphic mapping and also the role of satellite remote sensing in riverine engineering by citing some evidences from Rajasthan, Pakistan and parts of Tamil Nadu.

1. INTRODUCTION

Rivers are the backbones for any nation owing to its enriched water resources. This is the reason why all our earlier civilization was greatly restricted to the riverine systems. Whereas, when the population has exploded, the humans started spreading over to the plains. Even then the livelihood and their growth was greatly dependent on the water resources. In the process of their growth, the humans have started constructing dams and reservoirs. But in the aspect of site selection they did not take some of the important parameters like the fluvial geomorphology into consideration which stand as a testimony to the flood cycle and river dynamics. Hence, the present paper is dedicated to bringout the importance of fluvial geomorphology in riverine engineering like site selection for dams, reservoirs, bridges and culverts.

2. ROLE OF FLUVIAL GEOMORPHOLOGY IN RIVERINE ENGINEERING

2.1 General

Just like the humans, the rivers have also got three stages namely the youthful stage, the mature stage and the old stage. In youthful stage, the rivers normally occupy the high order mountain ranges and occur very much above the mean sea level/base level of erosion. Hence, the rivers will have enormous energy and as a result it causes extensive erosion in the youthful stage. Whereas in the mature stage, as the rivers flow in the plains, they are normally controlled by the geology and the structure of the area and in addition, as they occur little above the base level of erosion, they cause both eroison and deposition and possess only moderate energy. But they frequently show phenomenon of migration, submergence and also meandering pattern. In the old stage, the rivers flow along the coastal regions and hence they almost flow at the base level of erosion. Hence the rivers do not have any energy at the old stage and as a result they cause only deposition. Hence such types of fluvial geomorphic appraisals are very essential in planning the riverine projects, as such geomorphic rhythms directly document the flood cycle. In the aspect of such fluvial geomorphic mapping, the remotely sensed data supplied by the Polar orbiting satellites provide useful information as in their raw and digitally enhanced modes, they vividly express various facets of the geomorphic features. Hence, in the present study such a remotely sensed data has been extensively used.

2.2 Youthful stage

As discussed earlier, river cause extensive erosion in the catchment zone in the youthful stage. Hence the reservoirs get extensively silted thereby the capacity of the reservoirs will get lost quickly. Such a reservoir siltation can provoke flooding also in the adjacent areas, hence prior to the construction of any reservoir in the youthful stage of a stream, a careful study is warrented to map the pattern of erosion, regions of erosion and also the terrain features that are responsible for erosion and accordingly remedial measures, such as check daming, gully plucking, afforestation and silt trapping etc must be taken (Ramasamy et al. 1992).

The remotely sensed data is really a boon in mapping out such zones of catchment area erosion as they appear bright white and vegetation free in black and white images and in digitally processed colour coded data they occupy dominant bright colours. In the present discussion, the Vaippar area has been shown (Fig.1) as a documentary evidence how the remotely sensed data spectacularly shows the area of erosion.

2.3 Mature stage

In the mature stage, the rivers flow in 'watch and walk phase' cause erosion and deposition and also develop flood plains, palaeochannels, compressed meanders, meander loops and also submerging streams. These features need a deserving study in the event of constructing dams or reservoirs, bridges and culverts.

2.3.1 Floodplains

The rivers normally show floodplains in the mature stage. Sometimes such floodplains are very wide (Fig.2) or very narrow. While the narrow floodplains indicate that the quantity of water that has flown in the rivers has little bit reduced, whereas the wider floodplains suggest that the quantity of water has drastically come down since its origin. Such reduction in surface flow could either be due to reduction in rainfall in the catchment or excessive damming in the upper reaches. In addition, such wide floodplain could also indicate the sea level fall or the upliftment of the landmass. By the careful selection of sediment samples along a profile perpendicular to the river flow and the followed up C14 dating, the pattern and period of reduction in surface water flow can be brought out. This could also give a predictive model about the future trend in water quantity.

Fig. I LANDSAT 2 Imagery showing highly eroding (1) catchment in Vaippar, Tamil Nadu.

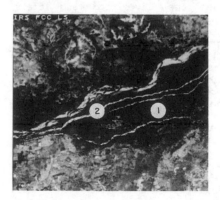

FIG. 2. IRS 1A Imagery showing a very wide floodplain (1) with a narrow river (2) in Cauvery-Colleroon rivers near Thanjavur.

However, as and when dams are decided in such areas, the dam axes must be considered to a full length completely covering the limit of floodplains, as the rivers may flood: on either sides if the dam is constructed only within the present river breadth.

The Cauvery-Coleroon river in the area north of Thanjavur is shown to explain the floodplain behaviour in the present discussion (Fig. 2). The present day Cauvery-Coleroon flow is having a width which is only 1/20 th of the total floodplain width. If dams are constructed only within the present day river breadth the river will definitely rejuvenate on either sides in the floodplain and the dam will become a failure.

2.3.2 Unpaired floodplains

The rivers sometimes develop floodplains only on one side of their courses. The Figure 3 shows similar phenomenon in River Indus, Thar desert, Pakistan. The river Indus is flowing in the western edge whereas it has developed a very wide floodplain to its South-East. This indicates that the river Indus has migrated towards Northwesterly. This feature is called as "unpaired floodplain". Adequate care must be taken in deciding the locations for the dams in such domains as the rivers are showing signs of preferential shifting towards one direction, in such regimes. Hence dams must be constructed with more length in the banks where the floodplain is not there. Such an extended length of the dam would take care of the floods during the process of preferential shifting of the rivers in one direction. Similarly bridges must also have an extended length in the banks where the floodplains are not found.

2.3.3 Migratory rivers

The rivers in the mature stage do migrate due to various reasons such as spinning of the earth, eustatic movements, isostatic adjustments, frequent floods, climate changes etc. Such phenomenon of migration must be studied before dams are constructed. Such migratory paths and also the buried river systems could be pickedup clearly from the satellite images as they show moistured and vegetative ribbons. The Figure 4 shows a set of major palaeochannels or an old river

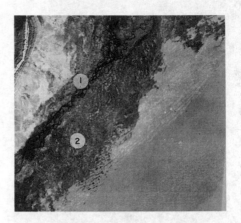

FIG. 3. LANDSAT Imagery of Thar desert, Pakistan showing Indus river in one edge (1) with an unpaired floodplain (2) to its South - East.

connecting Walajapet with Madras in Palar basin. This shows that the river Palar has once flown North-Easterly from Walajapet and confluenced the Bay of Bengal near Madras and subsequently migrated to its present tract (Vaidyanadhan 1971., Ramasamy et. al. 1992). In this case, if a dam is constructed in Palar river just at the down stream of Walajapet, then the river might probably rejuvanate through its old course and flow towards Madras and as a result the whole dam will become a failure. Hence, the regions of such migratory rivers must be avoided for the construction of dams and reservoirs and if situation compels such migratory patterns and behaviour must be studied in detail before constructing a dam.

2.3.4 Meander Systems

Similarly rivers will have a meander pattern and also will have meander loops in the mature stage and also in old stage. Such meander loops and also cutoff meanders can be mapped very clearly from the satellite pictures as they show similar signatures like palaeochannels with moisture and vegetation ribbons. The Figure 5 shows one such meander loop system in Banas river in the area east Bhilwara in Rajasthan. Such meander loops show that once the Banas river was flowing in a sinuous path and subsequently got straightend out. Now

5

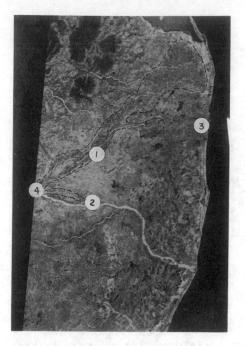

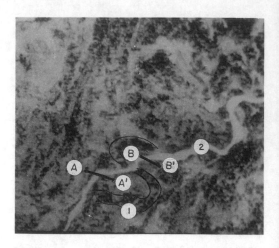

FIG. 5. LANDSAT imagery showing meandering loop (1) in Banas river (2) in the area East of Bhilwara, Rajasthan. (A-A', B-B' regions unfavourable for dams and reservoirs)

FIG. 4. LANDSAT Imagery showing palaeochannels (1) in Palar (2) in Madras (3) Walajapet area (4)

these fetures are very significant in the site selection process for river valley projects. If a dam is constructed in Banas river along A-A' or B-B' (Fig.5) then the river can migrate along such meander loops or cutoff meanders thereby the whole dam will become a failure. Hence such areas of meander loops and cutoff meanders must be avoided while sites are selected for dams and reservoirs.

2.3.5 Compressed meanders

The rivers both in their mature stages and in old stages show some anomalous compressed meanders i.e the river which is otherwise flowing very straight will show compressed meanders in a very restricted area which indicates that the area is seismotectonically active (Ramasamy et al. 1992). Hence, such areas of anomalous compressed meanders must be avoided from constructing dams or reservoirs. The Figure 6 is a Landsat imagery showing a part of Narmada-Tapati rivers in Central India. These two rivers which are otherwise flowing very straight for hundreds of kilometers show restricted compressed

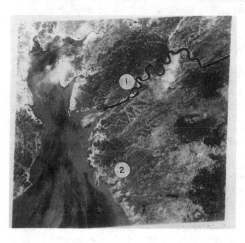

FIG. 6. LANDSAT imagery showing compressed meanders in Narmada (1) and Tapti (2) rivers along West Coast of Maharastra.

meanders only in the fringe of the continent. This phenomenon has been attributed to tectonically active faults (Ramasamy et al. 1992). Hence such fetures too must be strictly studied before the selection of sites for dams and reservoirs.

3 OLD STAGE

In old stages, the rivers normally develop deltas and also show enormous cutoff meanders and meander loops and migratory channels. Hence the old stage of a river is invariably a negative domain for constructing dams or reservoirs.

4 CONCLUSION

Thus the above discussion clearly shows that the fluvial geomorphology needs to be understood very thoroughly for selecting suitable sites for river valley projects and also for bridges and culverts. The above discussion also shows clearly that the art of remote sensing plays a very crucial role in fluvial geomorphic mapping.

5 ACKNOWLEDGEMENTS

The senior author is very grateful to the Department of Space, Government of India for sanctioning a research project entitled "River migration studies in Tamil Nadu (RIMIGTA), through which only the author has developed the concept for the present paper.

REFERENCES

Ramasamy,SM., Bakliwal,P.C., and Verma,R.P., (1992) Remote Sensing, River migration of Western India Int. Nat.Jour.Rem.Sens. Taylor & Francis Ltd.London : pp 2597-2609

Ramasamy,S.M., Balasubramanian,T., Thillai Govindarajan (1992) Remote Sensing aided integrated study and identification of erosion prone areas.T.S.Chauhan and K.N.Joshi (ed) Vol on Readings in Remote Sensing Applications Scientific publishers.Jodhpur : pp 439-454.

Ramasamy,SM., Venkatasubramanian,V., Riaz Abdulla and Balaji,S. (1992) The phenomenon of river migration in Northern Tamil Nadu - evidence from satellite data, Archaeology and Tamil literature, Man and Environment XVII(1) : pp 13-25.

Vaidhyanathan,R. 1971 Evolution of the drianage of Cauvery in South India, Journal of the Geological Society of India 12(1)

Techniques to assist the classification of fault activity and seismic hazard for engineering projects by remote sensing

Techniques de télédétection pour aider à la classification de l'activité des failles et des aléas sismiques

W. Murphy
Department of Geology, The University of Portsmouth, UK

ABSTRACT: With the expansion of large engineering projects the necessity to assess seismic hazard is becoming more widespread. Remote sensing provides a quick and cost effective method for the investigation of faulting and the classification of the degree of activity. This paper reviews the techniques available to the engineering geologist for the examination of faulting and proposes a method using remote sensing to classify fault activity for engineering purposes.

RESUMÉ: Avec l'augmentation dans le numéro des grands projets ingénieuries, la nécessité d'évaluer la risque sismique devient plus générale. La télédétection est un moyen rapide et pas cher pour l'investigation des failles, et pour classer la mesure de mouvement de la faille. Ce papier considère les techniques disponibles à l'ingénieur géotechnicien pour examiner les failles, et propose un moyen d'utiliser la télédétection pour classer le mouvement d'une faille. Cette information sera très utile pour un ingénieur géotechnicien.

1 INTRODUCTION

The expansion of large engineering projects in Europe has necessitated the evaluation of earthquake hazard. Even in relatively aseismic areas such as the British Isles, seismic hazard is often considered This has led to recent publications such as that by Gosschalk & Hinks, (1993), who proposed a code of practice for the assessment of seismic hazard for large dam projects in the United Kingdom.

Other regions of Europe have greater need for rapid methods of seismic hazard assessment. And while hazard is not solely concerned with the activity of faults, the movement of potentially seismogenic structures is of great importance in the estimation of seismic hazard.

Few studies have investigated the role of remote sensing in the delineation of active geological structures. Part of the problem in doing this is that the main satellite based sensors still have a coarse resolution by comparison to most engineering requirements. However, projects which truly require earthquake hazard assessment, normally take a regional overview where remote sensing may be helpful.

It is useful to consider the scales with which we would normally deal. Remote sensing in the assessment of hazard is most useful at the level of synoptic (1:100,000 and coarser, (IAEG, 1976) and medium scale mapping. The increased availability of SPOT panchromatic and multispectral images has increased the use of remote sensing in the field of engineering geology. With the prospect of new instruments such as HIRIS (High Resolution Imaging Spectrometer) the use of these techniques may become more widespread still.

The most commonly used remotely sensed data available, with the exception of aerial photography, is collected by the LANDSAT satellites. Of the instruments on LANDSAT 5, the Multispectral Scanner, (MSS), is severely limited by its 79 m ground resolution element (pixel). The Thematic Mapper, (TM), which was initially installed with LANDSAT 4 collects data at a ground resolution of 29 m. This is still coarse by engineering standards but it is a great improvement. The poor ground resolution is offset by the collection of data in the seven spectral bands shown in figure 1.

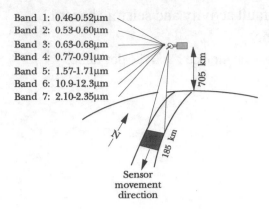

Band 1: 0.46-0.52μm
Band 2: 0.53-0.60μm
Band 3: 0.63-0.68μm
Band 4: 0.77-0.91μm
Band 5: 1.57-1.71μm
Band 6: 10.9-12.3μm
Band 7: 2.10-2.35μm

705 km

185 km

Sensor
movement
direction

Figure 1. The LANDSAT Thematic Mapper

Remote sensing is a general term which includes a variety of airborne and satellite based devices. It involves any kind of data acquisition which is collected from a distance. Aerial photography, satellite imagery and some geophysical techniques are all forms of remote sensing. The use of Global Positioning Systems, although not strictly remote sensing, has also seen increased applications in seismic hazard and crustal deformation studies.

This paper will explore satellite based, multispectral, remote sensing techniques.

2.1 Remote Sensing Platforms

Although the majority of space based imaging devices have too course a spatial resolution for actual site investigation, image interpretation can provide useful information on regional geological structure, and indicate more specific hazards based on terrain types. The great strength of remote sensing is as a compliment to field studies and can only be used with great difficulty as a substitute for ground based investigation programs. However these types of data can be useful not just because of the broad overview they provide but also because they can be processed to enhance and clarify different image features. Additionally, satellite image data are *raster* data, that is: they are in the form of discrete elements, called pixels, which represent an area on the ground. Therefore these data can be readily integrated with the other raster data using geographic information systems (GIS).

2.2 Image processing techniques

The potential to process satellite image data to enhance different ground conditions can be used to significant advantage in seismic hazard studies. Generally for the interpretation of geological structure some of the most simple techniques are the best.

There are a number of techniques which can be applied usefully to an image in order to enhance the ease of interpretation. In structural geological investigations these techniques normally are edge enhancement and contrast stretching. Band ratios and principal component analysis (PCA) are more complex techniques which require a broader understanding of the image data to allow a reasoned interpretation.

2.3 Contrast stretching

Often satellite data tend to be rather dark images as a result of atmospheric scatter of the reflected radiation. One of the main aims of image processing is to account for this effect. A fundamental principal in image processing is the image histogram.

If the data are displayed as a histogram with the digital number (DN), which is a measure of brightness, on the X- axis and the frequency of occurrence of that DN, on the Y-axis, a DN value of 0 (the minimum) can be set at some lower point in the histogram and a DN value of 255 (the maximum) can be applied at some point in the upper range. The data can then be stretched between these two points using a variety of mathematical techniques such as histogram normalization (which tends to enhance the lower and upper ends of the histogram), equalization (which tends to enhance the centre of the histogram), or, a linear stretch (which enhances the whole histogram by an average amount). For example, an image histogram has a distribution of data between DN values of 10 and 55. This would describe a dark image with poor contrast. In contrast stretching break points are inserted which map the DN value of 10 to 0 and the DN value of 55 to 255. The data are then stretched between these two values.

2.4 Edge enhancement

Another useful technique is that of edge

enhancement. This technique involves using varying mathematical processes (such as Fast Fourier Transform) to look for high frequency image features. This type of image processing works on the basis of rapid changes in image contrast (i.e. large variations in DN over a small number of pixels). An example of this would be the rapid change from bright pixels to dark pixels associated with shadows cast by topographic irregularities. These filters are described as kernels and take the form of a matrix over which the data are examined. These kernels can be general filters or directional filters which enhance lineations, (linear features on satellite imagery of varying origin), in only one orientation.

2.4 Complex image processing techniques

Given that geological materials are most reflective in two particular regions of the electromagnetic spectrum (bands 7 & 5, see figure 1). Data from these regions can be combined to into one display of data. Normally when an image is displayed one band is assigned a colour. When using the technique of band ratioing the user divides one band of data by another in order to gain detailed information about the spectral characteristics of the ground. For example, a band ratio employing LANDSAT TM bands 7 and 5 will reveal thermal anomalies which may indicate a prelude to volcanic eruptions.

Band ratioing is of limited use in structural geological interpretations since it is most frequently used to highlight spectral differences (i.e. between bands) in the geology. It subdues the effects of topography and therefore reduces the information available to the engineering geologist who is attempting a structural interpretation.

In general there is considerable redundancy between bands of image data (that is an area dark on one band will often be dark on another). When this is the case principal component analysis (PCA) may be useful.

Image bands are plotted against each other and the mean, variance, standard deviation, eigen values and eigen vectors are calculated. Using these statistics it is possible to sort the data into bands containing the variation in DNs in an image. The first principal component (PC1) contains the majority of the variation in the image, the second contains the next greatest variation and so on. The problem with this technique is that it the higher order principal

components have a very low signal to noise ratio and are therefore of limited use. For example, a principal component analysis is performed on an image using six bands of LANDSAT T.M. data (band 6 is not used because of the coarse, 120m, resolution). PC 1 contains the majority of variation in the image which is mainly due to topography and vegetation. PC 2 contains less variation. However PCs 4 through 6 are useless because of the noise in the image.

A full review of image processing techniques is beyond the scope of this paper. A good review of geological remote sensing techniques is given by Drury (1993).

2.5 Image interpretation techniques

The interpretation of image texture is one the principal methods of gaining information on geological structure. In general faults will display one, or more, of the characteristics shown below:

(i) Image lineations which may, or may not be, straight. Due to the interaction of topography and dipping structures, low angle faults areas of high relief will not show a straight outcrop. The degree of deflection of terrain contacts, and a knowledge of the terrain as indicated by the position of shadows, can give a useful indication on the type of fault. A consideration of the mechanics of faulting will show that strike-slip faults often dip steeply, while thrust faults are often shallow dipping.

(ii) As a result of comminution on the fault plane and the juxtaposition of different geological units the rocks in the fault zone may show markedly different mechanical properties to those of the local area. Such changes will result in differences in terrain which will be shown as a textural change on the image.

(iii) Major faults will also show a major change in topography which will often be reflected by a marked change in vegetation type, such as, grass dominated communities giving way to conifers etc.

(iv) Active faults are defined on the basis of the displacement of some recent phenomena such as housing, roads, rivers or recent sediments.

(v) Active faults will often show a scatter of seismic activity on, or around, the fault trace.

The best method of relating seismic activity to image features is through geographic information systems. Satellite images can be geometrically corrected to latitude and longitude projections using most image processing software. Once this is done earthquake epicentre data can be plotted on the image to make a seismicity map. These data can be derived from a number of sources such as the United States Geological Survey (on Compact Disk), the International Seismological Centre in Reading (on magnetic tape) or in published form (e.g. Postpischi 1985).

The user of these data should be aware however that with historical data the epicentral location could be in error by up to 50 km. This amount of error is very high by comparison to modern satellite images. Therefore the interpreter should be aware of this source of error.

(vi) Blind faults present a major difficulty in any stage of seismic hazard assessment which is employing remote sensing. It is possible that winter (low sun angle) imagery may be useful for the detection of these kinds of subtle image feature.

3 SEISMIC HAZARDS

The evaluation of earthquake hazard is a more substantial process than merely the estimation of earthquakes and their return periods. While the examination of faulting provides a useful start it is by no means the whole process.

3.1 Fault activity & earthquake recurrence

Regional geological structures must be investigated to determine which structures may cause earthquakes. Fault activity can be described by a variety of terms. The United States Atomic Regulatory Commission (USARC) defined to types of fault: *active faults* and *capable faults*. Both types of fault are classified on the basis of the occurrence of seismicity on the fault. An active fault being one in which at least one earthquake has occurred on the fault in the last 11,000 years, or, repeated movements in the last 35,000 years. Capable faults are those which have shown some evidence of earthquake activity in the last 350,000 years.

However fault activity remains a somewhat nebulous concept. If these figures are taken out of context then dangerous generalisations could be made resulting from the paucity of accurate records of seismicity rather than true estimates of active faulting. To illustrate this let us take two varying examples.

Records of seismic activity in the United Kingdom appear to have been reasonably reliable for the last 240 years or thereabouts. The United Kingdom is prone to small to moderate size earthquakes which normally cause only minor damage.

In the rather stable area of the British Isles the 11,000 year cutoff proposed by USARC for active faults is severely limiting. Return periods on faults in the British Isles could be in the order of millions, rather than, thousands of years. Therefore under the USARC scheme faults would be classified as inactive when they are capable of producing earthquakes over a very long time period.

Figure 2. shows a segment of the Wellington-Lower Hutt fault in New Zealand. Increasing urban development of the Totara Park residential area is now destroying the scarp created by at least one earthquake on this fault during the last 15,000 years (the fault movement caused the deformation of a 15,000 year old river terrace). However satellite imagery reveals the larger trace of the structure and therefore permits easy mapping of the fault. Despite the degradation of the scarp at Totara Park this has to be considered as an active fault.

Figure 2. A segment of the Wellington-Lower Hutt fault at Totara Park.

3.2 Evaluation of the design earthquake

Making estimates of a design earthquake invariably requires information on the seismicity of a region as well as some information on the pattern of faulting. Clearly there is the need to evaluate the level of seismicity on a fault or fault zone in order to define the maximum credible earthquake for that structure. However the estimation of site-specific ground motions requires more data than can be provided by satellite-based remote sensing methods and will not be discussed further in this paper.

4 FAULT CLASSIFICATION FOR ENGINEERING PURPOSES

Faults present a number of problems to the engineering geologist related to the extreme heterogeneity of material properties of soils and rocks in fault zones and enhanced permeabilities. The classification outlined below is specifically designed for seismic hazard and does not attempt to address these other engineering difficulties.

The scheme sets out to group faults into 6 classes based on the image interpretation, seismic activity and historical records.

Class I. A fault which, on the basis of satellite imagery, can be shown to deform recent sediments and geomorphic features. A class I fault should have an orientation consistent with the tectonic environment. When seismicity is plotted on this fault there should be a scatter of seismicity around the structure and focal plane solutions should indicate a primary, or, secondary plane consistent with the strike of the image lineation. This is an active fault.

Figure 3 shows a fault which falls into this category. The West Wairarapa Fault Zone in the North Island of New Zealand has given rise to large magnitude earthquakes (e.g. the Wellington earthquake in 1855, M_w=8.2 and the Wairarapa earthquake 1942, M_s=7.1). It is clear from examination of the image that there is a clear structure which strikes across rugged terrain without showing a great deal of deflection (hence indicating that the fault is very steep, field investigation confirms that this is a strike-slip fault). Historical and instrumented seismicity can be associated with distinct fault segments (figure 4).

Class II. A class II image lineament is a fault which shows evidence of recent displacement and has an orientation which is consistent with the tectonic environment. Seismicity is poorly defined in the region of the fault so that it is

Figure 3. LANDSAT Thematic Mapper image showing major faults in the west Wairarapa Valley (from Murphy and Bulmer, in press).

difficult to make convincing correlation. Historical information on isoseismal fields show a long axis orientation aligned around the fault. In terms of critical engineering structures this class of structure still should be considered as active.

Figure 5 shows a satellite image of the Messima graben in southern Calabria. The eastern boundary of the graben is shown in the image and is believed to have caused the catastrophic 1783 calabrian earthquake. The earthquake sequence is too poorly spatially to attribute to the boundary fault however the isoseismal field, and associated landslides, shown in figure 6, are consistent with the fault orientation.

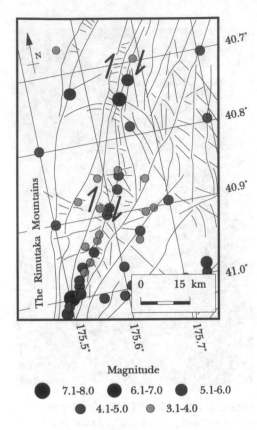

Figure 5. Satellite image of the Messima Graben, southern Calabria

Magnitude

● 7.1-8.0 ● 6.1-7.0 ● 5.1-6.0
● 4.1-5.0 ● 3.1-4.0

Figure 4. Interpretation of the satellite image shown in figure 3 (from Murphy & Bulmer (in press).

Class III. This category of fault is one which shows evidence of deforming sediments which represent a time period of less than 50% of the time for which the tectonic environment in which the fault is found has existed. A class III fault in the United Kingdom may show evidence of deforming quaternary sediments whereas in southern Italy a class III fault may only deform Devensian sediments. For critical structures this should still be considered as a potentially seismogenic structure.

Figure 7 shows a fault from southern Sicily at Santa Augusta. The fault is well defined in the centre of the picture however above this the plane is degraded by wave action indicating a number of periods of coseismic slip. On satellite image (figure 8) the fault outcrops as shown on the image. The fault appears irregular due to the degradation by coastal erosion processes.

Class IV. A class IV fault is one which shows

evidence of deforming sediments which were deposited during the current tectonic regime and has an orientation consistent with the tectonic environment. There is little, or no, evidence for historical seismicity being located along the strike of this fault.

Class V. These faults are those which are consistent with the current tectonic environment but show no evidence of deforming young sediments. The fault scarp will be eroded and incised streams will have cut down to the base level of the fault. There is no evidence for movement during the period of the tectonic regime in which the fault was formed. In engineering terms these structures are difficult and should be considered as being capable of producing earthquake events over long time periods. The lack of geological or seismological data makes assessment of hazard difficult since slip dependant models of seismicity and characteristic earthquakes cannot be used to indicate seismic risk because of the lack of data.

Figure 9 shows an interpretation of lineaments from LANDSAT MSS and TM imagery by Gutmanis *et al.* (1985). Image lineaments were classified on the basis of whether they were structurally controlled, or merely, an expression of topography (although the latter may also have a structural control). This interpretation contains

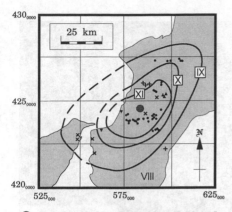

Earthquake epicentre + Mud or debris flow
× Liquefaction & lateral spreading ▲ Rotational slips
▼ Disrupted soil & rock falls • Coherent slides & slumps

Isoseismal lines
(Modified Mercalli Scale)

Figure 6 Isoseismal field of the 1783 Calabrian earthquake.

environment exists. There is no recorded, or historical evidence, to indicate movement, along such a fault. Slip on the fault can be shown to have occurred under a different tectonic environment.

SUMMARY & CONCLUSIONS

It is clear from the discussion above that there is a role for satellite based remote sensing in the assessment of seismic hazard. Perhaps the most useful is the investigation of regional faulting patters and seismicity. While site conditions can be evaluated by remotely sensed data, generally, the spatial resolution of this information is too course for most engineering projects.

Difficulty arises in assessing earthquakes of anything other than shallow focal depths (i.e. greater than 33 km) since these will generally not cause surface rupture. Additionally in areas like the United Kingdom where there are few earthquakes associated with surface faulting, and

class V and class VI faults. The image shown in Figure 10 is a LANDSAT T.M. image (band 7) for that area. It is clear that the interpretation is difficult. The authors comment that principal component analysis was used to enhance the image and ease interpretation.

Figure 8. Satellite image of the fault at Santa Augusta, Sicily, shown in figure 7.

Figure 7. Fault scarp at Santa Augusta, Sicily.

Class VI. A class VI structure is an image lineation which shows an orientation inconsistent with the current tectonic environment and shows no evidence of deforming sediments deposited within the period over which that tectonic

a wide scatter of seismicity, it is difficult to assign earthquakes to any individual fault. Clearly for these more difficult situations it is necessary to undertake detailed ground based investigations to investigate small surface structures or look for

15

evidence of palaeoseismicity (e.g Ringrose *et al.* 1991)

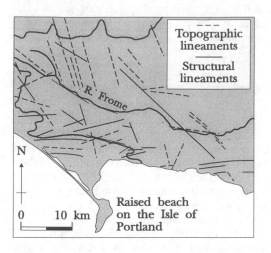

Figure 9. Interpretation of satellite imagery from Gutmanis *et al.* 1985

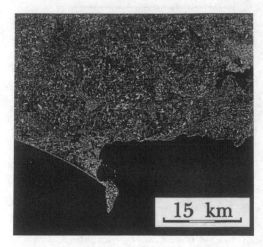

Figure 10. LANDSAT Thematic Mapper image of the area shown in figure 9.

However the use of remote sensing at the planning stage, and the incorporation of those data, into geographic information systems provides a useful tool for the assessment of hazards and allows more precise direction of more time consuming and more expensive techniques. Given the falling cost of archived satellite data which covers most parts of the earth the use of remote sensing in seismic hazard assessment is likely to become a major labour and cost saving tool.

5 BIBLIOGRAPHY & REFERENCES

International Association of Engineering Geology, 1976. *Engineering Geological Maps: a guide to their preparation,* UNESCO, Paris, 79pp

Drury, S. 1993. *Image interpretation in geology.* Allen & Unwin Press, 2nd edition.

Gosschalk, E. M. & Hinks, J. L. 1993. The preparation mof an engineering guide for seismic risk to dams. In *Soil Dynamics and Earthquake Engineering VI,* (Cakmak, A. S. & Brebbia, C. A., Eds) 445-460, Computational Mechanics, Southampton.

Gutmanis J. C, Holt, R. W. & Whittle, R. A. 1985. Satellite imagery applied to the interpretation of geological structure as part of recent seismic hazard studies in the Unite Kingdom. In *Earthquake Engineering in Britain,* Thomas Telford Press, London.

Murphy, W. & Bulmer, M. H. K. 1994. Evidence of pre-historic seismicity in the Wairarapa Valley, New Zealand, as indicated by remote sensing. *Proceedings of the 10th Thematic Conference: Geologic Remote Sensing,* 8-12, May San Antonio, Texas (in press).

Postpischi, D. (editor). 1985. *Consiglio Nazionale Delle Ricerche Progetto Finalizzato Geodinamica 1985 Catalogo dei Terremoti Italiani Dall' Anno 1000 al 1980.* Bologna.

Ringrose, P. S., Hancock, P., Fenton, C. & Davenport, C. A. 1991. Quaternary tectonic activity in Scotland. In *Forster, A., Culshaw, M. G., Cripps, J. C., Little, J. A., Moon, C. F. Proceedings of the 25th annual conference of the Engineering Group of the Geological Society; Quaternary engineering geology,* Engineering Geology Special Publications, Geological Society. 7, Sept. 10-14, 1989, 679-686.

Remote sensing investigation of slope instability occurrences

Investigation par télédétection des instabilités des pentes

Radmila Pavlović & Miroslav Marković
Faculty of Mining and Geology, University of Belgrade, Yugoslavia

ABSTRACT: On the right bank of Danube River in NE Serbia, Yugoslavia, between the dams "Djerdap-I" and "Djerdap-II", the significant occurrences of slope instability are developed. The activity of the area is strongly increased after the dam "Djerdap-II" was constructed. The instability endangers the main road and the settlements on the river bank.

For the investigation of the instability occurrences two techniques of remote sensing method were applied. Analysis of the Landsat satellite images enabled the determination of regional ruptural pattern, which controls the flow of ground water and the youngest, neotectonic activity of the wider region. By means of stereoscopic analysis of aerial photos the detailed ruptural pattern was investigated, and identification of instability occurrences was made. These occurrences were classified according to their genesis, time and intensity of development.

The collected data enabled classification of the slopes according to their instability, as a base for the further engineering geological projects.

RESUME: Le developpement d'importants phenomenes d'instabilite du sol a ete constate sur la rive droite du Danube, au nord-ouest de la Serbie, en Yougoslavie, entre des barrages "Djerdap-I" et "Djerdap-II". Une augmentation importante de l'activite du terrain a eu lieu apres la construction du barrage "Djerdap-II". L'instabilite presente un danger pour la route principale et les habitats sur la rive du fleuve.

Deux methodes de teledetection ont ete appliquees a la reconnaissance des instabilites du sol. L'analyse des images transmises par les satellites americains de la serie Landsat, a permis la detection de la structure regionale des ruptures, qui controle le mouvement des eaux souterraines ainsi que l'activite neotectonique, la plus recente, de toute la region. L'identification des instabilites du sol a ete etabli a l'aide de l'analyse stereoscopique des photographies airiennes. Elles ont ete classifiees d'apres leur genese, le temps et l'intensite de leur developpement.

Ces donnees ont permis la classification des versants, a partir du degre d'instabilite, qui servirait de base pour les projets utilitaires en geologie d l'ingenieur.

1. INTRODUCTION

After the dams "Djerdap-I" and "Djerdap-II" on Danube river were erected the ground water level was upraised. This caused development of numerous landslides of significant dimensions and occurrences of other forms of slope instabilities as well. The detailed investigation of those occurrences was carried out by application of remote sensing. Examined was right river bank area, where the direct influence of oscillation of river level on slope activity was obvious. The width of the area varies from 2 to 12 km., depending on its relief and geological composition.

Investigation of slope instability occurrences was aimed in two directions. One purpose was to determine regional and detailed ruptural pattern. The second was the recognition of processes, occurrences and forms of slope instabilities, their location and classification according to the intensity of activity.

The obtained results on ruptural pattern, and on locations and mutual relationships between the slope instability occurrences of different origin is presented on specially constructed maps. The maps represent the interpretation of bearing of slopes in the year of 1981.

2. METHODOLOGY OF INVESTIGATION

The methodology of investigation of slope instabilities of right bank area of Danube river comprehends the application of two remote sensing techniques. First is analysis of satellite images, the second stereoscopic analysis of aerial photos.

For obtaining the regional ruptural pattern the analysis of the Landsat satellite images was executed. The images scale of 1:1.000.000 to 1:200.000 were used. Examination of the satellite images was carried out as logical, or visual comparative analysis. Analyzed and correlated were the images of the same area, same scale, but from different spectral bands. The comparison of data obtained from the images of the same area, same spectral band, but presented in different scales was also made. The transfer of data from satellite images to the topographic base was done also visually.

Characteristics of regional ruptural pattern have direct influence on the flow of the surface and ground water. They are significantly controlling the youngest, especially the recent tectonic activity of the area. The ruptural structures are influencing the intensity of disintegration of rock masses. Thus regional ruptural pattern in general impels the slope instability.

Analysis of detailed ruptural pattern and the examination of slope instabilities were carried out by stereoscopic analysis of aerial photos, scale 1:26.000 to 1:50.000, from three different periods of time, i.e., from year 1957, 1970 and

1981. The three dimensional models of the terrain enabled determination of location and chronology of activity of slope instability occurrences. The analysis of their origin, and mutual relationships, as well as classification according to the time and intensity of activity, were also possible.

It was possible by means of study of big scale aerial photos to detect all important details of topography. On the other hand, the correlation of the data collected from the aerial photos made in different years enabled determination of changes of topography with the time.

3. THE REVIEW OF OBTAINED RESULTS

The area of the right river bank of Danube, between the dams "Djerdap-I" and "Djerdap-II" lies on the western margin of Dacian basin. The whole terrain is inclined to the central part of basin, situated to the east of the investigated area. The recent topography represents the complicated geomorphological complex. It comprehends active and abandoned geomorphological forms of different processes as are marine, limnic-marine, aeolian, colluvial and fluvial processes. These processes were replacing one another in time and expanse from Miocene up to now.

The highest influence on the shape of recent relief has fluvial process. Danube River with its tributaries makes well developed drainage pattern. The temporary dry gullies produced by proluvial process are included in this pattern. The erosional basis of the process is the level of Danube River. The proluvial process is consequently controlled by developing of fluvial process. The reinforced development of fluvial process, caused by upraising of Danube River level after the erecting of Djerdap dams, makes development of colluvial process more intensive. The old landslides are activated, and many new ones were produced.

3.1. Ruptural pattern

The results of ruptural pattern obtained by examinations of satellite images, also by stereoscopic analysis of aerial photos, are presented on the unique map, given as figure 1. All determined ruptures are classified according to their significance as of regional and of local importance. The other criterion of classification was the reliability of determination. Consequently the observed and supposed ruptural structures were distinguished.

In the northern part of the investigated area, locality named Dunavski ključ (the Key of Danube River), a small number of ruptures was registered. The two faults of regional importance are significant, both of NW-SE orientation. They are well expressed in the relief of Quaternary age, leaving no doubts about their neotectonic activity. The other faults in this locality, mostly distinguished as supposed ones, are of the same orientation, i.e., NW-SE.

In the other area, south of Brza Palanka, the density of ruptural structures is much higher. Ruptures are arranged in two systems. The first one, better developed, is oriented in NW-SE direction. The other has the orientation NE-SW. The regional fault with strike in N-S direction is the boundary between two tectonic units.

The ruptural structures of the explicit significance for the slope instability are the faults, or systems of close bigger joints, mutually parallel. They are separating the rock mass into blocks inclined to rock slump. If they are active during the neotectonic time the vertical movement of different blocks causes development of landslides.

3.2. Distinguished forms of relief

The data on slope instability occurrences are presented as a separate map (figure 2). The classification of distinguished forms of relief was based on genetic principles. All forms were connected with the geomorphological processes that generated them. The enormous reduction of original scale of the map effected that lot of details are lost.

(1). Colluvial forms

The forms of colluvial process in the investi-

gated area on the right bank of Danube River are results of the very slow movements of rock mass (delapsia). Such forms, identified as landslides, are classified according to their activity as active ones or as potential, temporary tranquillized landslides.

Recently active landslides have length and width of hundreds of meters. They are developed on slopes of valley of Danube River, or on the slopes of valleys of its tributaries and dry gullies. The lithologic composition of the terrain is presented by Pliocene sands and clays. The underlying beds are inclined towards the river and that fact increases the intensity of the movement. It is easy to recognize on the aerial photos morphological elements of landslides, such as scars, open cracks, etc. Identification is very quick, and highly reliable.

Temporary tranquillized landslides are defining the areas of unsteady stability. The morphologic signs of former movements, such as scars in topography and sudden changing of slope angle, here were observed. Some settlements, mostly small villages, are developed in areas of unsteady stability. Temporary tranquillized landslides are covering spacious surfaces of several square kilometres in the investigated area. They are mostly developed on the slopes of Danube valley.

The estimating of the depth of landslides was made on the base of stereoscopic analysis of aerial photos. According to it the depth varies from few meters to few dekameters. Field observation has confirmed the results of estimation (Marković & Pavlović, 1987, Pavlović 1990).

(2). Proluvial forms

The conditions for the development of proluvial process in the investigated area are very favourable. The rainfalls are often and intensive. The slopes are steep, and that permits high velocity of temporary running water. As the results of proluvial process the huge gullies and alluvial fans are formed.

The forming of gullies is well expressed. Gullies are cut in loess or clastic sediments. Their depth reaches a hundred of meters, their

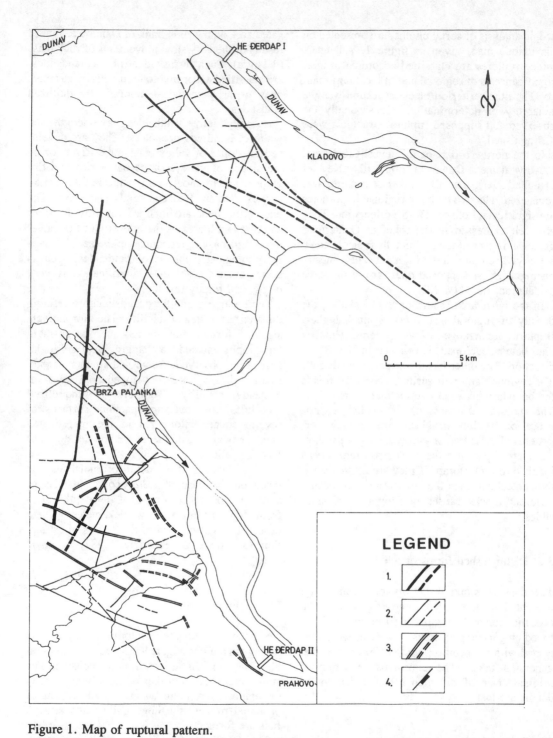

Figure 1. Map of ruptural pattern.
  Legend: 1. Regional ruptural structures, observed and supposed. 2. Local ruptural structures, observed and supposed. 3. Ruptures of direct influence on slope stability, observed and supposed. 4. Relatively downthrown block.

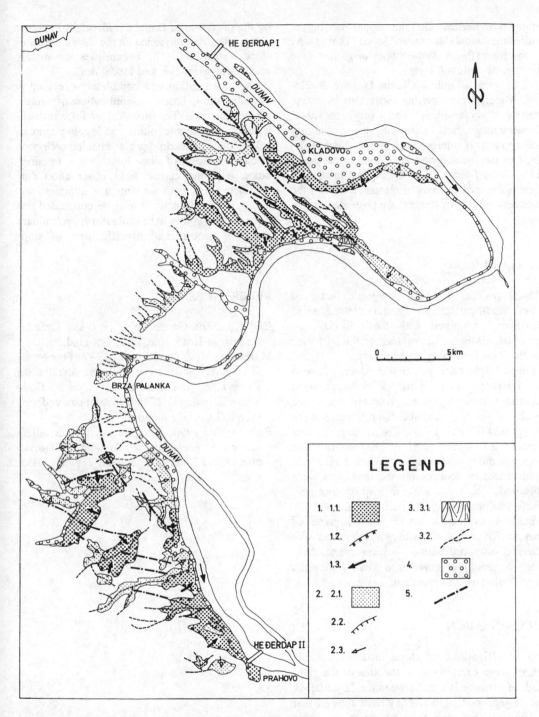

Figure 2. Map of slope instability occurrences. Legend: 1. Active landslide. 1.1. Landslide body. 1.2. Scar. 1.3. Direction of mass movement. 2. Temporary tranquillized landslide. 2.1. Landslide body. 2.2. Scar. 2.3. Direction of mass movement. 3. Proluvial forms. 3.1. Alluvial fan. 3.2. Gully. 4. Fluvial forms: river terraces and alluvial planes. 5. Ruptures significant for slope stability.

length kilometres. On the slopes of bigger gullies the landslides are developed. That is why on the map (figure 2) the slopes of gullies have the sign of colluvial forms.

At the ends of gullies alluvial fans are developed. Material constructing those fans is coarse grained, unclassified, and very movable. Consequently fans represent the instability occurrences. Numerous alluvial fans were registered in the investigated area. Due to the small scale of the figure 2 only the biggest fans, with dimensions of hundreds of meters, and with the thickness of tens of meters, are presented on the map.

(3). Fluvial forms

Fluvial process is main exogenetic factor of relief modifications in the investigated area. However, compared with the colluvial and proluvial forms the importance of fluvial forms for the lope instability is secondary.

On the right bank of Danube River, between the "Djerdap-I" and "Djerdap-II" dams, the most important fluvial forms are river terraces. Three levels of younger Danube river terraces were recognized (Cvijić, 1926). The surfaces of river terraces are stable parts of the terrain. The terrace's cliffs are unstable, and that is where occurrences of rock slump and landslides were observed. The small scale of map has not permitted the presentation of such forms.

Besides terraces a broad alluvial plane of Danube River and of its bigger tributaries was detected on aerial photos. Alluvial plane, constructed mostly of gravel and sand, represents very stable part of investigated area.

4. CONCLUSION

The investigations of slope instability occurrences were carried out in the area of the right bank of Danube River, between the "Djerdap-I" and "Djerdap-II" dams. They have showed that relationship between ruptural structures, their neotectonic activity and development of slope instabilities does exist.

Oscillation of Danube River level provoked

by the operation of Danube hydroelectric power plants has direct influence on the development of slope instabilities. The instabilities are mainly manifested as gullies and landslides.

The investigation was completely executed in the laboratory, based on application of remote sensing methods. The inventory of slope instability was very reliable, quick and highly economical. Applied methodology also enabled collection of data on causes of slope instabilities. In correlation with the classic field observation data obtained by remote sensing are quantitatively and qualitatively new. It may be concluded that remote sensing should be obligatory, preliminary phase in geotechnical investigations of slope instabilities.

REFERENCES

Cvijić, J. 1926. Geomorfologija, II knj. Državna štamparija Kraljevine SHS, Beograd.
Marković, M. & Pavlović R. 1987. Fotogeološka studija stabilnosti priobalja na desnoj obali Dunava, na potezu od brane "Djerdap-I" do brane "Djerdap-II". Fond Instituta za vodoprivredu "J.Černi", Beograd.
Pavlović, R. 1990. Kompleksna analiza reljefa kao metod geološkog istraživanja. Doktorska disertacija. Rudarsko-geološki fakultet, Beograd.

Development and application of slope stability assessment system by using satellite remote sensing data and digital geographic information

Développement et application du système d'évaluation de la stabilité des pentes utilisant les donnés de la télédétection par satellite et l'information géographique digitale

M. Kurodai & H. Kasa
Technical Research Institute, Hazama Corporation, Japan

S. Obayashi & H. Kojima
Faculty of Science and Technology, Science University of Tokyo, Japan

ABSTRACT: It is important to predict the time(period), the place(location) and the scale(magnitude) of the slope failure for disaster prevention. A technique for predicting the location of the slope failure using aerial photograph, satellite remote sensing data and digital geographic information, has been developed. The "Slope Stability Assessment System" based on statistical estimation was made using personal computer. The prediction model and the application results of the system to several natural slopes and construction sites were reported.

RESUME: Pour prévenir les désastres dues aux ruptures du talus, il est important de prévoir quand, où et sur quelle étendue s'engendrait une rupture du talus. Dans nos jours sont bien développées les technologies de prévision de la localisation du désastre sur le talus au moyen des vues aériennes, de l'analyse de données de télédétection par satellite ou de données géographiques numériques. En ce sens, les auteurs ont conçu le modèle de prévision en vue de localiser statistiquement une rupture du talus et en fin mis au point le système d'évaluation de la stabilité du talus basé sur ledit modèle et fonctionnant sur l'ordinateur personnel. L'objet de cet essai est donc de présenter le contenu du modèle de prévision et les résultats d'application du système d'évaluation de la stabilité du talus effectivement appliqué à de divers désastres naturels sur le talus ainsi qu'à ceux des travaux de construction.

1. INTRODUCTION

The Japanese Archipelago is located on one of the world's most active seismic belts and has very complex topographic and geological structures. This district is subject to frequent earthquakes and many volcanoes are active at present. The climate of Japan shows the distinct characteristics of a temperate monsoon climate, and average annual rainfall is from 1000mm to 2000mm. Slope failures are frequently caused from these conditions, and in particular landslide among these failures is located in Tertiary deposits and fractured zones. Most people live on the plains which constitute 30% of the country. Therefore the construction of roads, railways, dams and the other structures has continued on the hill close to the urban areas.

Construction in these areas, which are potential slope failure areas, tends to bring about slope failure directly. Because of this, it is important to predict slopes in danger of failure beforehand and to assess the slope stability appropriately. The range of area, predicted for slope failure caused by construction work, can be limited, and slope failure can be prevented during the pre-project investigation stage with suitable countermeasures. In short, prediction of slope failure makes it possible to prevent slope failure. If slope failure can be predicted in advance, the plan and the construction method can be changed and there will be a big merit in terms of costs.

Recently, satellite remote sensing data with geographic information is used in assessing slope stability[1],[2]. This method involves analysis of various slope characteristics using topographic and geological information in addition to information of land surface cover.

Under circumstances described above, the authors have constructed a system to accurately predict the areas in which slope failure is liable to occur, using statistical methods on the satellite remote sensing data and digital geographic information.

2. PURPOSE

The main purpose of the development is to build a slope failure prediction system that can be easily used on site. This report has the following three objectives:

1. To propose a "prediction model" which predicts the areas liable for slope failure using satellite remote sensing data and digital geographic information;

2. To construct a "Slope Stability Assessment System" (called "SSA-System" hereafter) which can make predictions from data input and project analyzed results, on the basis of the proposed "prediction model"; and

3. To confirm the prediction accuracy of the system by applying to several natural slopes and construction sites;

3. PROPOSITION OF "PREDICTION MODEL" [3],[4]

Fig.1 shows the flowchart of the "prediction model". The satellite remote sensing data and digital geographic information on the topography, geology, etc. are analyzed using statistical techniques described after, a Slope Failure Prediction Map based on the analyzed results is drawn up.

In general, the risk evaluation of the slopes was given points in proportion to the importance of various characteristics on slopes in danger of failure. However this process is very ambiguous because of the relative experience and subjectivity of the engineers concerned. The proposed model objectively assesses slopes susceptible to failure on the basis of the caluculated scores for each pixel (picture element : analysis unit) by the Quantification Method Type II, say, discriminant analysis with categorical data. In identifying potential failure areas, the analyzed area is divided

into a failure group and a non-failure group by the Mini/max second group discriminant Method (Fig.2). As shown in Fig.2, the "hit ratio" is utilized for the accuracy to divide into secondgroup. The objectivity of "prediction model" improves much by using "hit ratio", because there is no numerical index exactly to express the prediction accuracy until now.

Thus the Mini/max second group discriminant Method is mechanical method to predict slopes in danger of failure, but the knowledge and exprience of experts must not be ignored. In short, when the elements such as experts' knowledge, which are difficult to represent quantitatively, is broken out, the elements can be reflected as training data selected as standards to predict using the Quantification Method Type II. For example, the training data is selected by the experts as potentially dangerous slopes in study area, judging from the overall circumstances around the potential failure slope or the locations of actual slope failure. A description of each steps in the "prediction model" is as follows:

STEP1 : Areas in which slope failure prediction is to be performed are identified. Since the "prediction model" is designed for predictions on construction work and display area on colour monitor is limited, a single area is assumed to cover 3.0km × 1.5km. And wider areas will also be possible.

STEP2 : Soil, Subsurface geologic and Vegetation maps including endogenous factor data and training data are collected and classified. It is desirable that the scale of the maps is above 1:10000 for prediction accuracy.

STEP3 : Mesh mapping of the data obtained above is digitized for computer processing. The size of the grid squares used needs to be matched with the ground resolution of the

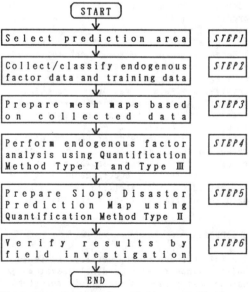

Fig.1 flowchart of "prediction model"

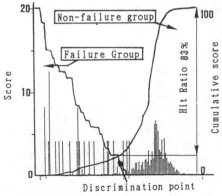

Fig.2 Mini/max second group discriminant Method

satellite remote sensing data(Landsat TM is 30m × 30m).

STEP4 : Wavebands to be suitable for the classification of the ground reflection characteristics based on satellite remote sensing data, are selected using the Quantification Method Type I , say, regression analysis with categorical data. The usefulness of the adopted endogenous factors in slope failure prediction is evaluated through the analysis using the Quantification Method Type III which is a dual scaling method, and useful endogenous factors in prediction is selected.

STEP5 : Slope Failure Prediction Map is prepared using the Quantification Method Type II and the Mini/max second group discriminant Method. The areas liable for slope failure are projected on a topographic map to aid engineers.

STEP6 : Based on the obtained Prediction Map, field investigations are conducted to evaluate the prediction accuracy and to plan for more detailed prediction.

4 . DEVELOPMENT OF "SLOPE STABILITY ASSESSMENT SYSTEM"
4.1 Hardware configuration

A personal computer, with configuration as shown in Fig.3, was used for easy operation. The hardware consists of 32bit-personal computer, hard disk drive unit for managing a number of data, frame buffer for advanced image projecting, numeric data co-processor for reducing calculating time etc.. As digitalizing endogenous factor data is requested for the rapid correspondence to the construction work, a digital scanner was used for the mapping.

4.2 Software configuration

The programming language is "C" on MS-DOS. As man-machine interface in the Japanese language was adopted, no special training is needed. The software configuration is shown in Fig.4.

1)Data Input Sub-System
1. Satellite Remote Sensing Data Processing Sub-System: Determining the "Land cover" and the "Vegetation Index" by processing the satellite remote sensing data.
2. Geographic Information Processing Sub-System: Determining the digital geographic information such as the "Slope gradient" from the Digital Terrain Model, the "Subsurface geology" and so on.

The endogenous factor data determined from these two sub-systems are managed for a data set group. This is because the confidence in and the objectivity of the prediction accuracy is assured, and because the data management is reliable. As certain knowledge is required to process satellite remote sensing data, processing and managing it at the analysis center is more effective than doing it individually on site.

2)Prediction and Analysis Sub-System
This sub-system is based on the "prediction model" shown in Section 3. The sub-system automatically processes in accordance with the menu number, software modules for each function are sufficiently considered for the flows of existing prediction techniques. Modification and improvement of the software and addition of functions become easier with this development. In particular, the training data can freely changed in accordance

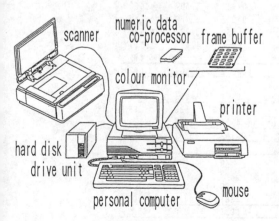

Fig. 3 Hardware Configuration

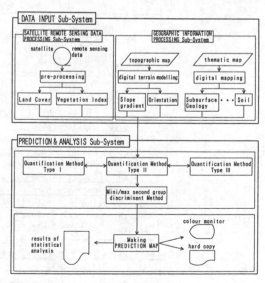

Fig. 4 Software Configuration

マウスカーソルの位置 (36. 29)
拡大表示位置 (11. 17)-(60. 42)

□ : points selected as training data
Fig. 5 Display image selection of training data

Table 1 Endogenous factor data of slope failure

Acquired Endogenous Factor Data	Endogenous Factor	Number of Categories
Vegetation map (1/50000)	Vegetation	12
Subsurface geologic map (1/50000)	Subsurface geology	8
Soil map (1/50000)	Soil	11
Topographic map (1/10000)	Orientation	17
	Slope gradient	7
Landsat-5 TM(May 21, 1987)	Land Cover	7

with each slope failures(Fig. 5). Thus, the great merit of this sub-system is the processing from various view points.

5. APPLICATION OF "SSA-System"
5.1 Application to slope failure[3]

1)Area of application
The area, where slope failures had occurred previously due to the earthquakein 1987, is 3.3km east-west × 1.5km north-south. The southwestern part is Alluvial soil, while the rest of the area consists of marine terraces in Diluvium. Slope failures frequently occur on the boundary between the two parts. The endogenous factor data for the area are shown in Table 1.

2)Analysis of endogenous factors and slope failure prediction
The results of the Quantification Method Type II and the Mini/max second group discriminant Method using the points where slope failures had occurred as training data, are shown in Table 2. The hit ratio is as high as 89.1% and the "Slope gradient" is considered to be closely related to slope failures in the area judging from its range value. In general, slope failure occurs when the angle of inclination above 20°. As the slopes in the area are above 30° and the results of the analysis coincided with the site investigations, it was confirmed that the analysis was correct.

Table 2 Results of statistical analysis

Evaluation Item		Calculated Score		
		Slope Failure	Landslide	Dam
Range	Vegetation	1.040	1.639	0.282
	Subsurface geology	1.979	1.066	0.932
	Soil	0.675	——	——
	Orientation	0.812	1.507	1.132
	Slope gradient	9.915	3.736	1.087
	Land Cover	0.660	3.277	0.277
	Vegetation Index	——	1.677	1.797
	Topographic features	——	——	5.692
Correlation ratio		0.072	0.007	0.097
Observed value	Number of failures	29	6	30
	Number of non-failures	5471	4994	4970
Predicted value	Number of failures	616	595	729
	Number of non-failures	4884	4405	4271
Discrimination point		-0.782	-1.261	-0.497
Observed value → Predicted vaiue				
Failure ⟶ Failure		26	6	26
Failure ⟶ Non-failure		3	0	4
Non-failure ⟶ Failure		590	589	703
Non-failure ⟶ Non-failure		4881	4405	4267
Hit Ratio(%)		89.1	88.0	85.2

Table 3 Endogenous factor data of landslide

Acquired Endogenous Factor Data	Endogenous Factor	Number of Categories
Vegetation map (1/50000)	Vegetation	11
Subsurface geologic map (1/5000)	Subsurface geology	7
Topographic map (1/10000)	Orientation	17
	Slope gradient	9
Landsat-5 TM (Jun. 16, 1985)	Land Cover	6
	Vegetation Index	10

Table 4 Endogenous factor data of unstable block

Acquired Endogenous Factor Data	Endogenous Factor	Number of Categories
Vegetation map (1/50000)	Vegetation	6
Subsurface geologic map (1/5000)	Subsurface geology	10
Topographic map (1/10000)	Orientation	16
	Slope gradient	10
Landsat-5 TM (Oct. 22, 1984)	Land Cover	5
	Vegetation Index	14
Aerial photograph (1/13500)	Topographic features	9

Fig. 6 shows the Prediction Map(3.0km × 1.5 km shown) for slopes susceptible to failure. The slopes are along the terrace scarp. According to site investigations in these slopes, trees with winding roots were observed. Winding tree's roots like this proved that slope failures had occured. Other small slopes observed to be liable to fail but not shown on the topographic maps or in the aerial photographs, were found in the predicted area which verified the accuracy of the Prediction Map.

5.2 Application to landslide[5]

1)Area of application
The area is a typical Tertiary type landslide. The landslide had been active for several years due to melting snow, but now it is controlled by various countermeasures. The area is 3.0km east-west × 1.5km north-south. It consists of Neogene deposits covered by Quarternary scrces. The Neogene deposits consist of mainly dark brown shale with some green tuff. The hydrated tuff is very viscous and acts as a lubricant for landslides. The endogenous factor data are shown in Table 3.

2)Analysis of endogenous factors and slope failure prediction
Landslide displacements were monitored by electro-optical survey for six months from September, 1985. The six points which had moved the most(about 30cm) out of the fourteen monitoring points were selected as training data, and the areas susceptible to landslides were predicted. Table 2 shows the results of the statistical analysis. The hit ratio is as high as 88.0%. The range values for the "Slope gradient" and the "Land cover" are large. On the other hand, the "Subsurface geology", which in general has a close relationship to landslides, is not so large.
The large value for the "Land cover" suggests that the satellite remote sensing data which provides the information on the sur-

face state, is effective in this application. Utilizing this data is effective for predicting landslides covering a wider area of activity than that of slope failures.
The Prediction Map for slopes susceptible to failure is shown in Fig. 7. The west slope of the main scarp, which is the boundary of the landslides, was shown very clearly to be a area in danger of failure. This fact proves that the prediction was correct. The predicted slopes are close in accordance with the large active landslide blocks found in another study[6].

5.3 Application to Dam construction site

1)Area of application
The area is 3.0km × 1.5km and covers the dam site and the reservoir area. The geology is Shimanto group sandstone and shale covered with Shirasu(sand and gravel) and some talus sediments. The endogenous factor data are shown in Table 4.

2)Analysis of endogenous factors and slope failure prediction
Slope failure prediction was carried out for the "unstable block", selected as training data from photo interpretation. The results are shown in Table 2. The hit ratio is 85.2% and the "Topographic features" has a close relationship to the unstable block.
The Prediction Map shown in Fig. 8 shows that the slopes in danger of failure grouped on the right bank of the river(center of Fig. 8). As these areas are very similar topographic feature to the areas selected as training data, prediction was carried out again excluding the "Topographic features". There were almost no changes to the predicted slopes. From this, it is clear that if selection of the training data is appropriate, even if one endogenous factor has a big effect on the prediction, there were no effects on the legitimacy of the prediction.

27

Fig. 6 Slope Failure Prediction Map(for slope failure)

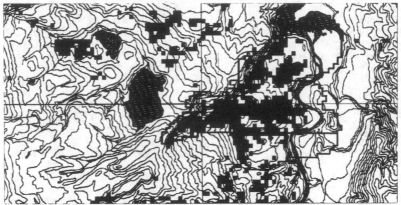

Fig. 7 Slope Failure Prediction Map(for landslide)

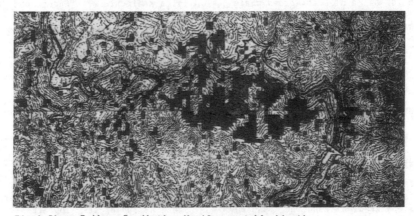

Fig. 8 Slope Failure Prediction Map(for unstable block)

LEGEND ■ : areas predicted as slopes susceptible to failure

Table 5 Applications for the "SSA-System"

OBJECT	GEOLOGY	EXOGENOUS FACTOR	RENGE OF AREA	PREDICTED RESULTS HIT RATIO	PREDICTED RESULTS SLOPES IN DANGER OF FAILURE	EVALUATION OF RESULTS
1 Landslide	Alluvial soil (Kanto loam)	Earthquake	3.3km×1.5km	87% ~ 89%	The slopes are distributed along terrace scarp	Propriety of constructed prediction model is comfirmed
2 Landslide Slope failure	Tertiary madstone	Typhoon	3.0km×1.5km	74% ~ 88%	The slopes are separated each type of failures	It is possible to predict for different failure's types in the same area
3 Colluvial slope	Alternation of sandstone and slate	Dam construction	1.8km×4.5km	87% ~ 93%	The slopes which had failed at the left bank of dam are clearly predicted	It is possible to apply to slope stability assessment under dam construction
4 Landslide Unstable block	sandstone and shale (Shimanto group)	Dam construction	3.0km×1.5km	85% ~ 89%	The predicted slopes are identical with dangerous slopes from site investigation	Propriety of method to select training data is comfirmed
5 Landslide Slope failure Unstable block	Tertiary tuff	Dam construction	2.25km×1.5km	80% ~ 87%	The slopes are grouped on the right bank of dam and identical with dangerous slopes from site investigation	It is possible to predict unstable and active blocks with virgin ponding
6 Landslide	Neogene shale	Melting snow	3.0km×1.5km	86% ~ 88%	Landslide blocks with different displacements are clearly predicted	It is possible to predict different landslide blocks by jointing results of electro-optical survey
7 Devris flow	granite	Typhoon	21.6km×17.6km	83% ~ 92%	The slopes in danger of devris flow are predicted each unit of geology	It is possible to apply to devris flow with wider affected region

The "Vegetation Index" is the next largest after the "Topographic features", and it is considered that monitoring changes in this index over a period will be an effective method to assess the slope stability around a reservoir.

The unstable blocks in this application were assessed to be where the growth of vegetation is good on north-facing slopes, indirectly suggesting that there is an abundant supply of underground water.

6. CONCLUSIONS

1) Conclusion of three applications
The "SSA-System" was applied to the areas where natural failures due to rainfall and melting snow had occurred, and be confirmed to be able to make an accurate estimate. There were only slight differences between the predicted results and the actual site situation, thus verifying the practical capability of the system.

Application to a dam construction site using the "unstable block" as training data, showed the possibility of appropriate prediction because of close correspondence between predicted slopes and site investigations. It is considered that the "SSA-System" is much effective to the prediction of slope failures.

2) Evaluation of practicability for the "SSA-System"
As the system easily to assess slope stability for a member of administrative organ and constructor is developed, it is very important in developing process to reduce the time inputting and analyzing data, not to limit the slope failure's type and application sites and to make planning of slope failure prevention etc.. Described above, the "SSA-System" can flexibly correspond to these requirement.

Applications for the "SSA-System" including three application reported in this paper are shown in Table 5. This table shows that the "SSA-System" can predict various slope failures' areas due to different exogenous factors on different topographic and geological conditions.

It is proved that the "SSA-System" can practically utilize for slope stability counterplan on natural failure prevention plan and construction work.

7. CURRENT PROBLEMS

The current problems facing the "SSA-System" can be summarized as follows:

1) Standardization of data selection
The three standardization methods described below should be examined.
1. Method to select satellite remote sensing data with consideration of the time when the data was taken;
2. Method to incorporate new endogenous factor data excluding natural conditions;
3. Method to select training data ;

29

2)Improving data management methods
A data management capability is necessary because of the great number of data sets stored. The "SSA-System" allows step-by-step growth and development of the system by providing capability for both data set changes and management, and the development of both the data set knowledge base and its management will be required from now.

3)Dealing with exogenous factor data
The "SSA-System" predicts slopes susceptible to failure by analyzing the endogenous factor data from various view points, but does not deal with the exogenous factor data directly. This system indirectly deals with slope failures due to earthquakes and rainfall, with the training data. A method to input information such as earthquake magnitude, precipitation and duration of rainfall should be developed in the future.

8. FUTURE PERSPECTIVES

Slope failure prediction is very difficult in general. Effective techniques for prediction have to be developed so that the solutions may be as correct as possible. Having divided prediction into prediction of either local "points" or wide area "planes", the "SSA-System" approach corresponds to the latter, and this approach gave the results expected. It will be important in the future to define the procedure for slope failure prediction, such as conducting "point" prediction on potentially dangerous slopes predicted using "plane" prediction, as a total system using both "point" and "plane" prediction.

It can be seen that the number of endogenous and exogenous factor data become larger, the greater total system is going to be developed. Therefore, it will be important to formulate a data base management system for all this data.

As the final stage of development, construction of a system enabling "real-time prediction" to be made from time to time on rainfall forecasts announced by meteorological organizations is greatly desirable. With the recent rapid development of electronics technology, the momentum for the development of scientific technology has increased rapidly. In the future it is hoped to complete the prediction system by introducing the techniques into other fields effectively.

REFERENCES

1)Lawrie. E. Jordan Ⅲ 1990. Integration of GIS and Remote Sensing, GIS SOURCE BOOK, Geographic Information System Technology in 1990, pp. 261 ～264.
2)General System Characteristics(Table 1) 1990. GIS SOURCE BOOK, Geographic Information System Technology in 1990, pp. 20 ～25.
3)S. OBAYASHI, H. KOJIMA, H. KASA 1990. APPLYING SATELLITE MULTI-SPECTRAL SCANNER DATA FOR LANDSLIDES PREDICTION, PROCEEDINGS OF JSCE, No. 415/ Ⅵ-12.
4)H. KASA & M. KURODAI, H. KOJIMA & S. OBAYASHI 1992. Study on landslide prediction model using satellite remote sensing data and geographic information, Landslides, PROCEEDINGS OF THE SIXTH INTERNATIONAL SYMPOSIUM ON LANDSLIDES, pp. 983-988.
5)M. KURODAI, H. KASA, K. FUJITA, S. OBAYASHI, H. KOJIMA 1993. Possibility of analyzing the causal relation between displacements and primary causes, PROCEEDINGS OF THE 28TH JAPAN NATIONAL CONFERENCE ON SOIL MECHANICS AND FOUNDATION ENGINEERING - JSSMFE, pp. 105-106.
6)T. TERAKAWA, S. NISHIDA, M. KONDO 1979. A Study on the Relationship between the Landslide and its Geological Conditions in Yachi, Akita Prefecture, Japan, Landslides - Journal of Japan Landslide Society, Vol. 16, No. 1, pp. 9-18.

Inductive electromagnetic methods in engineering geology: A case study

Méthodes électromagnétiques inductifs en géologie de l'ingénieur: Un étude de cas

F.A. Ferreira
EDP, Porto, Portugal

M.J. Senos Matias
University of Aveiro, Portugal

ABSTRACT: Conventional Electrical Methods have been applied to engineering studies, but their use may have been restricted by logistics and operation costs; however, the recent development of techniques that enable efficient and fast direct terrain conductivity measurements, using inductive electromagnetic methods, opens new possibilities for the use of Electrical Methods in Engineering Geology. In this work, these inductive electromagnetic techniques are tested and fully interpreted in an area where previous conventional resistivity and seismic refraction data were obtained and the advantages and disadvantages of the method are discussed.

RÉSUMÉ: Le développement et l'utilization courante des méthodes électriques classiques en Géologie Appliquée peuvent être feinés par des coûts opérationnels et logistiques; récemment, le développement des méthodes électromagnetiques permettant d'obtenir d'une manière rapide des mesures directs de conductivité a apporté un nouveau élan à l'utilization des méthodes électriques en Géologie Appliquée. Dans ce travail nous avons testé la méthode électromagnetique dans une zone où ont été effectués des travaux de resistivité et de sismique de refraction et discuté les avantages et désavantages de la méthode.

1. INTRODUCTION

Geophysical Methods play an important role in Engineering Geology studies, such as high-rise buildings, tunnels, dams, waste disposal, ground water studies and determination of engineering parameters. Commonly geophysical surveys are used to aid a rapid economical choice between a number of alternative sites for a proposed project or as a part of a detailed site assessment at a chosen location. Nevertheless, Geophysical Methods are always intended to supplement direct methods, Cratchley (1978). Resistivity and Seismic surveys have been the most used techniques, but the development of instrumentation and associated interpretation techniques led to the more recent application of other methods, Table 1 adapted from McDowell et al (1988). Conventional resistivity methods have been applied to Engineering Geology, either using profiling, mapping or sounding. However, its use requires considerable manpower, accurate and dense grid of the area under investigation, physical contact between the electrodes and the ground, data can be seriously affected by topographic

Table 1 Geophysical Methods in Engineering Geology; 4-excellent, 3-very good/excellent potential, 2-fair, 1 poor, 0-not used

METHODS / APPLICATIONS	Seis. Refraction	Seis. Reflection	Ele. Sounding	EM	GPR
Depth of the bedrock	4	2	4	3	2
Stratigraphy	4	2	3	2	3
Faults	3	1	2	4	2
Elastic Parameters	3	0	0	0	0
Ripability	4	0	1	1	0
Cavity detection	1	2	2	3	3
Buried objects	1	1	1	3	4
Hydrogeology	2	2	4	4	2
Buried channels	4	2	3	3	2

effects, ground inhomogeneities in the vicinity of the electrodes and some data processing is usually necessary to obtain ground resistivity values.

As it can be seen in Table 1, the application of inductive electromagnetic methods in Engineering Geology has an excellent potential, although its use is not fully developed yet. These techniques overcome most of the problems arising from the use of conventional resistivity, and recent development of equipment enables the direct ground conductivity measurement, McNeill (1980) and (1988).

Generally, a time varying magnetic field, with frequency in the range of 100 to 5000 Hz, is used as an energy source. To generate this primary field, H_p, an electric current is passed through a transmitter coil, Tx in Fig. 1. This magnetic field induces small currents into the ground and thus generate a secondary magnetic field, H_s, measured together with the primary field in a receiver coil, Rx in Fig. 1. This secondary field, H_s, is a function of the distance, s, between the coils, the frequency, f, and of the ground conductivity σ. For certain values of "s" and "ω=2πf" such function is very simple and can be shown to be, McNeill (1980):

$$\frac{H_s}{H_p} \propto \frac{i\omega\mu_0\sigma s^2}{4} \qquad (1)$$

where $i^2=-1$ and μ_0 is the permeability of the free space.

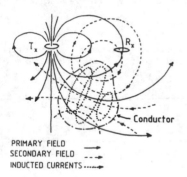

PRIMARY FIELD ⟶
SECONDARY FIELD ⇢
INDUCTED CURRENTS ⟶

Fig. 1   EM Transmitter - Receiver system

The ratio of the secondary field to the primary field is linearly proportional to the terrain conductivity, and thus it is possible to obtain this value once H_s/H_p is measured, that is:

$$\sigma_a = \frac{4}{\omega\mu_0 s^2} \left(\frac{H_s}{H_p} \right) \qquad (2)$$

The MKS units of conductivity are the mho,

or siemen per meter, or, more conveniently, the millimho per metre.

The equipment used can operate with distances between the coils of 10, 20 and 40 metre and operating frequencies of 6.4, 1.6 and 0.4 KHz, respectively. The coils can be used on the horizontal plane, thus using vertical fields, VD, or on the vertical plane, that is, horizontal fields, HD. The reported exploration depths are, McNeill (1980):

Table 2 Reported depths of exploration values for EM34

Exploration depths (m)	Horizontal fields (HD)	Vertical fields (VD)
Intercoil spacing "s"(m)		
10	7.5	15
20	15	30
40	30	60

The receiver, Rx, and transmitter, Tx, coils are connected by a reference cable and an automatic check of the distance between the coils is carried out, so that operation is only possible if that distance is the appropriate one. Thus ground gridding efforts are reduced considerably.

The Inductive Electromagnetic Method was tested, and the data were fully interpreted, in an area where previous conventional Resistivity Mapping and Seismic Refraction profiles had been carried out as part of a detailed site study for the future construction of a dam. This area is in the Mondego River right bank, central Portugal (lat 40° 24'N, long 7° 59'W).

2. GEOLOGY OF THE AREA

The area where the method was tested is located in a granitic valley. The local geology is described in Teixeira (1961), Ferreira (1990) and Rodrigues et al (1984) and consists of coarse grain porphyroid granites, weathered towards the surface. The area is covered by a granitic soil and near the river some alluvial deposits, up to four metre thick, can be found. Granitic boulders are also found frequently.

From the tectonic point of view there are three main sub vertical faults, Ferreira (1990), one N63E running parallel to the Mondego River approximately, another N48W roughly perpendicular to the river, and a less important one N36W also perpendicular to the Mondego River.

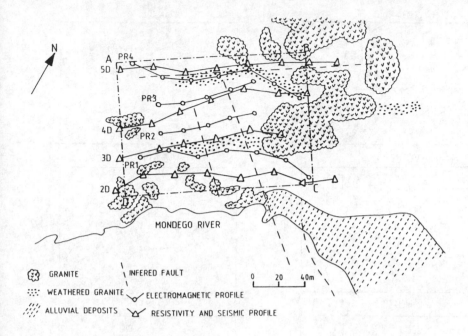

Fig. 2 Geology of the area

3. ELECTROMAGNETIC SURVEY

The electromagnetic data were obtained using the four profiles in Fig. 2, Ferreira (1992). These profiles were oriented perpendicular to the geological features and the use of both vertical and horizontal fields, at spacing of 10, 20 and 40 metres between the coils, and thus three different frequencies, enabled the construction of:
 - six apparent resistivity maps, in Fig. 3;
 - four electromagnetic sections, obtained from the interpretation of electromagnetic sounding data, Fig. 4.

3.1 Electromagnetic Maps

All the electromagnetic maps in Fig. 3, defined by the box ABCD in Fig. 2, display a central area of higher conductivity with an approximate direction NW-SE and a vertical or sub-vertical behaviour. This area must be related to the sub vertical fault N48W, that runs perpendicular to the Mondego River (Ferreira, 1990).

To the centre top of the maps, and specially in those corresponding to shallower investigations depths, there is another higher conducting zone. This zone corresponds to a highly weathered area in relation with the fault N63E, parallel to the Mondego River.

From the qualitative interpretation of the electromagnetic maps it is not possible to identify, or to discriminate, the two faults with directions N48W and N36W (Ferreira, 1990) perpendicular to the Mondego River.

3.2 Electromagnetic Sections

As six measurements were taken for each point, electromagnetic sounding interpretation was carried out using appropriate interpretation software, Interpex (1988), and results are shown in Fig. 4. In this figure, section PR1 is the nearest to the river while PR4 is the furthest one.

Analysis of Fig. 4 reveals that:
 - all the sections show a central deeper conductive zone, and thus are coherent with the mapping discussed previously;
 - as the profiles are closer to the river, the granitic bedrock is shallower; the depths of the bedrock vary from 10 metres, in section PR1, to 20 metres, in

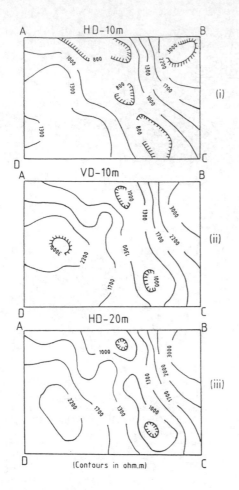

Fig. 3 Electromagnetic Maps

sections PR2 and PR3;
  - the thickness of the weathered layer is variable, although it is thinner towards the river, and varies between 10 and 15 metres;
  - section PR4 shows two conductive areas at depth, that may be associated with the faults N48W and N36W; it must be noticed that in this region the distance between the two faults should be the largest in the area under investigation and, in fact, electromagnetic maps VD-10 and HD-20 suggest the existence of these faults.

4. RESISTIVITY MAP

The area ABCD, in Fig. 2, was mapped using conventional resistivity methods. Thus, the resistivity profiles 2D to 5D, in Fig. 2, were carried out with a Schlumberger array (AB=65 metres, MN=5 metres) with a spacing of 25 metres between readings.

The resistivity map, in Fig. 5, shows clearly a structure with a NW-SE direction in the same area of the one revealed by the electromagnetic mapping. However, in Fig. 5, this structure is composed by two narrow conducting bands separated by a more resistive one. Thus it seems that conventional resistivity mapping discriminated the effects of the two faults N48W and N36W.

It is known that array geometry can induce "noise" in resistivity profile readings carried out over simple two dimensional structures, such as faults, contacts, dikes, Acworth and Griffiths

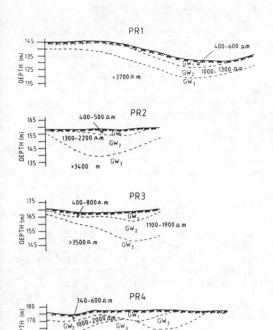

PR1

400-600 Ω.m

>3700 Ω m

GW₁
GW₂ 1000-1300 Ω.m
GW₃

PR2

400-500 Ω.m

1300-2200 Ω.m   GW₁
                 GW₂
>3400   m   GW₃

PR3

400-800 Ω.m

GW₁
GW₂   1100-1900 Ω.m
>3500 Ω.m   GW₃

PR4

340-600 Ω.m

GW₂ 1000-2000 Ω.m   GW₁   GW₂
                    GW₃
>3100 Ω.m

GW₁ - granitic soil
GW₂ - weathered granite
GW₃ - granite

0 10 20 30 40 (m)

Fig. 4 Electromagnetic Sections

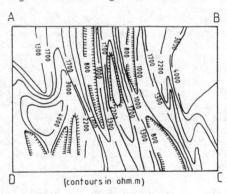

A   B

D   (contours in ohm.m)   C

Fig. 5 Resistivity Map

(1985). Therefore geometry array effects may account for the anomalies in Fig. 5 and thus they may be false anomalies and so a larger conductive area could be interpreted solely, as in the case of the electromagnetic maps of Fig. 3.

However, geological evidence and the overall EM interpretation favour the hypothesis of the two faults.

5. SEISMIC REFRACTION

Four refraction profiles , 2D to 5D in Fig. 2, were carried out in the area ABCD. Interpretation of these profiles is shown in Fig. 6.

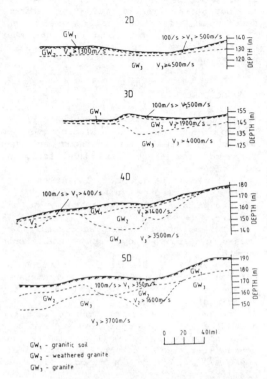

2D

GW₁   100/s >V₁>500m/s
GW₂ V₂>1300m/s
          GW₃   V₃≥4500m/s

3D

                100m/s> V₁500m/s
GW₁
          GW₂ V₂ 1900m/s
          GW₃   V₃ >4000m/s

4D

100m/s> V₁>400/s
          GW₁   V₁≥1400/s
V₂
                GW₂
          GW₃   V₃ > 3500m/s

5D

          100m/s > V₁ >350m/s   GW₁   GW₃
GW₂   GW₂
          GW₃   V₂ >1600m/s

          V₃ > 3700m/s

GW₁ - granitic soil
GW₂ - weathered granite
GW₃ - granite

0   20   40(m)

Fig. 6 Seismic Refraction Profiles

The location of electromagnetic and seismic profiles is not the same but comparison between the electromagnetic sections and the seismic sections reveals that the overall conclusions are the same in both cases, that is:

- profiles PR4 and 5D are topographically almost coincident and obviously their interpretation is very similar. However, the interpreted seismic depths are greater than those from the electromagnetic interpretation;
- profile PR3 is coincident with part of profile 4D and the interpreted depths in both methods are very similar;
- profile PR2 is located between profiles 3D and 4D and the part of the electromagnetic profile, that can be compared with the seismic profiles, provides interpretation depths very close to those obtained from the interpretation of seismic data;
- profile PR1 is located between the

seismic profiles 2D and 3D the interpretation is again identical and electromagnetic depths are between the seismic depths for profiles 2D and 3D.

However, it must be noticed that the interpreted depths are usually greater, two to four metres, in the case of the seismic sections but it should be remembered that the profiles are not coincident and that the methods deal with different physical properties.

6. CONCLUSIONS

The inductive electromagnetic method proved to be an excellent alternative to conventional resistivity methods and when compared to these has the following advantages:
- field operations, logistics and equipment transportation are easier;
- two persons are enough to carry out all the field tasks;
- can be adapted more successfully to areas of rough topography and dense vegetation;
- provides a far better compromise between depth of penetration and distance between coils than conventional resistivity soundings;
- no need for physical contact with the ground, enabling the operation in hard rock outcropping areas as well as in regions where electrical contacts are difficult such as very dry conditions, ice and permafrost;
- less dependent to lateral resistivity variations and to local ground inhomogeneities;
- enables continuous readings
- excellent operational versatility enabling its use both in general and detailed studies.

However the following disadvantages must be point out:
- its operation is difficult in noisy areas (industrial areas, power lines, metallic fences, buried electrical cables);
- inductive electromagnetic methods use A.C. currents and therefore depths of penetration are limited to the range of frequencies available;
- interpretation of electromagnetic soundings is somewhat unstable and the starting model must be of adequate accuracy;
- electromagnetic equipment is more expensive than conventional resistivity equipment.

It must be remembered that EM data gives an overall average of local conductivity values which can be advantageous or disadvantageous depending on the characteristics and aims of the survey.

From the previous discussion is clear that the advantages of inductive electromagnetic methods overcome the disadvantages, but some resistivity soundings should be carried out to provide the starting interpretation models. However, this procedure must be common to the application of any geophysical method in a particular project.

Finally it must be pointed out that all the inductive electromagnetic mapping and soundings, herein discussed, were carried out by two persons in just three days.

7. ACKNOWLEDGEMENTS

The authors wish to thank JNICT for the financial support of one of us (F. Ferreira), EDP that provided the seismic and resistivity data and Miss Trindade for helping out in the field.

REFERENCES

Acworth, R. and Griffiths, D., 1985. Simple Data Processing of Tripotential Apparent Resistivity Measurements as an aid to the Interpretation of Subsurface Structure. *Geophysical Prospecting* 33: 861-887.

Cratchley,C., 1978. *Manual of Applied Geology for Engineers*. Institution of Civil Engineering, London.

Ferreira, F., 1992. *Métodos Electromagnéticos de Prospecção aplicados em Geologia de Engenharia*. Tese de Mestrado, Universidade Nova de Lisboa, Portugal.

Ferreira,M., 1990. *Aplicação da Geologia de Engenharia ao Estudo de Barragens de Enrocamento*. Centro de Geociências da Universidade de Coimbra, Portugal.

Interpex, 1988. *EMIX-34, User's Manual, EM Conductivity Data Interpretation Software*. Interpex Ltd, Golden,Co, USA.

McDowell,P., Hooper,J., Barker,R. Darracot,B., Jackson,P., McCann,P. and Skipp,B., 1988. Engineering Geophysics, Report by the Geological Society Engineering Group Working Party. *Quartely Journal of Engineering Geology* 21: 207-271.

McNeill,J., 1980. *Electromagnetic Terrain Conductivity Measurements at Low Induction Numbers*, TN-6 Geonics. Geonics Limited, Ontario, Canada.

McNeill,J., 1988. Advances in Electromagnetic Methods for Groundwater Studies. In *Proceedings of the Symposium on the Application of Geophysics to Engineering and Environmental Problems*,

252-348, Golden, Co, US.

Rodrigues, L. e Oliveira, R., 1984. *Prospecção Geofísica em locais de barragem do Rio Mondego - Local de Midões*. Laboratório Nacional de Engenheiro Civil, Lisboa, Portugal.

Teixeira, C. et al, 1961. *Carta Geológica de Portugal - Nota Explicativa da folha 17C (Santa Comba Dão)*. Serviços Geológicos de Portugal, Lisboa, Portugal.

Acoustic emission in salt-work (Slanic salt mine)
L'émission acoustique dans les mines de sel (La mine de Slanic)

Cristian Marunteanu & Victor Niculescu
Bucharest University, Romania

ABSTRACT: Periodic measurement of acoustic emission activity using an AE analyzer is a method able to detect and to locate any significant stress - strain concentration generating stress waves in roofs, walls or pillars of mining works. The analyzer used for analysis of acoustic emission activity counts by an electronic timer the relative time of arrival at the transducer and the amplitude of the pulse reaching the transducer, 4 level amplitude distribution displayed. Registration of the acoustic emission both of the salt samples in compression tests and of the salt massif in mining works, correlated with the stress - strain curves and strain - time curves, allowed to define a model of relationship between the stress - strain in the salt massif and the acoustic emission. The peculiar evidence of acoustic emission curves gives a clear warning before rupture. This particular evolution of the acoustic emission curves enables not only to locate the stress zones and to watch their behaviour in time, but also to forecast and to warn the failure of roof or pillars in salt mines. The case of Slanic salt mine evolution is discussed.

RESUMÉ: La mésure périodique de l'émission acoustique utilisant un analyseur d'EA ç'est une méthode capable de détecter et de localiser les concentrations significatives des contraintes - déformations générant des ondes de contraintes dans les plafons, les parois ou les pilliers des travaux minièrs. L'analyseur d'EA utilisé pour l'analyse de l'émission acoustique compte à l'aide d'un "timer" éléctronique le temps rélatif d'arrivée au traducteur et l'amplitude de l'impulse atteignant le traducteur, distribuée sur 4 niveaux d'amplitude. L'enregistrement de l'émission acoustique, tant sur des échantillons de sel essayés à compression que dans les travaux minièrs d'un massif de sel, corrélé avec les diagrammes contrainte - déformation et déformation - temps, ont permis de définir un modèle de relation entre l'état de contrainte - déformation dans le massif de sel et l'émission acoustique. L'apparence particulière de courbes d'émission acoustique montre un avertissement distinct avant la rupture. Cette évolution des courbes d'émission acoustique permet non seulment de localisér les zones soumises aux contraintes et de surveiller leur comportement, mais aussi de prévoir et d'avertir la rupture des plafons ou des pilliers dans la saline. Le cas de l'évolution de la saline de Slanic est presenté.

1. INTRODUCTION

Most rocks produce acoustic emission activity when stressed. The release of energy internally stored in a structure (e.g. a rock mass) generates short impulsive stress waves or elastic waves that can be defined as acoustic emission (mechanical shocks and seismic phenomena are not included in this part of spectrum).

The source of acoustic emission (AE) are different processes emitting elastic waves: dislocation movements, phase transformations, friction mechanisms, crack formation and extension.

The signals emitted may be divided into two types: continuous emission (resembling with noise) and burst type emission. The acoustic emision signals generated by the crack formation are of the burst type which are often emitted at a very high rate in the case of rocks.

The AE analyzer used for analysis of acoustic emission activity counts by an electronic timer the relative time of arrival at the transducer and the amplitude of the pulse reaching the transducer, 4 level amplitude distribution displayed. The AE count rate (counts/time unit) vs. time curve provides an output proportional to the severity of AE activity.

Periodic measurement of acoustic emission activity using an AE analyzer is a method able to detect and locate any significant stress - strain concentration generating stress waves in roof, walls or pillars of mining works.

Registration of the acoustic emission both of the salt samples in compression tests and of a salt massif in mining works allowed to define a model of relationship between the stress - strain in the salt massif and the acoustic emission.

The particular evolution of the acoustic emission curves, correlated with the stress - strain curves and

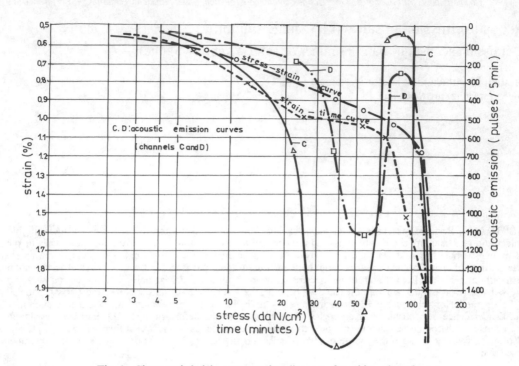

Fig. 1. Characteristic laboratory testing diagrams for white salt rock

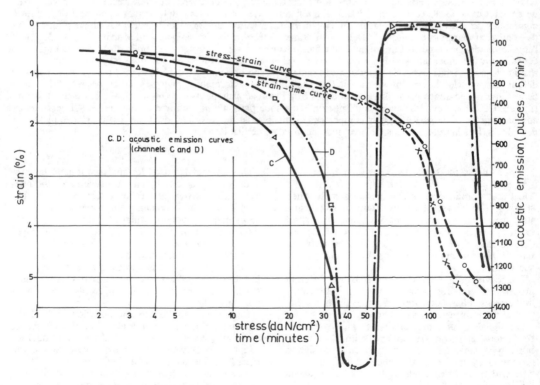

Fig. 2. Characteristic laboratory testing diagrams for stratified salt rock

40

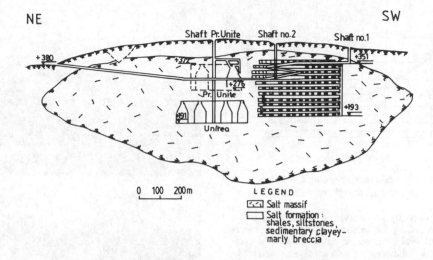

NE   SW

Fig. 3   Section draft through the salt-rock ore deposit
and the salt-mine works from Slanic

the strain - time curves, enables not only to locate the stressed zones and to watch their behaviour in time, but also to forecast and to warn the failure of roofs or pillars in salt mines.

The case of Slanic salt mine stress - strain evolution is discussed.

2. SLANIC SALT - ORE DEPOSIT

Slanic salt - ore deposit is situated in the southern part of the East Carpathians Bend, being provided by the Miocene formation of the Slanic syncline.

The salt carrying formation is the Salt formation (Middle Miocene), disposed on the tuffs and the Globigerina marls of Slanic tuffs. Salt formation is represented by shales, siltstones, sedimentary clayey marly breccia, salt and gypsum. The salt - ore deposit, developed as lense, is composed of white salt and grey stratified salt with thin marl intercalations.

The age of salt is considered by many authors to be Badenian (Upper Langhian) (Marinescu & Marunteanu, 1990).

The main physical - mechanical properties of Slanic salt are: volumic weight γ = 2.08 - 2.19 g/cm^3, compressive strength σ_c = 22.6 - 24.7 MPa for white salt and 11.0 - 13.5 MPa for grey salt, Young Modulus E = 16000 - 18700 MPa for white salt and 7350 - 16000 MPa for grey salt, Poisson coeficient ν = 0.21 - 0.38 for white salt and 0.12 - 0.37 for grey salt.

3. ACOUSTIC EMISSION IN COMPRESSION TESTS

Uniaxial compression tests on 10 x 5 x 5 cm samples of white and stratified salt from Slanic salt mine were carried out. The load was applied in steps of 16 daN/cm^2, an approx. 15 min. time for each step.

The strength of the tested specimens varies between 134 and 176 daN/cm^2. Different parameters were recorded: load, deformations, time, acoustic emission (number of pulses in the first 5 min. after loading).

Typical diagrams are presented in figures 1 and 2.

The stress - strain curves show a good correlation between stress and corresponding strain and an intensive increase of the deformation before rupture. The strain - time curves show also a rapid increase of the strain rate preceding rupture.

Despite of the relative rapid loading of the specimens, the strain - time curves of the white salt point out an interval with slow deformation rate similar to a secondary creep phase.

The AE activity (total number of pulses in a period of 5 minutes) during the compression tests was measured by a transducer connected to an amplifier and an AE pulse analyzer, 4 amplitude channels and different amplifying levels diplayed. The acoustic emission curves as a function of increasing stress in time were built using channels C and D recordings of peak output amplitudes over the noise peaks.

Figures 1 and 2 show the results. The AE activity increases when the load increases, reachs a maximum corresponding to a stress of 40 - 50 daN/cm^2,

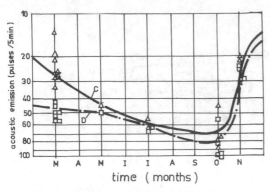

Fig.4 Acoustic emission cuves in Slanic salt-mine works (channels C and D)

but is followed by an interval of minimum acoustic activity preceding the intense activity before rupture. The interval of acoustic calm seams to correspond to the behaviour of salt specimens similar to secondary creep, especially for white salt.

This particular evidence of acoustic emission curves gives a clear warning before fracture.

4. ACOUSTIC EMISSION IN SLANIC SALT MINE

Acoustic emission measurements were also performed in Slanic salt mine. Developing workings and mining of salt have started at Slanic in the year 1974 by the room and pillar mining. The width of rooms is 16 m and the height 8 m with pillars 14 x 14 m (17.5 x 17.5 m for deepest levels). The thickness of the protective roofs is 8 m. 11 levels have been already performed (figure 3).

The AE activity has been measured periodically from March till November 1992 on the walls of the transport raise and on the pillars.Instability phenomena have permanently increased beginning from 1978: deformation of roof and floor, displacement fissures, exfoliation of pillars, slab collapse in the roof etc. Floor subsidence spread to ground surface, the rate of subsidence varying from 3 mm/month in 1985 to 14 mm/day in October 1992.

The AE measurements in some active zones in the salt mine during the year 1992 show a continuous increasing of the AE activity (figure 4), correlated with the rapid increasing of the instability: the mining roofs exhibited intensive fissuring and a growing magnitude of mine roofs displacement fissures.

In November, the last month when the measurements might be performed, the AE decreased warning the failure, similar to the behaviour of salt specimens under stress (figure 4).

The rapid evolution of the instability phenomena determined the closing of the mining activity in December 1992. Under these circumstances the roof strata fractured to surface, the salt mine collapsed and damage of ground surface was unavoidable : fracture and step breaks occur on ground surface (figure 5).

Fig. 5. Damage of ground surface above the collapsed Slanic salt mine

5. CONCLUSIONS

The results of the acoustic emission measurements on salt rocks both in laboratory and in mining works, correlated with stress - strain and strain - time curves, confirm the availability of the AE method in estimating significant stress - strain modifications and in warning the state preceding rupture in salt mining.

6. AKNOWLEDGEMENTS

The authors wish to aknowledge the permission and the support provided by Slanic Salt Mine Company and IPROMIN Bucuresti

REFERENCES

Marinescu Fl. and Marunteanu M. 1990. La paleogéographie au niveau du sel badénien en Roumanie. Geol. Zb.- Geol. Carpath., 41, 1, Bratislava.
Popescu B. and Radan S. 1976. East Carpathian Miocene molasse and associated evaporites. Guidebook serie, 15, Inst. Geol. Geofiz. Bucuresti.

Seismic measurement on Pontevedra dam

Mesures sismiques au barrage de Pontevedra

Pavel Bláha
Geotest a.s., Brno, Czech Republic

J. Knejzlík, Z. Kaláb & B. Gruntorád
Academy of Sciences of Czech Republic, Institute of Geonics, Ostrava, Czech Republic

ABSTRACT: Some new open cracks occurred on dam crest in Pontevedra during the dam reconstruction. One of possible way how the cracks came into being was the blasts of the near quarry. That is way the vibration measurements were materialized in the dam surroundings and the dam body. These measurements were fulfilled by using refraction and tomographic measurements. All these methods gave an image of dam body and dam subsoil conditions. Further necessary measurements and building arrangements were able to be taken on this basis of measuring results.

RESUME: Au cours de la reconstruction de la vieille digue á Pontevedra, les ruptures de la couronne du remblai ont été obsérvées. On a supposé que les ruptures sont provoquées par les explosions dans un carriére pas trop éloigné. Pour cette raison, les mesures des vibrations et les mesures séismiques ont été éffectuées. Sur la base des résultats, on peut juger l' état de la digue et ses fondations. On peut aussi recommander les mesures suivantes.

1. INTRODUCTION

The appreciating of the seismic effects is necessary not only for the new structure designing but there are some cases when it is necessary to investigate these effects on the completed constructions. An example of such measurement results is presented in this paper.

The new cracks appeared in the dam crest during the old masonry dam reconstruction. Because the assumption was expressed that those cracks came into being as a result of blasts (in the quarry at a distance of about 600 m) it was necessary to verify that assumption by the in-situ vibration measurements. The seismic waves were registered at three stations in three-component system of sensors. The acoustic waves due to blasting were monitored too. The storage lake emptied during our measurements.

Both the shallow seismic refraction and the seismic tomography measurement were performed in the dam to the estimate of the dam homogeneity as the basis for the processing of

seismic effects of quarry blast measurements. The task of those measurements was to give an idea of the dam body heterogeneity and to give the information for the proposal of the further test distribution that will be carried out in the dam.

The dam body in Pontevedra (Spain) that was the subject of investigation was constructed in the 1930s and 1940s. The dam is masoned from granite blocks so that the outside masonry is made from dimension stones. The dam inside is made from rubble stones and space among them is filled with the cement-lime grout. The height of the original dam was later increased to 7 m. Now the height of the dam reaches 23 m, the dam toe width is 18 m, and the dam crest length is 100 m. The concrete trough (width 35 m) is connected with the dam. The whole dam was constructed without any waterproof sealing elements. Therefore the big seepage occurred through the dam body after construction. That is why deterioration of the dam body was progressing. The dam was gradually discarded from the operation. The dam reconstruction began in the year 1992. During this reconstruction the parallel cracks with the long dam axis appeared.

The vibration measurement was carried out without any quarry co-operation so as not to change the usual blast technology with the intention of reducing harmful effects on the dam body. It was necessary to use such equipment that could record the seismic events in the triggered regime.

2. SEISMIC MEASUREMENTS

2.1 Refraction seismic

The seismic refraction results in the dam crest were anomalous. The dam body does not behave as the half-space in common geophysical investigation. Increasing velocity (or its stagnation) occurs with increasing distance receiver-source in case of usual geological bed position. The opposite effect occurs in the recorded traveltime curves in the dam crest. After arriving direct (V_0) and refracted wave (V_H) a slower wave V_1 arrives. Then it is replaced by the even slower wave (V_2).

The direct wave velocity responds to the longitudinal wave velocity in the concrete in the dam crest. The refraction horizon depth responds to the concrete cover bed thickness, too. It is obvious from V_0 velocity course that the concrete plate quality is different. It can be given by various intervals of cover concrete plate maturing in the dam crest.

The velocities on the refraction horizon are the highest ones in the dam (3.5 km/s). They respond to the velocity of the seismic wave spreading in the screen. The screen is a horizontal bed situated closely to the dam crest that occurs in that type of dam. Its compactness is increased. This is a response to the stress distribution in the dam body or to the crest dam traffic. The screen is very often separated from the dam body by a horizontal crack or at least by a discontinuity plane of mechanical properties. It is obvious from the refraction wave velocities that the properties of the screen along the dam crest aren't the same. The refraction waves going through that bed are of a high frequency (about 1000 Hz). The traveltime curve of this wave type disappears really to the source-receiver distance about 12 m. It is due to the big attenuation effect of a high frequency oscillation. Another interesting fact was the interruption of this type traveltime curves in one spot on the dam crest. It is possible to say that there must be a sub-vertical discontinuity in the dam body. Later one transversal crack was registered on this spot of the dam crest.

After disappearing screen wave 2.8 km/s velocity waves V_1 velocity appears. These waves have essentially longer running and it is possible to record them up to a distance of 35 m from their source. The wave can be characterized as a middle-frequency one. Mostly these waves are replaced to another type of waves with a lower frequency and velocity (V_2 wave). Its velocities are about 1.8 km/s. At first sight both last wave types seem to be of relatively small velocities (with consideration of the material the dam had been constructed from). We must realize that the dry or moist material was investigated. The small velocities were determined not only in the refraction wave results. 1.65 km/s velocities recorded in the dam body were also derived from the measurements of seismic effects of blasts. It is similar to the tomography velocities (see further). At first sight it is obvious that these velocities are in good accordance.

The inspection gallery that had been driven trough the dam gave us an opportunity to get to know the material the dam had been constructed from. We can see even the grout filling of the originally free spaces in the moment of seismic measurements. According to some marks on the inspection gallery walls it wasn't possible to eliminate the possibility of existence of some free spaces during the measured time. This idea can be approved by grout discharging on the dam downstream face during the consolidation grouting of the second sequence. The granite stones that make the essential part of dam inside are deteriorated with various degrees of weathering. Some stones could be scraped of by hand. All these facts suggest that the recorded velocities of longitudinal seismic waves are real.

2.2 Seismic tomography

A tomography cross section was run in the vertical plane perpendicular to the dam long axis. The seismic source and seismic receiver placements are shown in Figure1. Their interference strongly complicated the chance of tomography interpreting. The tomography processing results of the seismic radiography measurement are the seismic rays (Figure 1) and the velocity field (Figure 2).

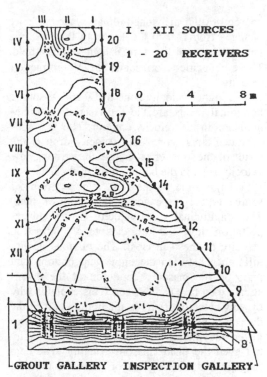

Fig.2 Pontevedra - seismic field from tomography processing.

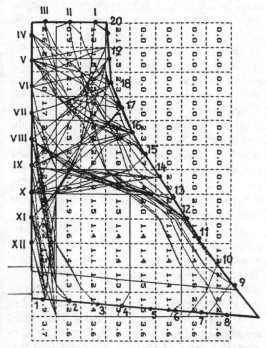

Fig.1 Pontevedra - seismic rays from tomography processing

It is unambiguously obvious on the tomographic rays how the rays avoid places with small velocities. They can be seen best at the large velocity minimum in the lower dam part. It is best proved by the rays from XI and XII sources to 10-14 receivers. The rays concentrate on the places with relatively high velocities especially in the middle of the dam, where the high velocity zone lies above the low one.

The velocity contours divide the dam body and the dam subsoil into four zones. The first ofthem is the low velocity zone under the dam crest (especially under I and II sources). At this place the example of the discontinuity under the screen occurs. In the case the rays go parallel with the dam crest, increasing effect prevails. In this case the velocity is increasing due to high velocity along the screen. If the rays must go perpendicularly to the screen, the effects of the open sub-horizontal crack under the screen manifests and it may increase the effect of other discontinuities under the screen. In this case the velocity drop occurs. Because most tomography rays go through the screen under obtuse angles, the velocity drop effect prevails. The velocity drop effect of the screen cracks is combined with the triangle extension dam block stress on the downstream face. The tensile stresses that work to this dam part result in further velocity drop. Therefore the velocity difference occurs between the cells under III and I+II sources. The horizontal velocity gradient between II and III sources gives evidence to the most tensile stresses at this place. These obviously reflected in the longitudinal cracks rise during the dam reconstruction.

In the further dam sequence, i.e., in V-VIII

source levels the longitudinal wave velocity changes from 2.0 to 2.6 km/s. In this case we can speak about the normal dam velocity field. These velocities correspond to V_1 refraction velocity.

The higher velocity zone in IX and X source horizontal levels are the third significant phenomenon in the velocity contour. This is a concentrated stress zone. The velocity drop is the result of the arch effect above the striking low velocity zone in the bottom third of the dam.

A large zone of the low velocity from XI source to the inspection gallery follows then. The longitudinal wave velocity is, beyond expectation, low and responds to V_2 refraction velocity of the dam crest. The obvious quality difference between the upper and bottom dam parts on the masonry condition of the downstream dam face can be seen. This explains why the low velocity occurs in the bottom part of the dam. It is possible to assume that the dam does differ not only in the outside masonry quality but even the inside masonry quality. The dam seepage and washing out of the bond material must have been bigger in the bottom part than in the upper one.

The last significant zone is a high velocity belt in the dam subsoil. At first sight it is obvious that the granite massif mechanical properties are better than those of the dam body. According to the absolute velocity values (3.6-3.7 km/s) it is obvious that the granite massif is weathered and that the subsoil rock doesn't excel in any good geotechnical properties.

We would like to say something about the validity of the individual velocities and about the incidental modulus calculation or about some other mechanical parameters derived from the seismic measurements. If we are to get to know the mechanical properties of any dam part (especially the upper one), it will be favourable for the calculation to use V_1 velocity (i.e. 2.6-3.4 km/s). This recommendation corresponds to the fact that the mechanical properties of rock massif of certain dimension must be in correlation with the wave range. If we are to characterize the whole dam it will be necessary for us to use V_2 velocity (1.4-2.4 km/s). It is possible that after finishing grouting and after maturing all concrete grout material sufficiently, V_2 velocity will level with the V_1 velocity.

3. MEASUREMENT OF QUARRY BLASTS SEISMIC EFFECTS

3.1 Apparatuses

The points for seismic station establishment were determined by the Spanish client to seismic effect of quarry blast measurements. S1 station was placed in the middle dam crest. It was to present information for the appreciation of the dam crest behaviour. S2 station was located in the grout gallery proximity on the toe slope. It was to give us information about the behaviour of the contact between the rock massif and the dam body. S3 station was on the granite outcrop closely by the dam. It was to give us a character of the rock massif behaviour in the dam proximity. Three electrodynamic seismometers were placed on these stations (vertical component - Z, perpendicular to the dam crest - T, parallel to the dam crest - L). The velocities of vibrations were registered by these seismometers. The station for acoustic wave measurement was placed the left dam site. The acoustic wave was measured by piezoelectric omnidirectional microphone.

A number of small quarry blasts was expected during measurement interval and their intensities and origin time were not known. Therefore it was necessary to use the apparatuses with triggered recording, high dynamic range of recorded signals and maximal reliability of record. The equipment and software used for data acquisition were developed in The Institute of Geonics of The Czech Academy of Sciences.

Block diagram of apparatuses is on Figure 3. Analog signals from seismometers (components Z, T, L) from S1-S3 stations are connected to inputs of PCM3-A apparatus. This apparatus records the seismic signals in frequency range 0.5-30 Hz with dynamic range 90 dB (MSB/LSB). The signals are recorded in analog form by UVO oscillograph in regime of automatic gain control. They are also recorded in digital form by three tape recorders (TR1-TR3). The sampling frequency of A/D conversion is 100 Hz. It is possible to reproduce the recorded signals from tape recorder in analog form by OSC oscilloscope or UVO oscillograph using PCM3-R apparatus. It is also

STATIONS:

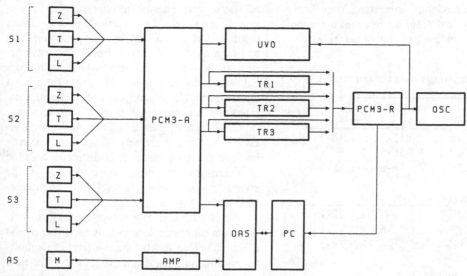

Fig.3 Block diagram of seismological apparatuses

possible to transfer data from PCM3-R to PC computer via RS232 serial port. PCM3-R apparatus allows continuous monitoring of signal quality during operation time.

PCM3-A apparatus also contains amplifiers and filters for data acquisition system DAS. DAS is realized as A/D board for PC computer and special software for record triggering. Signal from M microphone is amplified in AMP charge amplifier and is also connected to input of DAS. Frequency range of seismic and acoustic channel is respectively 0.5-100 Hz and 0.3-200 Hz. Sampling frequency of A/D conversion is 500 Hz, dynamic range is 72 dB (MSB/LSB).

The above mentioned system of apparatuses represents double reserved system of data record. High dynamic range analog records quickly give immediate information after the first blast, that make possible fast selection of right sensitivity of PC based recording system. The digital data have a good quality for the signal processing. Both systems record also 5 seconds data history before triggering.

Power sources of PCM3-A and TR1-TR3 apparatuses are battery buffered. It increases reliability of the recording system in the case of a mains interrupt.

3.2 Analysis of the recorded data

Eight blasts were recorded in the site during our measurement. In four cases (D, E, F, H blasts) the full quality digital record (on PC computer) was obtained. The other low seismic intensity blasts were recorded by PCM3-A only.

The software that was developed in The Institute of Geonics was used for the analysis of recorded seismic data. It makes possible ways of:
- the signal digital filtration
- the representation of seismograms in the suitable scale respectively their representation after the motion and acceleration conversion
- the wave group time picking
- the amplitude and prevailing frequency determination (in the chosen record sequence)
- the spectral analysis.

The Lennartz electronic GmbH software was used for particle motion analysis.

. According to the preliminary analysis we can say that E, F, H blasts had comparable intensities. They were described as "weak" by the construction workers. The seismic appearances of the D-blast that was described as "strong" were higher in the order. All PC recorded signals were filtered before interpretation by digital

47

low-band filter with cut-off frequency 90 Hz. The reading maximum amplitude from the velocity records and from the recounting records (acceleration, motion) are shown in Table 1.

Table 1. Monitoring measurement results

Ev	St	v_{max} [ms^{-1}]	A_{max} [m]	a_{max} [ms^{-2}]	P_{max} [Pa]
D	S1	T:2.4E-3	T:5.1E-5	Z:1.3E-1	
	S2	Z:2.8E-4	T:7.5E-6	Z:6.8E-2	
	S3	L:5.7E-4	T:7.1E-6	L:1.1E-1	
	Ac				63.0
	resonance frequency 8Hz				
E	S1	T:7.4E-4	T:1.5E-5	T:1.2E-1	
	S2	L:2.3E-4	L:4.7E-6	Z:5.7E-2	
	S3	L:4.8E-4	L:5.6E-6	L:8.2E-2	
	Ac				18.5
	resonance frequency 7Hz				
F	S1	T:6.5E-4	T:9.7E-6	Z:1.8E-1	
	S2	Z:2.1E-4	L:3.3E-6	Z:6.8E-2	
	S3	T:4.9E-4	L:4.2E-6	T:1.3E-1	
	Ac				6.3
	resonance frequency 8Hz				
H	S1	T:6.1E-4	T:8.9E-6	Z:1.9E-1	
	S2	Z:2.2E-4	L:3.0E-6	Z:6.5E-2	
	S3	Z:3.5E-4	L:3.7E-6	L:1.0E-1	
	Ac				37.6
	resonance frequency 7Hz				

Legend:
    Ev   seismic event (D,E,F,H)
    St   station (S1,S2,S3)
    Ac   acoustic pressure
    Z,T,L   record component
    v_{max} maximum velocity amplitude
    A_{max} maximum motion amplitude
    a_{max} maximum acceleration amplitude
    P_{max7} maximum amplitude of acoustic
                pressure

3.3 Analysis of D-blast seismogram

The D-blast seismogram was analyzed in the most detailed way and the result of these analyses is presented in next part of this paper. These data are best sampled and have the biggest signal/noise ratio. D-blast seismogram is in Figure 4. The uniform velocity amplitude scale is used in this figure.

The arrival direction of the seismic waves to S3 station was determined with the particle motion diagrams in two vertical planes. These planes were determined by the transversal (T) and longitudinal (L) components. The wave arrival in both planes is obviously oblique from below the station under a 70º angle from horizontal level. The diagram tail parts are strongly interfered and removed from the random motion. The starting particle motion in the horizontal level (P-wave arrival) responds to the azimuth to the blast place. It is distinctly possible to determine an explosive character of the seismic event focus-the arrival with the orientation outside from the focus.

The vertical arrival P-wave (cca 90º) is obvious from the vertical particle motion diagram at S2 station. That is why the P-arrival in the horizontal diagrams is difficult to identify. The next sequences are again very chaotic. The diagram character is similar to the S3 station.

The vertical arrival of the P-wave is obvious from the vertical diagram of the particle motion at S1 station. The resonance dam oscillation was determined in the horizontal level in the transversal direction after 160ms of P-wave arrival. The oscillation direction is not led to the focus, but it is perpendicular to the dam crest.Such an oscillation in the longitudinal plane is not observed.

The intensive resonance wave group is of 8 Hz frequency. The resonance wave character is of the driven attenuated oscillation. The maximum amplitudes of A_v registered velocity of all the blasts occur in this wave group. We suppose that it is the resonance response of the dam body. The oscillation character is obvious from Figure 4 (especially S1 T channel). The coefficient of the resonance oscillation attenuation $\alpha=2.05$ s^{-1} was calculated from D-blast record data. The particle motion oscillation is shown in Figure5. It is obvious according to ZT diagram (the plane determined by Z and T axes) that the particle motion has a significant linear character in the transversal direction. PM diagram for that ZL plane is shown in Figure 5 below. We can notice very well how the big difference between the oscillation in the longitudinal and transversal direction is.

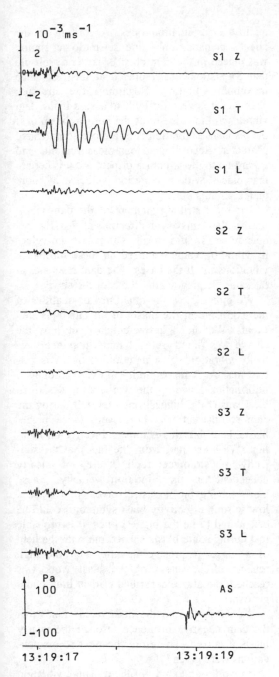

Fig.4 Seismic waveform from D blast

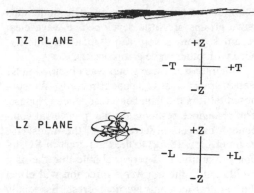

Fig.5 Particle motion of resonance waves

The amplitude spectra were calculated for the determination of the registered record spectral composition in the individual channels (including the acoustic pressure). We can state that there aren't any significant local maxima or minima in the spectra besides the transversal component of S1 station. The significant growth or attenuation of the some frequency zone doesn't take place then. 8 Hz significant frequency maximum is in the transversal component spectrum of S1 station.

The acoustic pressure record of D-blast is shown in Figure 4 and it is marked as SA. The acoustic wave has a strongly damping oscillation character. The acoustic pressure maximum values of the individual blasts are given in Table 1. 63 Pa acoustic pressure maximum amplitude was registered during D-blast. We can see in the records that the seismic and acoustic effects aren't in the same ratio. The acoustic effect depends on the blast work efficiency. For example the big acoustic effect and the small seismic one come into being by an imperfect packed charge.

It is interesting that the acoustic wave blast calls the resonance oscillation in the transversal direction (Figure 4, S1 T channel). The character of damping curve course is again exponential ($\alpha=1.55$ s^{-1}).

4. CONCLUSION FROM ANALYSIS

The wave image records of all registered blasts are of a similar character. It is possible to say that the seismic wave oscillation time, registered after the blast, is short and that the oscillation amplitude attenuates quickly. The individual wave groups are little developed. P-wave clear record section is very short, then the interference with other wave group occurs.

The intensive wave group was observed in S1 station transversal component records. We suppose that it is the dam body resonance response. The resonance response on the transversal component can be observed after the transverse hammer blow (8 kg) to the dam crest in S1 station proximity. It is very probable that the dam oscillation in the transverse direction will effect the car traffic along the dam crest. Especially the high speed truck entry to the road along the dam crest could call forth a dam oscillation about the relatively high velocity oscillation. The danger cases will particularly occur by the negative influence combination, i.e., in the case of the quarry blast and the truck movement or in the simultaneous truck entry from both directions.

The seismic effect evaluating on the dam belongs professionally to the dynamic study area of these structure types. The relevant standards, which are generalizing rules and experience, are used to the vibration admissible value determination. The specific nature conditions in the subject surrounding aren't therefore considered in these standards, for example geological, tectonic, hydrogeological ones and so on. The increasing amplitudes by the wave stacking can occur under the unsuitable conditions. Therefore the parametric and experimental seismologic measurements are necessary for the seismic influence determination to the structure.

We come out from DIN4150 standard appreciating this example. The structures are divided into three types. The dam belongs to the most sensitive structure category. The reality that this dam is without any strengthened elements leads us to this classification. No microcrecks rise may be admitted in the dam as the hydrotechnical work.

As the structure consists of two different parts (the masonry dam and the concrete trough) it is clear that the dam behaviour as a whole is very complicated during its vibration. The rock massif quality in the dam subsoil (the fault deformation in the left slope) influences substantially the vibration character. Therefore it is obvious that the first measurement results we

qualify as orientation ones. According to the client's demand, only one seismologic station was placed in the dam crest. It is very probable that S1 station wasn't placed at the resonance maximum vibration. Regarding the structure character it is very difficult to assess if the dam vibrates as the string or as the built-in beam or if the resonance vibration is of a more complicated character. According to our measurements and according to the structure inspection, it is certain that places with a vibration velocity of more than 2.4 mm/s exist.

The most striking element of the dam vibration in the transversal direction at 7-8 Hz frequency was in all records. The maximum velocity vibration value occurred on these resonance vibrations in all the cases. The dam resonates at the same frequency after the acoustic shock.

We didn't know the quantities or qualities of the explosive in the blasts of all our registered events. We didn't know either if or how the charge was timed. A_v=2.4 mm/s maximum velocity vibration value determined by us is getting near to 3 mm/s limit value. According to information given by the workers the described blast wasn't the biggest one they felt during the dam reconstruction. There were twice bigger blasts there, but this is most subjective information. We must start from the fact that the subjective physiological feelings are not linearly dependent on the vibration velocity. These blasts having big acoustic waves (it may be a low seismic effectivity blast symptom) could be considered to be the biggest ones. It is probable that during some blasts carried out with the help of a bigger explosive quantity and the high seismic effectiveness of the blast work was reached, the above described critical limit could get over.

We can say from the discovered values that the dam negative influence through the quarry blasts occurred. Therefore the next recommendations were given:

- to work out a study of the dynamic vibration effects to the dam
- to carry out further seismological measurements of a bigger number of blasts in the cooperation with the manager of the quarry
- to carry out a detailed measurement of the dam resonance vibration

-to work out a study of quarry blast technology optimization
-to secure a permanent dam seismological monitoring during its operation.

REFERENCES

Bláha, P., J. Knejzlík, Z. Kaláb & B. Gruntorád 1993. *Informe final de las mediciónes geofísicas para la presa Presa de Pontillón do Castro - Pontevedra.* Brno: Geotest Brno & HOÚ AVČR, (spain and czech), MS.

Bláha, P., J. Kottas & J. Sochor 1994. Geological inhomogeneities in seismic tomography. *Proc. 7th IAEG.* Rotterdam: Balkema.

Svoboda, B. 1992. General classification of construction damages. *Proc. XXIII General Assembly of the ESC: Vol. 2,* 459-462. Prague: Czechoslovak Academy of Sciences.

Toth,R. 1992. Koncepce programového vybavení pro zpracování dat ze Seismického polygonu Frenštát. *Proc. Sborník referátů z celostátní konference seismologů.* Ostrava: HOÚ, ČSAV.

Méthodes sismiques en forage appliquées à la reconnaissance des massifs rocheux en Belgique: Évolution et applications

Borehole seismic methods applied to rock mass investigation in Belgium: Evolution and case histories

D. Demanet, D. Jongmans & A. Monjoie
Laboratoires de Géologie de l'Ingénieur, Université de Liège, Belgique

RESUME: Les méthodes sismiques en forages constituent un outil privilégié pour reconnaître un massif rocheux et elles ont été abondamment utilisées en géologie de l'ingénieur dans une très grande variété d'applications. Jusqu'il y a peu, les techniques classiquement utilisées étaient les essais down-hole ou cross-hole qui permettent d'évaluer l'évolution des vitesses de propagation des ondes en fonction de la profondeur. Plus récemment, le développement des appareils d'acquisition de données et des ordinateurs a permis d'augmenter le nombre de signaux sismiques enregistrés et d'améliorer considérablement leur traitement. Cet article retrace l'évolution des méthodes en présentant des cas d'études en Belgique et montre l'intérêt grandissant des méthodes tomographiques.

ABSTRACT: Borehole seismic methods are powerful tools for rock mass investigation purposes and they have been extensively used in a wide range of applications in engineering geology. In the last twenty years, the standard techniques were the cross-hole and down-hole tests which are aimed at the determination of velocity values as a function of depth. More recently, the development of data acquisition systems and computers has allowed to increase the number of seismic records and to greatly improve data processing. This paper shows the evolution of borehole seismic methods by case histories in Belgium and emphasizes the interest of seismic tomography.

1 INTRODUCTION

Depuis de nombreuses années, la prospection sismique entre forages est couramment utilisée en géologie de l'ingénieur pour reconnaître les caractéristiques d'un massif rocheux. Par rapport aux méthodes de surface (sismique réfraction et sismique réflexion), elle présente l'avantage de déterminer directement des paramètres dynamiques à une profondeur donnée en se libérant des conditions de surface qui peuvent rendre difficile la pénétration des ondes. Les caractéristiques élastiques du massif généralement mesurées sont les vitesses de propagation des ondes P (Vp) et des ondes S (Vs) qui varient en fonction de la lithologie et de la fracturation. Ces grandeurs sont reliées aux modules de Young (E) et de cisaillement (G) par les relations suivantes:

$$G = \rho \ Vs^2$$

$$E = \frac{\rho \ (\ Vp^2 - 4 \ Vs^2)\ Vs^2}{(Vp^2 - Vs^2)}$$

où ρ est la densité.

2 METHODES CLASSIQUES EN FORAGE

Les méthodes sismiques classiquement utilisées en forage sont les essais de type cross-hole (figure 1B), down-hole et up-hole (figure 1A). Lors d'un essai down-hole, les ondes sont générées en surface et mesurées à différentes profondeurs dans le forage. La mesure des temps d'arrivée permet d'obtenir le profil de la vitesse (Vp ou Vs) en fonction de la profondeur. Dans l'essai up-hole par contre, la source est située dans le forage et les ondes sont enregistrées en surface. Enfin, l'essai cross-hole consiste à descendre la source dans un forage et à mesurer les ondes à la même profondeur dans un ou plusieurs autres forages. Connaissant la distance entre les forages, la mesure des temps d'arrivée aux différents géophones permet d'obtenir les valeurs des vitesses aux différentes profondeurs. La détermination correcte de ces paramètres nécessite cependant un choix adéquat de matériel et de dispositifs de terrain (Butler et Curro, 1981).

Ces méthodes ont été fréquemment utilisées en Belgique (Henriet *et al.*, 1986) pour la reconnaissance de massifs naturels dans le cadre d'ouvrages du génie civil : pont (Ben-Ahin), projets de métro (Anvers, Liège), centrales nucléaires (Doel, Thiange), ascenseur hydraulique (Strépy-Thieu)... De nombreux essais de type cross-hole et down-hole ont été réalisés pour le tunnel de Cointe qui est actuellement en construction au centre de Liège.

Il s'agit d'un ouvrage de 1 km de long qui recoupe des terrains carbonifères fortement tectonisés constitués d'une succession de couches de grès et de schistes avec des couches de charbon anciennement exploitées. Les travaux miniers ont contribué à altérer les caractéristiques des formations géologiques. Un exemple de données cross-hole est présenté à la figure 2 qui reprend l'évolution de Vp, Vs et E en fonction de la profondeur ainsi que l'interprétation géologique à partir des descriptions de forage. Les valeurs de vitesses et de module croissent régulièrement avec la profondeur entre les cotes 103 et 84 m, traduisant l'augmentation du degré de compacité. En-dessous de la cote 84 m, les valeurs restent en moyenne relativement constantes avec un niveau de valeur faible à la cote 76. Ces observations s'expliquent probablement par la présence d'une série gréseuse fissurée et/ou l'influence des zones de faille apparaissant sur la figure 2.

La méthode cross-hole a également été utilisée dans des terrains houillers non plissés (à Beringen dans le Limbourg belge) en vue de caractériser la détente du massif suite au creusement d'un bouveau à 800 m de profondeur. Trois forages horizontaux de 70 m de long ont été réalisés à respectivement 1,5 m (F1), 3 m (F2) et 5,5 m (F3) de la paroi de la galerie à partir d'une petite niche latérale. La figure 3 reprend la vitesse des ondes de cisaillement mesurées entre F1-F2 et F2-F3 en

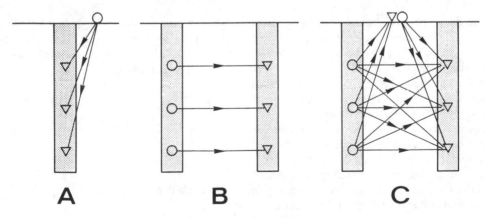

A   B   C

Figure 1 : Méthodes sismiques en forages. A: essai down-hole. B: essai cross-hole. C: panneau tomographique.

fonction de la distance à la niche. Jusqu'à 45 m de distance, les valeurs de vitesses à proximité de la galerie (entre F1 et F2) sont inférieures à 1000 m/s tandis qu'à l'intérieur du massif (F2-F3) les valeurs avoisinnent 2000 m/s. Ces résultats résultent de l'effet de la déconsolidation du massif à proximité de bouveau. Par contre, au-delà de 45 mètres par rapport à la niche, aucune différence notable n'apparaît entre caractéristiques proches et lointaines de la galerie. Ces observations semblent liées à la présence d'une zone fracturée mise en évidence par les forages horizontaux et dont les valeurs de vitesse ont peu évolué lors du creusement du tunnel.

3 METHODES DE TRACES DE RAIS

Dans les essais classiques de type cross-hole ou down-hole, les valeurs de vitesse observées ne sont supposées varier que selon une direction (la profondeur dans le cas de forages verticaux). En pratique, les valeurs de vitesse dans un massif peuvent varier selon les trois directions de coordonnées en raison de variations latérales de faciès, de changements d'épaisseur des couches et de zones faillées. Dans ces cas, les méthodes classiques ne permettent pas de déterminer la localisation de zones présentant des anomalies de vitesse. En considérant un plus grand nombre de données (plusieurs positions de récepteurs pour un même émetteur), il est cependant possible de localiser approximativement un corps de vitesse particulière pour autant que le milieu environnant soit relativement homogène. La technique consiste à ajuster un modèle théorique jusqu'à obtenir une bonne concordance entre les temps de propagation mesurés et calculés. Le calcul de temps de propagation théoriques nécessite la résolution du problème direct, généralement par une méthode de tracés de rais.

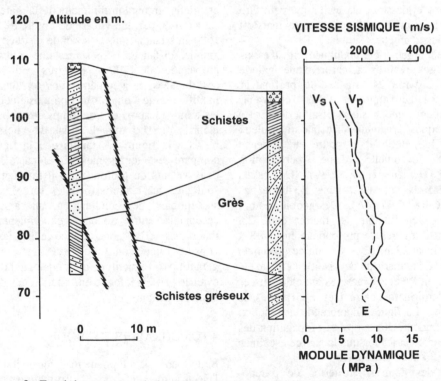

Figure 2 : Essai de type cross-hole réalisé pour le tunnel de Cointe. Evolution de Vp, Vs et E en fonction de la profondeur.

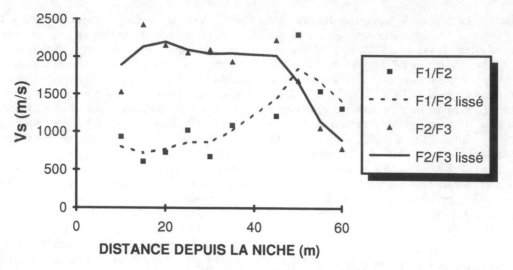

Figure 3 : Evolution des valeurs de Vs le long de la galerie. Mesures réalisées entre 1,5 m et 3 m (F1-F2) de la paroi de la galerie et entre 3 m et 5,5 m (F2-F3).

Contrairement aux méthodes tomographiques développées ci-dessous, la technique peut être appliquée à un nombre restreint de rais mais est limitée à des géométries simples.

Cette technique a été appliquée lors d'essais entre forages réalisés à partir d'une galerie creusée à environ 240 mètres de profondeur dans une couche d'argile de 100 m d'épaisseur. Cette campagne géophysique visait à définir les caractéristiques dynamiques du massif argileux entourant la galerie et à mettre en évidence l'influence éventuelle d'une perturbation thermique (Demanet *et al.*, 1989). Des essais ont été réalisés entre les forages R3-R6 et les forages R6-R8 (figure 4A). Des géophones ont été disposés en R3 tous les mètres entre 2,5 mètres et 9,5 mètres de profondeur et en R8 à 2,5 m , 6 m et 13 mètres. La source sismique, constituée d'un marteau de cisaillement, a été descendue en R6 et des mesures ont été réalisées tous les demi-mètres entre 1 et 5.85 mètres de profondeur. La figure 4B reprend, en fonction de la profondeur, les temps de propagation des ondes entre R6 et R3 lorsque la source est située à 1 m et 5,85 m de profondeur.

L'analyse des données en terme de pseudo-vitesses (distance source-récepteur divisée par le temps de propagation) semble indiquer que le massif est subdivisé en deux zones. La première, entourant directement la galerie est caractérisée par une vitesse de l'ordre de 1400 à 1600 m/s tandis que la seconde , située à plus grande distance, présente une vitesse de propagation de 1700 à 1800 m/s. Sur base de ces données, la géométrie des terrains a été modifiée jusqu'à obtenir un bon ajustement entre les temps mesurés et les temps de propagation théoriques. Le modèle final de vitesses est repris à la figure 4A tandis que la figure 4B compare les temps mesurés et calculés pour deux valeurs de profondeur. Il faut cependant souligner que d'autres modèles sont également susceptibles de conduire à un ajustement acceptable entre données expérimentales et théoriques. La concordance entre les deux constitue une condition nécessaire mais ne garantit pas l'exactitude du modèle, qui doit être évaluée par le géologue en fonction des données à sa disposition.

4 TOMOGRAPHIE SISMIQUE

Suite au développement considérable de l'électronique et de l'informatique , des méthodes plus complexes d'imagerie sismique (tomographie) sont apparues depuis une dizaine d'années. Ces techniques sont basées sur l'acquisition d'un grand nombre de données de temps de propagation entre sources et récepteurs placés en forages ou en surface et sur la reconstruction d'une image représentant le milieu par un de ses paramètres (généralement la vitesse de propagation des ondes P). L'inversion des données peut être réalisé par un grand nombre de méthodes dont les principes généraux sont donnés par Ivansson (1987). Parmi ces différentes techniques, les méthodes géométriques (Cote *et al.*, 1990) semblent particulièrement intéressantes en raison de leur relatif faible temps de calcul pour le grand nombre de données envisagé. Le domaine d'application des méthodes tomographiques, très vaste, couvre les reconnaissances détaillées du sous-sol, le contrôle des injections, la localisation des zones fracturées ou d'anciens ouvrages enterrés.... Un des avantages de la méthode est qu'aucune hypothèse préalable n'est faite sur la structure du milieu investigué. Par contre, pour une configuration cross-hole classique (émetteurs dans un forage, récepteurs dans l'autre), l'unicité de la solution n'est pas nécessairement garantie (Ivansson, 1987).

Une étude tomographique a été réalisée en Belgique sous le barrage de la Gileppe en vue de localiser des zones fracturées susceptibles de provoquer des fuites sous le barrage. Cet ouvrage de 270 m de long et d'environ 60 m de haut est implanté sur des formations siegeniennes consistant en une alternance de couches de grès, de schistes gréseux et de schistes. Les couches sont grossièrement

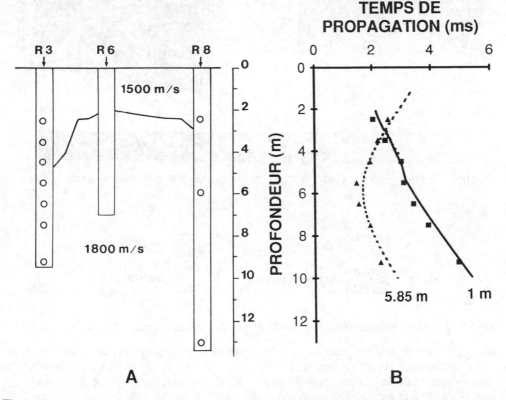

Figure 4 : A: Disposition des forages et modèle de vitesse obtenu. La position des géophones est indiquée par des cercles.
B: Courbes profondeur-temps de propagation ajustées aux données expérimentales pour deux profondeurs de source (1 m et 5,85 m). Les données expérimentales sont représentées par des symboles, les données calculées par des courbes.

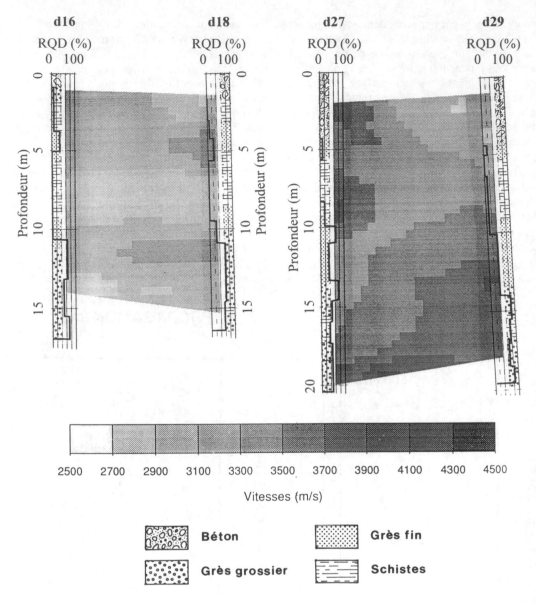

Figure 5 : Panneaux tomographiques d16-d18 et d27-d29 mesurés sous le barrage de la Gileppe.

orientées N70°E avec un fort pendage vers le Sud. L'ensemble de la région est parcourue par un réseau important de failles transversales orientées NNW-SSE et fortement redressées.

Quinze panneaux tomographiques (10 m × 15 m en moyenne) ont été réalisés selon l'axe du barrage entre des drains forés à partir de la galerie de drainage. Les drains, d'une longueur de 15 m, sont inclinés de 50 à 55° vers le sud.

Sur base des descriptions de forage, des RQD et des résultats de la prospection sismique, trois lithologies principales ont été distinguées :

• les grès grossiers caractérisés par un RQD moyen de 60 % et des valeurs Vp comprises entre 3500 et 5000 m/s dans un état peu fracturé.

• les grès fins dans lesquels le RQD est en

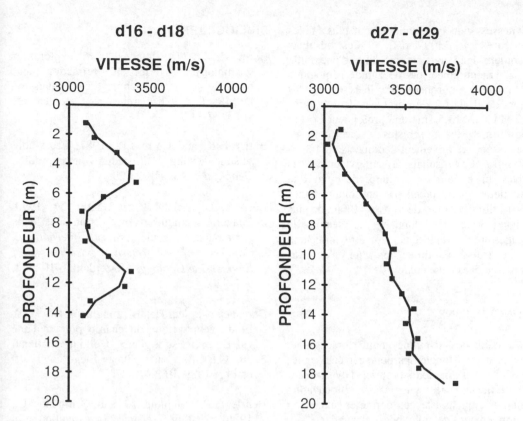

Figure 6 : Essais cross-hole correspondant aux deux panneaux tomographiques présentés à la figure 5.

moyenne égal à 40 % et Vp varie entre 3300 et 4000m/s (état peu fracturé).
- les schistes (éventuellement gréseux) dont le RQD est très faible à nul. La vitesse Vp est généralement comprise entre 3000 et 3500 m/s.

Les valeurs de vitesses sismiques mesurées vont être influencées à la fois par la lithologie et par le degré de fracturation. Deux exemples de panneaux tomographiques calculés à partir de la méthode développée par Cote *et al.* (1990) sont présentés à la figure 5. Sur chaque panneau sont également repris les descriptions de forages et les valeurs de RQD. La figure 6 présente les résultats des essais cross-hole classiques (Vp en fonction de la profondeur) obtenus pour ces deux panneaux.

L'examen des deux figures montre des différences sensibles existant entre les deux panneaux étudiés. La courbe vitesse-profondeur du panneau d16-d18 se présente sous la forme de deux maxima (correspondant à des couches de grès) encadrant une zone à vitesse plus faible liée à la présence d'une couche de schiste. L'image tomographique correspondante confirme cette structure caractérisée par une variation de Vp selon la profondeur et principalement influencée par la lithologie. Sur le panneau d27-d29 par contre, l'essai cross-hole montre globalement une augmentation de Vp avec la profondeur jusqu'à atteindre 3700 m/s dans les grès grossiers compacts. Dans ce cas, l'examen du panneau tomographique révèle des variations importantes de vitesse dans les grès avec des valeurs supérieures du côté du forage d29 confirmées par un RQD élevé (80%). A l'autre extrémité du panneau (d27) par contre, les vitesses sont systématiquement plus faibles (inférieures à 3300 m/s). Ces résultats concordent bien avec la présence d'une faille normale mentionnée lors de l'étude géologique et à laquelle pourraient être liées les faibles valeurs de RQD (40%) observées dans les grès entre 11 et 13,5 m de profondeur. Par opposition, les schistes surincombants apparaissent relativement compacts (3100 - 3500 m/s). Ces résultats ne sont évidement pas visibles sur la courbe d'évolution de Vp avec la profondeur qui ne prend pas en compte toutes les variations latérales de vitesse. Il faut de plus souligner que les données de temps de propagation peuvent également être influencés par la troisième dimension qui n'est pas considérée dans cette étude.

5 CONCLUSIONS

Si les méthodes sismiques entre forages sont utilisées depuis longtemps pour caractériser les massifs rocheux par leurs propriétés dynamiques (essentiellement les vitesses de propagation sismiques), on assiste ces dernières années au développement de méthodes d'inversion qui permettent d'obtenir une image continue du milieu à 2 ou 3 dimensions. Cette évolution a été illustrée par différents cas d'étude en Belgique qui montrent les améliorations remarquables des méthodes sismiques en forage durant ces dernières années. Lorsqu'il est possible d'entourer correctement le milieu à étudier par des capteurs, les méthodes tomographiques constituent à peu près les seuls outils capables d'obtenir une image complète du milieu sans hypothèse préalable.

REMERCIEMENTS

Nous remercions la Direction Générale des Voies Hydrauliques du Ministère Wallon de l'Equipement et des Transports qui nous a permis de présenter certains résultats de l'étude tomographique réalisée sous le barrage de la Gileppe. D'autre part, les essais entre forages à Beringen on été réalisés dans le cadre du projet de recherche n° 2.0003.88 financé par le Fonds pour la Recherche Fondamentale Collective.

BIBLIOGRAPHIE

Borm G., 1977, Methods from exploration seismology : reflection, refraction and borehole prospecting, Proceedings of DSMR 77, Karlsruhe, 5-16 september, vol.3, 87-113.

Butler D.K. and Curro J.R., 1981, Cross-hole seismic testing - Procedures and pitfalls, *Geophysics*, **46**, n°1, 23-29.

Côte P., Lagabrielle R. et Gautier V., 1990, Imagerie sismique: reconstruction à 2D dans le domaine du génie civil, *6th International Congress of the Internal Association of Engineering Geology*, Amsterdam, 6-10 août, 913-920.

Demanet D., Neerdael B., Jongmans D., 1989, Etude géophysique du champ proche d'une galerie en massif argileux, Colloque national du C.B.G.I., Sart Tilman-Liège , 17-19 octobre 1989, III.26-III.37 .

Henriet J.P., Monjoie A. and Schroeder C., 1986, Shallow seismic investigations in engineering practice in Belgium, First Break, **4**, N°5, 29-37.

Ivansson S., 1987, Crosshole transmission tomography, in *Seismic Tomography, edited by Guust Nolet, D.Reidel Publishing Company, Dordrecht,* 159-188.

McCann D.M., Baria R., Jackson P.D. and Green A.S.P., 1986, Application of cross-hole seismic measurements in site investigation surveys, Geophysics, **51**, 914-929

Shallow refraction processing in geology

Traitement des données de la sismique de réfraction peu profonde en géologie

Pavel Bláha
Geotest a.s., Brno, Czech Republic

ABSTRACT: Methods of shallow seismic refraction measurement are evaluated. Besides commonly used methods the article shows a new processing method, i.e., plus-minus method with a velocity penetration. Possible ways of using a new plus-minus method, i.e., a bigger processing accuracy and demonstrating velocity changes under the refraction horizon are presented using the examples of the real measurements. Figures from a dam site, landslide, deep weathered beds and fault cases are shown and show the advantages of the new procedure.

RESUME: L'article est consacré aux méthodes qui apprécient les mesures séismiques de réfraction peu profondes.Sauf des méthodes courantes on parle de la nouvelle méthode du traitement des mesures, c'est-à-dire de la méthode t0 à pénétration.Les possibilités de cette méthode, c'est-à-dire l'interprétation plus précise des données et l'illustration des changements de vitesse sous le horizon principal de réfraction sont montrées sur la base des exemples pratiques. Les images du lieu de la digue, de l'écroulement, des strates profondement désagrégées et des ruptures sont présentées. Les exemples montrent les avantages de cette nouvelle méthode du traitement de la séismique peu profonde.

1.INTRODUCTION

The shallow refraction method is used an the integral part of geophysical measurements for engineering geological applications, has been used for several years and given good results.

The classical methods of processing refraction measurements start from the theoretical assumption of a velocity constancy in the vertical direction. The other assumptions are the level boundary of small inclination. These assumptions are not fulfilled very often. The cover layers from the loam or debris mostly turn downward in an interrupted way to the weathered bedrock. The degree of weathered rock drops with the depth and nonweathered rock can be found in bigger depths. In these cases the vertical velocity changes are therefore slow without any violent jumps. It is difficult to determine a traveltime curve fracture in such areas. The

wrong determination of these fractures brings out an inaccurate interpretation of the depths and velocities of the seismic beds. The development of computer application in geophysics gives a lot of opportunities to look for new processing methods, and new theoretical procedures. Newly derived formulas are applied in the computer processing and so the results are of higher accuracy.

2.THEORETICAL SOLUTION

The original interpretation methods, i.e., the critical distances method and the plus-minus method proceeded from single theoretical assumptions. The critical distance method (also called the breaking point method) even supposes the refracted wave velocity stability along the refraction horizon. Another assumption is the linearity of the refraction horizon. The data about the refraction horizon depth and its velocity under the shot point can be obtained by using this method. The method gives information about several horizons (one under another). The refraction velocity is considered to be an average of the whole sections of the relevant traveltime curves. The advantage of this method is its facility and consequently the speed with which it is possible to give information about the investigated rock massif.

The plus-minus method already admits some fluent velocity changes along the refraction horizon. The small changes in the refraction horizon relief are possible, too. The radius of the rough boundary curvature must be bigger than the boundary depth. This method already gives information about the depth under each receiver that was used during the field measurement. The velocities are determined in the definite sections. Their sizes are connected to the velocity in the bedrock. Certain inaccuracies occur by its application if the violent changes occur along the refraction boundary.

Both shown methods assume that velocity is constant under the refraction horizon.

New interpretation ways try to remove the shown disadvantages of these methods. The contemporary trend goes in two ways: by removing errors originated by the refraction horizon roughness and the other by lowering errors originating by the velocity penetrations of seismic rays.

Palmer's method elects the former solution way. This method considers the environment formed of two beds. It succeeded in eliminating the influence of the big rough boundary curvature by a suitable solution and by the velocity election. This method is suitable for the application to thicker overburdens, and is believed to be suitable for use on overburden thickness greater 20 or 30 meters. The deviation influence that this method removes also depends on the relation between the direct and refracted velocities. If this relation grows then the rough boundary curvature influence drops.

The latter method results from removing errors caused by the velocity penetration of seismic rays to the bedrock. This processing method takes into account the velocity growth under the most significant refraction horizon. This growth is mostly given by the geological structures and to know it is essential for the solutions of engineering geological problems. The new method results set the thickness and the velocity in the upper bed, the refraction velocity and the velocity changes under the refraction horizon.

The velocity changes under the refraction horizon can be described by several velocity laws. We can consider the linear, exponential and parabolic laws. The velocity changes according to the linear velocity law can be expressed by the equation:

$$V(z) = V_0(1 + \alpha x + \text{ß}z) \tag{1}$$

where α is the horizontal velocity gradient

ß is the vertical velocity gradient.

Similar equations can be compiled for other velocity laws. These theoretical laws are more complicated and their program solutions are more complicated too. Therefore removing error occurring by velocity changes under the refraction horizon is carried out by the linear velocity law in real use. At the same time it shows that the influence of α coefficient, i.e., the errors caused by the noninstallation of corrections to the horizontal velocity changes are smaller than 5 %. The inaccuracies that occur in applying the vertical velocity gradient law are higher than 5 % in the violent vertical changes. In addition, very frequently during the field measurement the horizontal velocity changes quickly both in size and in sign. In practice introduction of these corrections is very difficult. Therefore αx was omitted from these corrections and the equation (1) simplifies to:

$$V(z) = V_0(1 + \text{ß}z) \tag{2}$$

The theoretical traveltime curves of the refracted waves are derived from this equation and then the interpretation procedures are derived, and used to compile the processing programs for the field measurements. It is possible to eliminate the disadvantages of the linear velocity law by using suitable measured system, i.e., using variously long receiver spread and using multiple overlap. Thus we can use depth velocity changes close to reality for the velocity contour design. The results of such an approach to the processing of shallow refraction measurements are described in the examples.

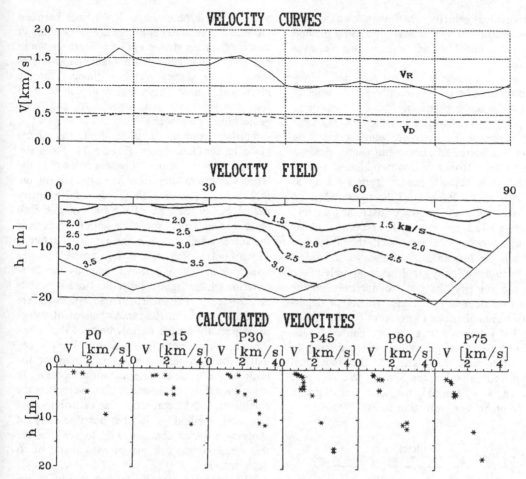

Fig. 1 Jihlava-complex results from velocity penatration method

3. EXAMPLES

The real outputs of the described way of shallow refraction processing are the determination of:

- the major refraction horizon depth
- the direct and refracted wave velocities including their changes along the measured profile
- the velocity changes under the refraction horizon.

These results are in practice shown graphically by drawing the refraction horizon and the velocity contours under this horizon. The velocity contours are drawn to the depths rather greater than the seismic wave penetration depth given by the calculation. These depths are not the same along the measured profile and therefore the lower limit of the velocity contours is

saw-toothed. Further visual outputs are the direct and refracted velocity graphs. In some cases we can complete these basic outputs by the deeper study of vertical velocity changes under the shot points or under middle of the receiver spread.

The first real example (Figure 1) shows the complex graphic output of such a processing of shallow seismic measurements. The example is taken from the Jihlava surrounding. From the geological point of view this place belongs to the Czech Moldanubicum Massif which is formed of gneisses. The upper part of Figure 1 shows the graphs of the direct and refracted wave velocities, the middle part, the refraction horizon depth and the velocity field under this horizon are drawn. In the lower part of Figure 1, the individual velocity values are shown (as they were determined by this new processing method under the middles and ends of the receiver spread).

The direct velocity changes minimally along the profile and we can see only a little velocity drop to the higher stationing, but the situation in the velocity on the refraction horizon is essentially different. This velocity whose penetration effect was corrected quite significantly shows a rock massif distribution into two blocks. The refracted wave velocity ranges from 1.4 km/s to 1.6 km/s on the left side of the figure while on the right side the velocity oscillates around 1.0 km/s. The velocity field gives a similar picture. We can see very well the difference between the left and right cross section sides. Unlike in the velocity graphs it is possible to see that the velocity changes are not only on the refraction boundary, but also in bigger depths (in this example from 15 do 20 m depth). In this case the velocity differences are caused by the different degree of gneiss massif weathering. In the lower part of Figure 1 the individual interpreted velocities are shown. We can well observe the difference in the behaviour of gneiss massif on both cross section sides.

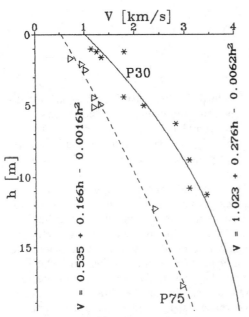

Fig. 2 Jihlava-relations between depths and velocities

In Figure 2 the extreme dependence between the velocity and the depth that are determined in the cross section shown in Figure 1 are shown in one graph. We can obtain mathematical relations for the vertical velocity changes for the individual points. These equations can be used for the velocity recalculation to the rock geotechnical parameters.

Further example is taken from the Genal place in southern Spain (Figure 3). From the geological point of view this area belongs to the Betic Cordillera. Phyllites and slates form the rock massif. Figure 3 shows how the velocity field looks in the case of the rock massif fault deterioration. Under the 85 m station we can see the velocity contours of a narrow fault. Between 115 m and 175 m station it is possible to see how rock massif velocities change by the deterioration of the significant fault. Here it is suitable to recall that the calculated velocities are not quite correct at the rapid changes of velocity, especially for the narrow belts of the velocity minimum or velocity maximum. We can say that at the high horizontal velocity gradient, the error of the velocity determination increase. In these cases it is not essential if the velocities are calculated 100 % correctly. The contribution of this new method is in the possible way of showing velocity changes, i.e., to see changes that occurred in rock massif as a result of its deterioration.

The opposite example to that caused by the velocity drop on the fault belt occurs in the case of velocity growth. Such a velocity field is shown in Figure 4. This example is taken from Žamberk surroundings. From the geological point of view this place belongs to the Czech Cretaceous Area. The bedrock is formed of marlites. A very interesting element in this cross section is the significant velocity maximum on the right side, caused by the strong silicification of the marlite sediments. This strong silicification was discovered in the hole cores in the narrow surroundings of this cross section. Using a more detailed analysis of the velocity contours we can prove that this belt is limited by the fault deterioration in its left side. The question remains whether silicification itself is the part of this fault and how old the silicification is. The most probable explanation is that the original fault was silicified and during the new move-

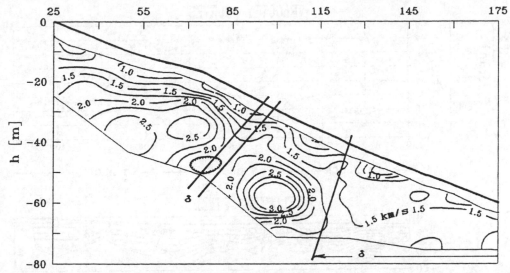

Fig. 3 Genal-velocity field in rock massif fault deterioration.

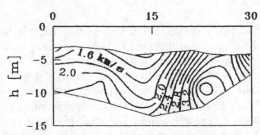

Fig. 4 Žamberk-velocity field in silicification zone

ments the rock massif deterioration occurred next to this silicification belt. Further geophysical measurements and further holes could give the correct answer.

The next two examples are from this method application for the landslide investigation. One is from the Genal landslide (Spain). The large slope deformation is not apparent in the field at first sight. Only the geophysical measurement called our attention to its existence. The example of the measurements on this slope deformation is taken from the cross section running in its bottom fifth. In Figure 5 we can see the difference between the velocity field in the not deformed slope (A block) and in the deformed massif by the slope movement (C block).

The velocity field divides the rock massif along the cross section into three blocks. There are phyllites and slates affected only by the usual weathering in A block. We can see in the velocity field that the usual velocity course was proved by evidence to 40 m depth. The most deteriorated massif is situated in C block. The velocity contours show a slower depth velocity increase than in A block. There are velocities only about cca 1.75 km/s at 30 m depth while in A block there are velocities about 3.25 km/s at the same depth, which is practically about 100 % more. Such low velocities in these depths show that the slide surface can be expected in the depths about 50 m. The next thorough engineering geological inspection proved the correctness of the geophysical interpretation. It is necessary to note here that the principal aim of the geophysical measurement was not to carry out a landslide investigation, but that slope deformation was discovered in the framework of the dam side investigation. Therefore the geophysical measurement working was not chosen to determine the slide surface depth.

The rock massif can be divided according to profiling methods (Figure 5, upper part). The curves of resistivity profiling and magnetic profiling and refracted wave velocity curve are shown here. T and ρ_a values are relatively quiet in the nondeteriorated massif (A block). In B block (transition block) these curves are mildly fluctuated. W10 curve in 145 m stationing shows a unambiguous anomaly. This anomaly is the most northern limit of the slope de-

65

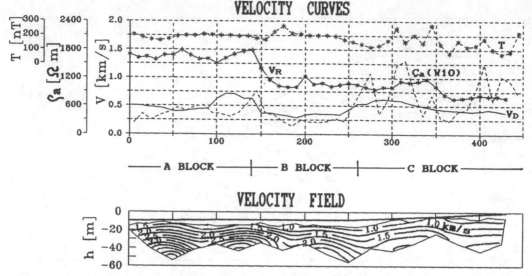

Fig. 5 Genal-velocity field in landslide

formation. The left side of the resistivity anomaly is accompanied by the direct velocity increase. Both anomalies give an evidence of the stress zone in the surface bed close the landslide. C block, as it has been said, belongs to the core of the slope deformation. W10 and T curves are of the biggest variability in this block and their courses are quite characteristic for the slope deformations. The direct wave velocity increase in this block is not caused by the stress increase in the surface bed (in comparison with A block), by the absence of soil horizons; the rock outcrops come out in a day. V_H curve course is interesting and characteristic as well. V_H values are the highest of all in Ablock and reach up to about 1.5 km/s. The velocity drop of the values cca 0.9 km/s in B block and the same values are in C block. This drop together with the other profiling curves give evidence to B block belonging to the slope deformation.

The other landslide example is taken from Luhačovice landslide (Figure 6). Here we intend to show what changes in the velocity field occurred in the landslide after the correction measures had been taken. The landslide area belongs to the Beskydy Flysh. The most striking element on the velocity curves is an anomaly at 20 m station. Close under this anomaly the sheet pile wall was placed into the landslide body. The velocity growth was significantly caused by

the stress concentration behind this wall. It is interesting that this growth is apparent not only on the direct wave velocity (dashed line), but also on the refracted wave velocity (solid line). The influence of stress concentration on the refracted wave was even more significant. The velocity growth behind the sheet pile wall can be seen not only on the velocity curves (Figure 6 upper part), but also in the velocity field (Figure 6 lower part). Among further anomalies the velocity minimum at 75 m station was interesting. The velocity drop was again apparent on both profile curves and in the velocity field, and the influence of tension zone was seen. That zone divided the landslide into two parts. Further knowledge (for example general V_H velocity drop to the slope direction) could be derived from the velocity curve courses. Their contribution was not, however, as significant as the first two events for the landslide investigation.

The last example (Figure 7) is taken from the geological interpretation of this method of processing seismic measurements. The measurement was carried out in the Čeladná place. The example is taken from the Beskydy Flysh again, this time from the place where sandstones prevail over claystones. The upper bed thickness is significantly greater than in the previous examples. The upper bed is formed of slope loams with about 0.5 km/s. The velocity in

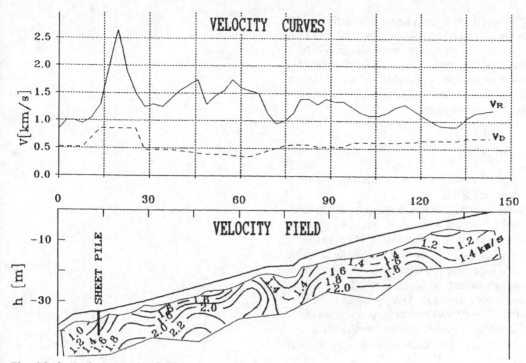

Fig. 6 Luhačovice-velocity field in stress zone

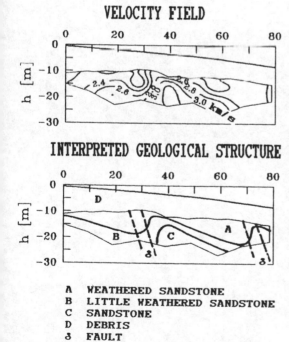

A WEATHERED SANDSTONE
B LITTLE WEATHERED SANDSTONE
C SANDSTONE
D DEBRIS
δ FAULT

Fig. 7 Čeladná-geological interpretation of velocity field

the nonweathered sandstones is about 3.0 km/s
in this area. Therefore the 3.0 km/s velocity
contour has been chosen as the basis for
drawing the nonweathered bedrock. On the
other hand the velocities under 2.5 km/s give
evidence of significant sandstone weathering.
We can draw a significant weathering basis
along the to 2.5 km/s contour. The space
between 2.5 and 3.0 contours then belongs to
the slightly weathered rocks. The main direction
of these geophysical boundaries corresponds to
the s-surfaces as they are known from the
outcrops in this area. The significant velocity
drops at greater depths suggest the rock
deterioration in the rock massif weakened zones,
e.i., in the fault place.

4. CONCLUSION

The plus-minus interpretation method with velocity penetration gives a new possible way of
increasing the accuracy of interpretation of the
result of shallow refraction seismic and the opportunity of velocity field investigation under
the refraction horizon is also very valuable. This

display provides a much better illustration of the geological structure of the investigated area than the other interpretation methods of shallow seismic processing. Unlike Palmer method, this method is especially suitable for small depths because the vertical velocity changes are the biggest in these depths.

REFERENCES

Nikitin, V.N. 1981. *Osnovy inženěrnoj seismiki.* Moskva: MGU.

Palmer, D. 1981. An introduction to the generalized reciprocal method of seismic refraction interpretation. *Geophysics* 46: 1508-1518.

Růžek, B. 1981. Interpretation of the medium with a general propagation velocity gradient using standard calculatin methods. *Acta Universitatis Carolinae - Geologica* 2: 185-201.

Skopec, J. 1989. Shalow seismic investigation ot the vertical velocity gradient media. *Acta Universitatis Carolinae - Geologica* 3:393-408.

Engineering geophysics for dam site selection

Géophysique de l'ingénieur appliquée à la sélection des sites de barrage

Pavel Bláha
Geotest a.s., Brno, Czech Republic

Otto Horský
Geoinza Madrid, Spain

ABSTRACT: This article shows the geophysical method facilities at the investigation stage of the dam site prospection. Especially good results are obtained in those cases when the complex of several methods is used for the geophysical investigation. The possible ways of this geophysical investigation complex for the dam sites are shown on the real examples in southern Spain and the Czech republic. Besides the basic geological structure evaluation we can obtain some information about further data necessary for the dam designing.

RESUME: Il est montré que les méthodes de la géophysique peuvent être appliquées pour la prospection géologique du profil de barrage. Les expériences nous ont convaincus que les bons résultats sont obtenus en utilisant plusiers méthodes pour la prospection géophysique. Les exemples pratiques(de l'Espagne de Sud et de la République Tchéque) montrent des possibilités d'applications de la prospection géophysique complexe dans le cas des barrages. On peut juger la structure géologique et en plus, il est possible d'obtenir des données complémentaires pour le projet du barrage.

1. INTRODUCTION

The basic requirement for the beginning stages of the dam site civil engineering investigation is its efficiency at the high quality observing. The mistakes at the beginning stages of the prospection return later and their removing is immensely difficult especially as to the time, skill and finance. Geophysics as science itself is able to bring plenty of information at relatively acceptable prices at this investigation stage.

The investigation stage of geophysical prospecting of dams is to give information in the next problem circles: the dam site prospection, the dam lake prospection, the prospection of the useful site for the construction material mining and the prospection for future dam structure sites. It is obvious that not all scopes in question are solved at the primary prospection stages. It happens very often that the entire emphasis is placed only on the first problem, i.e., on the dam site investigation.

The basic lithological types and their spreading in the investigation area are determined on the dam site first of all. The other problem scope forms the demarcation of the weak rock massif zones including identification of tectonic faults of rocks. Last but not least at this investigation stage it is suitable to assess the basic physical and mechanical properties necessary for water construction designing too. It is not even possible to omit the estimation of the rock massif permeability and the related problem of the courtain depth.

In investigating lake areas it is necessary to pay attention to the stability problems of the future lake slopes and to seeking places with the possible water leak from the dam lake. The former scope presents the existing landslide investigation and seeking for places of possible new slope deformation origin. It is necessary to determine places with the big thickness of the cover formation and places with the significant rock massif fault deterioration. The latter scope presents seeking for places with extreme permeability positions independent of their origins (lithologic or tectonic). The investigation for the incidental side and safety dams belongs to this sphere as well.

The construction of each dam needs ensuring the sufficient quantity of the construction material. The requirements for the construction material supplement differ according to the planned dam type. In any case this is the question of seeking for suitable material deposits, of discovering rock material deformation in future quarry places and of determining the construction material exploitation. Further it is necessary to find out the data about the ground water in the future exploited places.

In case the bases for the function structures are requested at the investigation stage it is necessary to ensure the answers to the questions on possible ways of founding of the requested structures. The tasks are similar to those in the dam site investigation, then.

A considerably wide circle of geophysical methods, namely sounding and profiling ones, is used for the questions mentioned. Their real composition is the question of the given dam site and of the problems that is being solved. It is necessary to put together requirements resulting from the following questions: designer's needs, rock massif and rock physical properties, earth's surface etc. Even thus it is possible to determine some common regularities in using individual methods. From the sounding methods, the shallow refraction seismic, Schlumberger sounding and ground penetrating radar are used when solving special problems the frequency sounding and the inducted polarization sounding can be used. The resistivity profiling with the symmetrical electrode arrangements and the electromagnetic profiling (especially at the very long frequency modification) are used most often. It is obvious that other geoelectric profiling methods are used too. These methods are completed either by the magnetic profiling or the inducted polarisation profiling or thermic profiling or radiometric profiling. The new hole logging or at least the measurements on cores cannot be omitted.

The basis for the complex interpretation of the geophysical measurements is all the geological data that are at disposal. They are all the archival data that are available. That means the geological maps of all types and scales, the archival hole data and of other subsurface works carried out in the area of interest. Further there are all the previous geophysical measurements that were materialized in the area of investigated construction. It is suitable to use the space photos and the results of the remote sensing. Last but not least there are the data that a skilled geophysist can obtain right in the field. These are the studies of the geodetic maps and the geological field inspections. Obtaining the basic idea of geological and geomorphological configuration of the area of interest is the purpose rather than the geological mapping. The determinations of the various physical properties of the outcrops belong obviously to that work. This enables to obtain the first idea of the rock behaviour and of its properties. All this mentioned above must be taken into account in electing geophysical methods and in their method working.

The result of such an approach is to obtain the bases about the attitude relations, rock deteriorations and geotechnical property estimates of the determined quasihomogenous blocks. Further, it is possible to obtain the bases for the weathering depth and the weathering degree determinations, the released stress zone depth and about the permeability. The abilities of the geophysical methods are shown on the examples of the dam site investigation in the Czech republic and in Spain.

2. SUPPLEMENT AREA INVESTIGATION

As it has been mentioned, the dam investigation includes not only the dam site solution itself but even other problem solutions. The first example is taken from the combined subject of Trsice dam in the Czech Republic (Figure 1).

The dam profile occurs in the area where the neogene basin projection lies on the culm rocks. The underlaying of the neogene basin is formed of claystones that weather very easily into the clays. The culm massif is formed of the greywacke and slate. The quarternary cover is formed of the loam on the slopes and of gravel and flood loam in the alluvial plain.

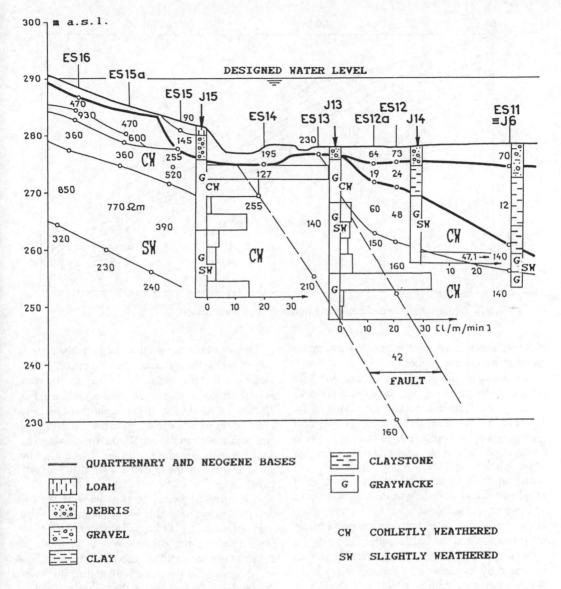

Fig. 1 Tršice - combined subject investigation

This example has been taken to show that the sufficient pieces of information can be obtained by using only geoelectrical methods. In this example the geoelectrical methods have been chosen on the basis of physical properties of the investigated rock massif. The whole geophysical investigation has been carried out only by the combination of the Schlumberger sounding and the symmetrical profiling. As this has been the question of the combine subject investigation, there has only been chosen a small distance between ES point. In the structure place and in its nearest surrounding, ES distance has been 10 m. In bigger surroundings the distance has been then about 20 m. The step of the profiling measurements has been 2.5 m.

All known lithological types have been successfully determined in the rock massif by the geophysical measurement. The weathering degree has been fixed in the rocks. The geological

71

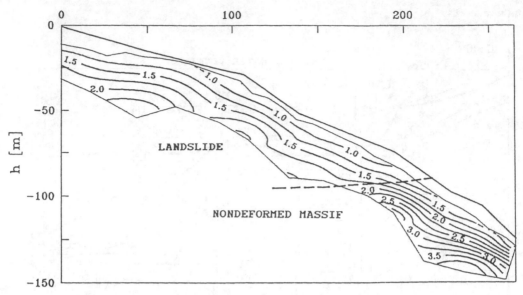

Fig. 2 Genal - velocity field in deformed and nondeformed rock massif

structure determined by the geophysical measurement is shown at Figure 1.

According to the geophysical work, the hole net has been proposed for the higher stage of investigation. The hole results are drawn in the Figure 1 to show how well the geophysical interpretation corresponds to the hole prospection. The hole prospection shows the basic lithological types, weathered degree and results of the pressure tests. If we compare the geophysical and hole results, it is necessary to add two remarks. One is a certain inconsonance in the gravel base asses between the geophysical interpretation and the hole results (Figure 1, the right side). According to our experience this fact is caused by the unsuitable drilling technology. When using this drilling technology the gravel fragments are pushed down to the bedrock clay. About 1 m gravel boundary pushing is typical for this area. The other remark is concerned with the culm rocks. Except the common difference of the rock weathered degree, the significant drop of the specific resistance can be seen on the right side of the fault. It is caused by carrying the clay from the overburden into the cracks of the culm rocks. Thus the conductive net comes into being and at the same time the rock resistance drop arises.

This example shows that the suitably chosen way of work, though using only one geophysical method, can bring valuable data about the rock massif structure under the dam structures. For all that we would like to point out that basically even better results can be obtained by using the combination of more geophysical methods. The geotechnical parameters can be determined by using seismic methods as early as at the investigation stage.

It is often necessary to sort out whether the designed slope is stable or whether the slope was deformed by sliding. Even here there are some opportunities of geophysical methods to help to the successful solution of this task. Figure 2 shows the possibility of getting to know the landslide according to the velocity field at the slope. The shallow refraction seismic was processed by the plus-minus method with the velocity penetration correction (Bláha 1994). The example is taken from a dam site at the Genal River (southern Spain). In the velocity field it can be unambiguously seen that the rock massif along the cross section can be divided into two parts. The bottom slope part is noted for the common velocities for the rock massif that is afflicted by the weathering only. In the upper slope part (roughly to 200 m stationing) the velocity field has quite a different character.

The low velocities reach as far as to the higher depths and they are about 50 % in comparison with those in the nondeformed rock massif. It is obvious that the velocity field itself cannot determine the landslide existence but it can warn of the slope deterioration, too. The deterioration origin must be solved by the geological interpretation of the geophysical measurement.

3. DAM SITE COMPLEX INVESTIGATION

The procedure described in the first chapter will be in a practice way shown on the example of the investigated stage of the dam prospection in the Genal Valley.

As it has been mentioned above, the valuable bases for the geophysical measurement interpretation can be obtained by the thorough study of the geodetic maps and the field morphology. The examples of such works are shown in Figures 3 and 4.

The morphological analysis of the area of interest came out from the geography map in the scale 1:50000. The analysis of river and brook stream directions served as the basic data for seeking faults. The mountain slope contour directions are taken into account in the further partial basis. The morphological analysis results are drawn in Figure 3. They show the basic direction of faults that can be expected in the investigated area. It is obvious that it is necessary to see the tectonic fault as the zone of the rock massif weakening in this case. This weakening can be caused by the tectonic massif deterioration but it can also correspond to the rock belts of a lower strength.

Perhaps the best distinguished departure from the geological map is the E-W direction in the southern half of the investigated area. It is, however, clear that this direction in Gaucin surrounding really exists. This direction is significant because the Genal River always changes its stream direction between those two lines. In these belts the river runs from north to south. The common direction of Genal stream is NE-SW or NNE-SSW (see the belts southern from P4 or northern from P3). The further reality that the E-W lines influence is the Genal stream character. The Genal River strongly winds in its valley to the north of the best significant E-W line (A line). Under this line its

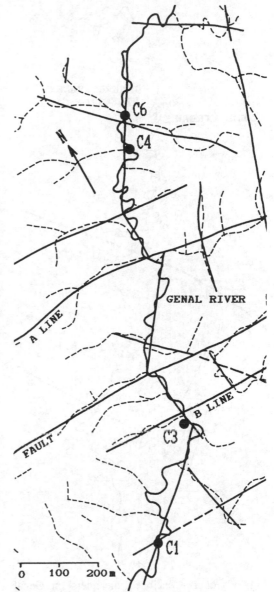

Fig. 3 Genal - morphological analysis

stream is straighter. We show this fact in Figure 4. There is shown the common direction of the Genal Valley and its real stream. In addition, the Genal River has a different stream inclination between A and B lines. If it overcomes the height difference of 10 m, from 300 to 600 m of the stream to the north from A line then between A and B lines from 700 to 1100 m of the valley length is necessary for the same drop. It is possible to say that just between

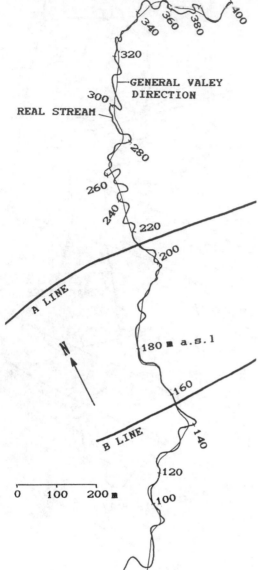

Fig. 4 Genal - enclosed meanders on Genal River

these lines there are tougher rocks and this area relatively rises in comparison with the northern and southern areas. In this way the Genal River is forced to wind in the northern part of the investigated area although it is the upper river stream. This type of the meanders is called the enclosed meanders because of its connection at the tectonic structure of the area of interest.

On the basis of the morphological analysis we can suppose that the geological structure is very complicated in the area of interest, especially the tectonic structure. The geophysical measurement fully certifies this assumption. It is apparent that the individual tectonic fault has a significant relation to the geophysical display of the rock massif.

The example of geophysical measurements for the investigation stage of the dam sites is taken from the Genal 1 dam site. The complex of the geophysical sounding methods consisted of shallow refraction seismic (SRS) and Schlumberger sounding (ES). SRS was processed by the critical distance method and the plus-minus method (with velocity penetration). One arrangement of the Wenner profiling with the depth radius of 10 m was used. It was completed with magnetic profiling (total vector). These real profiling methods are completed with the velocity changes on the refraction horizon along the cross section.

All the cross section measurements in the whole area of interest and their primary processing was followed by the first stage of geologic interpretation. On the basis of geophysical measurements and common geological laws the rock massif in the whole area was divided into six quasihomogenous blocks. A type is formed of the Genal River sediments. B type introduces the bed of the loam debris or the slope loam. The eluvium forms C type and D type is formed of the weathered rock massif. The nonweathered rocks are practically in E and F types. The rocks in the question are phyllites and slates.

The statistic analyses of the physical properties were carried out after that primary interpretation. The histograms of the specific resistivity are shown in Figure 5 and the refracted velocity histograms are shown in Figure 6. The specific resistivity histograms are calculated in the logarithmic scale. The histogram of the river sediments includes two basic types, i.e., the water-bearing or dry sediments. The boundary between them is about 500 ohmmeters. The lithological division into gravels and flood loams cannot be statistically determined. A very interesting distribution is in B rock type. The core of the overburden formation forms a scree loam debris with the average specific resistivity about 1000 ohmmeters. The lower values than those

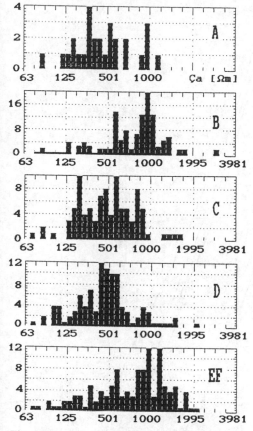

Fig. 5 Genal - specific resistivity histograms

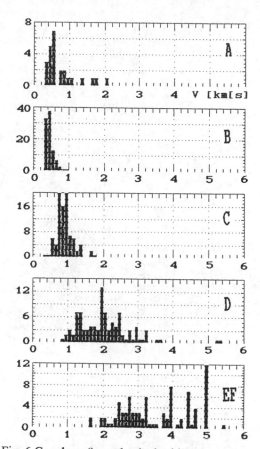

Fig. 6 Genal - refracted velocity histograms

responding to the logarithmic distribution are given by the transition of the scree loam debris into the pure loams. The specific resistivity histogram suggests a possible way of seeking the contribution into two subset even in this histogram. Since the absence of further data about eluvium occurred, we gave up the further dividing. D type forms the only large statistical set. The bedrock can be divided into two subsets that respond to two well-known lithological types in this place, i.e., phyllite and slate. It is very interesting to observe the specific resistivity changes between B, C and D types. In principle we can say that the specific resistance gradually drops. This drop is influenced by two facts. One is the depth moisture increase of the rock massif. The other is the abatement of the air drop content in fissures and cracks downward. The air is replaced with water or with clay mineral.

The statistic distribution of the velocities (in the linear scale) gives the further knowledge of physical properties of the rock. A set has the highest velocity representation close above the sound velocity. This velocity set belongs to flood loams. The velocity set from 1.0 to 2.2 km/s belongs to gravels. In B bed, i.e., in the bed of debris and slope loams, the velocities are pressed close above the sound velocity as well. The velocities of eluvium are higher and their middle is in the surroundings of 1.0 km/s. The further velocity increase (in the absolute velocities more significant) is in the weathered bedrock. An essential thing is that D bed does not crumble into two sets. It means that the weathered slates and weathered phyllites are of the similar mechanical properties. The velocity distribution at the nonweathered rocks is quite different. Two partial subsets are unambiguously apparent here. One of them has a mean

75

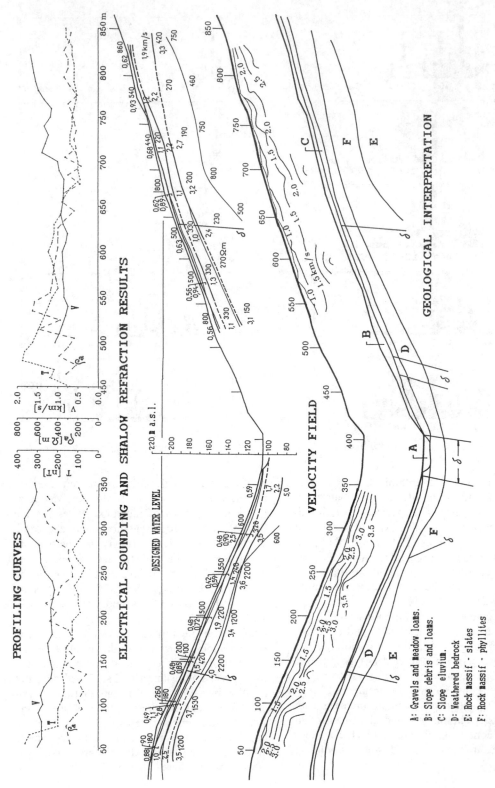

Fig. 7 Genal - complex interpretation on dam site P1

76

velocity about 3.0 km/s, the other has a chaotic distribution with the velocity from 4.0 to 5.0 km/s. It can partially be caused by the limited abilities of the critical distance method in the high velocity determinations. The tectonic deterioration of the rock massif is also influenced by the velocity distribution to a certain extent. It is essential especially at those slates where the velocity drop transfers the deteriorated slates into the phyllites. The velocity values cannot only be taken into the consideration when adjoining the lithologic meaning.

The construction of the engineering geological cross section is the most essential problem in dam site investigations. Figure 7 shows how to proceed by using geophysical methods. The bases for the construction of the main engineering geological cross section were the geophysical measurements on both declivity cross sections (in the higher part of the slopes) and the measurements on horizontal cross sections (in the lower part of the dam site). The results of the declivity profile measurements are shown in the upper part of Figure 7. The sounding results on the declivity cross sections processed by classical methods are drawn under the profile measurements. The velocity fields that are obtained by velocity penetration processing are shown below. The resultant engineering geological cross section along the whole dam site is shown in the most bottom part of this figure. The difference between the velocity fields on the left and right slopes is the most significant element of the whole geophysical measurement on this dam site. The velocity fields under the refraction horizon respond to the various lithological composition of the bedrock on both slopes. The left side bedrock (where the velocities are higher) is formed of slates. The right slope velocities are lower and therefore the bedrock is formed of phyllites. Only one fault is determined on the both slopes respectively. The faults were interpreted according to the velocity fields and the profile measurements.

The basic things for the engineering geologist, i.e., the main dam site cross section and the correlation scheme of faults can be designed after several geophysical cross section investigations. According to the classical interpretation methods we can determine individual geological beds. The beds with similar properties are mostly parallel with the surface. The geological beds differ not only by their properties but also by their thicknesses. The compiled main cross section is shown in the bottom part of Figure 7. All the lithological types and the main faults that were determined on this dam site are drawn in this figure.

The geological meaning of the geophysical beds was given according to both the individual physical property values and the common geological laws.

The next geotechnical types that were determined are shown on the cross section:

Quaternary:

A: Gravels and flood loams.

B: Slope debris and loams.

C: Slope eluvium. This bed can be characterized as a very loosened bedrock. Cracks can be filled with clay (in places with a low specific resistivity).

Bedrock:

D: Weathered bedrock. The conductive mineral beds (especially clay) can create the conductive nets that reduce the rock massif resistivity.

E: Nonweathered bedrock. The rocks are relatively little deteriorated and little weathered. From the lithological point of view this bed is formed of slates, locally of phyllites. The silicified slates can exist in places with the higher resistivity.

F: Nonweathered bedrock. The rocks are relatively little deteriorated and little weathered. From the lithological point of view this bed is formed of phyllites. The graphite slates can exist in some places.

The further parameters necessary for the dam designing can be determined on the bases of an interpretation like that and of common geological laws, but it is another problem described in many works (for example moduls, curtain depth, etc.).

4. CONCLUSION

This article shows the geophysical method facilities at the investigation stage of the dam site prospection. Especially good results are obtained in those cases when the complex of several methods is used for the geophysical investigation. The possible ways of this geophysical investigation complex for the dam sites are shown on the real examples in southern Spain and the Czech republic. Besides the basic geological structure evaluation we can obtain some information about further data necessary for the dam designing.

REFERENCES

Bláha, P. 1994: Shallow refraction processing in geology. *Proc. 7th IAEG*: Rotterdam: Balkema.

Bláha, P. & Horský, O. 1994: *Métodos de prospection geofízica para el estudio de las posibles cerrados en el anteproyecto o riabilidad de un embalse*. In print.

Kelly, W.E. & Mareš, S. (editors) 1993: *Applied Geophysics in Hydrogeological and Engineering Practice*. Monography: Developement in Water Sience. Rotterdam: Balkema.

Geological inhomogeneities in seismic tomography

Hétérogénéités géologiques en tomographie sismique

Pavel Bláha
Geotest a.s., Brno, Czech Republic

Jiří Kottas & Jan Sochor
Mathematics and Physics Faculty, Charles University, Prague, Czech Republic

ABSTRACT: Seismic tomography is one of the new geophysical methods. It can be applied not only in oil and structural geology but also in engineering geology and in the geology of small dimensions. This article shows some of its possible applications in studies of various geological inhomogeneities.

RESUME: La tomographie séismique est une des nouveaux méthodes de la géophysique. Les applications ont lieu non seulement dans le domaine de la géologie du pétrole et de la géologie des structures mais aussi dans le domaine de la géologie appliquée et de la géologie des petites dimensions. L'article montre des possibilités de la méthode pour les études des non-homogénéités géologiques.

1. INTRODUCTION

The contemporary development of construction design and building foundation needs better knowledge of the geological environment that is to be influenced by the structure. New methods for making a more perfect engineering geological investigation are required. Recently a new method of processing geophysical measurements has begun to appear in books and articles. This method is tomography.

Seismic tomography is a new geophysical contribution to the better knowledge of a geologic environment. It enables us to get a better idea of the velocity field between holes and galleries. In comparison to the classical assumption of seismic tomography, our method can be used not only in the "hole - hole" modification (or "gallery - gallery"), but also in the "hole - surface" modification. Good results have been produced with "surface" tomography. A much better description of the explored rock massif can be obtained than by classical methods.

Tomographic processing of geophysical measurements can be used in radiographic methods, i.e., in electromagnetic and seismic radiography. These two kinds of geophysical measurement need two different ways of processing. It is necessary to process electromagnetic radiography by considering straight rays of the spread signals among sources and receivers, which is simpler mathematically. Seismic radiography needs more complicated processing. The seismic rays spread in the quickest way, i.e., on curved lines. Tomographic processing of seismic radiography gives qualitatively better results than classical processing in fan diagrams.

In tomographic processing the whole measured area is built in a rectangle which is divided into rectangular cells. The cell dimension depends on two quantities, one of which is the number of measured rays, which must always be higher than the number of cells. For the inverse calculation it is necessary for the number of measurements to be greater than the number of unknown velocities (i.e., mathematically overdetermined). From the mathematical point of view it is suitable to use a multiple overdetermination, i.e., the number of seismic trajectories must be a multiple of the number of cells. From the civil engineering view point there is an antagonistic requirement, i.e., we need to have the number of places in which the velocities are determined as high as possible. Thus a geophysicist is forced to choose a compromise (often a task of geophysicists). The overdetermination is usually chosen between 1.5 - 2.0.

The other quantity that influences a cell number is the length of seismic waves and the connected frequency of the elastic wave. In real solutions there is no sense in selecting a cell dimension considerably smaller than the length of the used seismic wave. The inhomogeneities with dimensions considerably smaller than the wave length are of no consequence for spreading a seismic signal. The real output of tomography processing of the seismic radiography is the velocity field in the cells. It is possible to set up a velocity contour map using graphic programs. It is also possible to draw seismic rays with our program.

2. MATHEMATIC SOLUTION

2.1 Discretisation of the problem

Suppose that our domain is a rectangle which is divided into subrectangles. Suppose that points Z_i $i = 1,...,$ n (sources of a signal) and S_j, $j = 1,...,$m (receivers of a signal) are given. Finally suppose that an unknown velocity V is constant on each subrectangle so we have to find K-tuple of numbers $v_1,...,v_K$, where v_k is velocity of a signal in the k-th rectangle.

To do that we can use input data $T_{i,j}$ $i = 1,...,$m $j=1,...,$ n - minimal travel times of a signal from the source Z_i to receiver S_j. It is natural to suppose that mn > K. (A number of measurements is greater than a number of unknown velocities.)

The problem is solved by the iteration process where every iteration consists of two steps:

A. Direct problem

For given velocities v_k travel times t_{ij} corresponding to this velocities are evaluated.

B. Inverse problem

We compare evaluated travel times t_{ij} with measured travel times T_{ij} and then we choose a new value of velocities v_k to decrease a difference between them.

2.2 Direct problem

To evaluate times t_{ij} we have to know a curve minimizing the travel time from Z_i to S_j. Notice that velocity v_k is constant on each subrectangle and from this fact it immediately follows that the part of this curve in each subrectangle is a segment. We can choose a sufficiently dense set of points on boundaries of subrectangles and then try to find a piecewise linear approximation of a minimizing curve.

The problem to find a curve minimizing the travel time is transformed into a problem of graph theory - to find a minimal way in a graph. Our algorithm for solution of this problem is more effective than the general one. We use some natural assumptions coming from the physical background of the problem.

2.3 Inverse problem

Suppose that we know the old iteration of velocity v_k and we have to construct a new one. For this we use the following method: Define function F which in some sense expresses the difference between the real situation and our model with velocity v_k and we have to find v_k such that the corresponding value of F is minimal. For this we calculate gradient of function F where partial derivations are replaced by corresponding difference quotients. This gives us the direction of the steepest descent of function F and we can choose the new approximation of the velocities v_k.

The repetition of these steps gives us approximations of the solution of our problem.

3. TOMOGRAPHIC MODELLING

In the processing program that we have at our disposal it is possible not only to measure data to process, but also to produce mathematical models of various geological structures. This modelling is useful for several reasons. The first is to get an idea of the suitability of a seismic tomography application for various geological structures, the second is verify certain interpretation rules for real tomographic measurement processing, and finally the suitable system of field tomographic measurement for the real geological structure and the real subsurface work placement can be determined.

Figure 1 shows the real result of tomographic modelling of a deep glacial buried valley. The real model is shown in the upper part of Figure 1. We supposed that the valley is recessed in the rock massif which velocity 5.0 km/s V_1. The fault deformation of rocks under the valley middle reduces velocity to 3.0 km/s V_3. The buried valley is filled with clastogene matter of velocity 2.0 km/s V_2. As the complete exact analytical solution by oblique lines is very complicated, we chose a replacement of oblique lines by slightly broken lines (see Figure 1). The times of seismic waves spreading along the shortest time rays were calculated for this model.

Times thus determined were interpreted using our common program for tomographic processing. The whole area was divided into 15x13

cells with 2x3 m dimension. The seismic velocities in these cells were determined. The contour cross section (Figure 1 bottom part) was then constructed from the calculated velocities using the graphic program SURFER. Figure 1 shows that the fault position practically corresponds to the 3.5 km/s contour line. The fault width would be determined as about 50 % greater than it would be according to the common interpretation rules (for example, the inflexion spot rule). It is necessary, for the practice interpretation of tomographic measurement, to get the whole series of these models. Then the interpretation rules for the common works can be derived more exactly from those models.

The problem of determining a buried valley boundary is similar to that of determining faults. The actual results obtained corresponding to the model velocities of 2.0 km/s and 3.0 km/s were 2.2 and 3.2 km/s respectively. Local maximums resulted from interpreting the rock massif velocity (the 5.0 km/s). Two anomalies lie along both fault sides and another two under both valley margins. These anomalies appear by data processing on the graphic program. The determined velocity is 5.15 km/s instead of the 5.0 km/s.

We can see that mathematical modelling helps us to get some idea of the potential of the tomography method. We must realize that the final contour cross-section or map is influenced by both tomographic and graphic programs. It is very difficult to distinguish which inaccuracies are caused by which program. As we can see in the example given and as we know from our own experience, the total precision of the velocity and depth determinations is better than 10 %.

GEOLOLOGICAL MODEL

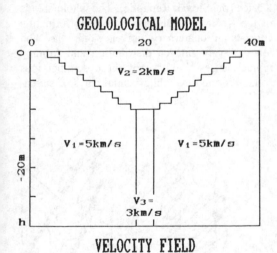

VELOCITY FIELD

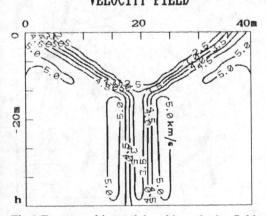

Fig.1 Tomographic model and its velocity field

4. FIELD EXAMPLES

Seismic tomography can be used in three systems according to the source and receiver placements in the space in engineering geology. The first system uses surface sources and receivers, the second uses a surface source and hole receivers (or the other way round) and the third uses all in holes or galleries.

The whole surface measured system is not used often because of the variability of surface bed thickness and velocity. This system can be .used only when these unfavorable influences can be eliminated or ignored them. The first variant is used if the surface-bed influence is really constant (the cover thickness and cover velocity are the same in the whole measured area). In these rare cases it is possible to install corrections and eliminate the influence of the low-velocity bed. Another opportunity occurs if the surface beds are mechanically stripped, so that the tomographic velocities approach to reality and are useful for the recalculation to geotechnical parameters of the investigated environment.

It is possible to ignore to surface-bed influence where the effect of measured time increase is small, i.e., in small thicknesses of overburden or over long distances. In these cases the tomographic velocities are not quite exact, but are a mixture of surface velocities and refracted ones. Nevertheless such a velocity field can give valuable information about the rock massif condition. These tomographic velocities cannot be used for subsequent recalculation to mechanical properties.

The second measuring system, surface - hole, is the most frequently used. Seismic sources are placed on the surface and receivers in the hole, or vice versa. The distances between receivers and between sources are both approximately 1-5 meters according to the conditions. Because velocity increases with the depth, velocity data from a depth greater than that of the hole can generally be obtained. The tomographic depth is 10 % bigger than the hole depth. Thus we can obtain data towards the side distance of 100 - 150 % hole depth in the horizontal directions. This variant gives an idea about the velocity field on both sides of the hole and also we can measure a few cross sections going through the hole. Usually two perpendicular surface cross sections are measured. Instead of the line information about the hole length h, a good idea of the geological structure in the cylinder of 3*h diameter and 1.1*h depth can be obtained. In the individual depth cross-sections radiograph there are not any rectangles or triangles (given by the last source, hole outset and hole bottom), but real areas between these extremes. This size depends on a geologic structure. Tomographic velocities taken from this measuring system are useful for calculating geotechnical properties of a rock massif.

The third system is the classical seismic radiography in the rock massif depth. As it has been introduced we can apply it in the gallery - gallery variant, the hole - hole one or in the combination of these subsurface works. The tomographic velocities can be used for recalculation to mechanical properties or to the stress field.

4.1 Surface system

The example of the real results of the surface measuring system is taken from the Budimír landslide in the eastern Slovakia. Open cracks had appeared during the roadbed construction. Geophysical measurements were made to explain the landslide structure and find the reason for the crack occurrence on the constructed roadbed.

The Budimír landslide lies 10 km north of Košice (the Slovak Republic). The whole area is of the Košice gravel formation. This formation consists of quaternary gravel, sand, clay and loam of very uneven distribution and forming typical lenses. Subbase of these semipervious soils is formed by subhorizontal clays of sarma-

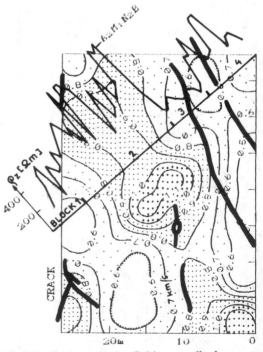

Fig.2 Budimír - velocity field on roadbed

82

tien age. The whole area is predisposed to sliding.

Surface tomographic measurement in the Budimír landslide was carried out in the nonline seismic measurement version. The source and receiver spread is shown in Figure 2. The distances between receivers were 3 m and between sources 7.5 m.

The seismic tomographic results show a rock massif deformation by tension cracks. Refraction results show that although the measurements were done on the roadbed, the refraction horizon still existed there. Therefore we must have in mind that the contours show only apparent velocities, but not real ones. The drawn velocities are the function of the velocity value of the direct waves (direct velocity), the velocity on the refraction horizon (cca 5 m depth) and the true depth of this horizon. This influence in the investigated area is approximately constant and it is thus possible to judge a relative size of the deformation or to judge relative mechanical stress according to the velocity value.

The comparison between tomography and symmetrical resistivity profiling (SRP) of P7 profile is the base for the geological interpretation of the seismic tomography. The course of the apparent specific resistance (on P7 profile measured) and the contours of the apparent tomographic velocities are shown on Figure 2.

Two striking elements can be seen on the curve of the apparent specific resistance. One is a violent variation of apparent specific resistance in places with sandstone superiority, the other is the fall in apparent specific resistances in places with clay superiority and the relatively quiet course of the apparent specific resistance in these places. Four bends of the various velocities can be determined from the apparent velocity contour map. They correspond well to the blocks from the resistivity profiling results on P7 profile.

By comparing all results it is possible to derive that the rock massif on the roadbed is situated in the tension stress area. The tension stress value is not the same in all places. It can approach normal values in places where the velocity reaches relatively high values (2/P7 block). This block is deformed with short discontinuous cracks (see the crack limits in Figure 2 center). The good contact between individual sandstone blocks gives relatively high velocities, and the presence of air in the cracks results in the high apparent specific resistances. Suffusion events arise in some cracks (see Figure 2, place x = 15 m a y = 11 m).

Another extreme is block 3. In this block a significant rock deformation accompanied by its alteration took place. The first tension crack rise in the anticlinal enabled intensive weathering of the origin rock material, which is gradually transformed into clay. Full alteration of the original rock mass is not necessary for reducing specific resistance, formation of a conductive clay net will be sufficient. This block has about five times higher specific resistance than clear clays. According to SRP (two arrangements) this block has a subvertical boundary (vice versa the sedimentation boundary between the sand - clay has a subhorizontal one). This block has not only a low specific resistance, but also a low tomographic velocity.

Blocks 1 and 4 are affected by the last type of deformation, tension stress. Some tension stress cracks are filled with clay. The apparent velocities that are significantly lowered (in block 1 more than in block 4) give evidence about the considerable tension deformation too. The apparent specific resistances vary according to the presence of tension cracks.

According to the seismic tomography results it can be determined that block 1 is similar to block 3. The apparent velocity values in both blocks really reach the same velocities, but it is of interest that no cracks are visible on the roadbed in block 1. This fully supports the theory of the rock mass alteration with tension stress moving to the relatively little deformed blocks. We can see in the left top corner of the measured area how the individual cracks limit a relatively nondeformed sandstone block. Also it is obvious how the crack in the middle area ends in front of the velocity maximum, i.e., in front of the zone with a relatively higher stress.

4.2 Combined system ("hole - surface")

The real case history is taken from the engineering geological investigation for the comprehensive structure of the Žilina dam. The structure foundation proposes opening 35 m depth pit. Seismic radiography was measured with

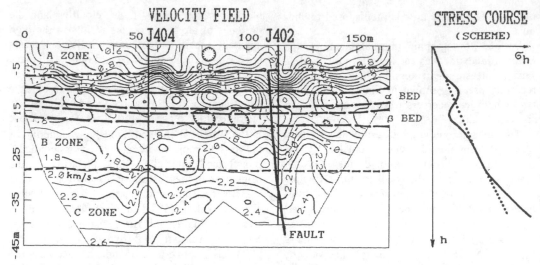

Fig.3 Žilina - velocity field and stress curves for hole - surface system

surface sources and hole receivers. These were step by step in JZ402 and JZ404 holes. This seismic radiography was processed as one seismic cross section. The results of these measurements are shown in Figure 3.

The area of interest lies on northwestern Slovakia. The overburden is formed from quaternary gravels of the River Váh. These gravels cover palleogenne flysh sediments, where claystone prevails on sandstone.

We can see from the contour map that no rectangle and no triangle is radiographed in this measured system. The radiographed area form depends on the real geological structure. From this measurement we can show very well how close is the relation between the longitudinal wave velocity and the mechanical stress.

According to contours we can divide the tomographic cross sections into three basic beds. Bed A consists of quaternary gravel. The gravel basis is determined according to contours with taking account to this bed limitation from holes. From the tomography modelling we know that the geological boundaries almost correspond to the beginning of significant velocity changes. A simple geophysical rule that the boundary lies in the inflexion spot or in its proximity does not hold true.

Bed B is the upper bed of palleogene flysh. The velocity field in this bed is the most interesting and it brings most information to the en-

gineering geologists and geotechnologists. All this bed consists of stress relieved rock massif. Inside this bed we can separate other thin beds. Beds α and ß are the most interesting of them. Bed ß responds to the dilated deformed bed of flysh complex. This deformation is very significant and this is no doubt about it especially in the figure right side. The velocity lowering of longitudinal waves occurs by the rock deterioration. This velocity drop is at least 0.5 km/s.

Vice versa the optical impression of the velocity field in bed α gives an idea of the concentrating stress over the dilated deformed bed ß. According to velocity contours it is very difficult to judge whether in this case it is the real concentration zone or the optical effect. The velocity growth shows the most probable solution. In the right side of the cross section some stress concentration can be proved, but in the left side it is only the optical deception. From the detailed study we can state that in the cases of more significant dilated deformation the partial contracted stress zone appears in the top wall.

The fault influence in cca 115 m manifests itself strongly in bed B. In this place α thin bed becomes extinct. The velocity lowering in ß bed is also bigger than in other parts of this bed. The fault influence is obvious in bed C too.

The velocity contours are relatively monotonic in bed C. The only exception is the small velocity drop at the fault. Higher velocities in the right side of the fault are caused by a higher proportion of sandstone in flysh rock (according to hole logging). We know from other tomographic cross sections in this area that in bed C the fault influence on the seismic velocities is small. This fact, and velocity depth growth give evidence about strong horizontal stress in this bed. This stress is most probably given by the existence of a first order fault about one kilometer northly from the measured place.

According to velocity curves we can assess the horizontal stress courses in the vertical cross sections. The scheme of the horizontal stress course in the middle of the cross section is shown in the Figure 3 right side. The dashed line shows the normal course of the horizontal stress. The solid line shows the derived anomaly course. It is very difficult to determine the absolute departure from normal stress value. The relationship between velocity and stress comes into existence here. Therefore it is necessary to understand the shown curve as model one. If it is necessary to know the absolute stress values, it is essential to carry out other measurements that would give the proper data for the relevant recalculations.

4.3 Subsurface system

The real example is taken from the geophysical measurement for the engineering geological investigation for the Dlouhé Stráně water pumping station. The original seismic radiation is of the old datum (1973). This measurement was processed in the classical way of fan velocities. Later this measurement could be processed again by seismic tomography. The comparison between the old method and the new one is obvious from Figures 4 and 5.

The area of Dlouhé Stráně lies in northern Moravia. The investigated area belongs to the Desná Arch system, and is formed predominantly of biotite paragneiss. In some places there are orthogneiss, amphibolite and pegmatite.

The radiation was performed between three galleries. The receiver distance was 10 m, the source distance was approximately 100 m. The spreading of sources and receivers is shown on Figure 5.

The classical processing by fan graphs (Figure 4) offers useful data about rock massif conditions during the measured time. These results give us an idea of the basic changes of rock massif velocities. They show that the velocities in the further inside rock massif are higher and tell about a possible way of founding the power station cavern in that place.

Tomographic processing (Figure 5) gives more evidence that this processing method improves the quality of seismic radiation results. At first sight a significantly higher rock quality inside the rock massif is obvious. Besides this basic knowledge, tomographic velocities give us a lot of further information. Knowledge about the fault deformation of the rock massif is basic for us. The most important fault in the area in question (the Desná Fault) causes a significant velocity drop. The gneiss deterioration and alteration are demonstrated in the gallery 2 in the Desná Fault. The thickness of the clay is as large as seven meters. Besides the most important fault, a further two longitudinal and two transverse faults can be proved. It is especially obvious in the second fault (the middle one) how significant the rock deterioration is. From the tomography velocity field, we can see that in the longitudinal faults velocity and mechanical properties both decrease.

Both classical geological and geophysical logging were made in the investigation galleries. The seismic measurement in the gallery bottom was one from many applied methods. The comparison between the logging velocity V_g and the tomographic velocity V_{ST} is shown in Figure 6. At first sight the harmony of both curves is obvious. The Vg curve is more articulated. It is given by the fact that this measurement investigates the gallery surrounding. Both velocity curves agree with rock massif deterioration (see geological logging). On this basis we can suppose that a similar good agreement exists in the inside rock massif.

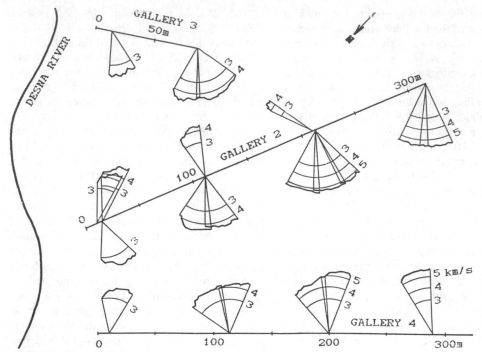

Fig.4 Dlouhé stráně - seismic radiation - velocity fan diagram

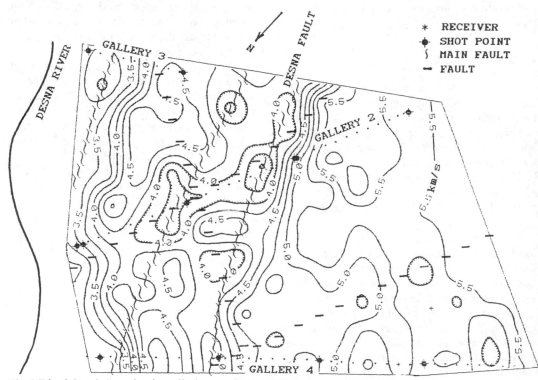

Fig.5 Dlouhé stráně - seismic radiation - velocity field from tomographic processing

GEOPHYSICAL MEASUREMENT

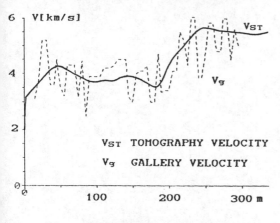

Vₛₜ TOMOGRAPHY VELOCITY

Vᵍ GALLERY VELOCITY

GEOLOGICAL LOGGING

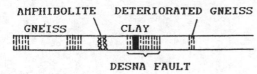

Fig.6 Dlouhé stráně - velocity curves in 2 gallery

5. CONCLUSION

Seismic tomography is a new geophysical contribution to the better knowledge of a geologic environment. It enables us to obtain a qualitatively new idea of velocity field among the holes and galleries. In comparison to the classical assumptions, it is possible to perform seismic tomography not only in the "hole - hole" modification, but also in the "hole - surface" modification. Very good results have been produced by surface tomography. Thus it is possible to obtain a much better description of the explored rock massif than by using classical methods.

Our program for processing seismic tomography results can also be used for mathematical modelling. Modelling enables us to obtain some interpretation rules and an opportunity to project the best measuring system of receiver and source spread during the field measurements.

This article sets out real examples from both mathematical modelling and field measurements. The examples are taken from all the three field possible measuring systems: surface, combine and subsurface.

REFERENCES

Bláha, P. and others 1994. Seismologic and seismic measurement on Pontervedra dam. *Proc. 7th IAEG*: Rotterdam: Balkema.

Bois, P. and others 1972. Well to Well Seismic measurements. *Geophysics* 37:471-480.

Ivansson, S. 1987. Tomographic modelling Used for Crosshole Data Analysis. In: Borehole geophysics for Mining and Geotechnical Application. *Geological Survey of Canada*: 167-171.

Worthington. M.H. 1984. *An Introduction for Geophysical Tomography*. First Break. Oxford: 20-26.

Problèmes de la prospection géophysique du karst

Problems related to geophysical surveying in karst

B. K. Matveev & R. P. Savelov
Université de Perm, Russie

ABSTRACT: The Typical physico-geological models of karst are described, principal geophysical tasks are formulated, possibilities of the electric and seismic methods are examined and data of karst study on the two geodynamic proving grounds. The results of investigations enable to choose a complex of the geophysical methods and to increase reliability of karst study.

1 INTRODUCTION

Karst est un phénomène complexe de la nature, lié avec le lessivage des roches par des eau superfitielles ou bien souterraines. En definitive les entonnoir, les ponores, les betoirs, les grottes et d'aurtes poches (cavités) karstiques sont formés. La karstification est plus forte dans les régions, ou les roches solubles se trouvent à faible profondeur. D'apres Maksimovitch (1963) la tiers de la terre ferme est couverte par les roches solubles et 45% de la terre ferme – à l'Oural. Il y a aussi le karst profond (karst hydrotermal) (Kropatchev, 1973; Koutirev, 1989 etc). On constate, que le mouvement neothectonique favorise le processus karstiques (Maksimovitch, 1963).

Il y a pres de 60 annés que les méthodes géophysiques sont appliqués pour l'étude du karst (Oguilvi, 1932; Golovitchin, 1935; Oguilvi, 1957; Karous, Faisulin, 1977; Chouvalov, 1985; etc) Dans les plus par des cas on l'étudie pour resoudre les problèmes de Genie Civil. Par ailleurs, on fait l'investigation pour les problèmes d'hydrologie et des recherches des gîtes de mineraux utiles (Gorelik, Saharova, 1951; Oguilvi, 1990).

On considere l'édude du karst comme l'étude des phénomènes physico-géologiques, géodynamiques,

hydrologiques ou bion géomorphologiques. Parmi les nombreux problèmes de l'étude du karst traités par géophysique, citons les principaux:
- étude du massif des roches, ou le karst se forme;
- recherche des zones fissurés et des roches karstifiables;
- étude et la prospection de l'espace du karst;
- étude de la dynamique des eaux souterraines;
- prospection et decouverte des gîtes karstiques des mineraux utiles;
- étude des karst protonds.

Les méthodes adoptées sont basées sur l'étude des anomalies géophysiques. Pour evaluer des anomalies et bien choisir l'ensemble des méthodes géophysiques il faut etudier des modèles physicogéologiques du karst. Les données, que nous avons, nous ont permis de généraliser nos idées sur les modèles physico-géologique (Matveev, Savelov, Gorbounova, Kataev, 1993), d'étudier les possibilités et proposer les méthodes géophysiques d'investigation du karst.

2 MODELES PHYSICO-GEOLOGIQUES ET CALCULS THEORIQUES

Les modèles physico-géologiques refletent en forme générale les éléments stucturels essentiels, qui se manifestent constamment en champs géophysiques. Selon l'etape d'investigation on fait les modèles generaux, particuliers, typiques ou bien concrets. Le modèle général est le modèle de la région karstique. Il peut se composer des modèles particuliers.

Les modèles typiques (apriori) et concrets (le modèle de travail) sont utilisés pour modèler dans le but du choix de la méthode des travaux ou pendant l'interpretation des données. En tout cas ils sont très variés et complexes. Cela est lié avec la genèse du karst.On sait qu'il existe les formations karstiques superficielles et profondes, qui sont distinguées par les indices de toute sorte.

Pour les géophysiciens les indices essentiels sont suivants:
- le type du karst, son afleurement, ou bien karst couvert, la dependence des conditions de formations etc;
- l'epaisseur, la composition et les proprietés physiques des roches recouvrantes;
- la profondeur des cavités karstiques et leur saturation par l'eau ou le comblement par les debris des roches;
- les proprietés physique des accumulations (remplissage du karst)
- la karstification des roches reconvrantes et leur fissurité.

Ainsi, le modèle typique du karst couvert peut être representé par la coupe à deux ou troix couches contenant la caverne karstique (fig.1). Audessus de la caverne il existe "l'aureole" des fissures et des trous et des roches recouvrantes plus recentes. Au fond du massif les roches karstiques etendent le calcaire lent, l'argile, les roches compactes representent le base de la corrosion. L'anomalie du champ géophysique resulte de la presence de la cavité et de "l'aureole". L'effet de "l'aureole" peut intensifier ou affaiblir l'anomalie. Cela depend de la position du karst, de la saturation et d'autres conditions differentes. L'eau minéralisée augmente la profondeur de la pénétration du courant et la possibilité de la prospection électique. En meme temps le courant, qui circule par les trous et les fissures, provoque le champ

magnétique et le champ de la polarisation spontannée.

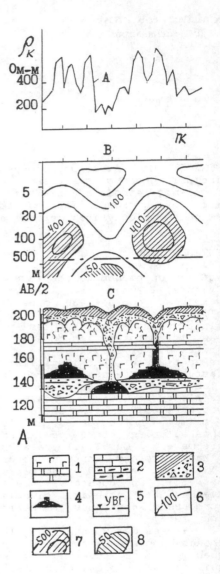

Fig.1. Modèle physico-géologique du karst:
a – profil de resistivité, b –coupe de resistivité, c – modèle de coupe; 1 – gypse et calcaire lent, 2 – calcaire compacte, 3 – argile et breche de karst, 4 – caverne, 5 – niveau de l'eau, 6 –equipotentielles de resistivité, 7,8 –anomalie de resistivité

90

Les premiers calculs et le modelé sont faits pour le milieu homogène avec le caverne du karst en
forme de sphère (Golovitcin, 1935;
Gorelic, Saharova,1951). Plus tard
on a calculé pour le milieu à deux
couches (Redosubov,1973; Serebrennikova, 1987). Sur la figure 2,a,b
on voit des courbes du profil
électrique (courbes 1,2), celles
de sondages électriques (courbes
3,4) et les cartes de resistivité
avec le modèle (5). En comparant
des courbes on peut remarquer
l'influence de l'épaisseur des
roches recouvrantes (épaissseur
des roches sur la fig.2,a est plus
faible que celle à la fig.2,b) et
de l'existence des cavernes
karstiques.

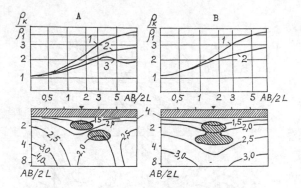

Fig.3. Exemples de sondages
électriques pour le milleu à deux
couches: 1 – sans caverne, 2 – un
caverne, 3 – deux cavernes.

3 TRAVAUX GÉOPHYSIQUES SUR LA GROTTE DE KOUNGOUR

La grotte de Koungour (glaciale) –
typique karst littoral de sulfate.
Elle est situé pres de la ville
Koungour (region de Perm, Russie),
au bord de la rive Silva. La coupe
de la berge est representé par les
roches permienes (gypse, calcaire
lent, calcaire compacte), qui sont
surmontées les terrains diluviems
de l'épaisseur de 5 à 15 m (fig.4
c). Dans le gypse il y a les
cavernes karstiques (leur longueur
ne depasse pas 60-70 m. leur
hauteur – 23 m) avec les defilés
entre eux. La lonqueur totale de
la grotte (partie etudié)est egale
à 2,5-3 km. Dans la grotte il y a
le lac, la nappe d'eau est liée
avec la rive Silva.

La grotte est un polygone assez
bon pour l'étude des possibilités
des méthodes géophysiques. Durant
quelques années (pres de 30) on a
fait l'investigation géodynamique
par la prospection électique (sondage électrique, profil de resistivité, méthode du gradient mi-
toyen), la sismique réfraction, la
biolocalisation, la radiometrie et
la gravimetrie. (Matveev, 1963;
Chmelevskoi, 1963; Chouvalov,1983;
Matveev, Savelov, 1993).
L'investigation est faite selon
les 9 profils paralleles (fig.4).

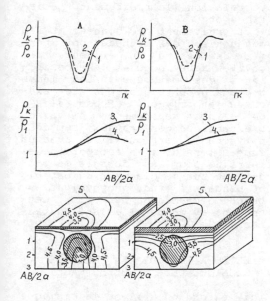

Fig.2. Champs électriques sur
les modèles du karst: 1,2 – profil
de resistivité; 3,4 – diagramme de
sondages électiques (3 – sans
caverne); 5(a,b) – modèles avec la
coups de resistivité

Les effets de l'aureole sont mo-
delés à l'aide des sphèroides et
des corps assimetriques. Les courbes de sondages électiques sont
reproduites à la figure 3. Elles
montrent l'influence des formes
karstiques et leur resistivité.
Les resultats du modèle ont été
utilisés lors de l'investigation
géophysique au polygone (champ
d'étude) de la grotte de Koungour.

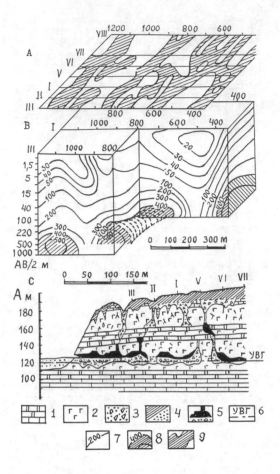

vités en ligne AB = 3, 50, 400, 500 m, profils de resistivité, les block-diagrammes. Sur la figure 4 on peut voir un exemple des résultats. Les anomalies positives (ρ_k>400 ohm-m) resultent de la presence des cavités. Toutefois la netteté avec laquelle apparait la remontée du champ électrique, depend de beaucoup de facteurs. Les exemples de ces phénomènes penvent être illustrés. Parfois les anomalies ne correspondent pas forcement aux cavitées karstiques. C'est "l'oreole" ou les conditions de site qui influent. C'est pourquoi nous avons fait les transformations differentes: calcul de la moyenne, calcul du paramètre , de la fonction d'auto-et d'intercorrelation, le traitement statistique etc.

Le traitement des données et la conjugaison des méthodes ont permis de mettre en evidence des cavités karstiques et les possibilités des méthodes. Les mêmes travaux géophysiques ont été executés en polygone "Predouralie", où le karst se forme en calcaire. L'experience des travaux sur deux polygones a été utilisé avec un succès lors de l'investigation des terrains pour les forages de production sur le gisement du petrole et pendant l'étude géophysique du tracé du chemin de fer pour mettre en evidence l'allure du bedrock.

4 ELABORATION DE LA METHODE SISMIQUE DE L'ETUDE DU KARST

Ici encore on utilese la sismique réfraction, mais pour fournir les résultats plus precis et étudier petit detail on propose la couverture multiple et la sismique 3D. Pour argumenter la possibilité de mettre en evidence les cavités karstiques par la sismique réflexion il faut évaluer le pouvoir de resolution vertical et horizontal. Nous avons fait le calcul theorique et, en utilisant les modèles physico-géologiques on a modelé le champ sismique. Les films syntetiques et les calculs nous ont montré, que le pouvoir de resolution ne depasse pas 5-10 m. Pour l'augmenter il faut élever la

Fig.4. Champs électrique sur la Grotte de Koungour: A - carte de resistivité, B - coupe de resistivité, C -coupe géologique (1- calcaire, 2 - gypse, 3 - breche, 4 - argile et sable, 5 - caverne, 6 - nappe d'eau, 7 - eqyipotentilles, 8,9 - anomalie de resistivité), I, II, III - profils de resistivité

La distance AB était egale à 3 ou 20 m pour étudier la fable profondeur et AB=200, 500 m pour avoir l'information sur la deuxieme couche des roches karstiques. Le sondage électrique est realisé avec le courant constant et alternatif et les électrades d'injection A et B distantés de 3 à 500 et parfois à 1000 m. D'apres les données de la prospection électrique on a tracé les cartes des resisti-

frequence du signal excité et perfectionner la méthodologie des travaux.

On a proposé de réaliser la sismique 3D avec le systeme d'observation regulié et irregulié (sur le profil curvilligne), où les géophones et les points d'excitation des ondes sont disposés sur les lignes orthogonales. Si le terain d'investigation n'est pas grand et il y a là une construction (la maison, l'ecole etc), l'observation peut être faite sur le carré où les géophones et les poit de tir sont disposés selon le perimètre.

Lors du traitement les sismogrammes sont groupés selon les points d'impact, les traces du sismogramme sont additionnées après les corrections correspondantes, les résultats reuvent être presentes sous forme des sections horizontales et verticales. On sait que pour le traitement de certains données géophysiques on utilise les transformations tomographyques. Nous les avons appliqué pour traiter les données reçues lors de l'observation sur le carré. Pour cela on a fait le programme qui était evalué lors du modelé. Les transformations tomographyques ont permis de mettre en evidence le cavité karstique sous l'ecole dans une ville (Savelov, Loukianov, 1993). La forage, qui a été executé après la sismique, a confirmé les estimations géophysiques.

L'étude par prospection sismique reproduite à la figure 5 a été executée à l'autoroute ou l'emposieu a été remarqué. On a realisé le systeme d'observation regulié, la distence entre les points de reception était egale à celle des points d'exitation, et egale à 5 m Le terrain d'étude est situé en zone du karst active. En comparant les sections des premieres arrivées des ondes sismiques (fig.5,A, B,C) on peut remarquer l'anomalie du temps sur la section avec x=24m. Ces sections sont composées des traces avec l'eloignement egal (x) entre géophone – point de tir. L'analise des résultats du modelé et des sections reçues ont permis de conclure, que cette anomalie correspond au temps d'arrivée des ondes directes en zone d'altera-

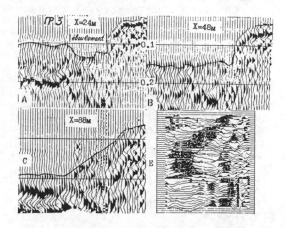

Fig.5. Exemples de l'étude sismique: A,B,C – les sections verticales, D – la section norizontale

tion. Sur les sections B et C, il n'y a pas de cette anomalie, parce que l'on y enresistre en premieres arrivées des ondes difractees ou bien de refraction.

Les données de la sismique 3 D permis de composer (former) les sections (coupe) horizontales au femps fixé. La coupe de la figure 5,E montre l'allure de la cavité karstique en plane horizale au niveux du temps t =50 ms (h ≈ 30 m). Les sections peuvent être reçues avec un pas egal à 2 ou 4 ms.

5 CONCLUSION

L'analyse de l'information et l'investigation faite nous a permis d'etablir, que les modèles physico-géologiques sont très variés et complexes. En forme générale le modèle physico-géologique du terrain karstique peut être representé par l'assise contenant les cavernes du karst. L'étude aux polygones typiques (Grotte de Koungour, "Predouralie") peut servir la base méthodique de la composition des modèles et l'elaboration des méthodes de l'investigation du karst et de l'interpretation des données. Dans le plus part de cas la conjugaison des méthodes electiques permet de resoudre les problèmes de l'étude du karst, mais la sismique fournit les résultats plus precis et donne une représentation en espace. Toutefois il faut appliquer le complexe des méthodes géophysiques en employant la conception de systeme basée sur l'analyse des données géologiques et géophysiques et sur la composition des modèles physico-géologiques. Les méthodes rationnelles peuvent être choisies à la base des modèles physico-géologiques.

Lors des traitements des données il faut faire l'attention de la représentation des résultats, de l'etablissement des cartes, des coupes differentes et des bloc-diagrammes.

REFERENCES

Gorbounova K.A., Andreichtouk V.N. Kostarev V.P., Maximovitch N.G. 1992. Le karst et les grottes de la région de Perm. Perm, edition de l'université.

Gorilik A.M., Saharova M.P. 1951. Application de la prospection electique en géologie de l'ingenieur. Moscou, Nedra.

Karous E.V., Faisulin I.S. 1977. Possibilité de l'application du sondage par ultra-sons en puits pour l'étude du karst. Moscou, la géologie et la prospection, N 2: 77-81.

Kropatchev A.M. 1973. Facteurs de migration et la precipitation des element en zone de l'hypergenese. Perm, edition de l'université.

Kouterev E.I., Michailov B.M., Lahnitchki I.S. 1989. Les gisement karstique. Leningrad, Nauka.

Maximovitch G.A. 1963. Les bases de la karsttologie. Perm, edition de l'université.

Matveev B.K. 1963. Les méthodes géophysiques de l'investigation des cavités karstiques en région de la grotte de Koungour. Perme, méthodologie de l'étude du karst, N 5: 3-24.

Matveev B.K., Savelov R.P. 1993. La conjugaison des méthodes géophysiques lors de l'étude du karst. Resumés de la conference de géologie, Perm: 69-70.

Matveev B.K., Savelov R.P., Gorbounova K.A., Kataev B.N. 1993. Modèles physico-geologiques du karst. Resumés de la conference geophysique, Perm: 9-10.

Oguilvi A.A. 1990. Bases de la prospection géophysique en Genie Civil. Moscou, Nedra.

Redosoubov A.A. 1973. Influence d'ecran de la couche conductice en étude des objets locals.Questions de la géophysique minerai, Sverdlovsk: 56-66.

Savelov R.P., Loukianov R.F., Loukianov S.R. 1993. La sismique 3 D en géologie de l'ingenieur. Resumés de la conference geophysique, Perme: 75-76.

Serebrennikova N.N. 1987. Calcul numerique du champs electrique dans le milieu lamelliforme. Revue de la géologie et de la prospection, Moscou: 109-114.

Chmelevskoi V.K. Méthode electique par courant alternatif de haute frequence en exemple de l'ourale et de Crimée. Perm, méthodologie de l'étude du karst, N 5: 25-34.

Chouvalov V.M. 1983. Investigation des régions karstiques et des cavités par les méthodes electriques. Perm, edition de l'université.

Chouroubor A.V., Chestov I.N. 1992. Question de la karstification a la grande profonder. Resumé du Congres international de la géologie de l'ingenieur. Perm:56.

Seismic tomography applied to iron mining and rock mass homogeneity assessment – Two case histories: The Santa Clara iron mine in Lazaro Cardenas Michoacan and the Huites Water reservoir dam project in Huites Sinaloa Mexico

Application de la tomographie sismique aux mines de fer et à l'évaluation de l'homogénéité des massifs rocheux

J.A. Ruiz-Reyes
ITESA, Mexico City, Mexico

ABSTRACT: Seismic tomography is one of the seismic prospecting methods in wich straight raypath analysis, between shotpoints and receptors, are used. Whit these travel time information is possible to create an image of the ground hardness distribution in a plane defined by shotpoints alignment and a second line of receptors. We present results of use of seismic tomographies in two sites, one of them applied to the mining industry and the other to the construction industry.

RÉSUMÉ: La Tomographie Sismique est une des méthodes de prospection sismique dans laquelle on emploie l'analyse de trajectoire rectiligne entre un point de tir et les récepteurs. Avec cette information sur le temps de voyage des ondes il est possible de créer une image de la distribution de la dureté du sous-sol dans un plan qui est défini par l'alignement des points de tir et une deuxième ligne de récepteurs. Nous présentons les résultats de l'application de la Tomographie Sismique en deux cas. Dans le premier, cette technique, a été utilisée dans le domaine de l'industrie minière. Dans le second il s'agit d'un problème qui concerne l'industrie de la construction.

1 INTRODUCTION

Seismic tomography is one of the seismic prospecting methods in which, straight raypath analysis, between shotpoints and receptors, is used to create an image of the ground hardness ditribution, along a plane determined by one line of receptors and other of shotpoints. These lines can be situated in two boreholes, two adits or one of them and the surface. This paper shows the results of two tomographies.

The first site is in the *Santa Clara* Iron Mine in south-west Mexico (figure 1). The method was used for identifying and quantifying mineral zone, as an aid in mining procedures. Making use of the fact that mineralization zones have wave velocity transmission in the order of 3500 m/s and the non mineralized zones are in the order of 2000 m/s.

The second site for tomography application is in *Huites* in north-west Mexico (figure 1). Where the contruction of a big water reservoir for irrigation and hydroelectric purposes, is in progress. The target here was to assess the homogeneity of a granitic

rock mass, on wich the dam is going to rest, using the fact that the rock hardness and wave velocity transmission are directly related.

FIGURE 1 SITE LOCALIZATIONS

In the first case, old inclined boreholes were used to place the shot points and lines vertically parallel to the boreholes on the surface were used to place the receptors. In the second case an old exploration adit was used to place the receptors and a line vertically parallel to it, on the surface, was used to place the shot points.

2 THEORY

The mathematics of method was originaly adapted, for this kind of job, by Vazquez (et al. 1988) and it is quite simple. First of all, the material between sources and detectors is separated as a grid made of rectangular cells, each one have its own seismic wave transmission velocity.

As is shown in figure 2, sources are placed along one line (not necesary straight) in one side of the searching area and in the opposite side receptors are placed in a similar way. For each signal produced by any shotpoint, it is posible to get as many travel times as receptor were placed. When all the shots are registered, with the travel times, first arrives of P waves, is possible to build a pattern of straight paths like the one shown in figure 2.

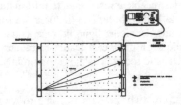

FIGURE 2 STRAIGHT PATHS

The measure is focused on the time that signal, produced at the source, needs to reach the receptors. If the material is homogeneous, the velocity calculated for any pair of points, *source-detector*, will be always the same, but this in not the normal case.

So it is necesary to find the velocity distribution in each cels in the material, by means of the straight path travel times (P waves) across them. Calculating, for any pair of source-receptor the travel time and making comparisons with the travel time recorded and minimizing differences between them.

A grid is proposed (figure 3), with an estimated velocity in each cell (initial model), to obtain the final velocity model.

Then all the paths, which cross a particular cell, are analized in terms of the travel distance inside the cell. The travel time is reconstructed with the sum of

distance and velocity in everyone of the cells crossed by the *"straight path calculated"*.

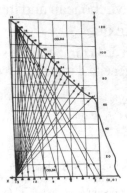

FIGURE 3 GRID OF CELLS

And error between the measured and calculated times for the whole path source-detector is obtain, as shown in figure 4.

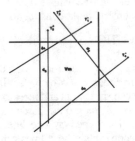

FIGURE 4 PATHS ACROSS ONE CELL

The algorithm used takes acount of the error sign as follows:

$$\Delta V < 0 \quad \text{if} \quad t_{ri} - t_{mi} > 0$$
$$\Delta V > 0 \quad \text{if} \quad t_{ri} - t_{mi} < 0$$

where

t_{ri} = Calculated time
t_{mi} =   Measured time
ΔV= Velocity change to be made in the cell to reduce error

Magnitude of ΔV is function of the lenght of each path in the cell and error magnitude in the same

96

path, as well as total number of paths crossing the cell.

If D1, D2, ..., Dn are the lenghts of the paths T1, T2,..., Tn who cross the cell m with velocity Vm, then for each path can be written the following proportionality:

$$\Delta V \propto d_i / D_i$$

where d_i is the part of the path D_i in cell m.

The velocity change, ΔV, must be proportional to the percent of error calculated with the last velocity model:

$$\Delta V \propto |t_{ri} - t_{mi}| / t_{ri}$$

Then can be said for any path, that:

$$\Delta V \propto d_i / D_i * |t_{ri} - t_{mi}| / t_{ri}$$

In every cell passed through by several paths, negatives and positeivs, the total contribution of the sums of both must be normalized by their number as follows.

Nt^+ = number of positive paths
Nt^- = number of negative paths

$$W_{1m} = \left[\sum_{i=1}^{Nt^+} \frac{d_i}{D_i} \frac{|t_{ri} - t_{mi}|}{t_{ri}} \right] / Nt^+$$

$$W_{2m} = \left[\sum_{i=1}^{Nt^-} \frac{d_i}{D_i} \frac{|t_{ri} - t_{mi}|}{t_{ri}} \right] / Nt^-$$

Sign for ΔV is given by *"sign"* :

If $W_{1m} > 0$ then   sign = -1
If $W_{2m} < 0$ then   sign = +1

Finaly to change the proportionality into an equality, a constant must be introduced:

$$\Delta V_m = sign * |W_m| * FAC$$

Where FAC determines the rate of change of ΔV.

By trial and error it has been found that FAC = 8000 is good enough for our purposes.

Some theoretical model was used in the original paper, where the behavior of the algorithm was tested using a computer program (Vazquez et al. 1988). The computer program was modified to accept irregular layout of sources and receptors. Like in figure 3, where the x axe, straight line in the lower part, represents the adit and the irregular line at the top (upper part) represents the topography of the ground surface, at *Huites* site.

3 APPLICATION CASES

3.1 Santa Clara Iron Mine in South-West Mexico

In this case the ore outcrops, give the idea of homogeneity in the subsurface and also underground. But when the area was drilled, surprisely the mineral was absent in many boreholes and in others it existed only in parts of the core extracted from de bore. In such cases to prepare the mining plan is very dificult and costly, because of the great amount of barren material to be removed. For these reasons it is desirable to have better knowledge of the ore distribution, as it was the goal in this project.

In old prospecting campaigns a lot of inclined boreholes were drilled to estimate and map the amount of ore. Many of them remain open, so in 1990 it was proposed to use them for the seismic tomographies.

It was known from lab tests that the compressional wave velocities in the ore are in the order of 3500 m/s and velocity in the barren zone was in the order of 2000 m/s, so it was expected a good velocity contrast in the tomographies. Using the boreholes to place the shotpoints and a line on the ground striking parallel to the borehole to place the receptors. An array of 24 receptor and 24 shotpoints were layedout to run the tomographic work, so a total of 24 x 24 paths were obtained.

Whit the geometric parameters of topograpy and travel times, read from the seimograms, the program was run. As many as 30 iterations were needed to reach the mean cuadratic error around 5 %. Which was considered small enough for our porposes. One of the Tomographies is shown in figure 5. Where the iron outcrops and mineralized zone, into the borehole, are marked with numbers (1, 2, 3, 4). After this tomographic interpretation, is quite obvious what limits of the ore can be marked to make a better estimation of the mineral reserves.

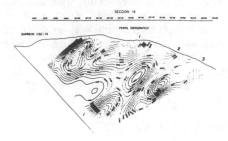

FIGURE 5 TOMOGRAPHY OF SANTA CLARA

3.2 Water reservoir Dam Project in *Huites*, Sinaloa

In this project was very important for the structural engineers to know the homogeneity of a granitic mass figure 6, on which the dam is going to rest.

FIGURE 6 GRANITIC MASS AT *HUITES*

A lot of previous exploration was already made in the site, among these, an adit of 162 m in length, was excavated in the granitic mass, this adit was used in our tomographic work to place the receptors and a line on the upper surface striking parallel to the adit, was used to place the shotpoints (sources).

The results of the tomographic work is shown in figure 7, and it was interpreted as follows:

Granitic mass is not as homogeneous as it was supposed.

The velocity (hardness) decreases outwards varies from a value > 5000 m/s to a value as low as 2500 m/s in the uper-right side.

An other decrease in velocity value is observed downward, being lowest between stations -122 and -132 at the end of adit. This is due to geothermical effects in the granitic mass. In that part of the adit the temperature is arround 55 °C and the granitic mass is altered to sand.

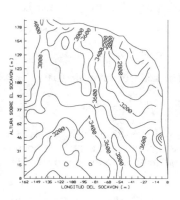

FIGURE 7 TOMOGRAPHY IN GRANITIC MASS

The meaning of these results in terms of construction is, by the moment, left to the civil engeneers in the project. It is because the seismic tomography is like a body radiography, in medicine. *It shows the problem but not the solution.*

BIBLIOGRAPHY

VAZQUEZ C.A., Aranda S.R., Benhumea L.M., "Tomografía Sísmica en la investigación de estructuras someras, utilizando un algoritmo iterativo", Boletin AMGE, V 28, N° 3 y 4, 1988.

Implementation of multivariate analysis of hydrochemical data and electrical resistivity sounding method in estimation of groundwater flow

Analyse multivariée des données hydrochimiques et de la résistivité électrique appliquées à la prédiction de la circulation souterraine

J. Ghayoumian
Graduate School of Science & Technology, Kumamoto University, Japan

S. Nakajima & H. Harada
Faculty of Engineering, Kumamoto University, Japan

ABSTRACT: In this article, the implementation of multivariate analysis of groundwater chemistry, in the prediction of groundwater flow, and its comparison with those results obtained through electrical sounding is reported. The Offset system of electrical resistivity sounding with a multicore cable was used for investigation of subsurface resistivity to predict groundwater level. Groundwater was sampled in the same area at different locations and was analyzed for water chemistry parameters. Multivariate analysis of hydrochemical data was performed to clarify the similarity of the groundwater and estimation of groundwater flow. The study proved that the multivariate analysis techniques of cluster analysis and principal component analysis can be used to elucidate the relationship among groundwater samples based on hydrochemical data. Also it was clear that the Offset system of electrical sounding is a fast and trustable method for resolving sub-surface resistivity and prediction of groundwater level for an area with simple subsurface structure.

RESUME : L'application de l'analyse multivariée de la chimie des eaux souterraines pour la prédiction de leur circulation, et sa comparaison avec les résultats obtenus par sonde électrique sont présentécs dans cet article.
Le système "offset" de la sonde à résistivité électrique munie d'un cablc à noyau multiple a été utilisé pour la détermination de la résistivité sub-superficielle pour prédire le niveau des eaux souterraines.
Les échantillons d'eaux souterraines ont été prélevés au même endroit et à des points différents puis analysés pour déterminer les paramètres de la chimie de l'eau.L'analyse multivariée des résultats hydrochimiques a été effectuée pour vérifier la similarité des eaux souterraines et estimer leur circulation.
L'étude montre que les techniques d'analyse globale et de l'analyse en composantes principales peuvent être utilisées pour élucider le rapport entre les échantillons d'eaux souterraines,basé sur les résultats hydrochimiques.Aussi , il a été clair que le système "offset" de la sonde électrique est une méthode rapide et fiable pour identifier la résistivité sub-superficielle et prédire le niveau des eaux souterraines.

1 INTRODUCTION

The understanding of flow and water quality in groundwater systems is a necessity at present for various purposes such as identification of recharge area, groundwater contamination, and water resources management.

For estimation of groundwater flow, direct measurement of water level in the observation wells or approximate determination of the depth of the water table using geophysical prospecting, which are calibrated with nearby borehole data, are the most popular methods. When the sub-surface geological structure is complicated, the above mentioned methods may not be successful for evaluation of water flow. Also in engineering projects such as dams and tunnels, the source of diverse waters which usually appear after the construction of the structures is one of the biggest problems which engineers are faced with most of the time.

We applied hydrochemical data as a tool for overcoming the previously mentioned problems. To compare and identify water masses, a large amount of water quality data may be needed. Traditional interpretive techniques (e.g. graphical methods) proved limited. In this research, multivariate statistical techniques are used as analytical tools to reduce and organize a large hydrochemical database into groups with similar characteristics. Multivariate analysis has been applied to hydrochemical data to a limited extent in the past [e.g., Reeder, et.al, 1972; Kiek, 1985; Eyles, 1988; Razack and Dazy, 1990; and Ruiz, et.al, 1990].

We have experienced availability of the method for prediction of seepage path through a rockfill dam (Nakajima et. al 1992). In this work, multivariate analysis of hydrochemcal data have been proposed for estimation of groundwater flow in Kumamoto Plain. The accuracy of the method is proved by performance of geoelectrical resistivity sounding method in the same area.

2 STUDY AREA

The study area is a part of Kumamoto Plain, in the south western part of Japan. The geology of the area has been studied by Watanabe (1980). The plain is mainly covered by upper quaternary unconsolidated deposits, particularly the pyroclastic deposits, which form significant aquifers in the plain. The pyroclastic deposits consist of four units: Aso - 1, 2, 3, and 4 in ascending order. The Aso - 1 and Aso - 4 are dated to be 360,000 years and 70,000 years respectively. The Togawa lava which consists of porous pyroxene andesite is also regarded as an excellent confined aquifer.

The Togawa lava's distribution is restricted to the southeast of the plain and is stratigraphically situated between the Aso - 1 and Aso - 2 (Koike, et.al 1990). Fig 1 shows a geological section of the study area.

3 IMPLEMENTATION OF MULTIVARIATE ANALYSIS OF HYDROCHEMICAL DATA IN PREDICTION OF GROUNDWATER FLOW

3.1 Sampling

Four times throughout the course of one month, water was sampled from wells and rivers in different locations (Fig 2) .

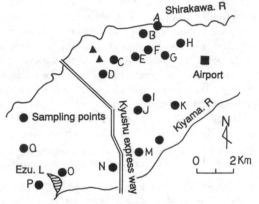

Fig 2. Study area including sampling stations

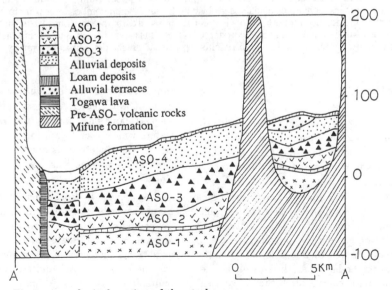

Fig 1. A geological section of the study area

Groundwater was sampled using manual equipment. The samples were analyzed for water quality parameters. The water quality parameters which were measured are temperature, pH, EC, Na^+, K^+, Ca^{2+}, Mg^{2+}, Cl^-, F^-, NO_3^-, HCO_3^-, SO_4^{2-}, and SiO_2.

Temperature, pH, and EC were measured in the field and the other parameters in the laboratory within half a day of sampling.

3.2 Cluster Analysis of Water Quality Data

Our initial objective is to classify the sampling points into different groups based on the similarities in the hydrochemical characteristics. For this purpose a pattern of recognition technique (e.g. cluster analysis) was used. For clustering of sampling stations based on water quality parameters, ordinary Euclidean distance was used.

Let r and s be two sampling stations with p variables (here p is the water quality parameters). The two sampling stations can be defined as the following vectors.

$$X_r = (Xr_1, Xr_2, \cdots, Xrp)$$

$$X_S = (Xs_1, Xs_2, \cdots, Xsp) \tag{1}$$

The Euclidean distance between the hydrochemistry of the two stations is defined as:

$$drs^2 = \sum_{k=1}^{p}(Xrk - Xsk) \tag{2}$$

where drs is the distance between the sampling stations s and r.

The Ward's agglomeration method was utilized for clustering and the out-put was displayed in the form of dendrograms. In this method all variables are given equal weight for the calculation of relative closeness for the quantitative definition of discrete water bodies (clusters). Eight water quality variables were considered for this analysis. To normalize the data a log transformation was applied to ion concentrations. The majority of the analyses revealed a dendrogram which classified the samples into four groups namely a, b, c, and d. The results of one of the cluster analyses is presented in Fig 3. Clusters b and c due to the low EC and site evaluation, were attributed to the shallow groundwaters which may not be

related to the other points. The points inside the clusters c and d were eliminated from further analysis. Clusters a and b were utilized for prediction of groundwater flow.

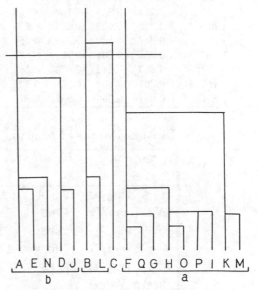

Fig 3. Cluster diagram for sampling stations

Principal component analysis which has been widely used in geoscience is applied in this study for recognition of the groundwater flow. Principal component analysis is essentially an analytical technique that permits identifications of mutually uncorrelated linear combinations of original variables.

The data from April to September in which the water shows higher electrical conductivity was used for analysis. The sampling points inside a group (cluster) demonstrates higher hydrochemical resemblance compared to those of different groups.

The sampling points inside a group were considered as waters with similar hydrochemical features which may be connected to one another. To find out about the relationship among sampling points inside a group, principal component analysis was performed. Table 1 shows the correlation coefficients matrix for group a.

It can be seen that the water quality parameters are heavily correlated, indicating the principal component analysis can be successful.

Table 1. Correlation coefficient matrix for the analyzed data

	pH	EC	Na^+	K^+	Mg^{2+}	Ca^{2+}	Cl^-	SO_4^{2-}	HCO_3^-	SiO_2
pH	1.000									
EC	-0.156	1.000								
Na^+	0.166	0.833	1.000							
K^+	0.261	-0.546	-0.169	1.000						
Mg^{2+}	0.618	0.008	0.358	0.630	1.000					
Ca^{2+}	0.077	0.879	0.749	-0.365	0.342	1.000				
Cl^-	0.482	-0.508	-0.185	0.885	0.794	-0.176	1.000			
SO_4^{2-}	0.633	-0.442	-0.023	0.862	0.785	-0.172	0.903	1.000		
HCO_3^-	-0.301	0.864	0.534	-0.725	-0.141	0.823	-0.574	-0.654	1.000	
SiO_2	-0.041	0.506	0.226	-0.879	-0.378	0.383	-0.703	-0.673	0.697	1.000

Table 2. Results of principal component analysis

variables	Mean	Variation	Stan. devi.	Eigen Vector PC1	PC2
SO_4^{2-} (mg/l)	19.514	62.996	7.937	0.995	0.029
Na^+ (mg/l)	9.762	5.670	2.381	0.082	0.571
K^+ (mg/l)	4.893	7.210	2.685	0.030	0.807
Ca^{2+} (mg/l)	13.629	1.932	1.390	0.029	-0.144
Mg^{2+} (mg/l)	6.883	0.623	0.789	0.046	-0.044
Eigen Value				63.629	8.827
Cum. Prop.				0.811	0.924

The water chemistry variables which were considered for principal component analysis were SO_4^{2-}, Na^+, K^+, Ca^{2+}, Mg^{2+}. The results of the principal component analysis for group a are given in Table 2.

The first principal component accounts for 81.1% of the variability and the second for 11.3%. The associated plane, which accounts for 92.4% of the variability is thus sufficient to interpret the results. The eigenvectors of the first principal component PC1 are positive and those of PC2 are positive for Na^+ and K^+, and negative for Ca^{2+} and Mg^{2+}. This reflects the fact that the dissolution phenomenon and ion exchange are in correspondence to the PC1 and PC2 respectively. A contact between sediments and flowing water causes dissolution of ions into the water. On the other hand, ion exchange between Ca^{2+} and Mg^{2+} in the water and Na^+ and K^+ in the clay minerals causes reduction of Ca^{2+} and Mg^{2+} and an increase of Na^+ and K^+ in the water, thus shifting from left to right indicates dissolution phenomenon.

This fact was utilized to estimate groundwater flow. For a certain group, for example group a, shifting from left to right indicate dissolution phenomenon. The principal component scores for group a is given in Fig 4.

The numbers in the figure indicate date of sampling. The groundwater flow can be extracted from the figure as follows: For example the water which exists in station H in May 25th flows toward station I and appears there in June 29th, and finally appears in station Q in June 27th.

Then the obtained water flow results were compared with the groundwater level in the sampling points, if the hydrochemical approach was confirmed by groundwater level method, the flow was accepted, otherwise it was not. The same procedure was performed for cluster d.

The results of groundwater flow using multivariate analysis are given in Fig 5.

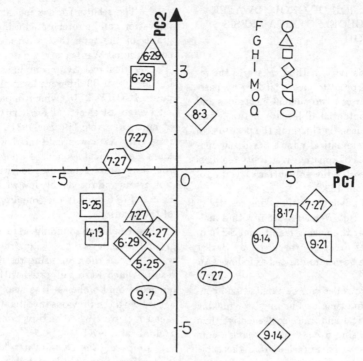

Fig 4. PC1 and PC2 score for group a

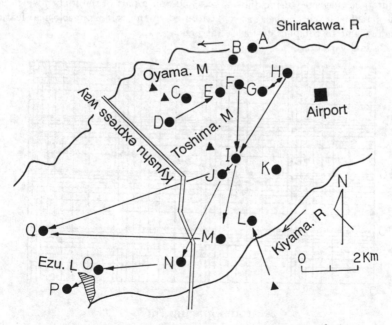

Fig 5. Prediction of groundwater flow using multivariate anlysis

4 INVESTIGATION OF GROUNDWATER FLOW USING GEOELECTRICAL RESISTIVITY METHOD

Geoelectrical resistivity method measures the resistance of the ground to electrical flow by passing a current between two current electrodes and measuring the potential difference between two potential electrodes. By changing the positions of the electrodes systematically for a particular configuration, a series of apparent resistivity readings is obtained to resolve the sub-surface layering of the ground (Auton; 1992).

The practical advantages of making resistivity sounding using offset Wenner system with a multicore cable rather than using traditional Schlumberger or Wenner array are discussed by Barker (1981); they have been summarized as follows (Auton; 1992).

1. All the electrodes can be planted and connected at the same time by one operator making the system practical and more cost-effective than sounding made using a conventional single core cable, which has to be unreeled from the cable drum for each electrode position.

2. Errors in the spacing of electrodes are eliminated.

3. By combining measurements from different electrode configurations (selected using the switching box), it is possible to check the consistency of the sounding and to compensate for effect of near-surface lateral variations.

4. The results from a modified Wenner configuration can be interpreted using computer programs of the type that are widely available for conventional Wenner array data.

Electrical resistivity soundings were performed in the area at 30 different locations, For this purpose MCOHM 2115, which implements the Offset system of electrical resistivity sounding with a multicore cable (Barker 1981), was used. The resistivity data was calibrated with the nearest known geological sequences from boreholes. The soundings were generally good with low observation errors and reasonably low with random Offset errors. Fig 6 depicts the sounding results for one of the locations.

The results of sounding were used to predict groundwater level. Fig 7 illustrates the groundwater flow in the area using resistivity sounding results which were calibrated with geological sequences from boreholes. It is clear that the results agreed with the previous investigation of the same area which were based on direct observation of the groundwater level.

Comparing the groundwater flow obtained using multivariate analysis with those estimated through geological sounding , a general agreement of the results is observed. However, there is a disagreement for a part of the area (e.g flow in the Northwest and South part of the Plain) which may be attributed to the complex geological features of the area.

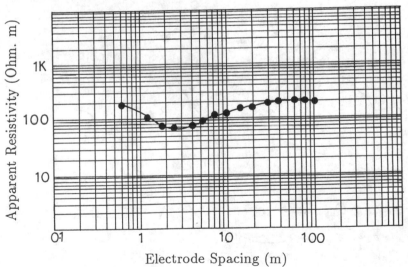

Fig 6. Sounding results for one of the stations

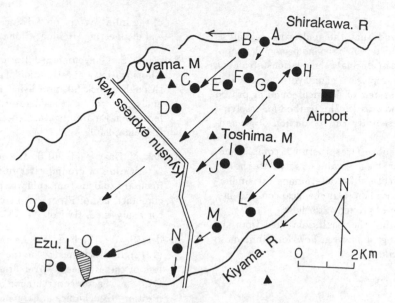

Fig 7. Estimation of groundwater flow using resistivity sounding

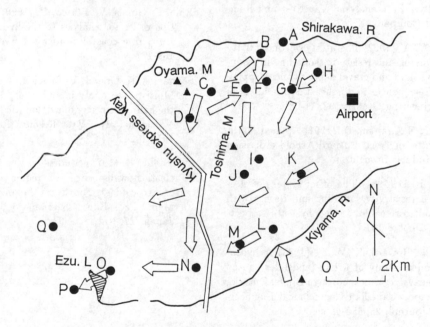

Fig 8. Groundwater flow for the study area

It can be concluded that when sub-surface is complex, prediction of groundwater flow using geophysical method or water table consideration can not be successful. Fig 8 illustrates the groundwater flow for the area which was estimated using the two previously mentioned methods.

.5 CONCLUDING REMARKS

In this paper, the implementation of multivariate analysis of groundwater chemistry in prediction of groundwater flow; and its comparison with those results obtained through electrical sounding are reported.

It was found that the multivariate analysis technique of cluster analysis and principal component analysis can elucidate the relationship among groundwater samples based on hydrochemical data.

The offset system of electrical sounding proved to be a fast and trustworthy method for resolving sub-surface resistivity and prediction of groundwater level.

When the subsurface structure is complicated, multivariate analysis of water quality data showed promise as a method for determining the similarity of waters based on their chemical constituents, and prediction of groundwater flow.

By comparing the results obtained from the two methods, groundwater flow for Kumamoto Plain was clarified.

REFERENCES

Afifi, A. A. Viginia, C. (1990) Computer-Aided multivari ate analysis. Van Nostrand Reinhold Company, NY.

Auton, C. A. (1992) The utility of conductivity surveying and resistivity sounding in evaluating sand and gravel deposits and mapping drift sequences in northeast Scotland. J. of Engineering Geology, 32, 11-28.

Brain, S. E Graham, D. (1991) Applied multivariate analysis. Edward Arnold a division of Hodder Stoughton.

Barker, R. D (1981) The offset system of electrical resistivity sounding and its use with a multicore cable. J. Geophysical Prospecting, 29, 128-143.

Eyles, N Howard, K. W. F. (1988) A hydrochemical study of urban landslides caused by heavy rain: Scarborough Bluffs, Ontario, Canada. Canadian Geotechnical Engineering Journal, 25, 455-466.

Kiek Steinhorst, R. (1985) Discrimination of groundwater sources using Cluster analysis, MANOVA, Canonical analysis and Discriminant analysis. Wat. Resour. Res. 21, 8, 1149-1156.

Koike, K; Omi, M Kaneko, K. (1990) Database system of geological information and its application to the engineering geology. Proc. of the Inter. Symp. on Advances in geological engineering, Beijing, China.

Nakajima, S., Ghayoumian, J Harada, H. (1992) Classification and estimation of seepage path through a rock fill dam using multivariate analysis. Proc. of the Mediterranean Conference on Environmental Geotechnology, Balkema, 93-99.

Razack, M Dazy, J. (1990) Hydrochemical characterization of groundwater mixing in sedimentary and metamorphic reservoirs with combined use of Piper's principle and Factor analysis. J. Hydrol. 114, 371-393.

Reeder, S. W; Hitchon, B Levinson, A. A. (1972) Hydrochemistry of the surface waters of the Mackenzie river drainage basin, Canada-I. Factor controlling in organic composition. Geochimica et Cosmochimica Acta, 36, 825-865.

Ruiz, F; Gomis, V Blasco, P. (1990) Application of Factor analysis to the hydrochmical study of a coastal aquifer. J. Hydrol. 119, 169-177.

Seyhan, A. A; Griend, V Engelen, G. B. (1985) Multivariate analysis and interpretation of the hydrochemistry of a dolomitic reef aquifer, Northern Italy. Wat. Resour. Res. 21, 7, 1010-1024.

Watanabe, K. (1978) Studies on the Aso Pyroclastic flow deposits in the region to the west of Aso caldera, Southwest Japan, Geology. Mem. Fac. Educ. Kumamoto University, Japan. 27, Nat, Sci., 97-120.

Long gravimetric profile: A new approach to determination of the regional variation

Profils de prospection microgravimétrique: Une nouvelle approximation pour la détermination de la variation régionale

Ch. Schroeder
Université de Liège, Laboratoires de Géologie de l'Ingénieur, d'Hydrogéologie et de Prospection Géophysique, Belgium

J. Fr. Thimus
Université Catholique de Louvain, Département Génie Civil, Belgium

I. Couchard & J. P. Wauters
TucRail, Brussels, Belgium

ABSTRACT: In the framework of the HST construction in Belgium, a microgravity survey has been performed on a 9 km length section, where the underground is composed mainly of chalk laying under a silty cover. The purpose is to detect, among other, eventual sinkholes or old underground quarries. The Bouguer anomaly profile shows several order of magnitudes of anomalies : some anomalies cover tenths of meters, others hundreds of meters. This constatation had led to determine the regional gravity variation on basis of specific signal treatment tools, with filtering at different "frequencies" (or wavelengths) corresponding to the observed anomalies. These "wavelengths" are correlated with geological parameters.

RESUME: Dans le cadre des travaux du TGV en Belgique, une campagne de prospection microgravimétrique a été réalisée sur un tronçon de 9 km à substratum crayeux dans le but de repérer entre autres des phénomènes de dissolution de la craie, ou d'anciennes exploitations de phosphates. L'examen des profils des anomalies de Bouguer fait apparaître des phénomènes d'amplitudes de différents ordres : anomalies de quelques dizaines de mètres à plusieurs centaines de mètres. Cette observation a amené la détermination de la variation de la gravité régionale sur base de techniques propres aux traitements de signaux avec filtrage à différentes "fréquen ces" (ou longueurs d'ondes) correspondant aux anomalies observées. Ces "longueurs d'ondes" sont à mettre en relation avec les phénomènes géologiques.

1 INTRODUCTION

The lay-out of the HST railway in Belgium, crosses an area where the underground is composed of quaternary silts, with thickness ranging from about 5 to 10 m, overlaying flintstone conglomerate and Cretaceous chalk. Between the conglomerate and the top of the chalk, one finds sometimes a phosphate bearing layer which has been locally extracted in the past.

In these conditions, there are real hazards of phenomenons like sinkholes due to the chalk dissolution and voids related to underground phosphate quarries.

A gravimetric survey has been decided on the most hazardous 9 km. The goal is the detection of lacks in density above-mentioned facts but also the variations of density and/or thickness of the superficial silts.

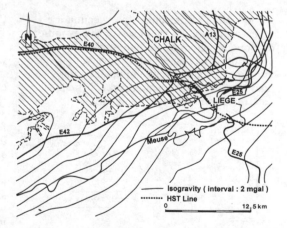

Figure 1 : Location of the studied area, sketch of the extension of the chalky underground and excerpt of the gravity map of Belgium

The survey consists of two parallel profiles, at 10 m interval. On each lines, one measure has been taken every 7.5 m. Perpendicular "antennas", 150 m long , have been made every 200 m. A total of about 3000 points has been measured, including the complementary and verification lines, which were performed in areas chosen according to the results of the first survey.

The measures were performed using a Lacoste-Romberg model D gravimeter, equipped with electronic filtering, which allows an accuracy of less than 5 µgal. The level of the measuring points is taken on the base plate of the gravimeter at the moment of the gravimetric measure in order to reduce the altitude errors. The landscape in the neighbourhood (200 m on each side of the profiles) has been modelised for the "topographic" corrections which are very important in the areas where the measuring profiles are very close (less than 10 m) of a 2 to 6 m height highway embankment. The "luni-solar" correction has been, of course, also taken into account as well as the "latitude" one which is not negligible in a 9 km long profile.

After corrections, the result consists of 2 "Bouguer anomaly" profiles. In this paper, examples of the determination of the regional gravity variations and residual anomalies will be given on the basis of one profile.

2 RESULTS OF PROSPECTION

The results of gravimetry prospection are given on figure 2. One observes the evolution of Bouguer anomaly for one profile. This evolution is characterized by several variations of significant amplitudes (more than hundred µgals) which make difficult the determination of regional anomaly.

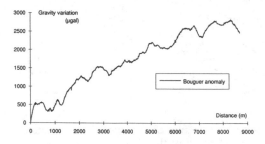

Figure 2 : Bouguer anomaly for 9 km profile

The technique generally used in civil engineering prospection, a linear regional anomaly, can not be retained.

The Bouguer anomaly curve shows clearly several orders of magnitudes of anomalies : some anomalies cover hundreds meters, others only about tens meters. This observation had led to determine the regional gravity on basis of specific signal treatments tools, with filtering at different "frequencies" corresponding to the wavelengths of observed anomalies.

Syberg in 1972 has used a similar technique, by analogy to magnetic fields interpretation, to separate the effects due to different sources. He has observed for all studied cases that the power spectrum have indicated a regional, a residual and a noise component. Camacho et al in 1994 have retained the same approach. After eliminating the regional anomaly using a polynomial fit, the residual anomaly is modelled by least-squares prediction. This method made it possible to model three signal components corresponding to three wavelengths, after filtering the final noise.

3 SIGNAL ANALYSIS

The processing principle is presented and the results are discussed. The detail of the signal analysis is reported in another paper (Demanet et Schroeder, 1994).

From this chapter, the word "signal" will be used in place of the gravity variations (Bouguer, regional and residual anomalies). The term "wavelength" (with its opposite "frequency") is related to the aspect of variations of the signal.

The first step consists of the determination of the frequency range of the global signal (Bouguer), using Fourier analysis. On figure 3, it appears that in addition to very low frequency components (wavelengths higher than 2000 m), the major components are at about 400 m, 300 m, 150 m and less.

Eliminating the low frequencies, the Fourier analysis has been conducted on a filtered signal in order to keep only the component whose wavelengths are between 500 and 50 m. The second curve on figure 3 shows the peaks at 500, 300 and 200 m. The signal was then processed with a low-pass filter at 125 m and the Fourier spectrum exhibits a peak at about 80 m. Taking this into account, it has

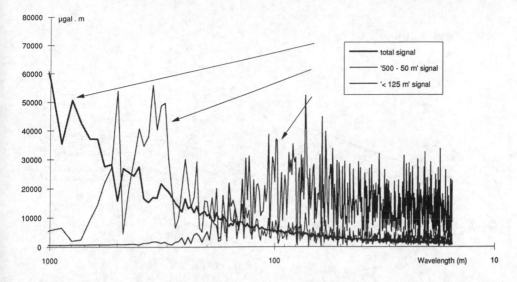

Figure 3 : Signal analyses - Frequencies ranges

been decided to consider 4 frequency ranges of the global signal and to choose filtering in order to keep the components of the signal whose wavelengths are :
- above 5000 m
- between 2500 and 500 m
- between 500 and 125 m
- less than 125 m.

4 INTERPRETATION

The low frequency component of the signal, wavelength greater than 5000 m, shows a regular increase of gravity (fig. 4).

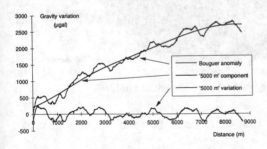

Figure 4 : Analysis of low frequency signal

This observation can be correlated with the results published by De Vos et al in 1993 for the map of Belgium of the Bouguer gravity anomaly determined from 373 gravity observations. One finds the same value of gravity gradient. It can be explained by the influence of the denser rocks of the northern Brabant Massif in comparison with the less dense rocks of the southern Ardennes area. Figure 4 shows also the signal after filtering of low components.

This analysis is also made for the 500-2500 m wavelengths range, for the 125-500 m wavelengths range and for wavelengths lower than 125 m. The results of this interpretation is given on figure 5 for the first 2000 m.

The 500-2500 m components indicate negative gravity variations between 650 and 1500 m. This observation can be correlated with a decrease of the thickness of the quaternary silts, 3 or 4 metres in this zone against 7 to 9 meters elsewhere.

The 125-500 m components give gravity variations which are more difficult for interpretation. The observed variations can may be explained by different thickness of silt and conglomerate beds, by different geomechanical characteristics of chalk layer or by decohesion phenomenons at the base of silt layer.

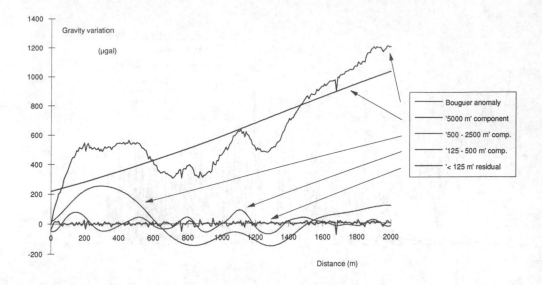

Figure 5 : Results of FFT analysis - Different components gravity variations

The residual components with wavelengths lower than 125 m show many anomalies of amplitude of about tens µgal. In a first step, one must subtract the noise level and the range of accuracy of measure estimated to average ± 5 µgal. It appears clearly several zones which must be verified by additional gravity measurements to extend the prospected zone or by other investigations as penetration tests, boreholes... These results are shown on figure 6. The amplitude of the observed anomalies can be reach 20 to 30 µgal.

The zone situated between 1940 and 1980 m has been investigated by 3 additional gravity profiles. The residual anomalies map is given on figure 7.

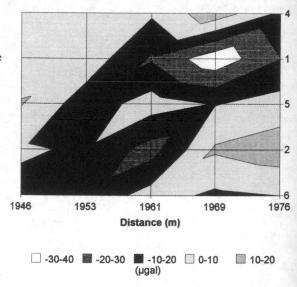

Figure 6 : Gravity variation for wavelengths lower than 125 m

Figure 7 : Residual anomalies for studied zone

This map shows a negative zone of anomalies crossing the HST profile which can be correlated with near underground phosphate quarries.

5 CONCLUSION

The use of specific signal treatment tools for determination of the regional anomalies proves to be conclusive. The filtering at different wavelengths permits to place in a prominent position anomalies of different natures : influence of geological phenomenons, variation of geotechnical characteristics of subsoil (thickness, quality,..) and possible existence of underground galleries or dissolution phenomenons. Nevertheless it appears also clearly that technique can be used only for profile enough long with a great number of gravity stations.

6 ACKNOWLEDGEMENTS

The authors would like to thank the technicians and engineers of L.G.I.H. and L.G.C., particularly D.Read, Ph.Jacques and P.Jaumain for the accuracy of the gravity measurements.

REFERENCES

Syberg, F.J.R. 1972. A Fourier method for the regional-residual problem of potential fields. Geophysical Prospecting 20, p.47-75.

Camacho, A.G., Vieira, R., Montesinos, F.G., Cuéllar, V. 1994. A gravimetric 3D global inversion for cavity detection. Geophysical Prospecting 42, p.113-120.

Demanet, D., Schroeder, Ch. 1994. Signal analysis for determination of regional gravity. 30th Annual Conference - Engineering Group of the Geological Society London. University of Liège, Belgium (to be published).

De Vos, W., Chacksfield, B.C., D'hooge, L., Dusar, M., Lee, M.K., Poitevin, C., Royles, C.P., Vandenborgh, T., Van Eyck, J., Verniers, J. 1993. Image-based display of Belgian digital aeromagnetic and gravity data. Professional paper 1993/5, N. 263. Service Géologique de Belgique.

Chacksfield, B.C., De Vos, W., D'hooge, L, Dusar, M., Lee, M.K., Poitevin, C., Royles, C.P., Verniers, J. 1993. A new look at Belgian aeromagnetic and gravity data through image-based display and integrated modelling techniques. Geological Magazine 130 (5), p.583-591.

Apport des méthodes de résistivités électriques pour caractériser l'état des conduites enterrées visitables

Contribution of the electrical resistivity methods to the characterisation of the condition of sewers

L. Cabassut, M. Frappa, M. Martinaud & C. Viguier
CDGA Université Bordeaux 1, Talence, France

J. D. Balades
CETE Sud Ouest, Bordeaux, France

H. Madiec
Lyonnaise des Eaux-Dumez, Bordeaux, France

RESUME: Il a paru intéressant au C.D.G.A. de développer une méthode géophysique électrique de diagnostic de l'état des collecteurs d'assainissement. Pour cela nous avons au préalable testé plusieurs dispositifs électriques (Wenner, tripôle, bipôle, unipôle) sur une dalle de béton recouvrant des cibles (vides, maçonnerie...). Les différentes géométries d'électrodes retenues ont ensuite permis de localiser les fissures d'un collecteur et de préciser la variabilité de l'état du bâti et celui l'encaissant géologique. Nous avons aussi prouver qu'il est possible de détecter des objets situés au delà de la conduite.

ABSTRACT: The C.D.G.A. has undertaken to develop a method of electric geophysical diagnosis about the state of the main sewers. With this aim of view we have tested as a preliminary some electrical arrays (Wenner, tripole, bipole and unipole) on a testing concrete slab covering some anomalies (cavities, stoneworks) We have then selected the best electrodes configuration that have permitted to locate the cracks and to give the variability of the framework and that of the geological encased on a sewer in Bordeaux. These devices have also revealed many objects set against the stucture simulating some cavities.

1. INTRODUCTION.

L'importance du réseau de collecteurs d'eaux usées, dont une partie est souvent vétuste, oblige les gestionnaires des métropoles urbaines a un examen et un entretien régulier de ces ouvrages. Les décisions concernant le type de réhabilitation ou le remplacement de ces conduites nécessitent une connaissance aussi précise que possible de l'état des matériaux de la structure (maçonnerie, béton...), de la nature du remblai et des terrains encaissants. La détermination des causes des désordres sera ensuite déterminée par l'étude du fonctionnement de ces ouvrages.

Pour atteindre ces divers objectifs nous avons proposé au gestionnaire du réseau de Bordeaux (Lyonnaise des Eaux-Dumez), d'utiliser la méthode des résistivités en courant continu et l'auscultation par ultrasons (Cabassut et al. 1993). Ce projet est soutenu par le Projet National REREAU 2.

De nombreuses autres méthodes sont susceptibles d'être utilisées: radar géologique, thermographie, impédance acoustique..., mais compte tenu du contexte géologique local: nappe phréatique peu profonde, encaissant argileux, l'adaptation de la méthode des résistivités faisant appel à la technique des pseudosections de résistivité (Sheriff 1989) est apparue comme la plus prometteuse.

2. RESISTIVITE EN COURANT CONTINU.

2.1. Principes.

Tout dispositif de mesure en courant continu nécessite l'utilisation d'un dispositif de quatre électrodes : deux pour l'émission du courant « C », deux pour la mesure de la différence de potentiel « P ». Suivant l'arrangement relatif de ces quatre pôles on distingue différents types de dispositifs.

Le choix d'un dispositif dépend des qualités intrinsèques de chaque arrangement possible ainsi que des facilités de la mise en oeuvre.

Pour un terrain non homogène plan, semi-infini, la mesure de V/I qui permet de calculer la résistivité apparente est fonction de cinq paramètres:

-le type de dispositif, caractérisé par le facteur géomètrique k de la relation $\rho_a = k.V/I$,

-les coordonnées (x,y) du point d'affectation de la mesure en surface,

-la coordonnée z=-a/2 du point d'affectation de la mesure, en profondeur. « a » est le paramètre d'espacement qui caractérise les dimensions du dispositif.

-l'orientation du dispositif.

Le choix d'un réseau d'électrodes conditionne tous les paramètres ci dessus. Des résultats représentatifs des résistivités nécessitent un réseau régulier, au mieux à mailles carrées, de points de mesures.

Dans le cas d'un collecteur dont surface interne n'est pas plane les électrodes peuvent être implantées selon deux possibilités simples:

- sur des cercles perpendiculaires à l'axe du collecteur, méthode C,

- sur des génératrices, méthode L.

Ces deux descriptions sont apparemment équivalentes; sur une surface déroulée, ces deux réseaux d'électrodes forment un plan à maille carrée ou rectangulaire.

Nous avons retenu la méthode L, qui consiste à réaliser depuis l'intérieur de la conduite, des pseudosections perpendiculaires à la surface de l'intrados, dans des plans passant par des génératrices. Dans ce but nous disposons selon une génératrice d'série de 25 électrodes régulièrement espacées constituées par des clous plantés à l'aide d'un pistolet. L'ensemble des électrodes est ensuite relié à un distributeur (scanner) permettant d'affecter à chaque clou, et successivement le rôle d'électrode de courant C ou de potentiel P.

2.2. Les différents dispositifs.

Sur ces bases, différents dispositifs à électrodes en ligne peuvent être envisagés: bipôle (pôle-pôle), tripôle (pôle-dipôle); quadripôle Wenner ou encore double dipôle. Ces divers dispositifs sont caractérisés par leur facteur géométrique k permettant de calculer la résistivité apparente:

$$\rho_a = k.V/I$$

- Quadripôle Wenner, c'est un dispositif tout à fait courant caractérisé par un espacement cons-

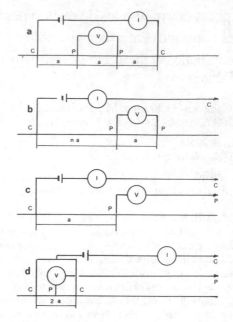

Fig. 1: Les divers dispositifs testés; a) Wenner, b) Tripôle, c) Bipôle, d) Unipôle.

tant entre chacune des électrodes: a; le facteur géométrique est $k = 2\pi a$; chaque point de mesure nécessite le déplacement de quatre électrodes (fig.1a).

- Le tripôle (fig.1b) implique l'utilisation d'une électrode C à l'infini. D'une mesure à l'autre trois électrodes sont déplacées, ici $k=2\pi an(n+1)$; (Dey et al. 1975; Brass 1981).

- Le bipôle (fig.1c) a deux électrodes à l'infini C et P, $k=2\pi a$, (Aspinall et Lynam 1970; Apparao et Roy 1973; Kumar 1974; Roy 1977; Martinaud 1990).

Toutes ces configurations utilisent comme source de courant une électrode ponctuelle ne présentant donc aucune focalisation. Des dispositifs focalisés, dérivés de ceux utilisés en diagraphie de résistivité ont été envisagés, nous n'avons testé parmi ceux-ci que l'unipôle (fig. 1d) (Gupta 1963; Apparao et Roy 1973, Brizzolari et Bernabini 1979; Jackson 1981), le nombre réduit d'électrodes mobiles nécessaires (3) étant plus compatible avec l'utilisation dans une conduite souterraine.

2.3. Représentation des mesures.

Pour un espace semi-infini, les résultats s'inscrivent

dans un espace à trois dimensions (x, y, z=-a/2). Pour un collecteur x est dans la direction des génératrices, y=-a/2 est la dimension perpendiculaire à la surface, la troisième coordonnée d'espace est l'azimut qui sera notée 1 à 7 de la rive gauche à la rive droite (fig.5).

Les quatre paramètres (trois d'espace et ρ_a) ne sont pas représentables ensemble graphiquement. Quelle que soit la méthode de déploiement des électrodes, C ou L, les représentations les plus synthétiques sont donc des séries de cartes à trois dimensions (représentation plane en courbes d'isorésistivité ou représentation en vue 3D pour faciliter la lecture):

-cartes déroulées à profondeur « constante »: elles intègrent les profils avec « a » constant en lignes ou en cercles.

-pseudosections longitudinales ou transversales qui intègrent respectivement les mesures effectuées sur une ligne (leur forme est rectangulaire) ou un cercle (leur forme est celle du collecteur). Le mode de report en profondeur couramment utilisé dérive de celui adopté en polarisation induite pour une configuration dipôle-dipôle (Bodmer et Ward 1968) (fig.2), le paramètre -a/2 représentant une pseudoprofondeur.

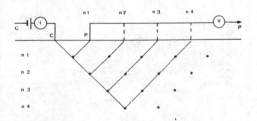

Fig.2: Construction de la pseudosection.

Des sondages électriques, représentation à une dimension: $\rho_a = f(a)$ sont aussi réalisables pour évaluer les valeurs des résistivités vraies et les épaisseurs des couches lorsque celles-ci peuvent être considérées comme parallèles à la surface.

3. ESSAIS SUR MODELE.

3.1. le modèle.

Des auteurs déjà cités ont étudié d'un point de vue théorique la réponse de ces dispositifs. Cependant dans la plupart des cas, les applications se rapportent à la reconnaissance sur des volumes plus importants

et dans un contexte bien différent. Nous avons donc testé ces différentes configurations sur un modèle (fig.3) constitué par une dalle de béton recouvrant un sol, dans lequel ont été disposées diverses cibles: vides, murs en maçonnerie de tailles variables.

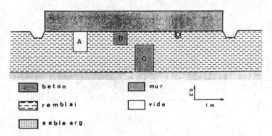

Fig.3: Schéma du modèle.

Pour augmenter le contraste, durant toute la durée des mesures, la saturation du sol est maintenue par une alimentation en eau dans les caniveaux bordant la dalle d'essai.

3.2. Choix du dispositif de mesure.

Les résultats obtenus avec les quatre dispositifs testés sont présentés sur la figure:4.

Ici l'intervalle entre clous et le pas sont égaux à 0,25 m. Les espacements utilisés pour chaque dispositifs sont les suivants:

- quadripôle: a=0.25; 0.5; 0.75; 1 m.
- tripôle: a=0.25, n=1; 2 ;3 ; 4; 6; 8.
- bipôle: a= 0.25; 0.5; 0.75, 1, 1.5, 2m.
- unipôle: 2a= 0.5; 1; 1.5; 2 m.

Les pseudosections obtenues sont sensiblement différentes .

On constate que :

-en Wenner et en unipôle seul l'objet C donne une anomalie nette,

-en bipôle et tripôle l'anomalie de C est déformée par l'anomalie de B; l'objet A est également détecté. Ces deux derniers dispositifs fournissent l'image la plus complète. Le bipôle « voit » plus loin pour une même valeur de a et par conséquent contracte l'axe des pseudoprofondeurs. Le tripôle par une moindre contraction permet de mieux différencier les divers milieux surtout en surface mais les anomalies sont allongées en profondeur et décalées latéralement.

En conséquence, pour l'étude des   collecteurs nous avons retenu en priorité le bipôle qui , de plus ne nécessite le déplacement que de deux électrodes. Pour obtenir une meilleure définition dans le béton on pourrait lui préférer le tripôle.

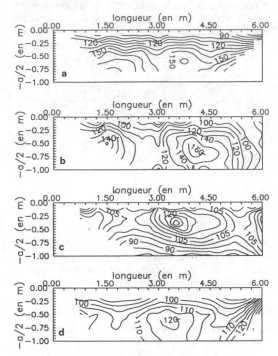

Fig.4: Pseudosections des 4 dispositifs testés sur le modèle a) Wenner, b) Tripôle, c) Bipôle, d) unipôle.

4. EXPERIMENTATION SUR LE COLLECTEUR DUPRE DE ST MAUR (BORDEAUX).

4.1 Le site.

Il s'agit d'une structure en béton non armée, de section ovoïde, coulée en place il y a·une cinquantaine d'années au moins, autour d'une forme gonflable (procédé S.A.T.U.J.O). L'encaissant géologique est une argile grise, trés plastique. Les pieds-droits du collecteur sont directement en contact avec le sol argileux non remanié. La tranchée a été remblayée sans précautions avec l'argile extraite ou avec des debris de maçonnerie. Cette conduite est parcourue sur presque toute sa longueur par trois fissures longitudinales, à la clef et au niveau de chaque pied-droit. Cet égout est aussi découpé par une série de fissures circulaires majeures, régulièrement réparties qui peuvent correspondre à des reprises de bétonnage. La clef du collecteur est située à une profondeur de moins d'un mètre, sous une chaussée à grande circulation de véhicules lourds (Zone industrielle). Cette situation pourrait être à l'origine des dommages.

Après les mesures, la destruction programmée de ce collecteur nous a permis, par contrôle visuel, de confirmer les résultats géophysiques

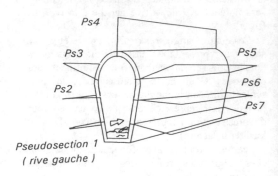

Pseudosection 1
(rive gauche)

· Fig.5: Position des pseudosection à la périphérie du collecteur.

4.2. Distance optimale entre clous.

Pour la déterminer nous avons réalisé 3 pseudosections bipôle sur la ligne n° 6 (x=30 - 49m) successivement pour une distance entre clous de 0,25m, 0,5m et 1m. (fig.6).

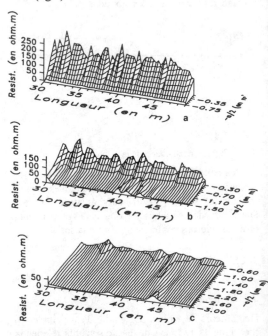

Fig.6: Pseudosection n° 6 réalisée avec un intervalle entre les clous de: a) 0,25m, b) 0,5m, c)1m.

L'intervalle de 0,5 m qui permet à la fois de caractériser avec précisions la structure et d'avoir une image des premiers décimètres de l'encaissant a été retenu.La détermination de la distance optimale entre clous doit être effectuée avant chaque campagne de mesures en fonction du but à atteindre et de la résistivité de la structure.

4.3. Test de deux configurations d'électrodes.

La figure 7 rend compte d'une partie des résultats pour la ligne de mesures n° 6.

4.4. Exemple de résultats.

Nous avons réalisé 7 pseudosections en bipôle sur tout le linéaire (7 x 96 m). La dépose ultérieure du collecteur nous a permis d'expliquer l'ensemble des pseudosections. Par exemple, la pseudosection n°4 (fig.8), à la clef, montre un remblai constitué exclusivement d'argile (x = 0 - 10 m) puis d'argile mêlée de bloc de béton (x = 10 - 30 m) à laquelle correspond les fortes valeurs de résistivités apparentes. Nous remarquons pour x=13 m. une valeur élevée de la résistivité due à la présence d'une buse au-dessus de la conduite.

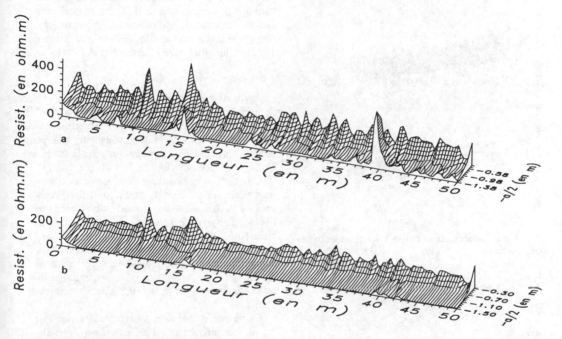

Fig. 7: Pseudosections ligne 6 de 0 à 50 m;
a) tripôle, b) bipôle.

Aux faibles pseudoprofondeurs correspondent des valeurs de résistivité plus élevées dues aux béton dont les variations traduisent sa qualité ou sa fissuration. Aux plus grandes pseudoprofondeurs correspondent les faibles valeurs de résistivité de l'encaissant argileux; les variations sont dues ici à des hétérogénïtés du remblai (x=40m) ou à la présence d'un piquage de bouche d'égout.

Le bipôle et le tripôle mettent bien en évidence, au niveau du bâti (x=10 - 15m) et dans l'encaissant, surtout le bipôle (x=15 - 40m), des zones de résistivités contrastées.

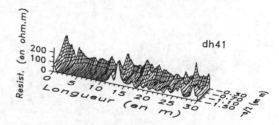

Fig.8: Pseudosection bipôle de la ligne 4 à la clef.

4.5. Effets des variations d'humidité.

Au point x=15 m.(fig.9), le piquage d'une antenne de bouche d'égout brisé en croix et surmonté d'une fissure circulaire majeure a imprégné d'humidité le béton environnant. En période humide, les pseudosections lignes 5 et 6 recoupant respectivement le piquage et la fissure montrent dans ce secteur une anomalie de faible résistivité.

La buse de l'antenne apparaît par contre en x=15m comme un objet de résistivité élevée.

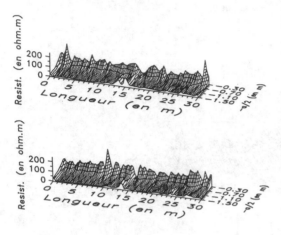

Fig.9: Pseudosections bipôle: 5 et 6 réalisées en période humide.

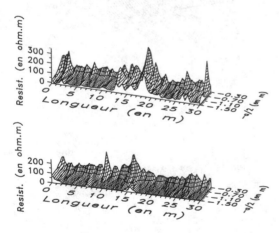

Fig.10: Pseudosections bipôle n° 5 et 6 réalisées en période sèche.

Pour illustrer l'importance des conditions climatiques et du niveau de la nappe sur les mesures électriques, la figure 10 présente les mêmes pseudosections, lignes 5 et 6 en période sèche (fig.10). La fissure et le piquage apparaissent alors comme des anomalies positives. Sur la pseudosection ligne 5 l'anomalie de la tranchée de pose de l'antenne, remblayée de sablon est alors très visible.

A la lumière de ces derniers résultats il est possible de faire le rapport entre les mesures réalisées sur le site durant une période sèche puis une période humide, ce qui est de nature à souligner les zones dont la résistivité présente de grandes variations, comme celà a déjà été réalisé pour évaluer la pénétration d'injections, en représentation de sections transversales (Erling and Lantier, 1992) ou longitudinales (Cabassut et al.1994).

4.6. Cartes développées:

La carte développée du collecteur pour des valeurs de résistivité apparente obtenues avec a=0,50 m montre la bonne homogénéïté de la structure en béton où les faibles valeurs de la résistivité, très ponctuelles, correspondent aux fissures circulaires et aux piquages défectueux (fig.11a, b et c).

Une carte sclérométrique développée du collecteur, pour des mesures de qualité du béton, réalisées à proximité de chaque clou confirme l'homogénéïté du béton dans sa partie superficielle (fig.11d).

La vitesse ultrasonique représentée de la même manière permet, par contre, de délimiter une zone réhabilitée il y a quelques années par coffrage interne sur l'ancien béton (x=10,5 - 28,5 m, figure: 11e).

En ne conservant dans cette représentation que les vitesses inférieures à 1800 m/s (fig.11f) les fissures circulaires de l'ancien béton et leur remontée dans la nouvelle couche sont soulignées.

La vitesse des ultrasons étant supérieure dans le béton récent, on ne peut pas mettre en évidence, à l'extrado, l'ancien béton très altéré (entre 18 m et 21 m). Cette zone, masquée par la couche de réfection, est par contre bien représentée sur la carte des faibles résistivités, inférieures à 60 ohm.m.(fig.11c).

4.7. Pose et détection d'objets entre.collecteur et encaissant.

Avant la dépose de ce collecteur, en accord avec l'entreprise gestionnaire, des cibles ont été

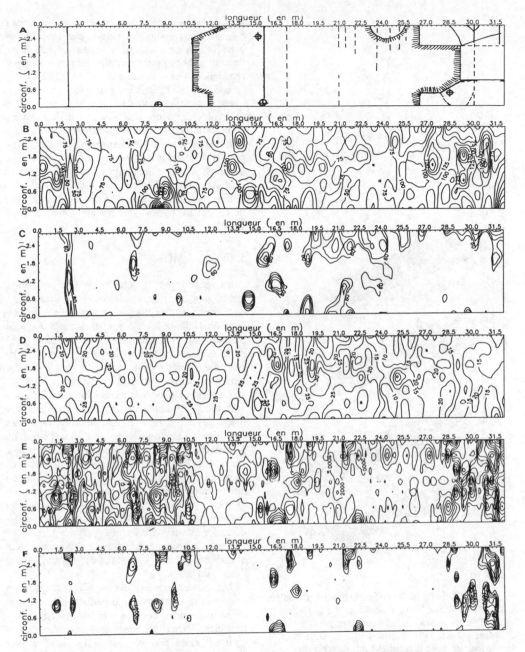

Figure 11: A: Carte développée des anomalies visibles; B: Carte développée des valeurs moyennes et supérieures de la résistivité apparente; C: Carte des faibles valeurs de résistivité (inférieures à 60 ohm.m).

D: Carte sclérométrique développée du collecteur; (en MPa) E: Carte développée des vitesses ultrasoniques; F: Carte développée des vitesses ultrasoniques (inférieures à 1800 m/s).

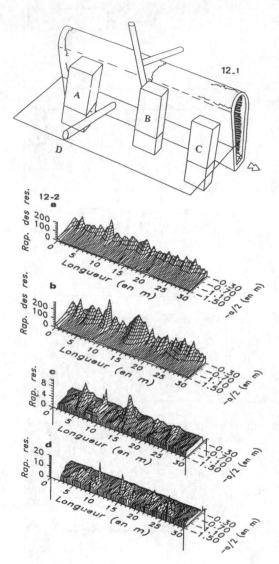

Fig.12.1: Position des differents blocs de dynaplast:
A: L=2.5m; l=1.5m; H=1.7m situé à x=5m; B: L=2m;
l=1m; H=1.7m situé à x=15.2m contre un piquage;
C: L=2m; l=1.5m; H=1.7m situé à x=25m; D: antenne
dont la jonction avec le collecteur est située à x=8m sous
le plan de mesure de la ligne 7.
       12.2: a: Pseudosection bipôle avant la pose des
objets; b: Pseudosection bipôle après la pose des objets;
c: Rapport des mesures après/avant la pose des objets
pour le bipôle; d Rapport des mesures aprés/avant la pose
des objets.

long de la conduite; pour simuler des vides dans l'encaissant nous avons fait enfouir des blocs de dynaplast (matériau plastique alvéolaire utilisé dans les travaux publics pour constituer en peu de temps des remblais ou des couches de forme) (fig. 12.1).

Les figures 12.2a,b,c,d présentent les résultats ~htenus avec les dispositifs tripôle et bipôle. Les .nomalies des trois objets sont parfaitement reconnaissables sur chaque pseudosection. Le rapport des mesures avant et après la pose des cibles, permet d'individualiser trés clairement chaque bloc de dynaplast. Sur les figures 12.2c et 12.2d, un piquage sur le collecteur est aussi trés visible en x=8m. La troisiéme anomalie (B), la plus éloignée du collecteur, (0,5m.), est comme prévu, la moins visible.

5. CONCLUSIONS.

Nous retiendrons de cette étude que les dispositifs électriques les mieux adaptés à l'auscultation des conduites souterraines sont le tripôle (pôle-bipôle) et le bipôle (pôle-pôle). Pour un espacement « a », le premier par sa profondeur d'investigation plus réduite permet de mieux détailler la structure de la conduite, le second, au contraire sera plus efficace pour analyser la struture et l'encaissant.

La série de pseudosections longitudinales réalisées dans ces conditions fournissent une bonne image des conduites et de leur environnement. Les conditions climatiques si elles sont sensibles n'affectent pas la représentativité des mesures. Ainsi, dans le cas du collecteur étudié la qualité du béton est bonne et constante, la fracturation est bien identifiée. En ce qui concerne le terrain encaissant, la dépose à permis de contrôler l'hétérogénéïté du remblai notamment, à la clef, déduite des mesures électriques.

Les mesures de célérité du son complètent avantageusement les mesures de résistivité pour l'étude du béton.

Un certain nombre de possibilités s'ouvrent pour rendre cette méthode performante. La modélisation rendra compte d'anomalies types susceptibles d'être rencontrées dans ce contexte.

Il est alors possible que le mode opératoire doive être modifié, suivant les solutions retenues. Par exemple une mesure en oblique par rapport aux génératrices devrait permettre de mieux appréhender la fissuration, quelle que soit son orientation.

REFERENCES

Apparao, A. & Roy, A. 1973. Fields results for direct curent resistivity profiling with two-electrodes ar ray. Geoexploration, 11: 22-44.

Aspinall, A & Lynam, J.T. 1970. An induced polarisation instrument for the detection of near surface features. Prospezioni Archeologiche, 5: 67-75.

Bodmer, R. & Ward, S.H. 1968. Continuous sounding profiling with a dipole-dipole resistivity array. Geophysics, 33,05: 838-842.

Brass, G. Flathe, H. & Schultz, R. 1981. Resistivity profiling with different electrode arrays over a graphite deposit. Geophysical Prospecting, 29: 589-600.

Brizzolari, E. & Bernabini, M. 1979. Comparaison between schlumberger electrode arrangment and some focused electrode arrangments in resisitivity profiles. Geophysical prospecting, 27: 233-244.

Cabassut, L, Frappa, M & Martinaud, M. 1994. Auscultation des collecteurs d'assainissement par méthode de résistivité électrique. A paraître dans « Géologues » UFG.

Dey, A. Meyer, W.H. Morrison, H.F. & Dolan,W.M. 1975. Electric fiel response of two dimensional inhomogeneities to unipolar and bipolar electrode configurations. Geophysics, 40-4:630-640.

Erling, J.C. & Lantier, F. 1992. Appraisal of electric and microgravimetric methods in underground works. No Trenchs in towns, Henry & Mermet (eds), Balkena.

Gupta, R.N. & Battacharya, P.K. 1963. Unipole method of electrical profiling. Geophysics, 28,4: 608-616.

Jackson, P.D. 1981. Focused electrical resistivity arrays: some theorical and practical experiments.Geophysical Prospecting, 29: 601-626.

Kumar, R. 1974. Direct interpretation of two electrrode resistivity soundings. Geophysical prospecting, 22: 224-237.

Martinaud, M. 1990. Intérêts du dispositif bipole C-P en prospection électrique non mécanisée. Rev. d'Archéométrie,14: 5-16.

Roy, K.K. & Rao, K.P. 1977. Limiting depth of detection in line electrode systems. Geophysical Prospecting, 25: 758-767.

Sheriff, R.E. 1989. Encyclopedic dictionnary of exploration geophysics. S.E.G. (eds) Tulsa.

Solution of specific engineering-geological problems at site of ancient Kremlin by geophysical methods

Résolution de problèmes spécifiques de la géologie de l'ingénieur dans le territoire de l'ancien Kremlin par des méthodes géophysiques

M.G. Ezersky & V.A. Jakubov

Geophysical Research and Survey Enterprise 'GEOTON' of Joint Stock Company 'Engeocom', Moscow, Russia

ABSTRACT: The results of geophysical (seismic and electrical prospecting researches at the site of th ancient Kremlin in the city of Ryazan (Russia) where stone buildings of XVII-XIX centuries have been deformed for a long time are described. Geophysical data made it possible without additional drilling of holes to define ancient lacustri-ne-palustrine deposits hidden under a technogeneous layer of soils and construct a three-dimensional geomechanical model of the foundation of the palace of prince Oleg.

RESUME: On décrit les résultats des recherches géophysiques (sismiques et electriques) sur le territoire du ancien Kremlin a Riazan (Russie), où les bâtisses en pierre de XVII-XIX ss. se deforment lors d'une période prolongée. Les statistiques géophysiques ont donné la possibilité de tracer les conturs des anciens de pots marecageux et de lac, caches sous une couche technologene des sols sans sondage complementaire, ainsi qu'ils ont donné la possibilité de construire un modèle volumineux géomecanique de la base du bâtiment du palais d'Olegue.

1 INTRODUCTION

Russian monuments of history and culture built as a rule on fill grounds with a large content of organic matters in the process of their history have been subjected to deformations connected with no-nuniform settlements of foundations (Gendel 1980). Provocative factors may be ground water of natural and technogeneous origin, karstic formations and construction and operation of subways in the cities (Legget 1973).

In the last decades in the process of studies of engineering-geological causes of deformations of architectural monuments geophysical methods of electrical prospecting "small seismics", different soundings such as nondestructive remote methods of studies of buildings, state and properties of soils and their variation with time have found expanding applications. One of the advantages of geophysical methods is the possibility to obtain a continuous data set along lines (profiles) with any specific detail. Ar-

ranging the profiles on the specific net the latter factor makes it possible to lay out maps of distribution of various indexes of soils in space (in depth, in planes, in volume), to reveal zones of structural disturbances and a different state (Liakhovitsky et.al. 1989) Possibilities of geophysical methods and seismic methods in particular expanded in view of digital computerized seismic equipment allowing sophisticated geophysical information to be obtained by simplified and shortcut methods directly in-situ.

In the present work geophysical researches and surveys were included in the complex of engineering-geological studies of the "deep protective" zone of the architectural monument of XVII-XIX centuries carried out by the leading scientists of the Moscow Geological Prospecting Institute - Dr. E. I. Romanova and A. G. Kuptsov. Stone buildings of the Ryazan Kremlin have been subjected to deformations for a long time the cause of which are treated by scientists differently. At the site of location of deformong buildings a series

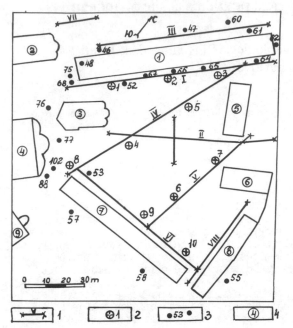

Fig. 1 Scheme of geophysical investigations on the territory of the Ryazan Kremlin; 1 - seismic profile and its number; 2 - point of vertical electrical sounding and its number; 3 - prospecting bore hole and its number; 4 - stone buildings; 1 - Oleg Palace; 2 - Christorozhdestvensky Cathedral; 3 - Archangel Cathedral; 4 - Uspensky cathedral; 5 - Choir building; 6 - Administration building; 7 - Consistory building; 8 - stables.

of prospecting bore holes were drilled (1948-1969) which did not produce an integral engineering-geological picture. At the present time for preservation of the architectural monument it was found expedient to limit the scope of drilling by the use of nondestructive geophysical methods.

Thus the purpose of geophysical studies is the refinement of engineering-geological conditions of the site of stone building the Ryazan Kremlin.

The task of investigations includes the following:

1. Refinement of an engineering-geophysical structure at the site adjacent to the Oleg Palace.

2. Determination of indexes of physical and mechanical properties of the soil mass and their distribution in space (in depth and laterally).

3. Construction of the soil foundation.

The methods of solution of these problems have been adequately developed and find use in surveys for construction of civil and industrial works.

The following methods were used in geophysical investigations:

seismic prospection refrection correlation method;

electrical prospection method of vertical electrical sounding.

The seismic method was used as the main one in the process of studies of the structure, state and properties of rocks and the electrical prospection method was applied as an auxiliary one for identification of soils by combination of geophysical parameters.

The scheme of location of seismic profiles and points of vertical electrical sounding is shown in Fig. 1. The site under study is an area (green lawn) adjacent to the Oleg Palace of about 100x100 m (10000 sq. m) limited from all sides by stone buildings: Archangel Cathedral, Consistory building, Stables and Choir building.

2. BRIEF INFORMATION OF ENGINEERING-GEO-
   LOGICAL CONDITIONS OF THE TERRITORY

The Kremlin hill is located in the central part of the city of Ryazan on the right bank of the Trubezh river which in its turn is the right tributary of the Oka river. The Trubezh river borders on the hill from the north; in the east the hill is limited by the Lebed river which is the tributary of the Trubezh river. The relief of the area is plain with a gradual low towards the Trubezh river.

From geomorphological standpoint the territory most probably is a morainic hill or an outlier of a fluvioglacial terrace which two alluvial terraces and a flood-plain are adjacent to from the north. According to our preliminary data the Oleg palace is situated at the joint of two terraces. Geologicic-lithological structure of the area is nonuniform. The soils encountered in the lithological cross-section are represented by sandy loams, loams, clays, lacustriue-palustriue, peat-silt-clay deposits of the Middle and Upper Quaternary Age. At some places deposits of the Dnieper glacial moraine are traced.

Hydrogeological conditions are characterized by presence of ground water classified as "irregular temporary water". The first aquifer was encountered at a depth of 18-20 m. "Irregular temporary water" was encountered in several bore holes ma-

Fig. 2 Deformed elevations of stone buldings of the Kremlin: southern elevation of the Oleg Palace with a crack in the western (a) and central (b) parts; c) bending of cornice due to settlement of of central and western portion of the Consistory building

inly in loam (occurred and stabilized water levels are shown in bore hole logs). Recharge of "irregular temporary water" is realized by infiltration of atmospheric precipitation in consequence of which its level is subjected to sharp seasonal variations. In spring-autumn periods "irre-

gular temporary water" is likely to occur near the ground surface.

3 CONDITIONS OF STONE BUILDINGS OF THE RYAZAN KREMLIN

For a long period of time buildings on the territory of the Ryzan Kremlin experienced settlements and deformations. The building of the Uspensky cathedral (1693-1702) was damaged, reinforced, cracked for a number of centuries; walls were broken loose from the floor arch and settled down behind foundations (Mikhailovsky 1972). In 1805 restoration was carried out and after that the building stood for about 100 years. Then the cracks were formed again in the building. In the early fifties a large crack split the cathedral from the middle of the western wall to the south-eastern corner through arches. In 1950 the opening of the crack on top of the western wall was 25 cm and in 1953 it reached 65 cm (partly because of fallen bricks). The cause of the settlements was a lake filled up in the end of the XVI century and mentioned in the chronicle. The peat-silt-clay deposits encountered by geological bore holes confirmed the location of the ancient buried lake at this place. The foundation of the cathedral was widened and deepened by 1.2 m (down to el. 112.0-112.5 m) and "rested" on underlying loams. After that the settlements discontinued.

The history of the Uspensky cathedral is typical of many buildings accommodated on the adjacent territory shown in the scheme of Fig. 1. The most serious deformations are acting on the Bishop house (Oleg Palace) shown in photographs of Fig. 2a and 2b. The Oleg Palace was build as a three-storeyed house (1653-1655).

The most ancient is a western section of the building. The eastern section of the Palace was added to the building in 1778-1780 and it was reconstructed in the XIX-century. Two large inclined throughout cracks running from the shed of the porch upwards in the western (fig. 2a) and eastern (fig. 2b) direction are observed from the side of the south elevation. The cracks have an opposite hand view on the north elevation of the building. Vertical cracks of up to 1 cm opening are observed at the joint of the building with the later eastern annex. Some cases of break-away of columns due to foundation settlement was observed. Brick masonry is

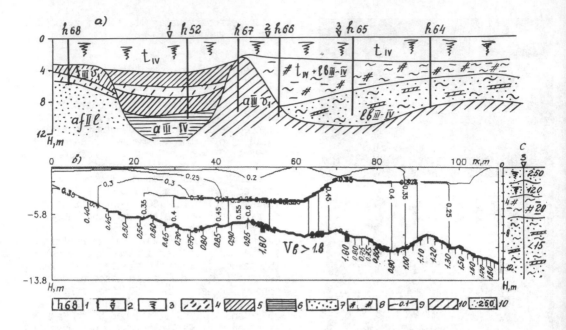

Fig. 3 Seimogeological section through profile 1: a) engineering-geological section;
b) velocity section; c) electrical section in point of vertical electrical sounding-3
(VES-3);   1 - No. of bore hole; 2 - No. of point of VES; 3 - technogeneous
soils; 4 - sandy loam; 5 - loam; 6 - clay; 7 - sand; 8 - lacustrine-palustrine deposits; 9 - isolines Vp, Km/s, 10 - zone of higher velocity Vp>0.6 Km/s; 11 - electrical resistivity of soil, Om*m; Vb - boundary velocity

heavily damaged, misalignment of brick
layers and falling down of separate
bricks may be seen.

The Consistory building is damaged to a
lesser extent. Here the most deformable is
the central settling section at the junc-
tion of two-storeyed and single-storeyed
buildings and the eastern aisle of the
administrative section of the building.
Settlements are clearly seen in photographs (Fig. 2 c) by bending of the cornice
line.

The stable house is heavily deformed.
The end (southern) wall is dessected for
the whole height by two vertical cracks
running from the foundation to the roof.
Buildings of the choir house and the Archangel cathedral were deformed and resto-
red at different time.

4 METHODS AND TECHNIQUE OF INVESTIGA-
TIONS AND DATA INTERPRETATION

Seismic investigations were carried out
by refraction method according to the se-
ven-point system of observations with the
impact excitation of elastic waves. CB-30
seismographs were spaced at a distance of

2m apart. The 12 - channel digital
engineering - geophysical station with
summing was used for recording. At each
arrangement (22m in length) impacts were
produced at seven impact points: by one -
at the ends of arrangement, one - in the
center of arrangement and at a distance
of 22 and 44m from each end of arrangement. This allowed the soil mass to be
studied to a depth of up to 16-20 m.

The materials of seismic investigations
were processed by the "SEISMIC-GEOTON"
program developed by V. A. Yakubov for PC
IBM for compatible computers in TURBO
PASCAL 5. 5. language. The distinctive feature of the program is the possibility
of construction of structural sections in
isolines of Vp velocities which in its
turn makes it possible to provide detail
partitioning by elastic properties. The
"SEISMIC-GEOTON" program package includes
the following programs:

text editor for input and editing of information on the place of work, coordinates and topography of seismic profiles
and time of wave arrival;

graphical look of seismic traces to
distinguish seismic waves and read out

the time of their arrival;

graphical editor of travel-time curves of seismic waves allowing travel-time curves to be looked and tied in on profiles;

program of processing of time-travel curves of seismic waves by the method on the basis of generalized solution of inverse kinematic problem of seismic prospecting which allows the problem to be solved in the class of leg velocity functions (Burmin 1980);

program of construction of three-dimensional model of the studied geological medium by collection of data obtained from seismic profiles.

All graphical constructions made by the "SEISMIC-GEOTON" program package are produced as output on the plotter or matrix printer. The program package is also provided with the utility adjusting the program to configuration of the computer in use.

On the basis of sections differentiated with respect to velocities Vp map-sections in lateral planes and on boundaries of layers were generated on a computer. Electric prospecting for determination of variation of electrical resistance with depth was carried out by vertical electrical sounding. Measurements were taken by AHЧ-3 low-frequency standard apparatus. Observations were carried out in 10 points located in 4 profiles of vertical electrical sounding. On the basis of results of measurements vertical electrical sounding curves of variation of apparent resistivity with depth were constructed by which later true resistivity of each layer were determined with the use of the computer. Electrometric data were processed with the help of the iteration program.

Multi-computer organization of seismic and electrometric parameters enabled the section to be dissected lithologically by depth with high reliability.

5 REZULTS OF INVESTIGATIONS

5.1 Structure of soil mass

The section of one of the main profiles N 1-R extending along the south elevation of the Oleg Palace is shown in Fig. 3. The seismic-geological section is characterized by the presence of two refracting boundaries. The first boundary is sharply defined in the central part of the profile and separates the soil with velocity Vp=0.25-0.30 km/s (upper layer) and with - velocity Vp>0.55 km/s (second layer). The second boundary (observed at a depth from 2 to 14 m) is vaguely defined in the initial part of the profile. Beginning from Sta. 0+30m the difference of velocities in the soil above and below the boundary starts increasing from 1.8 times (0.40 and 0.70 km/s) to 3 times (0.6 and 1.8 km/s) respectively in the central part of the profile; at the end part the difference in velocities above and below the boundaries is increased by 7 times (0.25 and 1.8 km/s). A dome-shaped high velocity structure in the central part of the profile (Vp=0.60-0.65 km/s) attracts interest which is identified with the presence of a domed structure composed of loams in the engineering-geological section. Section N 1-R is fairly typical of the site under study. To obtain a three-dimensional picture of a structure of velocity fields map-sections Vp in the lateral plane at different elevations on the refracting boundary and maps of the relief of different seismic-geological boundaries were constructed.

Fig. 4 shows a map-section in velocoties of seismic waves at el. - 5.0m (zero mark is the ground surface). A zone of higher velocities Vp= 0.65-0.85 km/s in the central part under the Oleg Palace building attracts attention. A decrease of wave velocity towards the periphery of the building is observed. Towards the left side of the building velocity decreases down to 0.4 km/s and in the right side the values of velocity Vp=0.25-0.30 km/s were recorded. The analysis of the given map shows the following main distinctive characteristics of the velocity structure of the soil mass at el - 5.0 m:

sharp velocity nonuniformity of the soil mass under the building with higher elastic characteristics in the center and lower values of Vp in the ends of the building (particularly in the right north-eastern aisle;

the present of a vast low-velocity area (Vp<0.35 km/s) in the north-eastern part of the Archangel cathedral.

A picture of a three-dimensional position of boundaries is produced by the map of the base of the low-velocity layer (by isoline V=0.35 km/s) (Fig. 5). The map displays a regular increase of the epth of a low-velocity layer from the west to the east. The bouday occurs at its highest (h<2.0m) under the western half of the Oleg Palace at it deepens regularly in the eastern direction. The maximum depth is reached by the low-velocity layer in the eastern section under the Oleg Palace (9-11m), under the eastern aisle

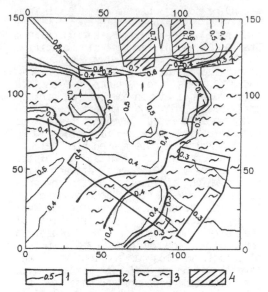

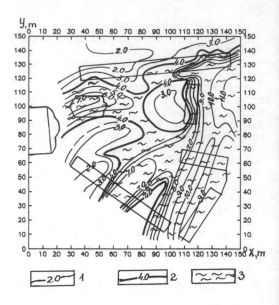

Fig. 4 Plan-section at el. 5. 0m from the surface in isolines of velocity Vp. 1 - isolines Vp km/s; 2 - isolines Vp=0. 35 km/s - boundary of zone of lacustrine-palustrine deposits; 3 - area of low velocity Vp< 0. 35 km/s - lacustrine-palustrine deposits; 4 - area of higher velocity Vp> 0. 70 km/s - loam

Fig. 5 Map of depth of foot of low-velocity layer Vp=0. 35 km/s - bottom of ancient reservoir, 1 - depth of isolines, m; 2 - isoline of maximum depth of filled layer - 4. 0 m; 3 - lacustrine-palustrine deposits.

of the Consistory building (8-10m) and along the western elevation of the Stables and under the central section of the Consistory building (h>7m) and under the Archangel cathedral (h>7m) as well.

5. 2 Indexes of physical and mechanical properties of rocks

On the basis of multi-computer organization of seismic and electrometric data and tie-in with real engineering-geological sections the following was done:
1. Development of criteia of identification of these or those types of rocks by combination of their geophysical values;
2. Establishment of connection between measured parameters (logitudinal wave velocity Vp physical and mechanical properties such as dynamic modulus of elasticity Ed , modulus of total Edef and continuous Et deformation, cohesion C, friction coefficient $tg\varphi$, voids ratio n, coefficient porosity e, density γ, and natural moisture content W.
As a result of the performed analysis

ranges of variability of geophysical parameters Vp, Vs, ratio Vp/Vs and associated dynamic Poisson's ratio γ, electric resistance φ were determined for each group of rocks. Main data are given in Table 1. As may be seen from Table 1 technogeneous soils and lacustrine - palustrine deposits with equal lowe values of elastic parameters Vp<0. 35-0. 40 km/s differ considerable (by 4 items and more) in values of electric resistance which are less than 15 Om×m for peat - silt -clay deposits and for technogeneous soils clays peat-silt-clay deposits at close values of electric resistance differ considerably (by two times and more) in Vp velocities. Similarly by combination of two parameters lacustrine-palustrine soils differ from other lithological varieties of rocks. These differences make it possible to single out unambiguously lacustrine-palustrine soils in the general engineering-geological section. It should be added that value Vp/Vs may provide a useful guide to identify the rocks and in this case this value is very much effec-

Table 1. Main indexes of elastic electrical and deformation properties of soils

NN	Description of rock type	Vp Km/s	ρ Om*m	γ Kg/m	e	Edef MPa	Ro 10 Pa
1	Technogeneous deposits	0. 15-0. 40* 0. 27	60-400	1700	0. 56-0. 97 0. 63	6-18 13	– 2. 6
2	Hard plastic clay	0. 40-1. 30 0. 79	10- 20	1390	0. 55-0. 85 0. 77	16-30 20	2. 5-4. 5 3. 3
3	Hard plastic loam	0. 35-0. 80 0. 43	15- 40	1700	0. 65-0. 85 0. 75	8-27 14	1. 5-2. 1 1. 8
4	Plastic sandy loam	0. 25-0. 70 0. 43	15-100	1700	0. 40-050 0. 45	26-50 32	2. 0-3. 0 2. 5
5	Fine-grained sand	0. 35-0. 55 0. 31	100-200	1660	0. 46-0. 50 0. 48	25-40 36	3. 0-4. 0 3. 5
6	Lacustrine-palustrine deposits	0. 20-0. 40 0. 31	5- 15	1380	0. 80-1. 12 1. 10	1. 5-3. 5 2. 5	1. 0-2. 0 1. 5

* Numerator-variation limits of index, denominator - average value of index

tive at watering of rocks. In our case the soils are in a natural-moist state.

5. 3. Analysis of obtained results

Using now the criteria of identification of rocks by geophysical parameters let us try to interpet the obtained geophysical data. The analysis will be performed using the plan-section in velocities Vp shown in Fig. 4. The values corresponding to the given depth in points of vertical electrical sounding and geological bore holes and rocks occuring at the depth of the section according to description of bore holes are plotted herein as well. A low velocity zone of rocks with Vp<0. 35 km/s attracts attention on the plan. In conformity with classification of Table 1 these may be both technogeneous soils and peat-silt-clay deposits. The former are characterized by values ρ =60-400 Om*m and the latter - 5-15 Om*m. Data of prospecting bore holes, "point" measurements of electrical resistance by vertical electrical sounding enable a low-velocity field to be identified as lacustrine-palustrine deposits. The boundary Vp=0. 35 km/s outlines the contour of the ancient lake.

In the light of expressed considerations the map of depth of the foot the low-velocity layer in Fig. 5 is actually a map of the foot of technogeneous and lacustrine deposits. The analysis of geological description of bore holes shows that a filled layer within the limits of the given territory is not more than 4 m in thickness. Consequently the isoline of depth 4. 0 m may be considered with a high degree of probability as the boundary between filled soils (h<4. 0) and lacustrine-palustrine deposits (h>4. 0). In the drawing this isoline gives also a rough idea of the contour of the ancient lake. The deposits of which nowadays together with a filled soil are the foundation of a part of buildings, in particular eastern sections of the Oleg Palace, Consistory building, Stables and Choir house. It is not incoceivable that there is a small ancient lake in the contour of isoline Vp<0. 35 km/s under a portion of the Archangel cathedral. This is attested by the presence of values ρ = 10-15 Om*m in the point of vertical electrical sounding-4. It should be noted that the outlines of the "lake " singled out on the depth map agree well with configuration of the similar boundary of low-velocity soils (Vp<0. 35 km/s in Fig. 4) which points to reliability of the obtained results. The difference in elastic wave velocities in lacustrine-palustrine deposits and enclosing soils is embodied in considerable difference in indexes of deformation properties i. e. moduli of deformation Edef. Thus assessment of moduli of deformation made by correlation relations given in paper (Methodics 1985) demonstrates that the layer of lacustrine-palustrine (peat-silt-clay) deposits stand-out prominently among the whole collection of soils by low (particularly deformation) indexes. Thus value Edef= 2.5 MPa for the given type of the soil is 6-15 times *less* than the similar indexes of other soils. A cultivated layer and loam have close average values Edef. = 13-14 MPa (in the upper part of the section these indexes are characterized by minimum values Edef. = 6-10 MPa). Hence long-time deformations of stone building of the Kremlin completely or partly constructed on lacustrine-palustrine deposits.

Observations of buildings indicate that deformation processes have not been completed until now. It is clearly seen by the nature of existing and newly appeared damages of the Oleg Palace, settlements of the Consistory building and Stables. Low values of elastic wave velocity in the stratum of soils of this type point to their high porosity and ability of further deformation and nonuniform porosity at the contact with more dense soils creates favourable conditions for deformations of those soils in the zone of contact with peat-silt-clay deposits.

6. CONCLUSION

The results of our investigations disclosed that seismic prospecting performed by the seismic prospection refraction correlation method with the use of the "SEISMIC-GEOTIN" program made it possible to detail the structure of the soil mass by indexes of physical and mechanical properties of soils and in particular by a state of soils both in depth, lateral planes and on the refraction boundary separating the soils with sharply different properties. The successful application of seismic methods in the given case results from a wide range of variability of longitudinal wave velocity in the upper portion of the section which is 0.25-1.80 km/s i.e. the difference between minimum and maximum values by an order of magnitude. It has been just the lower limit of values Vp (0.25-1.80 km/s) which is attributed to lacustrine-palustrine soils being the object of the search in the soil mass. Their distinction from technogeneous soils characterized by close values of elastic properties consists in a significant difference (more than 4 times) of electrical properties of these soils on which a success of comprehensive interpretation of data of investigations is based.

The results of the given work demonstrate that geophysical methods and in particular modified methods being used by us at a limited number of prospecting workings and minimum interference in natural environment hold much promise for the use on areas of location of monuments of history and culture and other objects limiting such interference.

Authors express their gratitude to Dr. E. I. Romanova, A. S. Kuptsov and L. D. Maksimova, director of the Ryazan Kremlin for useful, interesting and pleasant cooperation.

REFERENCES

Burmin V. Ju., 1980. Conversion formula discontinuos travel-time curves of refracted waves. Proceeding of the USSR Academy of Sciences. "Lithosphere physics" series, N6, 101-106 (in Russia)

Gendel E. M., 1980. Engineering work in restoration of monuments of architecture Moscow, Stroyizdat.

R. F. Legget. 1973. Cities and geology. McGrow-Hill Book Com- pany, New York

F. M. Liakhovitsky, V. K. Khmelevsky, Z. G. Yashchenko, 1989. Engineering geophysics. Moscow, "Nedra" (in Russian)

Methodics recommendations on determination of composition, state and properties of soils by seismoacoustic methods, 1985. VNII transstroy, M (in Russian)

E. Mikhailovsky, 1972. Conservation of the Uspensky Cathedral in Ryazan. "Theory and practice of restoration work", M. c. 64-74 (in Russian)

Application des méthodes tomographiques et microsismiques à la quantification des effets d'un tir aux explosifs
Application of the tomographic and microseismic methods to the quantification of the blasting effects

H. Heraud & J.J. Leblond
Laboratoire Régional des Ponts et Chaussées, Clermont-Ferrand, France

O. Abraham & P. Cote
Laboratoire Central des Ponts et Chaussées de Nantes, France

ABSTRACT: The carrying out of blasting with explosives always leads to important disorders beyond the theoretical mining area. Generally, the damages are not directly perceptible with the naked eye, and are expressed by a change of cracking and noticeably by the moving of pre-existing faults, which -in course of time- induce the weathering, the evolution and then the instability of rock walls.
A survey by means of microsonic diagraphy or seismic tomography, performed before and after a blasting with explosives, shows the gravity and the extent of the disorder.
The results appear as a decrease or an increase of velocities with changes reaching up to 40 % of the initial values.

RESUME: La réalisation d'un tir de mine aux explosifs induit toujours des désordres importants au delà de la zone d'extraction théorique. En général, les traumatismes ne sont pas directement visibles à l'oeil nu et se traduisent par une variation de la fissuration et surtout le rejeu de fissures préexistantes qui favorisent à terme l'altération, l'évolution puis l'instabilité des parois rocheuses. Une auscultation par diagraphie microsismique ou tomographie sismique effectuée avant et après tir aux explosifs montre l'importance et l'extension de la désorganisation. Les résultats se traduisent par une diminution ou une augmentation des vitesses avec des variations pouvant aller jusqu'à 40 % des valeurs initiales.

1 - PRESENTATION

Les études entreprises en France depuis une dizaine d'années sur les terrassements aux explosifs ont toutes montré que l'emploi des explosifs pour la réalisation d'ouvrages rocheux entraînait une relative désorganisation des matériaux restant en place.

Il est possible de limiter les effets arrière sur une paroi rocheuse, en adaptant les énergies et types d'explosifs au problème posé. Les gels allégés sont l'une des adaptations possibles (Rebeyrotte A. - Heraud H. 1989).

L'exemple présenté ci-après concerne le tir d'un trou avec un tel explosif (cisalite) et pour lequel les effets arrière ont été mis en évidence et quantifiés par mesures microsismiques et tomographie sismique.

2 - CARACTERISTIQUES DU SITE

Le site expérimental choisi est constitué d'un massif granitique porphyroïde à biotite.Les sondages destinés au tir et à l'auscultation ont été effectués sur une plateforme antérieurement terrassée dans un matériau rocheux sain (figure 1)

L'analyse de détail par lame mince montre que la matrice rocheuse n'est l'objet d'aucune micro-fissuration. L'analyse des discontinuités met en évidence des diaclases se répartissant en trois familles principales (figure 2)

N 135 E - famille la plus marquée la mieux représentée

N 185 43 W

N 67 S.

La répartition de la fracturation varie de ID2 à ID5 dans la classification AFTES avec 76 % des valeurs comprises entre ID3 (20 à 60 cm) et ID4 (6 à 20 cm).

Figure 1 : Vue générale du site. Les mesures sont effectuées dans un matériau rocheux sain

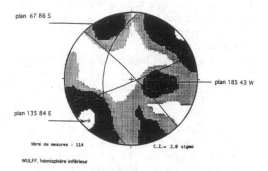

Figure 2 : Etude de la fracturation au front de taille.

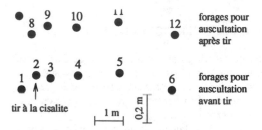

Figure 3 : Schéma de principe des sondages d'auscultation avant et après tir

3 - PROCEDURE EXPERIMENTALE

L'expérimentation a été réalisée à partir d'un certain nombre de sondages et d'un tir d'essai à la cisalite selon la procédure suivante (figure 3).

Réalisation d'une ligne de sondage à 10 m de profondeur - wagon-drill n° 1-2-4-5-6 carotté n° 3.
Auscultation microsismique et tomographique de l'ensemble de ces sondages (auscultation avant tir).
Tir à la cisalite dans le sondage n° 2 = bourrage gravillon sur les deux mètres supérieurs, cisalite entre - 2 et - 8 m.

Réalisation d'une seconde ligne de sondage à 10 m de profondeur décalés d'environ 0,50 m par rapport à la 1ère lignewagon drill n° 7 - 8 - 10 - 11 - 12 sondage carotté N9
Auscultation microsismique et tomographique de l'ensemble de ces forages (auscultation après tir).

4 - LE TIR

Le forage N°2 a fait l'objet d'un tir à la cisalite entre - 2 et - 8 m de profondeur. L'ensemble du trou a été bourré de gravillons et l'amorçage effectué au détonateur électrique sur cordeau détonant de 12 g.
La cisalite ou boudin de gel allégé (watergel) se présente sous forme d'une charge linéaire continue de 400 g au mètre et de diamètre 25 mm. La masse volumique est de 0,9 t/m3 pour une vitesse de détonation de 2850 m/s.
Les énergies piscine mesurées expérimentalement sont les suivantes :
Energie de choc : 653 cal./g - Energie de gaz : 1523 cal./g.
Energie totale : 2176 cal./g.
Il s'agit donc d'un explosif à relativement forte énergie de gaz par rapport à l'énergie de choc.
Précisons enfin qu'il s'agit, dans le cas de cette expérimentation au tir d'un seul trou en configuration parfaitement bloquée.
Toutes les conclusions doivent donc être ramenées au contexte de l'expérimentation : tir d'un seul trou avec explosif à relativement forte énergie de gaz et à faible vitesse de détonation.

5 - RESULTATS D'AUSCULTATION AVANT ET APRES TIR : MICROSISMIQUE ET TOMOGRAPHIE

5.1 Diagraphie microsismique : matériel et mise en oeuvre

La sonde, mise au point par le Laboratoire des Ponts et Chaussées d'Aix en Provence, est descendue à l'intérieur d'un forage de diamètre 76 mm minimum. Pendant la mesure, la sonde, comprenant un émetteur et deux récepteurs, est bloquée sur la paroi du forage par quatre microvérins pneumatiques (Les enrochements 1989).
Un choc est émis au droit de l'émetteur par un mini marteau pneumatique et le temps de propagation des ondes est réceptionné par deux capteurs de vibration.
On calcule ainsi la vitesse V1 entre l'émetteur et le

récepteur 1 et la vitesse V2 entre l'émetteur et le récepteur 2.

Le pas de mesure est généralement de 0,33 m et les résultats sont présentés sous forme de vitesse en fonction de la profondeur (figure 4).

5.2 Résultats microsismiques

Les profils de vitesse avant et après tir montrent que les vitesses avant tir s'échelonnent de 2000 m/s à 5500 m/s et après tir de 1500 m/s à environ 6 000 m/s.

La variation en pourcentage des vitesses microsismiques est calculée pour chaque sondage. Leur interpolation entre les forages conduit à une carte de variation de vitesse aux caractéristiques suivante (fig. 5) : dans la zone immédiate du tir (x = 50 cm), on note une diminution sensible des vitesses pouvant aller jusqu'à 40 %.

- dans la zone éloignée du tir (x>3 m), on remarque une augmentation importante des valeurs mesurées pouvant aller jusqu'à 20 %.

Cette dernière, très faible en tomographie sismique, est peut-être due à une erreur systématique de pointage des signaux sur le terrain avant le tir à la cisalite. En effet, la carte de vitesse microsismique est celle obtenue par tomographie sismique sont parfaitement corrélées après le tir.

5.3 Matériel tomographique et méthodes

La méthode de tomographie sismique employée ici s'attache à reconstituer la carte des vitesses sismiques de l'onde de compression entre forages à l'aide de mesures de durée de propagation (Côte 1988).

Elle repose sur un processus d'inversion géométrique itératif (Gilbert 1972). Elle présente l'avantage sur les méthodes matricielles (Bois et al 1971, Côte 1984) de minimiser le stockage des données et le temps de calcul.

Une de ses originalité est l'utilisation de rais analytiques qui accroît sa rapidité par rapport aux méthodes de tracé des rais dites physiques. Il est à noter que ces dernières restent tributaires de la discrétion du milieu et donc, par exemple, de la présence de blocs erronés (Côte 1988).

Dans cette application les pseudorais sont circulaires (fig.6). Dans un des forages remplis d'eau est immergée une flûte de neuf hydrophones récepteurs distants de 1 mètres et, dans un second forage, une source détonante et un dixième

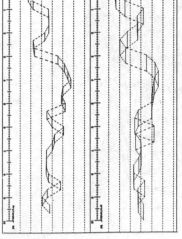

Figure 4 : Exemple de diagraphie microsismique effectuée après tir.

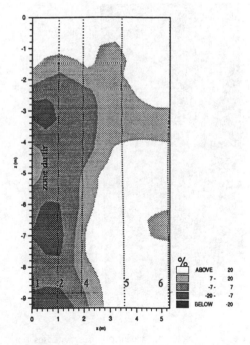

Figure 5 : Carte de variation des vitesses en microsismique

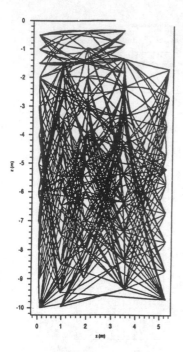

Figure 6 : Tracé des rais en tomographie

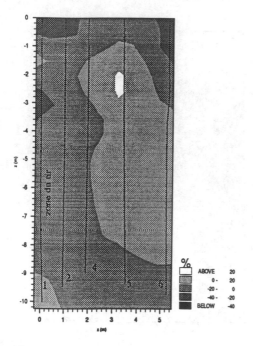

Figure 7 : Carte de variation des vitesses en tomographie

hydrophone sont conjointement déplacés tous les mètres. A chaque détonation de la source les informations sont stockées sur une centrale d'acquisition comportant 10 voies. Ces mesures sont répétées dans d'autres couples de forages afin d'assurer une bonne couverture en rais de la région à étudier.

Dans un premier temps une série de mesures est effectuée avant explosion dans les forages S1, S3, S4, S5 et S6 (fig 3). Une carte des vitesses caractérisant le terrain avant le tir à la cisalite est établie. Ensuite pour étudier l'impact de l'explosion, l'idéal est d'effectuer une deuxième série de mesures dans un contexte identique, i.e. dans les mêmes forages. La comparaison de la nouvelle carte des vitesses avec celle obtenue avant explosion conduit à une estimation de l'impact du tir su le terrain.

Pour le site de considéré la deuxième série de mesures est effectuée dans cinq nouveaux trous distants des précédents d'une cinquantaine de centimètres S7, S9, S10, S11 et S12 (fig 3). Deux cartes des vitesses dans deux plans voisins, sont donc établies.

5.4 Résultats tomographiques

Seuls les résultats des zones traversées par suffisamment de rais sont fiables. Aussi la zone superficielle, d'épaisseur inférieure à 1,5 mètres, et les bords du cadre de présentation des résultats sont écartés de la discussion (fig.6).

La reconstruction tomographique donne des vitesses s'échelonnant entre 2000 m/s et 6000 m/s confirmant le caractère sain du granite. A la suite de l'explosion, la vitesse moyenne du terrain, calculée avec tous les temps mesurés et les distances séparant les sources et les récepteurs, passe de 5231 m/s à 5 103 m/s. La carte de variation des vitesses (fig.7) met clairement en évidence le tir : la zone entourant le forage où l'explosion a lieu, aux alentours de $x = 0,5$ m. Dans cette zone, qui s'étend autour du forage sur environ 1,5 m, le matériau est détérioré et, globalement, le terrain est plus lent.

5.5 Comparaison diagraphie microsismique et tomographie sismique

Les deux méthodes peuvent être comparées en superposant les variations des profils sismiques au droit des forages.

134

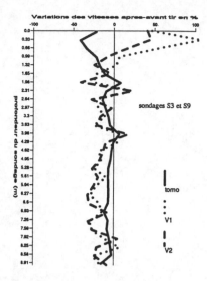

Figure 8 : profils de vitesse : comparaison des deux méthodes

La méthode microsismique, directement sensible à la fissuration et microfissuration de la paroi du forage fournit des renseignements précis et contrastés, mais ponctuels, alors que la tomographie, méthode beaucoup plus globale, intéresse un panneau entier et concerne l'ensemble d'un réseau de fractures.Mise à part la zone superficielle non représentative en tomographie l'allure générale des courbes de variation de vitesse sont globalement semblables même si elles sont systématiquement décalées (fig.8).

Sur la carte de variation de vitesse une augmentation de vitesse sismique est visible pour les x supérieurs à 2 m en microsismique (présence d'un décalage) mais reste minime en tomographie : l'influence de l'explosion reste localisée autour du forage où a lieu le tir.

6 - EVOLUTION DE LA FISSURATION

Afin d'essayer de vérifier les variations de vitesses mesurées en microsismique ou en tomographie, il a été effectué une comparaison de l'évolution de la fissuration par sondages carottés avant et après tir à proximité de la zone de tir. Les sondages carottés 3 et 9 réalisés à environ 0,50 m du tir permettent de reconstituer le nombre de fissures naturelles, fissures recristallisées ou fissures artificielles affectant le massif

	Fissures Natur.	Fiss. recrista.	Fiss. Artif	Total
Sondage 3 (avant tir)	34	6	6	46
Sondage 9 (après tir)	43	4	12	59

La longueur des éléments de carotte varie de 5 à 60 cm. La fissuration globale est plus importante après tir qu'avant.

Parallèlement une série de 7 lames minces effectuées sur le sondage 9 (après tir) montre qu'aucune micofissuration n'affecte la matrice rocheuse à 0,50m de tir.

Il semble donc que l'action de l'explosif, dans le cas précis de l'expérimentation réalisée, se manifeste au niveau de la densité de fracturation du matériau, et pas du tout sur la microfissuration. Cette observation est largement confirmée par les constatations faites par ailleurs sur d'autres tirs de mine : les blocs constituant les matériaux d'abattage sont très souvent délimités par des surfaces oxydées, ferrugineuses ou argileuses, témoins de fissures préexistantes qui ont été ouvertes par l'action des explosifs ; la matrice rocheuse proprement dite n'est que rarement affectée.

Les variations constatées à partir des méthodes géophysiques mises en oeuvre sont donc à rechercher essentiellement dans l'action des explosifs sur la fissuration

7 - CONCLUSIONS

L'évolution de la fissuration d'un massif rocheux liée à un tir à l'explosif peut être quantifiée par auscultation microsismique ou tomographique.

Le résultat des deux auscultations sont tout à fait comparables et traduisent des zones à augmentation ou diminution des vitesses plus ou moins sensibles selon leur position par rapport au tir.

Les deux méthodes géophysiques peuvent ainsi conduire à une meilleure connaissance de l'action des explosifs sur un massif rocheux et à une maîtrise des énergies nécessaires à la fragmentation des matériaux.

Références bibliographiques

Bois P. - Laporte M. - Lavergne M. Thomas G. (1971) - Essai de détermination automatique des vitesses sismiques par mesures entre puits - Géophy. - Prospect 19 = 42 - 83

Cote P, 1984 - Imagerie sismique : inversion en vitesse et facteur de qualité de mesures de transparence sismique entre forages - RR LCPC n° 132 - LCPC Paris

Cote P (1988) - Tomographie sismique en génie civil - Thèse d'Etat - Université Y. Fournier de Grenoble

Les enrochements (sept 1989) - document Laboratoire Central des Ponts et Chaussées - Ministère de l'Equipement, du Logement, des Transports et de la Mer - 107 p

Gilbert P - 1972 - Itérative Method for the three dimensional reconstruction of an object from projection - Journ. - Theor - Biol 36 = 105-117

Rebeyrotte A, Heraud H (1989) - Méthode d'extraction pour les terrassements rocheux. Découpage des talus à l'explosif. Evolution des effets arrière. Rapport des Laboratoires des Ponts et Chaussées GT 38, série Géotechnique - Mécanique des Sols - Sciences de la Terre, 154 p + annexes

Dielectric prediction of landslide susceptibility: A model applied to recent sediment flow deposits

La prédiction diélectrique de la susceptibilité aux glissements: Un modèle appliqué aux coulées de terres

Pekka Hänninen
Geological Survey of Finland, Espoo, Finland

Raimo Sutinen
Technische Universität München, Institut für Bodenkunde, Germany

ABSTRACT: A dielectric model for predicting landslide susceptibility is developed. Instead of previously reported polynomial expressions for water (θ_v) versus dielectric properties (ϵ) we consider the water-dielectric relation to be a four-step linear function, which allows us to determine critical dielectric range for sediment flow susceptibility. The model was tested for sandy/silty deposits at the margin of the Matanuska Glacier in Alaska, where *in situ* dielectric measurents were carried out using an electrical capacitance probe (f=35-55 MHz). The dielectric range of stabile (ϵ_{app}=6-16), unstabile (up to ϵ_{app}=28) and currently sliding (ϵ_{app}=31-36) materials matches that of the present model. Therefore, we suggest that dielectric monitoring is a viable method for predicting landslides in natural sediments as well as man made structures.

Résumé: Un modéle diélectrique pour prédisposition à l'éboulement a été mis au point. Au lieu des expressions polynomiales mentionnées précédemmet pour l'eau (θ_v) contre les propriétés diélectriques (ϵ), nous considérons que le rapport eau-diélectrique est une fonction linéaire à quatre phases qui nous permet de déterminer la gamme diélectrique pour la prédisposition d'écoulement de sédiments. Le modéle testé pour les dépôts de sable/limon au bord du glacier de Matanuska dans l'Alaska exécute les mesures diélectriques in situ en utilisant une sonde à capacitance (f=35-55 MHz). La gamme diélectrique de matiéres stables (ϵ_{app}=6-16), instables (ϵ_{app} jusqu'à 28) et ordinairement glissants (ϵ_{app}=31-36) correspond au présent modéle. C'est pourquoi nous suggérons qu'un contrôle diélectrique est faisable pour prédire des éboulements dans les sédiments naturels ainsi que dans des synthétiques.

1. INTRODUCTION

Soil moisture content is a very important factor in several disciplines such as agriculture, forestry and engineering geology. The measurement and monitoring of water content, however, has conventionally been rather time consuming. Electromagnetic (EM) techniques, such as ground penetrating radar (GPR), time domain reflectometry (TDR, Hoekstra and

Delaney 1974, Topp et al. 1980) and electrical capacitance probing (ECP, Selig and Mansukhani, 1975) provide effective indirect moisture information concerning subsurface materials. GPR-WARR (wide angle reflection and refraction) provides dielectric information from depths of several meters to tens of meters, while GPR-RSAD (radar surface arrival detection, Sutinen and Hänninen 1990) records surface dielectric data. Installation of TDR probes (parallel transmission lines) at desired depth allows continuous soil moistrure monitoring. The ECP method has the advantage of facilitating detailed dielectric logging.

Determination of soil moisture by the EM tecniques is based on the well known relationship between the (real) dielectric constant ϵ_r and volumetric water content θ_v. This is beacause the free water dielectric constant ($\epsilon_{fw}=81$) is significantly greater than that of solid soil particles and air. The ϵ_r/θ_v-relationship has been most commonly expressed as a polynomial function (Topp et al. 1980, Hallikainen et al. 1985). Several reports (e.g. Hoekstra and Delaney 1974, Topp et al. 1980) in the radio and microwave frequency range suggest that soil texture has only a very minor influence on this relationship. More recent laboratory tests in the microwave region, however, indicated a slight texture-dependence (Wang and Schmugge 1980, Dobson et al. 1985). The *in situ* measurements in the radio wave range do not, however, show soil type specific scattering (Sutinen 1992). This is presumably because the behaviour of free water in layered natural sediments tends to be largely controlled by hydraulic properties and unsaturated sediments tend to be in field capacity.

According to Tran Ngoc Lan et al. (1970) dielectric dispersion is significant at frequencies of less than 10 MHz, while the real dielectric constant (ϵ_r) is not frequency dependent in the range 25.6 MHz to 5 GHz for any soil saturation state between 0 and 100%. Similar results were obtained by Hoekstra and Delaney (1974) at frequencies between 100 MHz and 1 GHz. Since ϵ_r is significantly greater than the imaginary component (ϵ_i), the complex dielelectric constant, therefore, is effectively equivalent to the real dielectric constant.

As an atternative to polynomial equations, this paper describes a four-step linear model for the ϵ_r/θ_v relationship. The model allows us to dielectrically predict the landslide susceptibility of sediments. The model was evaluated by using dielectric data from recent sediment flow deposits at the margin of Matanuska Glacier, Alaska.

2. DIELECTRIC MIXING MODEL

The basic approach of the present four component model is similar to that of Wang and Schmugge (1980) (1.5-5 GHz) and Dobson et al. 1985 (1.4-18 GHz). The major difference, however, is the input grain size frequency data in the three classes: d>0.6 mm, d=0.6-0.006 mm and d<0.006 mm and the four-step linearity of the ϵ_{app}/θ_v model.

The volume fractions (V) of the four dielectric mixture components, rock particles (ϵ_{rock}), bound water (ϵ_{bw}), free water (ϵ_{fw}) and air (ϵ_{air}), yield an apparent dielectric constant as follows:

$$\epsilon_{app}=\epsilon_{air}V_{air}+\epsilon_{fw}V_{fw}+\epsilon_{bw}V_{bw}+\epsilon_{rock}V_{rock}$$

The dielectric constant of air $\epsilon_{air}=1$ is considered to be a constant. The dielectrics of solid particles are variable depending on the rock porosity and density (Parkhomenko 1967). We therefore determined ϵ_{rock} (Fig. 1) using specimens of dimension representing of six igneous rock types, by applying an open ended coaxial probe (HP 85070 dielectric probe kit).

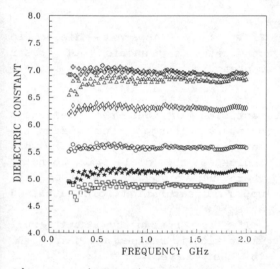

Fig. 1. Dielectric constant (real part) of igneous rock types as a function of frequency.

Between the measured frequencies, ranging from 200 MHz to 2 GHz, the real dielectric constant is practically independent of frequency (Fig. 1), reminiscent of what Tran Ngoc Lan et al. (1970) described for soils. The ϵ_{rock} varies between about 4.5 to 7, while ϵ_i is close to zero. These values are similar to those of solid matricies of soils and limestone obtained by Alharti and

Lange (1987) at 250 MHz. Based on the laboratory test, the average $\epsilon_{rock}=6$ has been used in the present dielectric mixing model.

Water is considered to have the following phases on solid particles and filling the voids in the mixture: an electrically bounded, 0.25 μm thick layer of water with $\epsilon_{bw}=3.5$, electrically free but physically bound water forming a layer up to 0.50 μm layer (Polubarinova-Kochina 1962) with an $\epsilon_r=81$ and free water with $\epsilon_{fw}=81$ filling the pore spaces, finally saturating and oversaturating the mixture.

Wang and Schmugge (1980) took the changing magnitude of bound water into consideration by changing the dielectrics of ϵ_{bw}. In the present model bound water is consistently assumed to behave like ice, although the increasing volume of bound water decreases the dielectric volume fractions of the other volume components of the mixture. Therefore, it is necessary to perform iterations between the volume fraction of water and other volume fractions of the soil.

The input parameters of the program are: dielectric constants of rock, bound water and free water, densities of rock particles and water, volume of rock particles in the beginning, maximum thicknesses of electrically and physically bound water, grain size frequencies and form factors in the three classes, d>0.6 mm, d=0.6-0.006 mm and d<0.006 mm. The coarsest class, has in practice, only an insignificant amount of bound water, where as the finest grain size class, due to their relatively large surface area, contain the predominant portion of the bounded water in the mixture.

Particles are assumed to be spherical, but a factor describing form could be used as an input parameter. However, natural sediments, particularly unstratified till, contain a wide variety of differently shaped rock and mineral components, so that no particular particle shape (see Ulaby et al. 1986) is representative of the whole sediment matrix.

3. GEOTECHNICAL SIGNIFICANCE

The apparent dielectric constant (ϵ_{app}) of a mixture containing no water is always small compared to that of solid particles due to $\epsilon_{air}=1$, i.e. $\epsilon_{app}<4.5$ (see Fig. 1). In terms of natural sediments, however, such a mixture is not possible because of water adsorption on the particle grains.

When water is added to the dry air-rock particle-mixture, all water will presumably be firmly bound onto the particles. This bounding is assumed to be electrical ($\epsilon_{bw}=3.5$), and relates to the hygroscopic coating (≈ 0.25 μm thick) on the particles. The increase of water volume will slightly change the volume fractions of solid particles and air in the mixture. In spite of the added water the apparent bulk dielectrics ϵ_{app} still increase very slowly (Fig. 2), since the presence of air $\epsilon_{app}<\epsilon_{rock}$, makes this kind of firm material geotechnically very resistant. Since the mixture lacks free water, these materials are not susceptible to frost either.

Assuming, that the bound water from 0.25 μm to 0.375 μm, half from hygroscopic coating to whole physicall coating (Polubarinova-Kochina 1962), the cohesion strength is large enough to separate solid

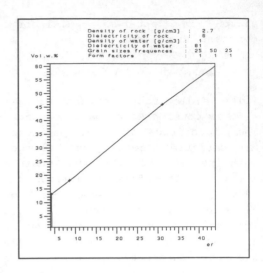

Fig. 2. Apparent dielectric constant as function of water content.

particles, we get the dielectric boundary range for stablile materials (see Fig. 2). This indicates the dielectric boundary where the voids will begin to fill up, full saturation being attained at $\epsilon_{app}=30$.

It is noteworthy that full saturation is reached again at $\epsilon_{app}=30$ in the curves of fig 3. This evidently corresponds to the dielectric limit for full saturation/ oversaturation and it seems to be independent of texture.

In the second step, the increasing volume of water is still physically bound to the particles and is able to separate them, the water is electrically free, i.e. $\epsilon_r=81$. Consequently, ϵ_{app} will increase rapidly with increasing amounts of water, because the relative portion of the volume fractions of the low dielectric components , ϵ_{air}, ϵ_{bw}, ϵ_{rock}, will decrease. According to Figs. 2

and 3, this dielectric range is from $\epsilon_{rock}=6$ to $\epsilon_{app}\leq16$ depending on the textural composition of the sediment. The general geotechnical quality is good for construction purposes. However, the degree of resistancy will decrease as a function of the ϵ_{app}.

After all the particles are coated with the physically bound and still electrically free water, excess water will start filling the voids in between the particles. The dielectric range of this model step three extends to the $\epsilon_{app}=30$. The pore spaces will gradually be filled with physically free water, thus both decreasing the resistance and increasing frost susceptibility of the material.

At the fourth step, beyond $\epsilon_{app}=30$ all the voids are filled with free water and the material is unstable. Particularly on slopes the excess of water will result in loss of shear streng thus increasing the risk of landslide. These may be induced by rainfalls or seismicity .

4. FIELD EXPERIMENT

The experimental data for testing the dielectric mixing model was collected at the margin of the Matanuska Glacier, Alaska. The supraglacial sediment flow deposits, previously described by Lawson (1979), are currently sliding by creep, but rotational lanslides are common, too. The sliding supraglacial debris, therefore, contains variable amounts of water released from the ice and are exposed to rainfall as well. Whenever the water content in the sediment mixture exceeds shearing strength, material will slide (Fig. 4). Also, the ice surface gradient tend to change with time.

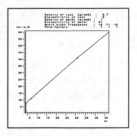
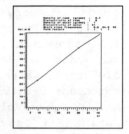
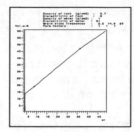
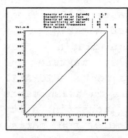

Fig. 3. Model curves for the sediments AK5, AK8, AK11 and AK12.

Fig. 4. Supraglacial landslide at the margin of Matanuska glacier, Alaska.

The ϵ_{scd}-measurements were performed using ECP (manufactured by the ADEK Ltd Co, Saku, Estonia) with an operating frequency range of 30 to 60 MHz. The bulk dielectric properties are determined by means of change of fringe capacitance using the equation $\epsilon_r=\Delta C/Ca+1$, where ΔC is the change of the capacitance due to probed sediment and Ca is changing (active) part of the probe capacitance.

Sediments were also sampled for the determination of gravimetric and volumetric water contents and particle size analysis (wet sieving and wet dispersed laser diffraction) (Table 1). The tested materials AK1-AK9 represent supraglacial flow sediments varying in texture and dielectric water content. AK10-AK11 represent proglacial ouwtwash sands and AK12 proglacial lacustrine silty sand. AK14 is a reference marine fine sand from the Turnagain Inlett, Alaska. Model curves have been presented for AK5, AK8, AK11 and AK12, as an example in Fig. 3.

Table 1. Particle frequences in classes > 0.6mm (1), 0.6-0.006mm (2), < 0.006mm (3) and ϵ_{app}.

Case code	Classes(%)			obs. ϵ_{app}		
	1	2	3			
AK1	14	59	27	10	15	25
AK3	17	67	16		13	14
AK5	0	75	25	27 28	28	31
AK6	3	66	31		22	28
AK7	29	63	8		14	14
AK8	19	71	10			7
AK9	31	59	10		6	7
AK10	80	20	0			4
AK11	82	18	0			4
AK12	0	65	35			36
AK14	0	92	8	25	26	27

Materials AK8-AK11 are dry according to the measured dielectric properties and water is in a bound state. Since the proportion of water-bearing fine-grained particles is only 0-10 %, excess of water will

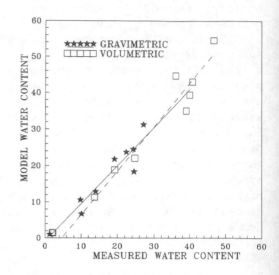

Fig. 5. Model water content vs. measured water content of supraglacial flow sediments, proglacial and marine sediments from Alaska.

be removed by gravity. The $6<\epsilon_{sed}<16$ indicate the materials AK1, AK3 and AK7 to be stabile, except for one observation of $\epsilon_{sed}=25$ recorded from AK1. All the measured ϵ_{sed}-values from AK5, AK6, AK12 and AK14 indicate near full saturation and/or oversaturation and the instability is evident. The water contents measured in the laboratory and model water contents show a good correlation of R=.91 for gravimetric and R=.92 for volumetric water contents (Fig. 5). The most evident sources of error were the lack of representative sediment sampling due to layered structure and moisture evaporation from the containers.

5. CONCLUSIONS

The dielectric mixing model developed here, along with the input grain size frequency data can reveal the following critical geotechnical sediment features. The $\epsilon_{app} < \epsilon_{rock}$ is typical of the first step in the model, where all water is electrically bound to the particles. Material in this phase are stabile. The apparent dielectric range from ϵ_{rock} to $\epsilon_{app} \leq 16$ contain water which is electrically free, but physically bound to the particles. The material is geotechnically safe and the upper ϵ_{app} may refer to the wilting point in soil science. When the ϵ_{app} ranges from about 16 to 30, porespaces in the material will gradually become filled with free water, reaching full saturation at $\epsilon_{app} = 30$. Consequently, geotechnical resistance will decrease linearly, but frost susceptibility will level out near full saturation. Beyond this critical point the increase in ϵ_{app} is very great along with the increase of θ_v. The sediment mixture will be unstable and susceptible to sliding.

REFERENCES

Alharti A. and J. Lange. 1987. Soil water saturation: dielectrric determination. Water Resources Research, 23, 4, 591-595.

Ansoult M., De Backer L.W. and M. Declerq. 1985. Statistical relationship between apparent dielectric constant and water content in porous media. Soil Sci. Am. J., 49, 47-50.

Dobson M.C., Ulaby F.T., Hallikainen M.T. and M.A. El-Rayes. 1985. Microwave dielectric behaviour of wet soil-Part II. Transactions on Geoscience and Remote sensing, GE-18, 1, 35-46.

Hallikainen M.T., Ulaby F.T., Dobson M.C., El.Rayes M.A. and L.-K. Wu. 1985. Microwave dielectric behavior of wet soil-Part I: Empirical Models and Experimental Observations. IEEE Transactions on Geoscience and Remote sensing, GE-18, 1, 25-34.

Hoekstra P. and A. Delaney. 1974. Dielectric properties od soils at UHF and microwave frequencies. Journal of Geophysical Research, 79, 11, 1699-1708.

Lawson D.E. 1979. Sedimentological analysis of the western terminus region of the Matanuska Glacier, Alaska. US.CRREL Report 79-9.

Parkhomenko E.I. 1967. Electrical properties of rocks. (Transl. and Ed. by G.V. Keller) Plenum Press, New York.

Polubarinova-Kochina P. 1962. Theory of ground water movement. Princeton Univ. Press, New Jersey.

Selig E.T. and A.M. Mansuhhani. 1975. Relationship of soil moisture to the dielectric property. Proc. Am. Soc. of Civ. Eng. 101, GT8, 755-770.

Sutinen R. 1992. Glacial deposits, their electrical properties and surveying by image interpretation and ground penetrating radar. Geol. Surv. Finl. Bull. 359.

Sutinen R and Pekka Hänninen. 1990. Radar profiling and dielectrical properties of glacial deposits in North Finland. Proc. 6th Int. IAEG Congress, Balkema, Rotterdam, 1045-1051.

Topp G.C., Davis J.L. and A.P. Annan. 1980. Electromagnetic determination of soil water content: Measurements in coaxial

transmission lines. Water
Resources Research, 16, 3 574-582.

Tran Ngoc Lan, P. Chaigne and A.
Philippe. 1970. Mesure des teneurs
en eau des sols par les méthodes
électriques. Rapport de recherche
N° 5. Laboratoires des Ponts et
Chaussées, Paris.

Ulaby F.T., Moore R.K. and A.K.Fung.
1986. Microwave remote sensing.
Active and passive. III. Artech
House.

Wang J.R. and T.J. 1980. An
empirical model for the complex
dielectric permittivity as a
function of water content. IEEE
Transactions on Geoscience and
Remote Sensing, GE-18, 4, 288-295.

Nonlinear acousto-elastic properties of grainy media with pore filling

Propriétés acoustiques-élastiques non-linéaires des milieux granulaires aux pores remplis

I.Yu. Belyaeva & V.Yu. Zaitsev
Institute of Applied Physics, Nizhny Novgorod, Russia

ABSTRACT: Nonlinear properties of media consisting of randomly packed spherical granules are investigated. The equation of state for such a medium is derived. The results of this work may be useful as a base new methods of seismo-acoustic probing.
RESUME:On examine les propriétés non lineaires de milieu qui est composé des granules spheriques. Les équations d'état de ce milieu sont obtenues.Les résultats de ce travail peuvent être utilisés comme la base des nouveaux méthodes de sondages seismiques.

1. INTRODUCTION

Recent years indicated a growing interest to the connection between medium structure and its nonlinear acoustical properties. It is realized now that the high level of nonlinearity of some materials is closely associated with the complexity of their inner structure, for example, with the presence of such inclusions as pores, cracks, grains, etc. Hence, the observation of nonlinear vibro-acoustical effects can be used in principle for judging on the structural and mechanical characteristics of the material under investigation. An essential step for the realization of such interpretation of the experimental data is the design of theoretical models connecting medium structure with, for example, its nonlinear "stress-strain" relation, the latter being reconstructable using vibro-acoustical measurements.

Let us briefly discuss the terminology generally adopted for the description of the nonlinear properties of the materials. Usually, in acoustics the nonlinear parameters are associated with the dependence of sound velocity c on acoustic pressure. Often used as the measure of nonlinearity for gases and fluids is the parameter

$$\Gamma = 1 + \rho_0 c_0 (dc/dp)_s \qquad (1)$$

where s is the entropy, ρ - density, p - pressure, and index 0 stands for the unperturbed state of the medium. Rather often, only the quadratic nonlinear term in medium governing equation is taken into account assuming the linear relation between the sound velocity c and the pressure p in acoustic wave: $c = c_0 + \Gamma p /\rho c_0$, where the parameter Γ is not dependent on pressure, namely, $\Gamma = 1 + B/2A = (1+\gamma)/2$, where A and B are the coefficients of linear and quadratic terms in the Taylor expansion for the equation of state $p=p(\rho)$, γ is the polytropic index. Similarly, at longitudinal strains in a solid medium the stress tensor component, $\sigma = \sigma_{ii}$ is related to the strain $\varepsilon = \varepsilon_{ii}$

$$\sigma(\varepsilon) = M(\varepsilon - \Gamma \varepsilon^2) \qquad (2)$$

where M - the elastic module of longitudinal deformation.
Typical values of Γ for homogeneous media range from 1 to 10 (specifically, $\Gamma = 3.5$ for water, $\Gamma = 1.2$ for air, and $\Gamma \approx 5$ for metallic monocrystals.) (Zarembo 1966). Meanwhile, Γ sharply increases in microinhomogeneous media, for example, Γ can reach 10^4

in fluid with gas bubbles (Kobelev 1980, Naugolnykh 1990). About the same values of the nonlinear parameter are observed, for example, in rubber-like media with spherical or cylindrical pores (Ostrovsky 1991, Belyaeva 1991) and polycrystalline metals (Nazarov 1991).

In this paper we consider the nonlinearity of granular media as another important group of microin-homogeneous materials. The nonlinear acoustic properties of these media are interesting not only as one more example of strong "nonlinear influence" of the microstructure but also as a base of possible diagnostic applications to various construction materials as well as earth soils, etc.

As the model of such granular material consider a medium consisting of elastic randomly packed identical spherical particles the porous space between them being filled with a fluid. Similar models was considered by many authors for describing the linear elastic properties of earth soils (Deresewicz 1958, White 1979). The main theoretical conclusions (the linear elastic constants, sound velocities) have been verified experimentally using the model packages of spherical particles, the field experiments also shown that these conclusions are applicable to the description of the linear acousto-elastic properties of nonconsolidated media (White 1979).

The nonlinear properties of granular media were investigated, for example, in (Nesterenko 1983) where the propagation of nonlinear elastic waves in a chain of spherical particles (producing a soliton) was considered theoretically. Results of the laboratory experiments with a one - dimensional chain, which substantiate the theoretical conclusions, are reported in (Lazaridy 1985). The data on nonlinear properties of a single contact are given in (Zosimov 1991).

As it is well known , acoustical properties of the isotropic elastic medium in a linear approximation can be described by two independent parameters, for example, by the bulk deformation module and the plain one which relate, respectively, the perturbation of pressure p to the relative variation of volume ε_v in the case of all-round

compression and the stress σ to the relative variation of linear dimension ε_L (strain) in the case of pure plane deformation.

As shown below, to describe the nonlinear properties of the medium we can obtain the nonlinear analogies to these relations both for plane deformation that corresponds to the longitudinal wave propagation in an unbounded medium and for all- round compression.

2.THEORY

Consider the sample composed of a large number N of particles of radius R, the properties of which are described by the constants K_s (the bulk compression module) and E_s (the Young module). Let α be the fraction of empty (porous) space per unit volume of the aggregate and $\bar{n}$ the average number of contacts between particles. The values of α and $\bar{n}$ for a system of randomly packed identical particles have been found experimentally: $\alpha = 0.392$, $\bar{n} = 8.84$ (Deresewicz 1958). (These values range from $\alpha = 0.477$ in the case of simple cubic packing to $\alpha = 0.23$ in the case of tightest packing of identical granules). Assuming that the porous space between particles is filled with a fluid under pressure P_f we find that the total volume of the "particle + fluid" system amounts to

$$V_t = \frac{4}{3} \frac{\pi N R^3}{1-\alpha} \qquad (3)$$

Suppose that the whole aggregate of N particles is subjected to quasistatic compression due to pressure P_{ext} that causes particle deformation. Therefore, the contact point transforms to a circle and the particle centers draw closer to a distance $2 \cdot \tilde{\Delta}$, which is related to the force F, acting at the contact, by the Hertz formula (Landau 1986):

$$\tilde{\Delta} = \left(\frac{3(1-\nu_s^2)F}{4 E_s R^{1/2}} \right)^{2/3} \qquad (4)$$

Here, ν_s is the Poisson coefficient

of the particle material. The variation of total volume V_t to the first order of $\tilde{\Delta}$ is equal to:

$$\delta V_t = - b \frac{4\pi N R^2 \tilde{\Delta}}{1 - \alpha} \quad (5)$$

where $b = 1$ in the case of bulk compression and $b = 1/3$ for longitudinal one.

This variation includes the decrease of the porous volume as the particle centers draw together, the decrease of the spherical granule volume due to the all- round compression by fluid, and at last the particle volume variation in the contact region. It can be shown that the last correction is of the order of $\tilde{\Delta}^2$ (while other items have the order of $\tilde{\Delta}$) and is therefore negligible.

To ascertain the relation between the force F, acting at the contacts between particles and the given external pressure P_{ext} we make use of the energy balance equation:

$$W_{ext} = W_f + W_s + W_c \quad (6)$$

where W_{ext} is the energy required for the quasistatic compression of the aggregate by external pressure, W_f is the energy of pore fluid , W_s the elastic energy of the particle matter due to the bulk compression by the fluid, and W_c the energy stored by the particles on the contacts.

The energy spent for reducing the aggregate volume by δV_t is determined by the expression:

$$W_{ext} = \int_{\delta V_t} P_{ext} dV_t \quad (7)$$

The energy accumulated by a pore fluid is given by:

$$W_f = \int_{\delta V_t} (p_f + \delta p_f)\, dV_f \quad (8)$$

where δp_f is the excess pressure due to the variation of the fluid volume δV_f.

The energy stored in the volume of fluid- compressed granules is

given by:

$$W_s = \int_{\delta V_t} (p_f + \delta P_f) dV_s \quad (9)$$

where δV_s is the variation of th spherical granule volume under the action of compressing pressure P_f from the pore fluid side.

The energy due to the work of deformed particles on the contacts can be written as:

$$W_c = b \bar{n} N \int_0^{\tilde{\Delta}} F dr \quad (10)$$

where the relation between $\tilde{\Delta}$ and δV_t is given by (5). The quantity b in this formulae describes the effective number of "active" (those that accumulate energy) contacts. It is obvious that in the case of all- round compression the particle centers are shifted with respect to each other only along the normal to the contact surface. Note that all contacts have "equal rights" owing to the isotropy of the problem and one should put $b=1$ in (10). In the case of plane deformation granule centres are also shifted along contact normales (i.e., in radial directions). However, due to the lack of granule centre shift along two coordinate axes (which are orthogonal to the plane deformation axis), because of the isotropy in contact orientation, and independence of each contact deformation work the average , effective share to the elastic energy of contact w_c is paid by one third part of the total contact number. Therefore one should put in (10) the factor $b = 1/3$.

The pressure and volume variations of the fluid and the granules are related via elastic constants by:

$$\delta p_f = - K_f \frac{\delta V_f}{V_f} \quad (11)$$

$$\delta p_f = - K_s \frac{\delta V_s}{V_s} \quad (12)$$

Thus, for the unknowns δp_f, δV_f and δV we have set of linear equations (6), (11), and (12) that gives the expressions of the unknowns via $\tilde{\Delta}$:

$$\delta p_f = \frac{\dfrac{3\tilde{\Delta}}{R}\, bK_f}{\alpha + (1-\alpha)\eta} \qquad (13)$$

$$\delta V_f = -\frac{\alpha b}{\alpha + (1-\alpha)\eta} \cdot \frac{4\pi N R^2}{1-\alpha}\,\tilde{\Delta} \quad (14)$$

$$\delta V_s = -\frac{\gamma b}{\alpha + (1-\alpha)\eta}\, 4\pi N R^2 \tilde{\Delta} \qquad (15)$$

where $\eta = \dfrac{K_f}{K_s}$.

In these expressions $\tilde{\Delta} = (\Delta - \delta R_s)$, where $2\cdot\delta R_s$ is the variation of the distance between the granule centres due to the all- round compression of granules by a pore filler (note that $\delta R_s / R_s \simeq (\delta V_s / V_s)/3$) i.e., from (15)):

$$\tilde{\Delta} = \Delta\left[1 - \frac{\eta b}{\alpha + (1-\alpha)\eta}\right] \qquad (16)$$

Substituting these expressions into the energy balance equation (6) and bearing in mind that the resultant equality should be valid at any arbitrary $\tilde{\Delta}$ (therefore, we can equate the integrand to zero) we obtain:

$$\frac{4\pi R^2}{1-\alpha}\left\{P_{ext} - P_f\right\} - \pi F - $$
$$- \frac{12\pi R K_f b\tilde{\Delta}}{(1-\alpha)(\alpha+(1-\alpha)\eta)} = 0, \quad (17)$$

where F, according to (4), is equal

$$F = \frac{4E_s R^{1/2}}{3(1-\nu_s^2)}\,\tilde{\Delta}^{3/2} \qquad (18)$$

Eq. (17) relates in fact the effective pressure in the medium and the strain of the system under the action of this pressure. Hence,

assuming $b = 1$ in the case of bulk compression and linearizing (17) with respect to strain (Δ/R) variations we can obtain the expression for the linear module of bulk compression.

Consider in more detail the case of pure longitudinal strain, which corresponds to the longitudinal wave propagation, and put $b = 1/3$ in (17). Introducing the effective strain $\sigma_{eff} = P_{ext} - P_f$ and the longitudinal strain $\varepsilon = \delta l_{11}/l_{11} = \delta V_t / V_t$, we obtain the "nonlinear Hook's law" for a fluid saturated granular medium:

$$\sigma_{eff} = \frac{K_f}{\alpha + (1-\alpha)\eta}\,\varepsilon +$$
$$+ \frac{\bar{n}(1-\alpha)E_s}{3\pi(1-\nu_s^2)}\,\varepsilon^{3/2} \qquad (19)$$

We emphasize that the nonlinear term is exclusively due to the presence of Hertz contacts rather than the nonlinearity of the aggregate component materials which we neglect in our consideration.

Expanding the dependence $\sigma(\varepsilon)$ into a series in small strain variations $\tilde{\varepsilon} = \varepsilon - \varepsilon_0$ in the vicinity of the initial static strain ε_0 we obtain:

$$\sigma(\tilde{\varepsilon}) = \sigma(\varepsilon_0) + \sigma'_\varepsilon(\varepsilon_0)\left[\tilde{\varepsilon} + \right. \qquad (20)$$
$$\left. + \frac{1}{2!}\frac{\sigma''_{\varepsilon\varepsilon}(\varepsilon_0)}{\sigma'_\varepsilon(\varepsilon_0)}\,\tilde{\varepsilon}^2 + \frac{1}{3!}\frac{\sigma'''_{\varepsilon\varepsilon\varepsilon}(\varepsilon_0)}{\sigma'_\varepsilon(\varepsilon_0)} + ..\right]$$

The coefficient of the linear term defines the longitudinal strain module of the aggregate $K_{\|agr}$. Thus, in accordance with (19) we have:

$$K_{\|agr} = \sigma'_\varepsilon(\varepsilon_0) = K_f\left(\frac{1}{\alpha+(1-\alpha)\eta} + \right.$$
$$\left. + \frac{\bar{n}(1-\alpha)}{2\pi(1-\nu_s^2)}\frac{E_s}{K_f}\,\varepsilon_0^{1/2}\right) \qquad (21)$$

148

According to (19), (20), the nonlinear parameter defined in (1) is given by:

$$\Gamma_{agr}^{(2)} = \frac{\sigma''_{\varepsilon\varepsilon}(\varepsilon_0)}{\sigma'_{\varepsilon}(\varepsilon_0)} =$$

$$\frac{\dfrac{\bar{n}(1-\alpha)E_s}{4\pi(1-\nu_s^2)}\,\varepsilon_0^{-1/2}}{K_f\left(\dfrac{1}{\alpha+(1-\alpha)\eta} + \dfrac{\bar{n}(1-\alpha)}{2\pi(1-\nu_s^2)}\dfrac{E_s}{K_f}\varepsilon_0^{1/2}\right)} \qquad (22)$$

In the case of small static initial strains and much less compressible pore filler compared to the contact compressibility, the second term in the denominator can be neglected. Thus, the expression for the nonlinear parameter is simplified:

$$\Gamma_{agr}^{(2)} = \frac{\bar{n}(1-\alpha)(\alpha+(1-\alpha)\eta)}{4\pi(1-\nu_s^2)}\frac{E_s}{K_f}\frac{1}{\varepsilon_0^{1/2}}$$

Using the same procedure we introduce the cubic nonlinearity parameter. We thus obtain:

$$\Gamma_{agr}^{(3)} = \frac{\sigma'''_{\varepsilon\varepsilon\varepsilon}(\varepsilon_0)}{\sigma'_{\varepsilon}(\varepsilon_0)} =$$

$$\frac{\dfrac{\bar{n}(1-\alpha)E_s}{8\pi(1-\nu_s^2)}\,\varepsilon_0^{-3/2}}{K_f\left(\dfrac{1}{\alpha+(1-\alpha)\eta} + \dfrac{\bar{n}(1-\alpha)}{2\pi(1-\nu^2)}\dfrac{E_s}{K_f}\varepsilon_0^{1/2}\right)} \qquad (24)$$

In the case of weakly compressible (in the above defined sense) liquid we get a simplified expression:

$$\Gamma_{agr}^{(3)} = \frac{\bar{n}(1-\alpha)(\alpha+(1-\alpha)\eta)}{8\pi(1-\nu_s^2)}\frac{E_s}{K_f}\varepsilon_0^{-3/2} \qquad (25)$$

For a granular medium without a pore filling Eq.(17) has the simpler form:

$$\frac{4\pi R^2}{1-\alpha}\,P_{ext} - \bar{n}F = 0 \qquad (26)$$

and in (16) and (18) we should put $\tilde{\Delta} = \Delta$. Consider the propagation of a longitudinal wave in such a medium. Introducing the stress $\sigma = P_{ext}$ and the relative longitudinal strain $\varepsilon = dV_t/V_t$, we write (33) in the form of the "nonlinear Hooke's law":

$$\sigma = \frac{\bar{n}(1-\alpha)E_s}{3\pi(1-\nu_s^2)}\,\varepsilon^{3/2} \qquad (27)$$

Linearizing this equation with respect to the strain variation $\tilde{\varepsilon}$ near the static initial compression ε_0, we obtain an expression for the elastic module of pure longitudinal strain:

$$K_{\parallel s} = \frac{\bar{n}(1-\alpha)E_s}{2\pi(1-\nu_s^2)}\,\varepsilon_0^{1/2} \qquad (28)$$

In a manner similar to that in (29) we write the quadratic nonlinear parameter $\Gamma_s^{(2)}$:

$$\Gamma_s^{(2)} = \frac{\sigma''_{\varepsilon\varepsilon}(\varepsilon_0)}{\sigma'_{\varepsilon}(\varepsilon_0)} = \frac{1}{2\,\varepsilon_0} \qquad (29)$$

where can be connected initial strain ε_0 with the value of external pressure P_0, using (28), then for $\Gamma_s^{(2)}$:

$$\Gamma_s^{(2)} = \frac{1}{2}\left(\frac{3\pi\,(1-\nu_s^2)P_0}{\bar{n}\,(1-\alpha)E_s}\right)^{2/3}$$

For the cubic nonlinearity parameter:

$$\Gamma_s^{(3)} = \frac{\sigma'''_{\varepsilon\varepsilon\varepsilon}(\varepsilon_0)}{\sigma'_{\varepsilon}(\varepsilon_0)} = \frac{1}{6\,\varepsilon_0^2} \qquad (30)$$

Thus, the nonlinear parameters of a granular medium without a pore filling is exclusively determined by the preliminary static strain.

3.EXPERIMENT

For experimental verification of the theoretical results we used the model aggregates of lead short and tuff grains (as an example of earth soil) of different dimensions both in the form of fractions of identical size particles, and as a mixture of different size granules (some results of the experiments with lead short are reported in (Belyaeva 1992). The quadratic and cubic nonlinear parameters were measured by the method of the second and third harmonic generation in the samples of such media. The material filled the cup with the bottom fixed on a massive platform. The cup was compressed from the above by a massive piston tight to the cup walls. The piston oscillation amplitude at the fundamental frequency was assigned by the vibrator.From the accelerometer fixed on the piston, the signal corresponding to the sample reaction to harmonic generation entered the spectrum analyzer. The experiments confirmed that the nonlinear parameters of granular medium are independent of the grain material and size but depend on the initial strain values. Thus, the quadratic nonlinear parameter was measured to range from $1,3 \cdot 10^2$ to $1.5 \cdot 10^3$,while the cubic nonlinear parameter was from $4 \cdot 10^4$ to $5.5 \cdot 10^6$ at the preliminary strain from $3 \cdot 10^{-4}$ to $3 \cdot 10^{-3}$, that is in a good agreement with the theory.

Note that the measurements were fulfilled also for samples consisting of the mixture of different size tuff granules. The values of linear plane deformation module obtained in these experiments were 15..20 percents larger (that should be expected for such more dense package) than that defined by Eq.(28), which is valid for less dense random package of identical grains. At the same time the measured values of the nonlinear parameters $\Gamma_s^{(2)}$, $\Gamma_s^{(3)}$ happened to be in good agreement with Eqs. (29),(30) relating this values with the initial stress. This is also the expecting result as the "nonlinear Hooke's law" for the case of multi-

size grains mixture should differ from the one for the case of identical grains only in quantitative factor before $\varepsilon^{3/2}$ (that is essential for the linear module value), but this factor does not alter the expressions (29),(30) for $\Gamma_s^{(2)}$, $\Gamma_s^{(3)}$. Therefore, we have certain reasons to use these expression to estimate the nonlinear characteristics of the real nonconsolidated (grainy) materials, which usually are also a mixture of different size grains.

4.CONCLUSION

The above results clearly demonstrate the high efficiency of the "structural" mechanism by which the nonlinearity of a medium may increase sharply if the material inner structure contains components (inhomogeneities, inclusions) with contrasting elastic properties. In the case considered such components are the elastic matter of the grains and the soft contacts between them. The derived relations connecting nonlinear medium properties with its structure, interparticle filler presence, prestrain value may be used for judging on the latter by experimental observations of nonlinear vibro-acoustic effects (combination harmonic generation, cross- modulation, etc.).

Similar "structural" mechanism can be seemingly revealed in other important classes of media (polycrystalline metals and minerals, cracked earth soils, etc.), in which high values of the nonlinear parameters were also observed [6,20]

Authors are grateful to Dr. E.M. Timanin for his participation in the laboratory experiments.

REFERENCES

Zarembo L.K.& Krasil'nicov V.A. 1966. "Introduction to 'nonlinear acoustics". (In Russian). Nauka, Moscow.

Kobelev Yu.A. & Ostrovsky L.A. 1980. "Models of gas-liquid mixture as nonlinear dispersive medium." (in Russian).In: Nonlinear acoustics. Gorky, IAP Acad.Sc. USSR, 143-160.

Naugol'nykh K.A. & Ostrovsky L.A. 1990. "Nonlinear wave processes in acoustics". Nauka, Moskow, (to be published by Cambridge University Press).

Ostrovsky L.A. 1991. Wave processes in media with strong acoustic nonlinearity. JASA, vol.90, No 6,p.3332-3338.

Belyaeva I.Yu. & Timanin E.M. 1991."Experimental investigation of nonlinear properties of porous elastic media".Sov.Phys.Acoustics, vol.37, No 5,p.1026-1029.

Nazarov V.E. 1991."Influence of copper structure on its acoustic nonlinearity".(In Russian).Physics of metals,vol.3,p.172-178

Gurvich I.I.& Nomokonov V.P. 1981. Seismo-prospecting. Reference book of geophysicists. .(In Russian).Nedra,Moscow.

Deresewicz H.A. 1958. Review of some recent studies of the mechanical behavior of granular media. Appl. Mech. Rev., 11, p.259-261.

White J.E. 1979. Underground Sound. Application of Seismic Waves. Elsevier, Amsterdam,Oxford,New York.

Nesterenko V.F. 1983. "Propagation of compression nonlinear pulses in grainy media". (In Russian). Jurn.Appl.Mech. and Tech.Phys.,No 5,p.136-148.

Lazaridy A.N. & Nesterenko V.F. 1985. "Discovery of new type solitary waves in the one-dimensional grainy medium". (In Russian) Jurn.Appl.Mech. and Techn.Phys., No 3,p.115-118.

Zosimov V.V. & Panasuk A.V. 1991."Acoustical effects of contact nonlinearity".(In Russian).In: Abstracts of the XI Acoust.Conf.,USSR, Moscow, p.43-46.

Landau L.D. & Lifshits E.M. 1986. "Theory of elasticity". Pergamon Press.

Belyaeva I.Yu., Ostrovsky L.A. & Timanin E.M. 1992."Experiments on harmonic generation in grainy media". Acoustic Lett., Vol.15, No 11, p.221-224.

On a method of the elastic nonlinear parameter logging using the effect of seismo-acoustic waves nonlinear backscattering

Une méthode de diagraphie des paramètres élastiques non-linéaires par l'effet de 'backscattering' des ondes sismo-acoustiques non-linéaires

V.Yu.Zaitsev
Institute of Applied Physics, Russian Academy of Sciences, Nizhny Novgorod, Russia

ABSTRACT: A scheme of profiling of Earth rocks nonlinear parameter is proposed, the effect of nonlinear sound backscattering being used as the physical base of the method. Estimations of the method feasibility are performed.

RÉSUMÉ: On propose un schéma du profilage du paramètre non-lineaire des roches sur la base d'effect de la dispersion non-lineaire des ondes. Les estimations d'applicabilité de cette méthode sont accomplies.

1. INTRODUCTION.

Conventional methods of seismo-acoustic Earth sounding are based on measurements of linear acoustic characteristics of the medium under investigation. Usually such parameters of probing waves are measured as the signal travel (delay) time and/or the amplitudes of reflected waves. Using this data one may determine acoustic parameters in the sounded region, the main characteristic being reconstructed is the profile of sound velocity along the path of the probing wave. The observed variations of sound velocity are attributed to the changes of the rocks inner structure and to the variations of the material inner stresses (loading conditions). Seismo- acoustic probing methods have proved their usefulness (R.E.Sheriff, E.Geldart 1982), though one can mention that such geological conditions can occur, when the medium structure is sufficiently inhomogeneous, whereas the corresponding changes of sound velocity are hardy noticeable.

It is realized now that very useful information on Earth rocks structure can be obtained from measurements of their higher- order elastic parameters that are associated with nonlinear acousto- elastic properties. It is proved also that the contrast of a medium non-linear properties can be much greater in comparison with the contrast of accompanying changes of the linear sound velocity (Nikolaev 1987). For example, it seems to be rather typical than unusual for Earth rocks, when the changes of the nonlinear parameter in a region reach several orders of magnitude, whereas the variations of linear sound velocity are about several percents (see also the report "Estimation of crumbly rocks nonlinearity on the base of a granular medium model...". by I.Yu.Belyaeva at this conference').

Therefore nonlinear parameters are of-ten much more sensitive to the changes of a medium inner structure or loading condi-tions. Due to such high sensitivity the determination of nonlinearity of a medium surely can provide rather useful information on its structure. An essential point in this connection is how these nonlinear parameters can be measured in practice. Various methods are known in nonlinear acoustics, but they are mostly res-tricted to laboratory conditions. This report considers the possibi-lities of such a me-thod, which seems to be feasible in field con-ditions typical of the problems of en-gineering geology, the physical base of the method being the effect of so called nonlinear sound backscattering.

2. THE EFFECTS OF NONLINEAR SOUND FORWARD RADIATION AND BACKSCATTERING

The discussions of the use of nonlinear acoustic wave interactions for probing problems sounds somehow exotic for seismics, though e.g. in hydroacoustics the nonlinear sound generation (or the parametric radiation, as it is usually called) has been already exploited in some commercial types of relatively short-range sonars, and a successful tests have been carried out of the long-range (up to 10^3 km) propagation of the nonlinearly generated sound (Dybedal 1993, Kalachev et. al. 1993). Certainly, due to the very nature of the nonlinearly generated sound waves, their amplitude is much more small to compare with the level of primary linear waves, usually called the pump waves. Thus, to increase the resultant signal level, the case of collinear wave interaction (or, in other words, the forward nonlinear radiation that accumulates the nonlinear signal along the interaction path) is used in hydroacoustical nonlinear sonars.

In nonlinear acoustics there is also an effect of nonlinear sound backscattering. The effect essence is the appearance of back-radiated nonlinear signal under propagation of powerful primary wave. Unlike the effect of spatial accumulation of the signal under collinear interaction, in the case of backward radiation from travelling nonlinear virtual sources, there is no such spatial accumulation, and the only nonlinear virtual sources contained within less than a $\Lambda/4$ layer (Λ is the characteristic wavelengths of backward radiation) are in phase. Due to this fact, the nonlinear backscattering intensity is usually much less than that of forward radiation.

The first observations of the nonlinear backscattering therefore related to the case of highly nonlinear scatterers, for example, the clouds of gas bubbles in hydroacoustics (Sandler, Selivanovsky 1981, Ostrovsky, Sutin 1983, Doskoy et.al.1984). It is essential to note that the nonlinearity of bubble-water mixture is much greater to compare with the nonlinearity of pure water. For example, the so called quadratic nonlinear parameter that equals 3.5 for pure water may be of the order 10^3 for bubbly liquid. Such a large value of the nonlinear parameter is connected with the inhomogeneous structure of bubbly liquids. For "ordinary" homogeneous material (such as glass or pure liquids) the nonlinear parameter is about several units. In the case of such a material as Earth rocks, their nonlinear acoustic parameter also often proved to be by several orders greater in comparison with nonlinearity of "common" materials (Nikolaev 1987).

Extremely high level of Earth rocks nonlinearity approves the possibility of observation of nonlinear effects at sound propagation. Possible schemes of such measurements may be based on the observations of transmitted signals ('two-sided' methods, see e.g. the report "On the possibility of seismo-acoustic nonlinear parameter tomography" by I.Yu.Belyaeva, A.M. Sutin, and V.Yu.Zaitsev at this conference) or of the back-radiated signal ('one-sided' methods). One-sided methods, if practically feasible, seem to be more attractive as being often more convenient. The following sections contain essential theoretical analysis of the possibility of the use of nonlinear sound backscattering as a base of a new remote sounding method.

3. GENERAL PROPERTIES OF THE NONLINEAR BACKSCATTERED SIGNAL

The amplitude of the secondary wave produced by a powerful primary wave can be found in the frames of the small perturbation method. The general expression for the secondary wave sound pressure can be written in the following form (Kustov et.al 1986):

154

$$P_{\bullet} = \frac{1}{4\pi\rho c^4} \times$$

$$\int \frac{\varepsilon(x,y,z)}{r} \frac{\partial^2 <P_i^2(t-r/c)>}{\partial t^2} \, dV \quad (1)$$

where c, ρ are the unperturbed sound velocity and medium density respectively, ε is the nonlinear parameter, r is the distance between the integration volume dV and the reception point, and the $<P_i^2(t)>$ is the primary wave pressure squared, averaged over the high-frequency period $2\pi/\omega$. This expression gives both amplitude of forward - radiated signal as well as the backward-radiated one, the latter being of the main interest for the purpose stated above.

Let us discuss some general properties of the secondary field.

It follows from (1) that, in the case of bi-harmonical primary waves, the nonlinear source amplitude is proportional to the difference frequency F squared. The corresponding formulae for forward and backward radiation in the case of a monochromatic source and a nonlinear layer with a sharp border and bi-harmonical primary radiation are derived by Kustov, Nasarov, Sutin (1986).

It can be shown that, when the curvature of the nonlinear virtual sources front across the primary pump beam can be neglected, the corresponding expression (Kustov et.al 1986) shows that the forward radiation pressure $P_{\bullet}$ is proportional to the difference frequency F squared and to the whole length of the interaction volume.

The backscattered signal amplitude is determined by the effective length of the interaction region, which, in this case, is approximately $(\Lambda/4) \propto 1/F$ (due to the phase mismatch between the counter-propagating primary wave and the backscattered one). Thus, in the case of a plane front of the virtual nonlinear sources (i.e they are in-phase within the whole cross section of the primary beam), due to the mentioned above frequency dependencies of the effective interaction length ($\propto 1/F$) and nonlinear sources amplitude ($\propto F^2$), the resulting dependence of back-

ward radiation $P_{\bullet}$ on F is linear, despite the quadratic dependence for forward signal.

In the case of a curved front of the effective nonlinear sources of the back- radiated signal, the backscattered wave amplitude is determined predominantly by the in-phase nonlinear sources within the first Fresnel zone of the insonified spot, and not by the total spot square. As the Fresnel zone square is proportional to the difference frequency wavelength $\Lambda \propto 1/F$ (Clay & Medwin 1977), the amplitude of the backscattered signal in the case of a curved primary wave front should not depend on F.

The amplitude relation between pressure of forward (P^f) and backscattered (P^b) nonlinear signals is predominantly determined by the ratio of interaction region effective lengths. In the case of the forward - radiated signal this length is determined predominantly by the absorption of the pump waves (that is about $10^2 \div 10^3$ m or more), and in the case of the backward signal this length is less than the wavelength of secondary field (that in practice may vary from fractions of one meter to several meters depending on the signal characteristic frequency). As the absorption distance is usually much greater than the secondary signal wavelength (by the factor of the order $10^2 \div 10^3$), the amplitude of forward signal should be correspondingly higher than the amplitude of the backward wave. Therefore, in hydroacoustics, as it was mentioned above, nonlinear backscattering can be easily observed only in the presence of highly nonlinear scatterers - gas bubbles. In seismics, the nonlinearity of Earth rocks often is even higher than the nonlinearity of bubbly water. Therefore, in the case of powerful enough primary waves and appropriate signal processing (such as it it used in hydroacoustics), the nonlinearly backscattered sound should be quite observable in Earth rocks.

To check up the validity of our theoretical representations, we have carried out some small-scale laboratory experiments on the nonlinear backscattering of ultrasound (Belyaeva & Zaitsev 1993). The observed frequency and ampli-

155

tude signal properties were in good agreement with the dependencies stated above.

4. ON THE POSSIBILITY OF THE NONLINEAR PARAMETER PROFILING USING TEMPORAL FORM OF THE NONLINEAR BACKSCATTERED SIGNAL.

Let us turn directly to the discussion of the possibilities of the nonlinear parameter profiling using the effect of nonlinear backscattering. Consider Eq.1 in the case of quasimonochromatic primary wave with a stepwise envelope $P_i \propto \exp(i\omega t)\,\theta(t)$, where $\theta(t)=0$ when $t<0$, and $\theta(t)=1$ when $t \geq 0$. The nonlinear parameter distribution is supposed to be quasi layered along the primary beam axis $\varepsilon(x,y,z)=\varepsilon(x)$. Thus for a high directivity (quasi-plane) primary beam the backscattered field amplitude appears to be proportional to the $\varepsilon(x)$ spatial derivative $\varepsilon'_x(x)$:

$$P_s(ct,0,0) = \frac{1}{2\rho c^2} \left(\int P_{io}^2(\rho)\rho d\rho \right)$$
$$\times \frac{\varepsilon'_x(ct/2)}{ct} \qquad (2)$$

where $P_{io}(\rho)$ is the transverse primary wave distribution in the vicinity of the transmitter.

In the case of a spherically diverging primary beam, the temporal form of the backscattered signal is proportional directly to the local value of the nonlinear parameter:

$$P_s(ct,0,0) = \frac{1}{2\rho c^2} P_{io}^2(0) \times$$
$$\frac{\varepsilon(ct/2)}{(ct)^2} \qquad (3)$$

where $P_{io}(0)$ is the primary wave axial amplitude normalized to 1m distance.

These properties of nonlinear backscattered signal temporal dependencies can be used for nonlinear parameter profiling. The longitudinal spatial resolution of the image is determined by the scale of the primary signal envelope front.

5. SOME ESTIMATIONS OF THE NONLINEAR BACKSCATTERED SIGNAL AMPLITUDE.

As it can be seen from the previous sections, the suggested method rather originates from nonlinear hydroacoustics that from conventional methods of seismic sounding. Correspondingly, it supposes more high signal frequencies than those used in traditional exploration seismology. Namely, this means the frequencies of primary waves from about 1 kHz up to $10 \div 20$ kHz, and about $100 \div 1000$ Hz for secondary signal. Traditional vibration sources developed for exploration seismology do not allow to generate such signals, but this can readily be done by using hydroacoustic sources adapted for radiation into Earth rocks thickness. These devices have the acoustic radiation power up to $10 \div 20$ kW and more, the transducer area of the order $1 m^2$, and beamwidth about 0.1 rad (J.Dybedal 1993). Because of strong dissipation of such high frequency signals the probing range of the method would be evidently restricted to several hundreds meters (up to 1km under favorable conditions).

To give an impression on the possible level of nonlinearly backscattered sound, let us accept that the source acoustic power W=10 kW, the rocks density r=2500 kg/m^3, the sound velocity c=1000 m/s. Suppose, that the acoustic nonlinear parameter of rocks is about $\varepsilon \sim 10^3 \div 10^4$ (A.V.Nikolaev & I.N. Galkin 1987, Johnson P.A., Shankland T.J. 1990), and the derivative $\varepsilon'_x \sim 100$.

Then, for the case of a plain wavefront, according to (2) we obtain for 500m sounding depth the sound pressure $p_s \sim 1$ N/m^2. For the case of a spherically diverging wavefront and 500m sounding depth, the expression (3) yields $p_s \sim 1$N/m^2 again. Such sound pressure is comparable with the level of noise in the $\Delta F=500$ Hz bandwidth at the spectral noise density about 90 dB/Hz$^{1/2}$ (re 10^{-6}N/m^2). Therefore, rather moderate coherent signal accumulation during $\Delta T=100$ sec. would provide signal to noise ratio about 10 (note that, when $\Delta T/\Delta F$=const., the signal to noise ratio is also constant).

156

The expressions (2) and (3) do not take into account the sound dissipation, and consequently the above estimations are somehow over optimistic. The influence of sound dissipation can cause the signal level decrease by tens times, though sound absorption varies sufficiently depending on the rocks type. The effect should be most significant in crumbly sedimentary rocks. Under such unfavorable conditions the sounding range may be shortened up to several dozens meters.

CONCLUSION

Extremely high nonlinear elastic properties of Earth rocks, on the other hand, and the successful experience in applications of nonlinear sound generation in hydroacoustics, on the other hand, allow one to believe that, in the case of powerful enough primary wave and appropriate signal processing, the technique based on nonlinear sound backscattering may became a new useful method for short-range probing in engineering seismics. The above estimates of the nonlinearly backscattered signal level may be considered as confirmation of the possibility of backscattering based methods of imaging of the acoustic nonlinear parameter of Earth rocks.

References

I.Yu.Belyaeva, V.Yu. Zaitsev 1993. Acoustics Letters, v.16, No 11, p.239-242.
C.S.Clay, H.Medwin. Acoustical oceanography: Principles and applications. Wiley- Interscience Publ. New York - London- Sidney- Toronto. 1977.
J.Dybedal 1993. TOPAS:Parametric end-fire array used in offshore applications. In: Advances in noninear acoustics (Proc.13th ISNA), World Scientific, p.264-269.
D.M.Donskoy, S.V.Zamolin, L.M. Kustov and A.M.Sutin 1984. Acoustics Letters, v.7, N 9, p.131-134.
Johnson P.A.,Shankland T.J. 1990. Nonlinear elastic wave interaction in rocks.In:Frontiers of nonlinear acoustics (12th ISNA), Elsevier,London, p.553-558.
Kalachev A., Scharonov G., Sokolov A., Sutin A. 1993. Long-range sound propagation from parametric array. In: Advances in noninear acoustics (Proc.13thISNA), World Scientific, p.259-263.
L.M.Kustov, V.E. Nazarov, A.M. Sutin 1986. Akust.Journ., v.32, No 6, p.804-810, (in Russian).
A.V.Nikolaev & I.N. Galkin (eds.) 1987. Problems of nonlinear seismics, Nauka Publ., Moscow (in Russian)
L.A.Ostrovsky, A.M. Sutin 1983. In Ultrasound diagnonstics, Gorky: Inst. Appl. Phys., p.151.(in Russian).
B.M.Sandler, D.A.Selivanovsky, A.Yu.Sokolov 1981. Doklady Acad. Nauk SSSR,v.259, N 6, p.1474-1476 (in Russian).
R.E.Sheriff, E.Geldart 1982.Exploration Seismology, Cambridge University Press, Cambridge, Englang.

Estimation of crumbly rocks nonlinearity on the base of a granular medium model – Application to engineering seismics and seismoprospecting

Analyse des paramètres non-linéaires des roches désagrégeables sur la base d'un modèle de milieu granulaire – Application à la sismique de l'ingénieur et à la prospection sismique

I.Yu. Belyaeva
Institute of Applied Physics, Nizhny Novgorod, Russia

ABSTRACT: Comparative analysis of elastic linear and nonlinear parameters variability depending on material structure and initial stress is fulfilled on the base of a model of granular medium with fluid pore filling. It is demonstrated that the nonlinear parameter may be used in seismo-acoustic probing problems as much more sensitive informative characteristic to compare with conventionally determined linear modules.

RESUME: L'analyse comparative de la variabilité des paramètres, lineaires et non-lineaires dans la dépendance de la structure du matériel et de la pression initiale est accomplie sur la base du model du milieu granulaire avec le remplissage des pores. On montre que un paramètre non lineaire peut être utilisé pour la solution des problèmes du sondage seismique comme une caractéristique plus sensible en comparaison des modules lineaires qu'on utilise habituellement.

1. Introduction.

Conventional seismo-acoustical methods used in exploration seismology are based on measuring of a medium linear seismo acoustic parameters, namely, sound velocity and coefficients of wave reflection and decay in the medium (White 1979, Sheriff 1985). During last years some investigations of nonlinear effects at elastic waves propagation in Earth soils and rocks were performed, the character of such effects being dependent on nonlinear elastic properties of the media (Groshkov 1991, Johnson 1990, Nesterenko 1983, Dunin 1989). Measurements of nonlinear elastic modules of natural rocks and soils (which always contain large amount of defects, discontinuities, etc.) has shown that their values may exceed by two or three orders the ones for similar homogeneous materials (for instance, glass, most metals), while there was not so drastic differences between linear characteristics. Such results persuade one

that the nonlinear parameters are often much more sensitive to the changes of a medium structure and presence of defects than the linear ones. Thus it seems worthwhile to estimate the possibilities of nonlinear parameters use as informative characteristics for creation of new methods of exploration seismology (Belyaeva 1992). In this connection it is important to investigate the origin and magnitude of medium nonlinearity and its connection with the material inner structure not only empirically, but to design some theoretical models of the phenomenon.

This report presents the results of comparative linear (longitudinal sound velocity) and quadratic nonlinear parameter calculation based on the analysis of the model of granular material (Belyaeva 1993) with fluid filling and the known field data (White 1979, Sheriff 1985, Nomokonov 1981). The aim of this consideration is t demonstrate strong variability of nonlinear parameter depending on the geological conditions (the influence of initial precompression, presence of fluid filling) that may be used for creation of nonlinear seismo-acoustical exploration methods.

2.NONLINEAR PROPERTIES OF GRANULAR MEDIA.

Consider the granular medium as a con tinuous one (e.g. them waves or vibrations in the medium are supposed to be long wave or low frequency enough). For the case of random packing of grains the mate rial can be treated as an isotropic one that makes it possible to characterize the longitudinal deformations by scalar stress-strain relation $\sigma=\sigma(\varepsilon)$, which often can be represented by a power expansion:

$$\tilde{\sigma} = \sigma_\varepsilon'(\varepsilon_0)\cdot\tilde{\varepsilon} + \frac{1}{2!}\sigma_{\varepsilon\varepsilon}''(\varepsilon_0)\tilde{\varepsilon}^2 +$$

$$\frac{1}{3!}\sigma'''(\varepsilon_0)\tilde{\varepsilon}^3 + \dots \qquad (1)$$

where $\tilde{\sigma}=\sigma(\varepsilon)-\sigma(\varepsilon_0)$, $\tilde{\varepsilon}=\varepsilon-\varepsilon_0$, ε_0 is initial prestrain. Using the expansion (1) the following linear and nonlinear parameters are usually defined:

$$M=\sigma_\varepsilon'(\varepsilon_0) \qquad (2)$$

- the linear elastic longitudinal modulus, which is related to the value of sound velocity $c_p = [M/\rho]^{1/2}$ (ρ is the medium density), and quadratic and cubic nonlinear parameters:

$$\Gamma^{(2)} = \sigma_{\varepsilon\varepsilon}''(\varepsilon_0)/\sigma_\varepsilon'(\varepsilon_0) \qquad (3)$$
$$\Gamma^{(3)} = \sigma_{\varepsilon\varepsilon\varepsilon}'''(\varepsilon_0)/\sigma_\varepsilon'(\varepsilon_0) \qquad (4)$$

that determine nonlinear vibroacoustic effects in the material.

The medium is supposed to consist of a random packing of spheres of radius R; the grain material is described by density ρ_s, Young modulus E_s (or volume compressibility coefficient K_s), Poisson coeffi-

cient ν_s. The packing is characterized by void content α and mean number of grain contacts $\bar{n}$. For description of the pore fluid filling the following parameters are chosen: fluid compressibility coefficient K_f and fluid density ρ_f.

Using the approach (Belyaeva 1993) based on energy balance equation that equals the work of given external pressure p_{ext} to the energy accumulated at intergrain contacts and in the volume of grains material and pore fluid one can derive stress- strain relation $\sigma=\sigma(\varepsilon)$ for such a grainy medium with fluid pore filling:

$$\sigma_{eff}(\varepsilon) = \frac{K_f}{\alpha + (1-\alpha)K_f/K_s}\varepsilon +$$

$$\frac{\bar{n}(1-\alpha)E_s}{3\pi(1-\nu_s^2)}\varepsilon^{3/2} \qquad (5)$$

Note that in the case of fluid presence the left-hand side of stress-strain rela tion is written (as it is used in such a case (White 1979, Nomokonov 1981) in the form of effective stress equal to the dif ference of external pressure and pressur e of the pore fluid:

$$\sigma_{eff} = p_{ext} - k\cdot p_f \qquad (6)$$

(where $k=1$ in the considered case of to tally non-consolidatedn media). For grainy rocks (which contain some amount of inter- grain concrete like in sandstone) there are empirical data showing that the unloading coefficient k belongs to the range $0.85 \leq k \leq 1$ (Nomokonov 1981) In accordance with definitions (2)-(4) the following expressions for linear and nonlinear parameters are derived from stress- strain relation (5):

$$M = \frac{K_f}{\alpha + (1-\alpha)K_f/K_s}\varepsilon +$$

$$+ \frac{\bar{n}(1-\alpha)E_s}{3\pi(1-\nu_s^2)}\varepsilon^{3/2} \qquad (7)$$

$$\Gamma^{(2)} = \left[\frac{\bar{n}(1-\alpha)E_s}{3\pi(1-\nu_s^2)} - \varepsilon_0^{-1/2}\right]/M \quad (8)$$

$$\Gamma^{(3)} = \left[\frac{n(1-\alpha)E_s}{8\pi(1-\nu_s^2)}\varepsilon_0^{-3/2}\right]/M \quad (9)$$

Point out that the appearance of nonli near terms is not due to material nonli nearity of grain volume or fluid compo nent, but is connected with the presence of highly nonlinear "soft" intergrain con tacts. For a particular case of dry me dium these expressions for nonlinear para meters and the stress-strain relation it self take very simple form (Belyaeva 1993):

$$\sigma(\varepsilon) = \frac{\bar{n}(1-\alpha)E_s}{3\pi(1-\nu_s^2)}\varepsilon^{3/2} \quad (10)$$

$$\Gamma^{(2)} = 1/(2\varepsilon_0), \quad \Gamma^{(3)} = 1/(6\varepsilon_0^2) \quad (11)$$

that is the $\Gamma^{(2)}$, $\Gamma^{(3)}$ values are determi ned only by the initial prestrain ε_0. It is obvious that at small ε_0 (of the order of 10^{-3} and less) the nonlinear parameters of grainy medium reach rather high magni tudes much more than the ones for homoge neous grain substance, the statement being corroborated by laboratory experiments (Belyaeva 1992).

3. COMPARATIVE ESTIMATES OF LINEAR AND NONLINEAR PARAMETER CHANGES.

Consider some examples demonstrating high sensitivity of nonlinear parameter value to the changes of geological conditions. As a first example choose the crumbly sand-like sediments with water layer beginning at depth H_0. The calculations based on the derived formulas (5),(6) were performed under the following model parameters: $\alpha=0.2$, $E_s = 5.3 \cdot 10^{10} \text{N/m}^2$, $\nu=0.2$; $\rho_s=2.65 \cdot 10^3$ kg/m^3; $K_f=2.2 \cdot 10^9 \text{ N/m}^2$, and it was kept in mind that p_{ext} value in (6) is determined by the weight of overla-

ying medium layer. It was shown that the sound velocity variation at the boundary H_0 of water layer is about 50-70%, while the nonli near parameter $\Gamma^{(2)}$ decreases by $4 \div 5$ times (both of the changes are connected with the increase of medium compressibility modulus due to the appearance of fluid filling). With H_0 increase the value of both characteristics changes decreases. Anyway the contrast of nonlinear parameter changes is significantly higher than that of sound velocity, though the difference is not so drastic as in the following example.

Much more abrupt variations of the nonlinear parameter can be observed at the fluid boundary when it is under anomalous pressure (Sheriff 1985). The discovery of such fluid layers covered by nonpenetrable domes is very important e.g. for oil exploration. Consider the case of oil collector similar to the real one in Ker-Girs oil field in Aserbaijan (Dobrovolsky 1991). The oil boundary depth $H_0 \approx 500$m. To model the anomalous liquid pressure at the horizon H_0 suppose that it is determined by the overlaying rock layer of thickness $H_1 \geq H_0$ (that is $H_1 - H_0$ is the effective collector thickness), so in the range $H_0 \leq h \leq H_1$ one can obtain for initial fluid pressure:

$$p_{f0}(h) = (1-\alpha)\rho_s \cdot g \cdot H_1 - \rho_f \cdot g \cdot (H_1 - h) \quad (12)$$

where g is acceleration due to gravity.

To find the initial strain ε_0 the stress-strain equation (5),(6) were numerically solved using expression (12) for $p_{f0}(h)$. The resulting magnitude of ε_0 is rather critical to the choose of unloading coefficient k in expression (6) for σ_{eff}, as existence of anomalous fluid pressure at $k=1$ provides total unloading of the grain "skeleton" ($\varepsilon_0=0$) and consequently leads formally to infinite values of non linear parameters according to (8)--(9). So we chose as more close to real conditions the unloading coef-

161

ficient values $k<1$. Other parameters under the calculation were as follows: intergrain void content in the collector $\alpha=0.38$; liquid compressibility $K_f=1.2\cdot10^9$ N/m^2 and density $\rho=0.85\cdot10^3$kg/m^3(that is typical of oil), the effective collector thickness H - H =50m. The corresponding results demonstrated that at $k=0.88$ the sound velocity changes are hardly seen, and at $k=0.96$ sound velocity variations are about 5%, while the corresponding relative chan ges of nonlinear parameter are by several orders higher. Presence of gas component may be also taken into account, and it can be shown that it leads (even for rather small cubic content about $10^{-3}\div10^{-4}$) to additional nonlinear parameter increase up to dozens of percents without significant influence on the sound velocity.

4.CONCLUSIONS.

The above analysis of elastic properties of fluid-filled grainy media demonstrated that the nonlinear parameters can be rat her useful informative characteristic for exploration seismology problems. Thus the nonlinear parameters profiling, especially by means of remote methods that has been successfully tested in medical and indus trial applications (Sato1990), opens new possibilities in resource prospecting.

REFERENCES

White J.E. 1979. Underground Sound. Application of Seismic Waves. Elsevier, Ams terdam, Oxford, New York.
Sheriff R.E., Geldart L.P. 1985. Exploration seismology, Vols.1,2. Cambridge University Press, Cambridge, London.
Groshkov A.L, Kalimulin R.R and Shalashov G.M., 1991. "Cubic nonlinear effects in seismics". (In Russian). Doklady Akad. Nauk SSSR, 65-68.
Johnson P.A., Shankland T.J. 1990. Noninear elastic wave interaction in rocks. Frontiers of nonlinear acoustics: Proc. 12thISNA, London, p.553-558.
Nesterenko V.F. 1983. Propagation on non linear compression pulse in grainy media. (In Russian).J. Prikl.Mech.Teor.Fiz. N 5, p.136- 148.
Dunin S.Z. 1989. Attenuation of finite amplitude wave in grainy media. (In Russian). Izv. Akad. Nauk SSSR, Fizika Zem li, N 5, p.106-109.
Belyaeva I.Yu., Sutin A.M., Zaitsev V.Yu. 1992. Seismo-acoustic tomography of the Earth soils nonlinear parameter. Res. of Int. Symp. "Frontiers in fundamental seismology". Strasbourg. p.33.
Belyaeva I.Yu., Ostrovsky L.A., Zaitsev V.Yu. 1993. Nonlinear acoustoelastic properties of grainy media. (In Russian). Akust. Jurn. N1.
Seismo-prospecting. (Handbook of Geophysics). 1981.Eds. Gurvich I.I., Nomokonov V.P. M. Nedra.
I.Yu.Belyaeva, L.A. Ostrovsky and E.M.Timanin. 1992. "Experiments on harmonic generation in grainy media". Acoustic Lett., v.15, N11,p.221-224.
Dobrovolsky I.P., Preobrazhensky V.P. 1991. Oscillation of pressure in the oil collector. Izv. Akad. Nauk SSSR, Fizika Zemli, N 6. p.103-106.
Sato T. 1990. Industrial and medical applications of nonlinear acoustics. Frontiers of nonlinear acoustics: Proc. 12th ISNA, London, p.98-112.

On the possibility of seismo-acoustic nonlinear parameter tomography

Sur la possibilité de la tomographie avec un paramètre sismique-acoustique non-linéaire

I.Yu. Belyaeva, A. M. Sutin & V.Yu. Zaitsev
Institute of Applied Physics, Nizhny Novgorod, Russia

ABSTRACT: A phase - impulse scheme of acoustic nonlinear parameter tomography is proposed for seismic applications. Estimations of the scheme feasibility are performed.
RESUME: On propose un schema phase-impulse de tomographie pour des applications seismiques. On fait les estimations d'applicabilité de ce schema.

1. INTRODUCTION

Common methods of active seismo-acoustic Earth sounding are based on measurements of linear acoustic characteristics. Recent years indicated a growing interest to application of non- linear acoustical and vibrational methods in diagnostical problems of different nature. These new parameters may provide useful information additional to the one obtained by conventional linear techniques. There are now a number of experiments and some theoretical models showing the close connection between medium nonlinearity and its inner structure. The measurements of this nonlinearity value can allow one to obtain information on the inner structure (presence of pores, grains, cracks, fluid presence, etc.) of earth rocks that may be utilized e.g. for mineral resources prospecting or detection of pre-earthquake changes of rocks stresses. In nonlinear acoustics it is accepted to characterize medium nonlinearity e.g. by so called quadratic nonlinear parameter defined through the sound speed derivative with reflect to applied stress (Ostrovsky 1992):

$$\Gamma = \frac{1}{\rho V_p^2} \frac{\partial V_p}{\partial \sigma} \qquad (1)$$

Consider some quantitative examples of nonlinear parameter value and its variations in different situations. Data of some field and laboratory experiments performed with samples of rocks, concretes, crumbly soils gave the value of Γ up to $10^3 ... 10^4$ (to compare with $\Gamma - 5..10$ for homogeneous materials such as glass, silicon etc.) (Johnson 1990, Groshkov 1990).

There are also results demonstrating much more high sensitivity of the nonlinear parameter value to the material structure changes in comparison with variations of its linear module, e.g. for polycrystallic metals with different structure the linear sound velocity changes can be about several percents while nonlinear parameter variations appeared to be 10..100 times greater (Nazarov 1987).

Some reasons allow one to consider that the appearance of numerous defects in earth soils rocks caused by high stresses preceding the earthquake can also give rise to dramatic variation (most probably increase) of this "structure originated" nonlinearity, this variation should be much more intensive than the above mentioned variations of linear elastic module. It can be elucidated having as an example the typical dependence of seismic longitudinal velocity V_p on the external pressure (Rikitake T 1976, Scholz 1973)(such dependencies are used in the dilatation model of pre-earthquake rock properties changes). The estimation of nonlinear parameter Γ as defined by (1) using the measured dependence of sound velocity v.s. stress gives the variation of Γ about 10^2 while the sound velocity variation in the same region is about 10^{-1}. Thus the manifestation of this nonlinearity variations induced by pre-earthquake stresses seems can be much more noticeable sign of earthquake coming in comparison with the known indications based on measuring of sound speed variations.

As the nonlinear parameter value can be sufficiently more sensitive to the changes of media structure the development of nonlinear parameter measuring methods seems to be very important in the connection with mineral resources seismo-acoustic prospecting and the detection of regions of pre-earthquake high stresses forecasting. The known methods (Ichida 1973, Sato 1990) of nonlinear parameter measurement based on the registration of combination harmonics of cw signals allow one to obtain the nonlinear parameter value everaged along the path between sound transmitter and receiver.For seismological and seismic applications it is rather important to provide a possibility of spatial resolution of nonlinear parameter distribution. We believe that the method of phase - impulse "nonlinear" tomography can be the one matching the above mentioned requirements for seismic applications.

2.PHASE - IMPULSE SCHEME OF NONLINEAR PARAMETER TOMOGRAPHY

The physical essence of the method is similar to the one discussed for medical and industrial problems, but there are some significant features of its seismological application connected with es-sential pump-wave divergency and, consequently, zero pump pulse square that are considered below. The informative characteristic in the scheme is the probe high-frequency wave phase variation caused by nonlinear sound speed variations under the influence of powerful pulse (usually called the pump wave). If the strain and stress amplitudes in the pump wave equal ε_p and σ_p the corresponding variations of longitudinal wave sound speed can be represented as:

$$\frac{\Delta V_p}{V_p} = \frac{\Gamma \sigma_p}{\rho_0 V_p^2} = \Gamma \varepsilon_p \ll 1 \qquad (2)$$

where Γ is defined in (1), ρ_0 is the unperturbed media density. The most simple expression for phase modulation of the probe wave can be written for the case of smooth enough inhomogeneities when the geometrical approximation is valid for the signal wave field. The relation between phase variation $\Delta\varphi(t)$ of the probe wave propagating along the ray path R_L with the nonlinear parameter distribution $\Gamma(l)$ can be written in the simplest form if the wave fronts of both waves are quasi-plane.

$$\Delta\varphi(t) = \int_{R_L} \frac{\omega \Delta V_p}{V_p^2} \, dl \cong$$

$$= \frac{\omega}{V_p} \int_0^L \Gamma(l)\varepsilon(l,t)dl \qquad (3)$$

where l is the ray coordinate counted along the ray trajectory, $\varepsilon(l,t)$ is the spatial-temporal structure of pump wave strain, L -is the length of the whole path. The equation (3) is the main one connecting probe wave phase variations with the nonlinear parameter distribution and the pump wave characteristics. In general case of $\varepsilon(l,t)$ being an arbitrary (but given) function of its parameters the unknown distribution $\Gamma(l)$ can be reconstructed by a computational tomographic algorithm using set of (3)-type integrals. But the problem solution can be significantly simplified when the pump wave $\varepsilon(l,t)$ is a pulse one counter- propagating the probe wave. In such set the nonlinear interaction takes place along the ray path of probe wave, the interacting region being moved along the trajectory providing the pulse time gating (without special computational processing) of the $\Gamma(l)$ distribution. In this case one can obtain from eq. (4) the following expression for the phase shift (in the receiver location point $l=0$) of the probe wave front that has $l=l_0$ when $t=0$ and $l=l_0 - V_p t$ when $t=(l_0-l)/V_p$:

$$\Delta\varphi\left(\frac{l_0}{V_p}\right) = \frac{\omega}{V_p} \int_0^L \Gamma(l)\ \varepsilon\left(1,\frac{l_0-l}{V_p}\right) dl \qquad (4)$$

Among the merits of the method we can mention that the pulse character of the pump wave is essential not only for the simplification of the spatial resolution process, but also allows to use powerful pulse sources (e.g. of explosive type) that is necessary for the increasing of the scheme working range.

In some cases the integral (4) can be calculated analytically, for example, for short pulses when the nonlinear parameter changes are small within the pulse length and this variations can be approximated by the first item of Taylor expansion. That is one can represent $\Gamma(l + dl)$ in the following form: $\Gamma(l + dl) \approx \Gamma(l) + (l \nabla r) dl$; the initial pulse form $\varepsilon(l=1m,t) = \varepsilon_0 f(t)$, where ε_0 is the pulse amplitude counted to the 1 m and the spatial pulse scale Λ is supposed to be less than the characteristic scale L_Γ of $\Gamma(l)$ variations. Then the expression (5) takes the following form (when $l\gg\Lambda$):

$$\Delta\varphi\left(\frac{l_0}{V_p}\right) = \frac{\varepsilon_0}{l_0}\ \omega\ \Gamma\left(\frac{l_0}{2}\right)Q +$$

$$+ \frac{\varepsilon_0}{l_0}\ \frac{\omega\ V_p}{2}\ \frac{\partial\Gamma}{\partial l}\left(\frac{l_0}{2}\right)\ R \qquad (6)$$

where $Q = \int_0^\tau f(t)dt$ is the pulse square and $R = \int_0^\tau t\ f(t)dt$. In the far zone (the region of spherical divergency) of the pump wave the pulse square $Q=0$ as it is well known, and the first item in (6) vanishes, so the probe wave phase variation is defined by the second item proportional to the derivative $\partial r/\partial l$. If one can have at the disposal a pump source with lengthy project zone where it is possible to create a pulse with $Q\neq0$ the direct measurement of the nonlinear parameter distribution would be realizable. However, having in mind the characteristics of the existing quasi-point pulse sources which emit pulses with $Q=0$ one should fulfill the estimations of the possible phase shift values on the base of the second (gradient) item in (6). If one assume the following explosive pulse parameters: pulse duration about 10^{-2}s., $\varepsilon_p=2\cdot10^3$m (that correspond the TNT source mass 45 kg), and the local value of $\partial r/\partial l \sim 10^2$ (when the r variation of the order 10^3 takes place along the length about 10 m) for the sounding distance $l = 1000$ m, $f=\omega/2\cdot\pi = 200$ Hz, $V_p = 3000$ Hz the value of $\Delta\varphi$ estimated from the second (gradient) item of (6) is equal to $3\cdot10^{-2}$ radn. This value can be reasonably considered as the measurable one (note that the $\Delta\varphi$ decreases with probe range as l^{-1}).

Let's remind that expression (6) was derived in the supposition that the pulse spatial scale is much less in comparison with the one of $r(l)$ variations ($\Lambda \ll L_\Gamma$). However for the achievement of sufficient range of remote sensing it is necessary to increase the pump pulses power that simultaneously leads to the growth of pulse scale Λ. For example an explosive source with TNT mass about 10^2kg emits pulse with characteristic duration about $(1\div3)\cdot10^{-1}$ s. with the spatial length up to $0.5\div1.5$ km. In such case the condition $\Lambda \ll L_\Gamma$ may be broken and even take the inverse

165

form: $\Lambda \gg L_\Gamma$ (that is the inhomogeneity of Γ distribution can be characterized by the value $\Delta\Gamma$ of nonlinear parameter jump having spatial position l). Under this condition the expression (5) for probe wave phase variations may be transformed to the following one:

$$\Delta\varphi\left(\frac{l_0}{V_p}\right) = \frac{\varepsilon_0}{l_0}\,\omega\Delta\Gamma\int_0^\tau f(t)dt +$$

$$+\,\frac{\varepsilon_0}{l_0^2}\,\omega V_p\int_0^\tau \Gamma\left(\frac{l_0 - V_p t}{2}\right) t f(t)dt \qquad (7)$$

where $\tau = (l_0 - 2l)/V_p$. The estimation of the relative shares of the two items in (7) gives the following condition when the first item (originated from the sharp jump of the nonlinear parameter) prevails: $\Delta\Gamma/\Gamma \gg \Lambda/l$. Further we will suppose that this condition is valid. Note also that the temporal dependence of phase variations defined by eq.-(6) (for "short" pulses) corresponds the spatial structure of nonlinear parameter gradient $\partial\Gamma/\partial l$. In the case of "long" pulses (eq.(7)) the $\Delta\varphi(t)$ dependence is proportional to the integral of pulse form function $f(t)$ over time, the moment of the nonlinear response beginning being defined by the distance l of the nonlinear parameter jump from the receiver.

For the estimation of $\Delta\varphi$ value in the case of "long" pulses let us consider the explosive TNT source with mass 100 kg which produces pulses with the duration of $(2\div3)\cdot10^{-2}$ s. and strain amplitude $\varepsilon_0 \sim 10^{-3}$ m ; the value of $\Delta\Gamma$ is supposed to be of the order 10^3. The phase shift value under this conditions is equal $(3\div6)\cdot10^{-2}$ radn. at $l \sim 10^3$. The value is measurable one, so such type pulse scheme of nonlinear parameter tomography seems to be able for work in seismic applications.

REFERENCES

Ostrovsky L.A. 1992. Wave processes in media with strong acoustic nonlinearity.- JASA, vol.90, N 6,p.3332-3338.

Johnson P.,Shancland T. 1990. "Nonlinear elastic wave interaction in rock". Frontiers of nonlinear acoustics.: Proc.12th ISNA, Elsevier, London, p.553-558.

Groshkov A.L., Kalimulin R.R., Shalashov G.M. 1990.Nonlinear inter bore sounding using sound modulation by seismic waves, Doklady AN SSSR, V.313,N 1,P.63-65 (in russian).

Nasarov V.E., Ostrovsky L.A., Soustova I.A., Sutin A.M. 1987. Investigations of anomalous nonlinearity in metals, In: Problems of nonlinear acoustics, Ed. Kedrinskii V.K.,Proc.XI-th ISNA, Novosibirk (in russian).

Rikitake T. 1976. Earthquake prediction. Elsevier Sc. Publ. Comp., Amsterdam-Oxford-New York.

Scholz C.H., Sykes l.R., Aggarval Y.P. 1973. Earthquake prediction: a physical basis, Science,181, 803.

Ichida N., Sato T., and Lin zer M. 1983. Imaging the nonlinear ultrasonic parameter of a' medium.-Ultrasonic Imaging N 5, p.295-299.

Sato T. 1990. Industrial and medical applications of nonlinear acoustics, In: Frontiers of nonlinear acoustics: Proc. 12th ISNA, London, P.98-112.

Geological analysis of dam site by radar tomography and well loggings combination

Analyse géologique de site de barrage par une combinaison des méthodes de tomographie par radar et analyse des écoulements d'eau

Yasuhito Sasaki, Atsuki Fujii & Yasuo Nakamura
Public Works Research Institute, Tsukuba, Japan

ABSTRACT: Radar tomography is conducted in combination with various well loggings such as rock resistivity, water resistivity, neutron porosity, gamma rays, and borehole TV. Electromagnetic constants (dielectric constant and resistivity) are then calculated by tomographic data of cells around boreholes and compared to the logging data.   Properties of core and gallery samples (dielectric constant, resistivity, porosity) are also examined. The investigated area consists of granite and diorite into which porphyrite sheets sporadically intrude. The granite and diorite are highly fractured with faults and joints, and also slightly altered by hydrothermal processes. Compared logging data with tomographic data, low velocity and high attenuation zone corresponded to high porosity zone where highly altered and fractured rock was distributed. Finally, a rock classification profile was drawn up using radar tomography, regarded as more reliable than core investigations only.

RESUME: La tomographie par radar est appliquée en combinaison avec plusieurs méthodes d'analyse des écoulements d'eau, telles que résistivité des roches, résistivité des eaux, porosité neutronique, rayons gamma ou caméra dans les trous forés. Puis les constantes électromagnétiques (constante diélectrique et résistivité) calculées d'après les données tomographiques des zones autour des forages sont comparées avec les résultats d'analyses. Par ailleurs, les propriétés des échantillons liés aux forages et galeries sont examinées, par exemple les constantes diélectriques, résistivité, porosité. La zone étudiée est composée de granite et diorite dans lesquelles quelques couches minces de porphyrite sont observées. Les granite et diorite sont fracturés par de nombreuses failles et joints, de même qu'il sont un peu altérés par les phénomènes hydrothermiques. En comparant les résultats d'analyse avec les données tomographiques, une zone de vitesse lente et de forte atténuation correspond à une zone de forte porosité avec des roches très altérées et fracturées. Enfin, une courbe de clasification des roches a été établie sur la base de la tomographie par radar. Elle peut être considérée comme plus fiable que la précédente uniquement fondée sur les études par carottages.

1 INTRODUCTION

Radar tomography is a preferred method for geological analysis of dam sites. But difficulties in intepreting and verifying observations have prevented higher use of radar tomography. We conducted radar tomography in combination with various well loggings such as rock and water resistivity, neutron porosity, gamma rays and borehole TV, and compared the logging data with tomographic cell data (velocity, attenuation, and so on) around boreholes. We also examined the geological properties of core and outcrop samples, including electromagnetic properties, porosity and density and rock forming mineral content.

2 STUDY AREA

The dam site is located in the Chugoku region of Japan (Figure 1).

The study area is composed of granite and diorite into which porphyrite sheets sporadically intrude. Boring core and bore hole TV showed granite and diorite highly fractured with faults and joints, and also slightly altered by the hydrothermal process (Figure 2).

Figure 1. Study area

3 RADAR TOMOGRAPHY

Radar tomography was conducted between boreholes A and C using RAMAC with 20MHz antennas. Field data was processed by the CG method. Figure 3 shows the results, slowness and attenuation ratio. The results of slowness and attenuation ratio are similar. The main findings were,

(1) The geological structure is layered with dip is about 30 ° E.

(2) Two fault−like structures inclined at 28−48 ° W cut the layered structure with reverse trend.

(3) The shallow area is relatively slow (low speed) with high attenuation ratio.

(4) The porphyrite sheet area is relatively slow.

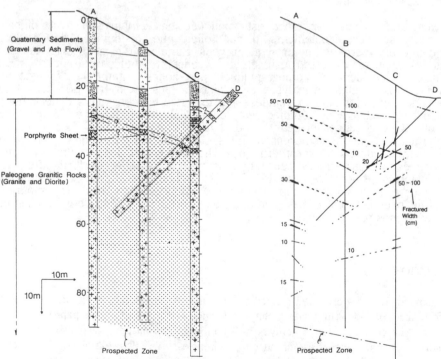

Figure 2. Geological section and the distribution of main fractures in prospected zone

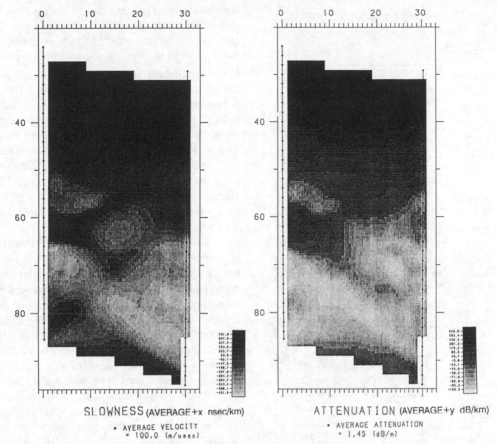

Figure 3. Radar tomography findings
(slowness and attenuation ratio)

4 COMPARISON OF RADAR TOMOGRAPHY, CORE OBSERVATIONS AND WELL LOGGINGS

Figure 4 shows radar tomography (dielectric constant ε and resistivity ρ), core observations, and well loggings around borehole A.

ε and ρ are calculated by the following equations,

$$\varepsilon =(\omega^2 - v^2 \alpha^2)/ \mu v^2 \omega^2 \qquad (1)$$

$$\rho = \mu v/2 \alpha \qquad (2)$$

Where, ω =angular frequency, v=velocity, α =attenuation ratio, μ =magnetic permeability.

Neutron porosity has higher correlation with ε or ρ than at other well loggings. This is consistent with the fact that the dielectric constant of water, at 81, is higher than rocks(4–20),or air(1) by far.

According to core observations and borehole TV, the variation in porosity in the area is due to the reciprocal action of the primary rock type, faulting and alteration.

Primary rock type leads to differences in crack density by faulting. Diorite has more cracks and faults than aplitic granite. This is most likely because the uniaxial strength of aplitic granite is about 1.5 times diorite. Furthermore, as the alterations occur along cracks or faults by the hydrothermal process, diorite becomes more porous.

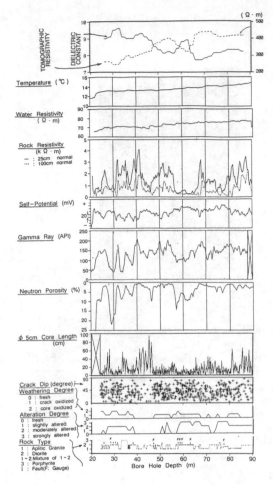

Figure 4. Results of radar tomography, core observations, and well logging around borehole A

5 COMPARISON OF RADAR TOMOGRAPHY AND ELECTROMAGNETIC PROPERTIES OF CORE

As mentioned above, radar tomography is determined primarily by the porosity of the rock mass, and the variation in porosity by the reciprocal action of primary rock type, faulting and alteration.

If pores form within the rock fabric, then strength or permeability would be affected by the properties of the rock itself. But pores form in cracks or faults, the latter will have a

significant effect on the strength and permeability of the rock. Therefore, it is necesary to confirm which pattern is dominant in this field. If the bulk of the pores are formed in the rock fabric, electromagnetic properties of the rock will be similar to rock mass. But if pores are concentrated in cracks or faults reflecting electromagnetic waves in the rock mass, then electromagnetic waves will be lower than as estimated by rock pieces. As a result, electromagnetic properties of the rock mass will have lower values than the rock pieces. The electromagnetic properties (ε and ρ) of the rock pieces are then examined and compared to the rock mass by radar tomography.

Electromagnetic properties are measured by impedance analyzer (YHP HP4194A) with test fixture (HP16451B). Rock samples are taken from borehole A. Fault gauge and inflow clay samples are taken from galleries around the site. ε and ρ are measured at 15MHz, the frequency of radar tomography electromagnetic waves after propagation, and also the highest frequency on the measurement system. Rock samples are saturated with tap water, while fault gauge and inflow clay samples are examined in their natural state of water content.

Figure 5 shows the relationship between ε and ρ for rock pieces at 15MHz. Granite has the lowest ε and highest ρ in all rock types. Diorite and porphyrite show similar values. Fault gauge and inflow clay has very high ε and low ρ . A considerable reflection ratio R of -0.42 for granitc (ε_1 =8) and fault gauge (ε_2 =50) is obtained from the following equation. It means that 42% of electromagnetic waves will be reflected by a thick fault.

$$R=(\sqrt{\varepsilon_1} - \sqrt{\varepsilon_2})/(\sqrt{\varepsilon_1} + \sqrt{\varepsilon_2}) \quad (3)$$

Figure 6 shows the relationships between ε and ρ of the rock mass by radar tomography. Though ε and ρ of the rock pieces in Figure 5 have a broader distribution than rock mass by radar tomography in Figure 6, the patterns are similar. The numerical gap between rock pieces and rock mass in one area is about 30%. Therefore most pores in the rock mass are formed in the fabric, with some exceptions. For example, depth 45m in the borehole A is obviously lower ; ε is higher and ρ is lower than expected by piece. In this area there are no major faults, and also core length is greater than anywhere else. As several open cracks were observed by borehole TV, the loosened area probably affects the electromagnetic properties of the rock mass. Based on core observations, this area is apt to be judged highly due to the long core. However the area loosened by open cracks can be estimated accurately using radar tomography.

Figure 7 shows the relationship between ε and porosity of rock pieces. Though granite and diorite have similar porosity (2–4%), ε

differs markedly, due to the difference rock–forming minerals. Diorite and porphyrite have roughly the same ε and porosity. The porosity of fault gauge and inflow clay ranges from 40 to 60 %.

Figure 8 shows the relationship between ε and neutron porosity of the rock mass. Compared with Figure 7, there is a positive correlation. Diorite has particularly high ε and higher porosity than granite. But the variation in numerical values is narrow and the correlation relatively low because of the exceptional area, high ε and low porosity. This area corresponds to depth 40–45m at borehole A, where open cracks are found.

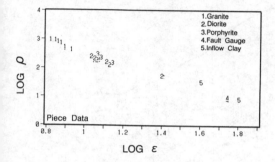

Figure 5. Relationship between ε and ρ of rock pieces at 15MHz by impedance analyzer

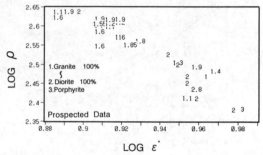

Figure 6. Relationship between ε and ρ of rock mass by radar tomography around borehole A

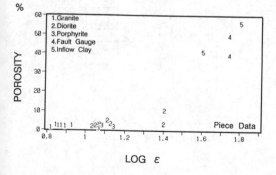

Figure 7. Relationship between ε and porosity of rock pieces

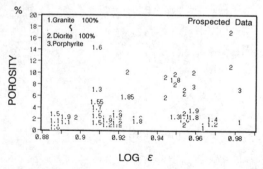

Figure 8. Relationship between ε and neutron porosity of rock mass

6 ROCK CLASSIFICATION BY RADAR TOMOGRAPHY

Radar tomography is strongly affected by porosity and/or rock type in this area. Porosity is closely linked to the extent of rock deterioration by faulting, alteration, or loosening, and correlates negatively with rock classification rank. Our rock classification profile was based on radar tomography.

Figure 9 shows rock classification profiles, the left side based on borehole and outcrop investigations and the right on radar tomography and well loggings such as borehole TV. Here, rock classes are compared with rock classification data at borehole A, class B: $\varepsilon < 8$, CH : $\rho \geq 270 \ \Omega \cdot m$, $\varepsilon \geq 8$, CM : $\rho < 270 \ \Omega \cdot m$, $\varepsilon < 9.5$, CL : $\rho < 270 \ \Omega \cdot m$, $\varepsilon \geq 9.5$.

According to the left profile, rock class should deteriorate along steep faults. But the right profile shows that gentle faults have a greatereffect on rock class.

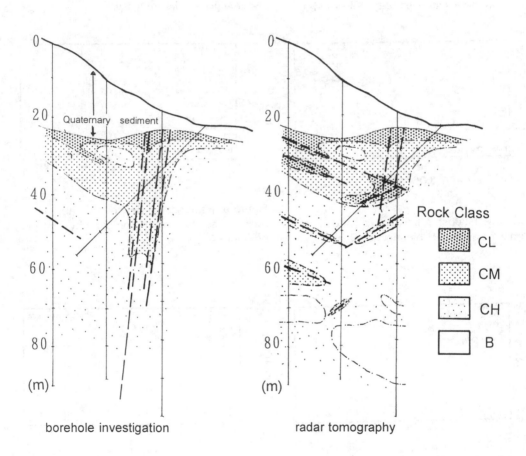

borehole investigation   radar tomography

Figure 9. Rock classification profiles (left: borehole investigations only, right: radar tomography)

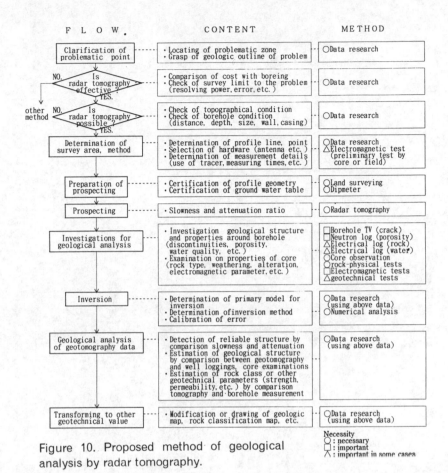

FLOW	CONTENT	METHOD

Figure 10. Proposed method of geological analysis by radar tomography.

7 GEOLOGICAL ANALYSIS BY RADAR TOMOGRAPHY AND WELL LOGGINGS

Figure 10 shows the proposed procedure for geological analysis using radar tomography. The most important difference from normal procedure is the need for geological interpretation of tomographic data. This investigation encompasses core observations, core tests (porosity, ε, ρ, and so on), and well loggings (neutron porosity, borehole TV and so on). In this investigation, it is necessary to clear the relationship between electromagnetic parameter of rock mass and geological properties, such as rock type, faults and cracks, degree of weathering or alteration, porosity and water content. On soft rock masses, this can be achieved via core tests. But hard rock masses require well loggings, since faults or open cracks affect rock mass properties significantly. The properties of these discontinuities can only be assessed accurately through by well loggings.

8 CONCLUSIONS

The comparison of tomographic data and borehole data enables us to transform tomographic data to other data forms used in dam design. At the same time, these procedures also give us information on resolution and observational errors in radar tomography. Geotomography and well logging combinations are likely to become a standard method of dam site investigation in the future.

Geophysical methods for the study of landslides: Some applications

Application des méthodes géophysiques à l'étude des glissements: Quelques exemples

M.L. Rainone & P. Signanini
Department of Territorial Planning, Ancona University, Italy

N. Sciarra
DiSSAR, 'G. D'Annunzio' University, Pescara, Italy

ABSTRACT: Due to economical reasons, geophysical prospectings are used more frequently permitting simpler applications than traditional methods (boreholes, geothecnical tests, etc.). Moreover geophysical methods could provide informations on underground bodies geometry and physical parameters of the investigated soils. In this note we show and discuss some applications of geophysical methods (refraction and reflection seismic profiles, G.P.R.) on landslides in different geological settings. The application and results are explained. In addition, results obtained with the classical P-waves and with Sh-waves high resolution reflection seismic profiles are compared, showing that the latter give a better resolution. Finally we try to define the limits in the application of such methods to the geological problem of interest.

RESUME : Quelques exemples de l'application de la méthodologie géophysique à l'étude des glissements de terrain sont présentés. En particulier l'illustration et la comparaison des résultats obtenus de la prospection sismique (réflection) à haute résolution (avec les ondes P et SH) et la prospection Ground Penetrating Radar utilisés dans diférentes situations géologiques sont réalisés, mettant en évidence les limites et les champ d'applications face aux problematiques géologiques recontrés.

1 INTRODUCTION

Geophysical prospecting are used more frequently for the landslides investigation and, in general, for studies regarding stability slope analysis. These methods (seismic, Ground Penetrating Radar, etc.), indeed, could provide informations on underground bodies geometry, physical parameters of investigated soils, (presence of subsoil discontinuities), etc. Moreover geophysical methods are often used due to economical reasons, permitting simpler applications than traditional methods like boreholes and geothecnical tests.

In the present note some examples of the seismic (rifraction and high resolution reflection profiles) and Ground Penetrating Radar prospecting, applied to study of landslides in different geological setting of instable areas in Italy (fig.1), are shown and discussed.

Fig. 1 Investigated areas: 1) Falconara (AN), 2) Caramanico Terme (PE), 3) Felder (BZ), 4) Sirolo (AN), 5) Grottammare (AP)

used in the esecutions of the seismic and G.P.R. prospectings.

2 APPLIED GEOPHYSICAL METHODS: THEORICAL ASPECTS

In succession we explain the main criteria

2.1 Seismic prospectings

Refraction seismic prospecting, using P

and SH waves, have been carried out for physical and mechanical characterization of the eluvial-colluvial deposits and bed-rock, and to compute the elastic-dynamic modulus of investigated litological units. P-waves and SH-waves high resolution reflection seismic prospecting, have been used to investigate the litostratigraphic characters and the geometry of landslides.

The use of transversal horizontal polarized waves (SH) for high resolution land seismic survey, begun to circulate in the last decade (Mc Cornack et alii, 1984; Stumpel et alii, 1984; Palestini et alii, 1988; Gasperini et alii, 1990), is preferable to the use of P-waves for several reasons and in particular:
- contrary to P and SV waves, SH-waves generate only SH;
- at the same frequency generated by the seismic source, S-waves, because of their lower speed, generally possess a smaller wavelength than P-waves and therefore have a higer degree of definition in thin layers;
- in a vertically anisotropic medium, the propagation problem can be exatly solved only for SH-waves (Blair & Korringa,1987).

For the seismic surveys illustrated in this paper (excepted seismic prospectings carried out Caramanico landslide), we used an EG & G 2401 seismograph with I.F.P. amplifiers, 15 bit A/D converter, series of 12 vertical component geophones for P waves, and of 12 horizontal geophones for SH waves (both series with 10-14 Hz natural frequency range). Moreover the energy system was different for the two types of body waves, an hammer of 10 kgs on a metal slab for the compressional waves, and on the two sides of a squared metal box for the shear waves (Gasperini and Signanini, 1983). The intergeophonic distance was 5-10 metres for refraction profiles and 2 metres for reflection profiles.

2.2 Ground Penetrating Radar prospectings

The G.P.R. (Ground Penetrating Radar) is a geophysical prospecting thechnique that utilizes radio frequency electromagnetic pulses. The measurement system is composed by two antennas (trasmitter and receiver) and a recording equipment (fig. 2).

When the emitted pulse reaches an interface between two media with different electric properties (e.g. resistivity, dielectric constant) part of the energy will be reflected back and caught by the receiver, while the rest will continue to penetrate through the second medium. The equipment will measure the travel-time needed

between transmission and reception of the reflected wave. The measurements of the received and recorded pulses permit to obtain a continuous signal series that can be observed on a screen or a film like that of a single-channel reflection seismic survey. In this way it is possible to discern the variation of the conductivity and the dielectric constant. Moreover, if we know the propagation velocity of an emitted pulse through the investigated medium, we can transform the travel-time into the depth of the reflecting layer.

The velocity of single layer can be obtained utilizing the WARR survey method (Wide Angle Reflection Refraction): one of the two antennas, for instance the transmitter, is kept in the same position while the receiver is moved on the profile.

The penetration depth capacity is negatively influenced by the following two parameters: high conductivity of the investigated medium (e.g. wet clay), high frequency of the electromagnetic emitted waves.

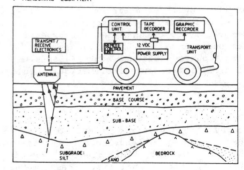

Fig.2 Radar equipment and the measuring process (from Hanninen, 1992)

3 SOME APPLICATION EXAMPLES

The following examples deal with some landslides and instable areas in Italy (fig. 1) in which seismic surveys or Ground Penetrating Radar prospecting have been applied. These examples are representative of different geological, geomorphological and structural settings.

3.1 An instable area in Falconara (Marche, Italy)

The main geological and geomorphological elements of the investigated area, situated between the city of Falconara and the city of Ancona, on the Adriatic sea, are

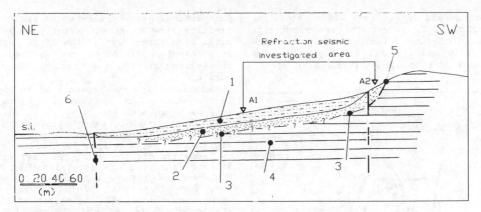

Fig.3 Morphological and structural sketch of Falconara slope: 1) eluvio-colluvial deposits, 2) marly clays with arenaceous layers, 3) alterated zone of substratum, 4) marly clays of substratum , 5) landslide scarp edge, 6) ipothetical fault

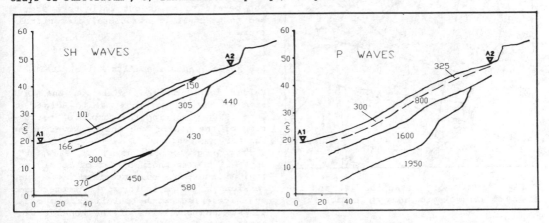

Fig.4 Seismic profiles obtained using SH and P waves along Falconara slope (velocities in m/sec)

recorded in fig 3. The stratigraphic succession is of the plio-pleistocenic period, and is represented by stratified marly clays and arenaceous sandy beds overlain by eluvio-colluvial deposits.

The structural settlement is characterized by a monocline dipping NE (15-20° of inclination) affected by faults with appenninic and transversal direction.

The slope (average gradient 15%), is affected by processes and forms linked to the action of the gravity. In particular it is possible to recognize several superficial mass movements in the whole area, interesting the eluvio-colluvial deposits and, probably, the bedrock too.

A programme of refraction seismic profiles, using P and SH waves, have been carried out to investigate the physical and mechanical properties of the lithotypes and to try to define the depth of

the mass movements. In spite of the very slight contrast of pyhysical properties of the investigated lithological units, seismic refraction surveys have given a good characterization of the superficial deposits and of the bedrock. The seismic profiles obtained using SH-waves (fig.4) show a better resolution than P-waves profiles; both, however, are too little resolutive to define the landslide geometry and the depth of the movements, in particular, when affect the bedrock.

3.2 Caramanico landslide (Central Italy)

In October 1989, after a period of intense rainfall, a landslide disrupted over 700 m of an important road at the southern periphery of Caramanico Terme. This event represents a remobilization of the deposits of an old and much larger slide (Buccolini et alii, 1991, 1992). The rocks outcropping in the area include Early Pliocene age marly clays and overlying calcareous breccias of probable Villafranchian age.

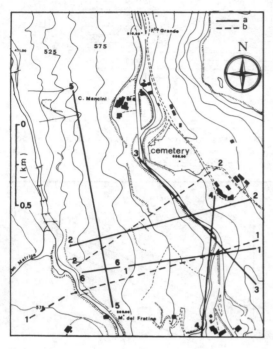

Fig.5 Location of refraction (a) and reflection (b) seismic surveys in Caramanico

The preliminary results of the geotechnical investigations show the rather complex nature of the fenomenon due to the heterogeneity of the materials involved (megabreccia blocks, calcareous and clayey debris mixed together, and marly clays), and the particular structural setting of the area related to the previous large scale gravitational movements.

A geophysical survey (fig.5), with refraction and reflection seismic methods using P-waves, has been carried out to drow further information about:
-the location of the bed-rock top;
-a better settlement of geometry and volume of the slide down bodies;
-the determination of the thickness of the potentially running ground because of the peculiar lithological and structural characteristics;
-a better reconstruction of deep lithostratigraphic ratios;
-to verify the possible existence of tectonic discontinuities (conjectural fault).

The seismic profiles using refraction prospectings are represented in fig.6. The profiles show a seismic stratigraphy composed with three layers: the first two are respectively attributed to aerated and sub-aerated ground, the third to the substrate.

Unfortunately, the presence of an important discontinuity has not been located, like we hopened to try.

The seismic lines using high resolution reflection prospecting are represented in fig.7. The proposed line is located quasi in superimposition with the refraction seismic line. For the acquisition an EG & G Geometrics seismograph and 24 geophons (one for trace) with 50 Hz natural frequency were used. The intergeophonic distance was 5 metres and the samples-rate/trace was 0.5 msec.

The analysis of the seismic sections resulted quite difficult, because these do not permit to point out reflectors ascribed to tectonic or structural discontinuities.

3.3 Felder landslide (Alto Adige, Italy)

Felder landslide took place on the left side of the Aurina valley (Alto Adige) on June 1990, producing the complete obstruction of the bed of the Aurino River along about 50 metres.

From a geological point of view, the area, at the southern border of Tauri tectonic window, is characterized by a succession of metamorphic rocks represented by calceschists with ophiolites and micaschists and phyllites of the Pennidi tectonic unit (Lias-Lower Cretaceous ?) overlain by alluvial-glacial clastic deposits outcropping in some strip of river-glacial terraces. The metamorphic rocks show a high level of fractures due to tectonics strains and to a strong schistosity.

The landslide has concerned both the bedrock and the overlaing clastic deposits and can be classified as a multiple retrogressive slides (Skempton & Hutchinson, 1969). Refraction and high resolution reflection seismic profiles, using P and SH waves (fig.8), have been carried out for stratigraphic and geotechnical characterization of the area, to define the depth and the geometry of the landslide and, moreover, to obtain a good definition of the mottled bodies. Fig. 9 shows a geomorphological section along a transversal profile of the investigated area. The seismic prospections were imposed along this transversal profile to verify the geomorphological evidence.

Fig.10 shows a stack seismic section obtained with P-waves; fig.11 shows the same profile obtained with SH-waves.

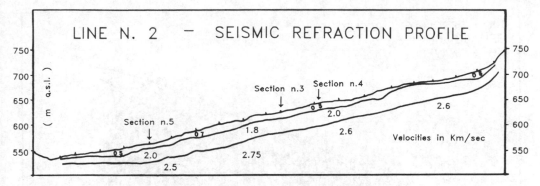

Fig.6 Caramanico: seismic profile using **refraction methods** (line n.2)

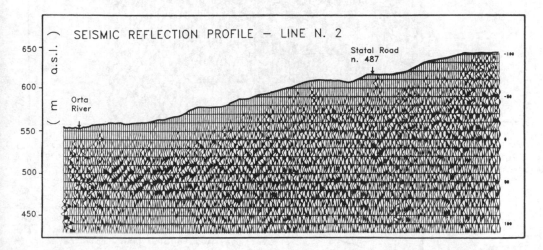

Fig.7 Caramanico: seismic profile using **reflection methods** (line n.2). No reflector is individuable

On the vertical the two-way time scale in msec and, on the horizontal, the space in metres are represented.

Confronting the seismic sections with final reflection in SH and P waves, we can clearly note a higher resolution obtained with the transversal waves in respect to the section obtained with the longitudinal ones.

In the P-waves section of fig.10, three strong reflectors are individuable. As seen, the deepest reflector is reconducible to the substratum not involved in the landslide. Regarding the structure we can note, rather clearly, only four vertical discontinuities, probably related to shear failures or scarps of the landslide.

In the SH-waves section, reported in fig. 8, the resulting situation is much more articulated and detailed. Infact, the number of individuated reflectors is superior as the vertical discontinuities,

which renders more complex the profile's reading and interpretation.

In any case the system of the individual fractures on the seismic profile is in very good agreement with the geological and geomorphological survey. The steps relative to the detachment niches inside the landslide body form an excellent coincidence with the surveyed discontinuities of the seismic section in SH reflection.

Some of these fractures show a downhill immersion, therefore they probably represent some of the rotational sliding surface. These are secondary in respect to the landslide true movement. The SH seismic section evidences further discontinuities which were not surveyed in the geomorphological investigation. The detail relative to this structure is, without question, inferior in the P waves section, while the probable sliding surface appears more distinct in it.

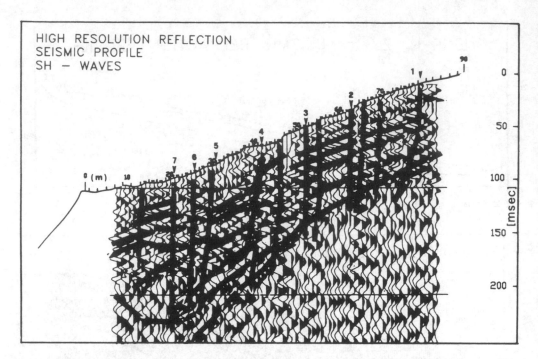

Fig.8 Felder landslide: High resolution reflection seismic profile using SH-waves; it is possible to note some interpretations of vertical and lateral discontinuities: 1-7) see simbology of fig.9

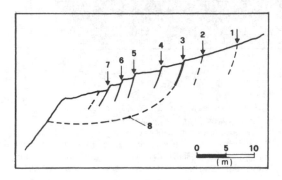

Fig.9 : Geomorphological sketch of Felder landslide: 1-2) probable landslide scarp edges, 3) main landslide scarp edge, 4-7) secondary landslide scarp edges 8) hipothetical rupture surface

3.4 Villages of Sirolo and Grottammare (Marche, Italy)

Sirolo and Grottammare are two interesting medieval villages situated on the Adriatic coast that are characterized by complex landslide phenomena affecting numerous buildings and infrastructures (Dramis et alii, 1988; Monna et alii, 1989).

The geological setting of Sirolo is represented by calcareous marls, sometimes with sandy levels, generally in thick layers or massive beds and, more rarely, in medium-thin layers (Schlier Formation, Lower Miocene) affected by intense bedding cleavage. From a structural point of view the urban area is characterized by a monocline dipping SE and with an inclination of the layers ranges from 20° to 30°. This monocline is affected by faults and joint systems, generally subverticals, trending E-W, NE-SW. The fig.12 shows a geological section of the urban area. The village of Grottammare is located on plio-pleistocenic regressive deposits represented by marine conglomerates overlaing sandy-silty clays; these lithological units are affected by a system of joints. The structural setting is characterized by a monocline gently dipping to NE.

Geological and geomorphological surveys and a preliminary analysis of surface effects on the constructions have shown that the damages are in correlation with the joints system.

Geophysical prospecting, using Ground Penetrating Radar method, has been carried out in both villages to investigate the structural discontinuities of the bedrock.

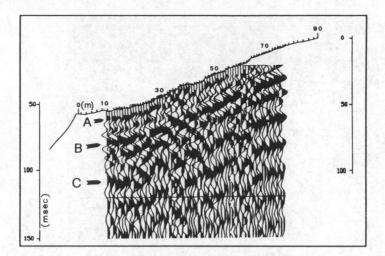

Fig.10 Felder landslide: stack seismic section obtained with P-waves: A-B-C) pointed out strong reflectors

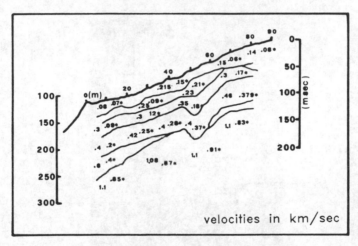

velocities in km/sec

Fig.11 Felder landslide: same section of fig.10 but obtained with seismic refraction method with SH-waves. The values of velocity with a star have been obtained with seismic reflection method (velocity analysis spectra and panels methods)

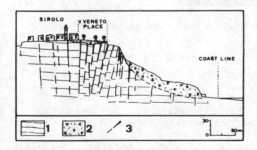

Fig.12 Sirolo: geological section, 1) Schlier formation, 2) landslide scree tongue, 3) joint

Figg.13 and 14 show the obtained results. It is possible to recognize several joints strongly interesting the bedrock.

The instability phenomena in Sirolo and in Grottammare are connected to differential subsidence and rotation of bedrock blocks.

4 CONCLUSIONS

The obtained results by the geophysical prospectings for investigation of instable areas or landslides confirm, in general, the applicability of these methodologies

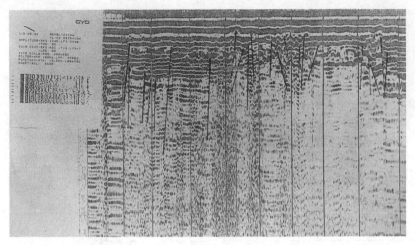

Fig.13 Grottammare: G.P.R. profile

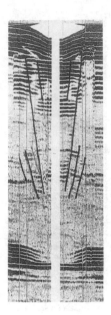

Fig.14 Sirolo: G.P.R. profile executed in the ground to study the presence of some structural fractures noted on a house

in different geological setting.

Refraction seismic survey, using P-waves and SH-waves, if correctly applied, could provide informations on physical and mechanical properties of investigated soils; meanwhile the limits of this method (mute layers, "disparition d'un terrain mince";

Astier, 1971) could give not good results, in particular, if the slip surface is located in the bedrock.

The utilization of high resolution reflection surveys offers a better detail, in particular concerning SH-waves, for their short lenght, compared with P-waves under the same frequency.

We suggest, in these kinds of research, to use an intergeophonic distance not more 2 metres to avoid the problems verified in Caramanico. Furthermore, we suggest to operate in split configuration central shot to obtain the minimum distance shot-last geophon.

About G.P.R. prospectings, we have noted that with these methods is possible to get to the best result to detect lateral discontinuities in solid rocks, in the same time we have found problems to determinate the geometry of the landslides.

REFERENCES

Astier J.L. 1971. Geophysique Appliquè a l'Hydrogeologie. Masson & C. Editeurs. Paris.

Buccolini M., Fiorillo F., Lollino G., Sciarra N. & Wasowski J. 1991. La frana di Caramanico Terme: risultati preliminari dell'indagine geologico-tecnica. Atti I Conv. Naz. dei Giov. Ric. in Geol. Appl., Gargnano (BS)

Buccolini M., Crescenti U. & Sciarra N. 1992. La frana di Caramanico dell'ottobre 1989: nota preliminare. Boll. Soc. Geol. It., 111, 181-191, 12 ff.

Dramis F., Garzonio C.A., Leoperdi S., Nanni T., Pontoni F. & Rainone M. 1988. Damage due to landslide in the anciet village of Sirolo (Marche, Italy): Preliminary analysis of risk mitigation on the historical site. Engineering Geology of Anciet Works, Monuments and Historical Sites, Marinos & Koukis, Balkema, Rotterdam

Gasperini M., Ligi M., Migliarini S., Signanini P. & Tombolini F. 1990. Esempio di prospezione sismica con onde trasversali polarizzate orizzontalmente. Atti IX Conv. G.N.G.T.S., Roma.

Gasperini M., Monna D., Pellegrini F., Piazza F., Rainone M.L. & Signanini P. 1993. G.P.R. data collected and processed as multichannel reflection profiles. XVIII General Assembly E.G.S., Poster Session. Wiesbaden 3/7 maggio 1993.

Hänninen P. 1992. Ground Penetrating Radar. Geophysical Research Methods. The Finnish Geotechnical Society, Tampere

Mc Cornack M.D., Dumbar J.A. & Sharp W.W. 1984. A case study of stratigraphic interpretation using shear and compressional data. Geophysics, 49, 5, 509-520

Monna D., Signanini P. & Tombolini F. 1989. Sull'applicazione di alcune metodologie geofisiche alle problematiche della stabilità dei centri abitati. Atti del Convegno "Studio Centri Abitati Instabili", 197-203, Portonovo di Ancona (Italy)

Palestini R., Signanini P. & Tombolini F. 1988. Esempio di prospezione sismica a riflessione in onde di taglio a ridotta profondità. Atti VII Conv. G.N.G.T.S. Roma.

Skempton A.W. & Hutchinson J. 1969. Stability of natural slopes and embankement foundation. Proc. VII ICSMFE, Mexico, State of the Art Volume, 291-340.

Stùmpel H., Kàhler S. & Meissner R. 1984. The use of seismic shear waves and compressional waves for lithological problems of shallows sediments. Geophysical Prospecting, 32, 662-675

A sensitivity study of subsurface-to-surface resistivity methods (Ω-methods) for monitoring groundwater contamination

Études de la sensibilité des méthodes de résistivité souterrain-surface (Ω-méthodes) pour délimiter la contamination des eaux souterraines

Satyendra Narayan & Maurice B. Dusseault
Department of Earth Sciences, University of Waterloo, Ont., Canada

ABSTRACT: Resistivity (Ω) measurements are most often used to delineate and monitor groundwater contamination because liquid contaminants usually have either lower resistivity (inorganic phases) or higher resistivity (for organic contaminations) than the surrounding groundwater. Although Ω methods are widely used, few parametric studies of sensitivity and measurability exist, although these studies can be instrumental in design and interpretation. Based upon such studies, we introduce a subsurface-to-surface Ω-method using a fixed borehole transmitter and surface receiver electrodes at the surface. A theoretical approach using adjoint solutions and reciprocity is developed to study the sensitivity of surface-to-surface Ω-methods. Several numerical model results are used to verify the sensitivity of the method, and these results demonstrate unequivocally that sensitivity is related to the amount of power dissipated across the region of interest. This study also demonstrates that subsurface-to-surface Ω-methods using the approach discussed here may give a better approach to long-term monitoring of migrating subsurface contamination.

RESUME: Si les éléments salins ou organiques dans les nappes phréatiques génèrent des propriétés électriques qui contrastent avec les couches avoisinantes, les mesures de résistance peuvent servir à délimiter les zones de contamination souterraines. Des études numériques sont nécessaires pour estimer la sensibilité et la détectabilité des cas typiques. Néanmoins, il n'existe que de rares études paramétriques dans la littérature. Basé sur ce genre d'études, nous suggérons une disposition d'émetteurs (dipôles) souterrains, couplés avec des récepteurs (dipôles ou pôles) placés en surface. Des études théoriques démontrent la supériorité de cette méthode, et sa sensibilité est démontrée sur des cas synthétiques. Entre autre, notons que la puissance des signaux est fonction de l'intensité électrique proche de la zone contaminée. Avec les émetteurs souterrains, les variations du champ électrique sont amplifiées à la surface par un ordre de grandeur. Ceci permet un suivi continu et détaillé de la pollution des nappes phréatiques.

1 INTRODUCTION

The integrity of landfills, mine tailings ponds, impoundment structures, slurry cement walls, sheet pile walls, and so on, must be monitored; it cannot be predicted nor guaranteed with existing technology. Other cases where subsurface mass transport is of environmental interest include saline plume tracking under waste halite embankments (potash tails in New Brunswick and Saskatchewan); thermal plume migration from hot fluid injection (steam injection clean-up) or enhanced oil recovery processes; groundwater migration under transportation ministry salt stockpiles; and so on. Providing that a sufficient resistivity (Ω) contrast is generated, these can be monitored using electrical techniques with a permanent electrode array. As new mathematical tools and better strategies evolve, Ω-methods become more economical and useful for tracking pollution migration problems, monitoring structure integrity, and evaluating clean-up effectiveness.

Electrical resistivity has been used as a petro-physical property to predict lithology and nature of pore fluid. Data are obtained from logs, surface electromagnetic data induced by plane remote-source waves, surface magnetotelluric response to a varying dipole excitation, or electromagnetic tomography methods. Surface methods generally lack penetration depth and resolution, and inhomogeneities or highly conductive or resistive overburden distort and obscure responses from deeper structures.

However, resolution can be improved with borehole or borehole-to-surface measurements, and penetration depth enhanced with dc (low) frequencies or current step functions. Furthermore, Δ-monitoring, where time differences in response ($\Delta V/\Delta t$) are analyzed in terms of resistivity changes ($\Delta\Omega/\Delta t$), reduces inhomogeneity effects and problem size, giving higher resolution of plumes or anomalies. In sequential Δ-monitoring, a complete inversion is not required at each analysis step; an efficient Dynamic-Adaptive Inversion Algorithm (DIA) can be developed to give rapid solution. We have conducted mathematical investigations based on reciprocity principles and perturbation analysis to study Ω-method sensitivity, and in this article, using a number of numerical results, we will explore the sensitivity of borehole-to-surface Ω-methods

2 ENVIRONMENTAL MONITORING

Various geophysical methods are used in environmental problems; a brief review is presented.

Because surface waters are of low salinity, inorganic contaminants generally have a lower Ω, and most organic contaminants lead to a higher Ω. Thus, Ω-methods may be used environmentally when targets have a lower or higher Ω compared to the host strata. Applications abound: Cartwright and McComas (1968) mapped leachate zones at sanitary landfills, Hackbarth (1971) monitored spent sulphite liquor dispersal from a pit, Plivas and Wong (1975) detected salt storage depot leakage into a sandy aquifer, and Rogers and Kean (1980) mapped leachate issuing from a fly-ash disposal site. On the development side, Cartwright and Shermen (1972) described ambiguities associated with classical Ω-methods for plume detection, and Merkel (1972) introduced borehole measurements for accuracy improvement. Nevertheless, Warner, 1969; Klefstad, 1973; Urish, 1983; Greenhouse et al., 1985; Mazac et al., 1990; Frantisek et al., 1990; and others have demonstrated the usefulness of Ω-techniques using conventional electrode arrays.

Monitoring methods can be classified into two broad groups: absolute measurements, and relative temporal measurements (Δt changes only). Absolute Ω-methods are used for conductivity distributions of geo-environmental targets, usually contaminated aquifers (Warner, 1969; Berk and Yare, 1977; Gilkeson and Cartwright 1983; Mazac et al., 1987). Classical Ω-methods have also been tried in storage impoundment liner integrity assessment (Shultz et al., 1984; Parra, 1988). Using relative measurements ($\Delta\Omega$) to detect small changes over time ($d\Omega/dt$) at a landfill site, Greenhouse and Harris (1983) analyzed effects of formation water and surface layer Ω changes, as well as water table depth changes. Similarly, Greenhouse and Monier-Williams (1985) detected Δ-$d\Omega/dt$ contaminant related anomalies by comparing with absolute Ω measurements. Bevc and Morrison (1991) introduced multi-dimensional Ω-methods for salt water invasion mapping. Kaufman et al. (1981) used permanent electrode arrays to monitor Ω changes in the vadose zone beneath an evaporation pond, as did Van et al. (1991) to detect saline water leaks at the Mohave Generating Station in Laughlin, Nevada.

Based on numerical analysis, scaled models, and field verification, Ω-methods can be shown to be limited by low Ω contrast, by deep small target geometry, by the obscuring effect of shallow layers of low Ω, by extreme heterogeneity, by poor penetration depth, and so on. To overcome these, borehole-to-surface Ω-methods (Asch and Morrison, 1990) and impedance tomographic techniques (Wexler et al., 1984; Wexler and Mandel, 1985; Tamburi et al., 1985; Daily and Ramirez, 1992) have been developed for environmental monitoring. Barker (1990) reported an offset Wenner array for vertical Ω sounding to reduce spurious effects on results from near-surface lateral Ω variability.

Other electrical techniques, particularly electromagnetic methods, have been tried for mapping lateral and vertical Ω distributions in environmental problems. When the region of interest is conductive, inductive electromagnetic (EM) and galvanic methods can be successful. Success with frequency-domain electromagnetic (FEM) methods for mapping is well known (LaBrecque et al., 1984). Goldstein et al. (1990) used low-frequency EM methods for mapping a plume of saline groundwater. Transient electromagnetic methods (TEM) have also been used for geo-environmental problems (Buselli et al., 1990). However, EM instruments are sensitive to natural and cultural noise, making frequency coupling necessary. EM methods are considered to be difficult and sophisticated in interpretation. EM data repeatability is usually poorer than in dc Ω-methods, and are also more severely affected by topography and surface inhomogeneities.

Ground penetrating radar and seismic methods can be used to delineate hazardous waste sites and zones of contamination (Olhoeft et al., 1992). Seismic response is insensitive to ionic properties ($\Delta\Omega$), but pore pressure changes without Ω changes affect wave velocities and attenuation. Radar techniques are depth-limited and most sensitive to saturation effects, but only second-order effects arise from pore fluid Ω changes. These methods are more costly than electrical techniques, are affected by surface inhomogeneities, and shallow case repeatability for Δ-monitoring applications is poor because sensors are not permanently installed, as suggested for the Ω-method. Evidently, EM, seismic, and radar methods are less well suited for continuous environmental monitoring than the Ω-method proposed.

As we will show, resolution can be improved using borehole or borehole-to-surface measurements, and penetration depth enhanced with dc (low) frequencies or current step functions. It is now possible to repeat dc Ω measurements with an accuracy of 0.1% (Morrison and Fernandez, 1986). Furthermore, as mentioned above, Δ-monitoring and development of a dynamic-adaptive inversion algorithm can minimize the inherent imprecision of absolute inversions. Using a permanent array of optimally placed deeply buried electrodes (5-15 m), combined with a step current change as the dc monitoring impulse, will enhance signal quality and facilitate interpretation. No quantitative study of these benefits has been found, but our preliminary work demonstrates that any Ω-mapping or monitoring problem can be more easily resolved with subsurface-to-surface Ω-methods along with Δ-monitoring approaches.

3 PHYSICAL PARAMETERS IN MONITORING

Intrinsic resistivity depends on pore fluid nature and salinity, mineralogy, porosity, pore shape and interconnectivity, and degree of saturation. Extrinsic factors such as temperature, stress, and pore pressure also affect resistivity. The Ω-method exploits electrolytic conduction involving charge carrier concentration and mobility in solution. The conductivity is proportional to charge carrier concentration, the charge in each carrier, and ionic mobility, which is proportional to temperature. Porosity and pore structure do not change as significantly as salinity and saturation in most geo-environmental problems, therefore we focus on salinity and saturations. Using Archie's Law, the effect of these two factors on resistivity ratio can be expressed as:

$$\frac{R_{t2}}{R_{t1}} = \left[\frac{C_1}{C_2}\right]\left[\frac{S_{w1}}{S_{w2}}\right]^n \quad (1)$$

Our preliminary conclusions, based on an in-depth review and new modelling, are that large Ω changes occur in many environmental issues, and these changes can be best delineated using Δ-monitoring and Ω-methods. Improvements are, however, necessary in both the general monitoring approach and in the mathematical analysis.

4 THE Ω-METHOD: THEORY, SENSITIVITY ANALYSES, IMPROVEMENTS

4.1 *Theory*

The equations governing the response to a point current source $I_s(x,y,z)$ are given by:

$$\nabla\phi(x,y,z) = -\rho(x,y,z)\boldsymbol{J}(x,y,z) \quad (2)$$

and

$$\nabla\cdot\boldsymbol{J}(x,y,z) = I_s(x,y,z), \quad (3)$$

where $\phi(x,y,z)$ is the electrical potential, $\boldsymbol{J}(x,y,z)$ is the current density and $\rho(x,y,z)$ is the three-dimensional (3-D) resistivity distribution. Combining these with the concept of reciprocity, perturbation analysis, and introducing the generalized Green's identity, surface potential field changes arising from a block Ω change in discretised space is given by (Narayan, 1990):

$$\delta\phi = \int_{\tau}\delta\rho\,(\boldsymbol{J}\cdot\boldsymbol{J}')\,d\tau. \quad (4)$$

Using (4), the sensitivity of the surface potential field to Ω changes can be expressed by the sum of individual block sensitivities. For a data set, potential field sensitivity over a 3-D discretised body becomes:

$$\delta\phi = \sum_{i=1}^{N}\delta\rho_i\int_o^{\Delta x}\int_o^{\Delta y}\int_o^{\Delta z}(J_xJ_x'+J_yJ_y'+J_zJ_z')_i\,dxdydz \quad (5)$$

Current densities can be obtained for each block by a forward scheme, thus this equation becomes:

$$\delta\phi = \sum_{i=1}^{N}\delta\rho_i\,(J_xJ_x'+J_yJ_y'+J_zJ_z')_i\,\Delta x\Delta y\Delta z. \quad (6)$$

The physical insight derived from this analysis is that surface measurement sensitivity is directly proportional to the amount of power dissipated (current density) around the region of interest.

4.2 *Surface-Surface Ω-Method Resolution*

Using synthetic analysis, a forward algorithm (Madden, 1971). Narayan (1992) showed that conductive anomalies cannot be resolved by conventional surface dipole-dipole Ω-methods when limiting depth-to-width ratios of 3:1 are exceeded with Ω contrasts of 10:1. Narayan (1992) also showed that a change of 15% in the Ω data for different n-spacing of the potential dipole for a 2-D conductive body at 3 units depth is not sufficient to permit inversion in terms of a 2-D structure. An inescapable conclusion is that surface-to-surface Ω-methods lack depth of penetration and resolution for deep structures.

4.3 *Excitation Electrode Placement*

Current density is a function of depth, electrode spacing, current magnitude, and strata Ω properties. To probe a zone of interest at greater depths, larger current and sampling electrode separations are required, and restrictions on power and geometric dispersion place inherent limitations on the Ω method applied uniquely at the surface. We must maximize current dissipated in and around the anomaly at depth to cause perturbations in the electrical field large enough to detect reliably at the surface (Narayan, 1990). Even at moderate depth, it is difficult to have enough current flowing through the plume or leachate body if only a surface configuration is used, but by placing current electrodes at the right depth, sufficient current flux through the Ω-changing zone can be achieved.

4.4 *Sampling Electrode Placement*

Across fixed sampling dipoles, potential field variations ($\{\Delta V\}$) from t_1 to t_2 ($\Delta V]_{t1-t2}$) during monitoring depend on changes in current flux magnitude and location in the zone of interest. Because of geometrical dispersion, it is best to have sampling dipoles near the regions in which Ω changes are concentrated to maximize ΔV across the sampling dipole length ($\partial V/\partial l$). Ideally, one would have deeply buried and densely spaced sampling electrodes, but such a 3-D array is costly; a permanent surface array with only a few deep electrodes will be more economically tractable in general. Note that possible $\partial V/\partial l$ sample points go up rapidly with electrode numbers: three electrodes allow three sampling dipoles; four electrodes allow six dipoles, and N electrodes allow $(N/2)\cdot(N-1)$ sampling dipoles. Thus, adding a few strategically placed electrodes near the affected zone greatly increases resolution by providing many voltage gradient dipole measurements in the process near-field.

5 MODEL RESULTS

Our studies indicate that surface measurement sensitivity is related to the power dissipated in the zone of interest at depth. In other words, to maximize $\Delta\Omega$ detectability, one must maximize the power dissipation in and around the zone. This is

the basis for us to recommend subsurface-to-surface resistivity measurements. In the simplest configuration, a single dipole is placed in a non-metallic strategically located borehole and an optimized surface array is installed using electrodes buried 5-15 m deep to reduce seasonal and anthropogenic effects. (Many enhancements are feasible, including multiple downhole electrodes along the axis of a single hole, a number of holes for transmission and reception , integrating tomographic approaches, and so on.) The transmitting dipole location and spacing with respect to the zone size and proximity controls power dissipation through it. Several two-dimensional (2-D) model results are used to verify the recommendations, and the suggested subsurface-to-surface Δ-monitoring Ω-method for geo-environmental cases is based on these studies.

A numerical network solution (Madden, 1971) is used to compute $\Delta(dV/dl)$ response over the various 2-D Ω-models discussed herein. In all the model results described herein, the reception pole is located on the surface and the transmission dipole is in borehole, as shown in the 2-D cross-section of the model. We start with a 2-D model consisting of a 10 ohm-m body of 1x1 unit size embedded in a 100 ohm-m halfspace at a depth of 5 units (Figure 1a). Note that although this body can not be detected by using transmission and reception dipoles at the surface, it is quite detectable using borehole-to-surface resistivity measurements. An homogeneous as well as a dipole-pole response over the model (Figure 1a) are plotted in Figure 1b. This shows that the sensitivity of surface measurements is enhanced as the transmission dipole is moved closer to the anomalous body. This is because the proximity of the current dipole increases the power loss in the body, giving a potential field perturbation, which in turn leads to a better sensitivity. The horizontal detectability and system sensitivity may be enhanced further by plotting response difference with respect to the baseline measurements. In our case, this is the difference between the homogeneous response and the response with the body, and an example is shown in Figure 1c. From the normalized differenced response over the model, the anomaly peak is clearly on the top of the body. This shows that by plotting the $\Delta(dV/dl)$ field response from one time to another an accurate tracking of the propagation of a zone of different resistance is achieved. Thus, contaminated groundwater plumes possessing Ω-contrast can be followed spatiotemporally.

In the results presented, locating the current dipole close to the body gives an anomaly of about 30 mV/A. A fundamental question arises: can we detect this anomaly on the top of background response? To assess measurability of this anomaly,

Resistivity Model

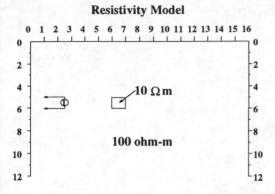

Figure 1a. Cross-section of 2-D resistivity model.

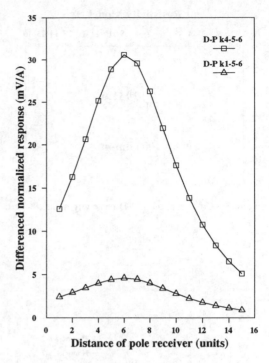

Figure 1c. D-P differenced normalized response for different current dipole locations over the model.

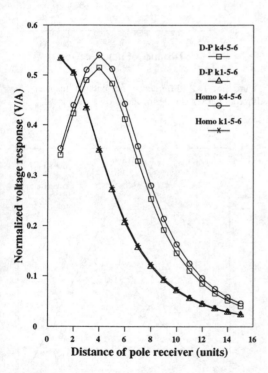

Figure 1b. Dipole-pole (D-P) normalized voltage response over the model shown in Figure 1a. ("k1-5-6" refers to location of current electrodes at depth 5 and 6 units below surface node 1).

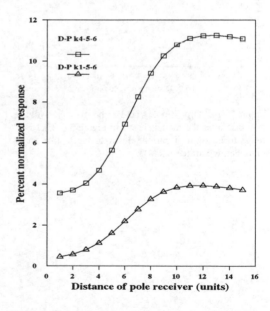

Figure 2. Normalized percent response over the model.

Resistivity Model

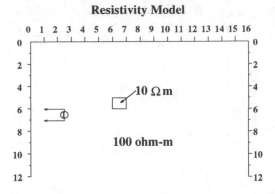

Figure 3a. Cross-section of 2-D resistivity model.

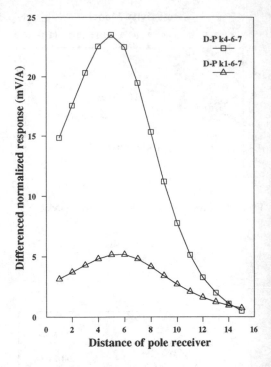

Figure 3c. D-P differenced normalized response for different current dipole locations over the model.

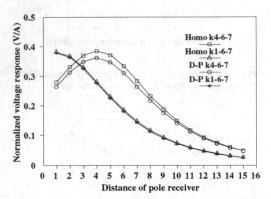

Figure 3b. Dipole-pole (D-P) normalized voltage response over the model shown in Figure 3a. ("k1-6-7" refers to location of current electrodes at depth 6 and 7 units below surface node 1).

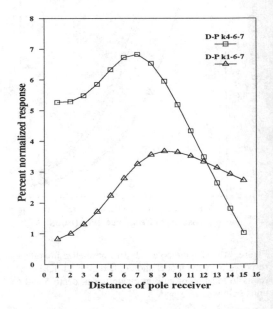

Figure 4. Normalized percent response over the model.

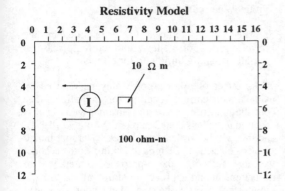

Resistivity Model

Figure 5a. Cross-section of 2-D resistivity model.

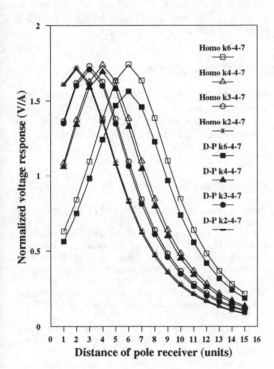

Figure 5b. Dipole-pole (D-P) normalized voltage response over the model shown in Figure 5a. ("k2-4-7" refers to location of current electrodes at depth 4 and 7 units below surface node 2).

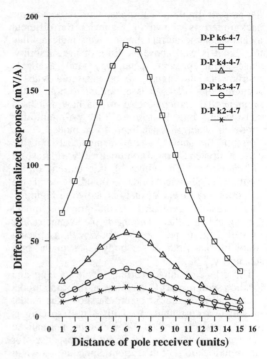

Figure 5c. D-P differenced normalized response for different current locations over the model.

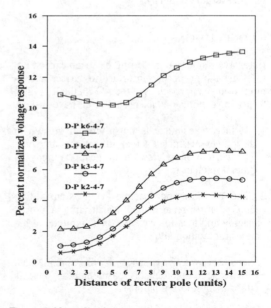

Figure 6. Normalized percent response over the model.

the percent dipole-pole response with respect to the background is shown in Figure 2 for different locations of the current dipole. This indicates that the signal is well above the limit of measurability, since with present technology and available instruments, the signal can be measured with an accuracy of 0.1%). It is important to note at the moment that geologic noise effects have not been considered, but because of the Δ-monitoring approach taken, they are likely to be modest.

Figure 3a shows a similar model except the current dipole is located asymmetrically with respect to the body. The effect of this current dipole location is shown Figures 3b and 3c, and the measurability of the anomaly is shown in Figure 4. These results, compared to the previous model results, indicate that a current dipole symmetrically located with respect to the body with a spacing equal to the width of the body leads to the best sensitivity.

The effect of longer current dipole spacing and location with respect to the body on the same model is studied in Figure 5. Figure 5a shows the model in which a current dipole with 3 unit spacing is moving towards the body. The absolute normalized voltage response and differenced response for different locations of the current dipole are shown in Figures 5b and 5c respectively. The measurability of the anomaly is shown in Figure 6. These results indicate that a current dipole with a bigger spacing located close to the body increases the magnitude of the measurable anomaly.

6 CONCLUSIONS

Extensive numerical modelling has been carried out to assess the proposed subsurface-to-surface and Δ-monitoring approach to the Ω-method. The following conclusions and observations have ensued:

1. Surface measurements sensitivity are related to the power dissipated across the anomalous zone. The current must flow in or near the body in order to detect it.

2. The location and spacing of the transmitting dipole control measurement sensitivity. A proper spacing and location increase the magnitude of the measurable anomaly.

3. Sensitivity may be enhanced by differencing data with respect to baseline measurements. This is vital for monitoring purposes.

4. Differenced response gives good horizontal detectability with the peak response above the zone, a vital observation for successful tracking of

groundwater contamination fronts.

5. The major sources of error are in the measurements of voltage and current, seasonal variations, and telluric current effects.

The effect of telluric noise may be reduced by using a step current function for excitation. The contribution from noise or seasonal variations is reported to be less than 3 percent (Van *et al.*, 1991), and this will be further diminished by proper buried surface electrode design. Taking into account various factors, the proposed method of measurement and approach to temporal analysis (Δ-monitoring) gives voltage gradient anomalies well above the seasonal noise level.

In summary, based on numerical studies, we propose difference-monitoring of geo-environmental Ω-changes with optimally located borehole excitation dipoles, along with a fixed grid of buried potential electrodes at the surface over the zone of interest. This will maximize detectability while reducing noise effects.

7 ACKNOWLEDGEMENTS

We thank the Alberta Oil sands Technology and Research Authority for research support, as well as the Natural Sciences and Engineering Research Council of Canada and the Government of India for scholarship support.

8 REFERENCES

Asch, T. and Morrison, H.F., 1990. Mapping and monitoring electrical resistivity with surface and subsurface electrode arrays. Geophysics 54:235-244.

Barker, R.D., 1990. Improving the quality of resistivity sounding data in landfill studies. In S.H. Ward (ed.), Geotechnical and Environmental Geophysics, v. 2, p. 245-251. SEG, Tulsa, Oklahoma.

Berk, W.J. and Yare, B.S., 1977. An integrated approach to delineating contaminated ground water. Ground Water 15:132-144.

Bevc, D. and Morrison, H.F., 1991. Borehole-to-surface electrical resistivity monitoring of a salt water injection experiment. Geophysics 56:769-777.

Buselli, G., Barber, C., Davis, G.B. and Salama, R.B., 1990. Detection of groundwater contamination near waste disposal sites with transient electromagnetic and electrical methods. In S.H. Ward (ed.), Geotechnical and Environmental Geophysics, v. 2, p. 27-38. SEG, Tulsa, Oklahoma.

Cartwright, K. and McComas, N., 1968. Geophysical surveys in the vicinity of sanitary landfills in northeastern Illinois. Ground Water 6:23-30.

Cartwright, K. and Shermen, F.B., 1972. Electrical earth resistivity surveying in landfill investigations. Proc. Ann. Eng. Geol. Soils Eng. Survey, Moscow, Idaho, 10:77-92.

Daily, W., Ramirez, A., LaBrecque, D., and Nitao, J., 1992. Electrical resistivity tomography of vadose water movement. Water Resources Research 28: 1429-1442.

Frantisek C., Mazac, O., and Venhodova, D., 1990. Determination of the extent of cyanide contamination by surface geoelectrical methods. In S.H. Ward (ed.), Geotechnical and Environmental Geophysics, v. 2, p. 97-99. SEG, Tulsa, Oklahoma.

Gilkeson, R.H. and Cartwright, K., 1983. The application of surface electrical and shallow geothermic methods in monitoring network design. Ground Water Monitoring Review 3:30-42.

Goldstein, N.E., Benson, S.M. and Alumbaugh, D., 1990. Saline groundwater plume mapping with electromagnetics. In S.H. Ward (ed.), Geotechnical and Environmental Geophysics, v. 2, p. 17-25. SEG, Tulsa, Oklahoma.

Greenhouse, J.P. and Harris, R.D., 1983. Migration of contaminants in ground water at a landfill - A case study, 7 DC, VLF, and inductive resistivity surveys. Journal of Hydrology 63:177-197.

Greenhouse, J.P., Faulkner, L.A., Pehme, P.E. and Wong, J., 1985. Geophysical monitoring of an injected contaminant plume: Experiments with a disposable E-log. Proc. NWWA Conf. on Surface and Borehole Geophy. Methods in Ground Water Investigations. Feb. 12-14, Texas, 64-82.

Greenhouse, J.P., and Monier-Williams, M., 1985. Geophysical monitoring of ground water contamination around waste disposal sites. Ground Water Monitoring Review 5:63-69.

Hackbarth, D.A., 1971. Field study of subsurface spent sulfite liquor movement using earth resistivity measurements. Ground Water 9:11-16.

Kaufman, T. Gleason, A., Ellwood, R.B., and Lindsey, G.P., 1981. Ground-water monitoring techniques for arid zone hazardous waste disposal sites. Ground Water Monitoring review 1:47-54.

Klefstad, G.E., 1973. Limitations of the electrical resistivity method for detecting landfill leachate in alluvial deposits. MSc Thesis, Dept. of Geology, Iowa State University, Iowa, Indiana, 66 pp.

LaBrecque, D.J., Weber, D.D., and Evans, R.B., 1984. Comparison of the resistivity and electromagnetic methods over a contaminate plume using numerical modeling. Proc. of the NWWA conference on Surface and Borehole Geophysical Methods. Feb. 1984, San Antonio, Texas, 316-333.

Madden, T.R., 1971. The resolving power of geoelectric measurements for delineating resistive zones within the crust, in the structure and physical properties of the earth crust. J.G. Heacock (Ed.), Am. Geophys. Union, Geophys. Monogr. 14:95-105.

Mazac, O., Benes, L., Landa, I., and Skuthan, B., 1990. Geoelectrical detection of sealing foil quality in light-ash dumps. In S.H. Ward (ed.), Geotechnical and Environmental Geophysics, v. 2, p. 113-119. SEG, Tulsa, Oklahoma.

Mazac, O., Kelly, W.E., and Landa, I., 1987. Surface geoelectrics for groundwater pollution and protection studies. Journal of Hydrology 93:277-294.

Merkel, R.H., 1972. The use of resistivity techniques to delineate acid mine drainage in ground-water. Ground Water 10:5.

Morrison, H.F., and Fernandez, R., 1986. Temporal variations in the electrical resistivity of the earth's crust. J. Geophys. Research. 91(B1): 11618-11628.

Narayan, S., 1990. Two-dimensional resistivity inversion. M.Sc. Thesis. University of California, Riverside, California.

Narayan, S., 1992. Vertical Resolution of two-dimensional dipole-dipole resistivity inversion. Sixty second Annual International Meeting & Exposition of Society of Exploration Geophysicists at New Orleans. October 25-29, p. 431-434.

Olhoeft, G.R., 1992. Geophysical detection of hydrocarbon and organic chemical contamination. Proc. of the Symposium on the Application of Geophysics to Engineering and Environmental Problems 2:587-595.

Parra, J.O., 1988. Electrical response of a leak in a geomembrane liner. Geophysics 53:1445-1452.

Plivas, G. and Wong, J., 1975. Resistivity investigations of the salt-water contamination of a sandy aquifer in the township of Ancaster. In: Case Histories in the Application of Geophysics to Ground Water Problems, Ont. Min. Environ., Toronto, Ont. Water Resour. Paper 5, 89 pp.

Rogers, R.B., and Kean, W.F., 1980. Monitoring ground-water contamination at a fly-ash disposal site using surface electrical resistivity methods. Ground Water 10:472-478.

Shultz, D.W., Duff, B.M., and Peters, W.R., 1984. Electrical resistivity technique to assess the integrity of geomembrane liners. Technical Report of U.S. Environmental Protection Agency, Contract 68-03-30331, SWRI Project 14-6289.

Tamburi, A., Allard, A., and Roeper, U., 1985. Tomographic imaging of ground water pollution plumes. Proc. Second Canadian/American Conference on Hydrogeology, Banff, Alberta, p. 164-169.

Urish, D.W., 1983. The practical application of surface electrical resistivity to detection of ground water pollution. Ground Water 21:144-152.

Van, G.P., Park, S.K., and Hamilton, P., 1991. Monitoring leaks from storage ponds using resistivity methods. Geophysics 56:1267-1270.

Warner, D.L., 1969. Preliminary field studies using earth resistivity measurements for delineating zones of contaminated ground water. Ground Water 7:9-16.

Wexler, A., Fry. B., and Neuman, M.R., 1984. An impedance computed tomography algorithm and system. Applied Optics 24:3985-3992.

Wexler, A., and Mandel, C., 1985. An impedance computed tomogaphy algorithm and system for ground water and hazardous waste imaging. Proceedings of the Second Canadian/American Conference on Hydro-geology, Banff, Alberta, p. 156-161.

One-dimensional geotechnical modelling by seismic methods of Lisbon alluvial basins

Modélisation géotechnique unidimensionnelle de bassins alluvionnaires de Lisbonne par des méthodes sismiques

Rogério Mota
Laboratório Nacional de Engenharia Civil (LNEC), Lisbon, Portugal

ABSTRACT: An experimental seismic microzonation was executed in Lisbon, in 1991. In this paper the results obtained in alluvial basins are compared, by spectral analysis, with those from one-dimensional (1D) nonlinear models. Geotechnical parameters are estimated for some sites in each basin, such as layer thickness, soil type (clay, sand or rock), mass density and shear wave velocity. SHAKE program, with 1969/2/28 Lisbon's earthquake accelerogram, was used to optimize theoretic models. Average relationships between dynamic shear modulus and damping ratios of soils, as functions of shear strain, were considered.

With this analysis it was possible to verify the existence of great soil amplifications at low frequencies depending on alluvium thickness at each site.

RESUMÉ: Cet article présente la comparaison des résultats expérimentaux à partir d'un zonage sismique réalisé en 1991 et des résultats théoriques obtenus par des modèles unidimensionnels non-linéaires de l'analyse spectral. Des paramètres géotechniques sont estimés en plusieurs points de chaque bassin, notamment: épaisseur des couches, nature du sol (argile, sable et roche) poids spécifique et vitesse des ondes sismiques SH. Les résultats ont été exploités à partir du programme SHAKE, en utilisant l'accélérogramme du séisme de Lisbonne daté du 28/2/1969. Cet analyse permet de vérifier l'existence de grandes amplifications du sol, pour des basses fréquences, en tenant compte l'épaisseur alluvionnaire de chaque endroit.

1 INTRODUCTION

Portugal's capital, Lisbon, has suffered since the beginning of times the effects of several earthquakes, usually divided in two categories: interplate (Eurasian and African plates (Fig. 1)) and intraplate source event.

The strongest event which affected Lisbon's region belongs to the first group - the 1755 earthquake (estimated magnitude $M_L = 8.5$); while the 1909 Benavente earthquake ($M_L = 6.7$), is an intraplate event.

The 1755 earthquake caused the death of about 10% of Lisbon's population and destruction or damage in most buildings. However these damages weren't of the same dimension all over town - in the East zone the intensities were IX and X, while in the West the intensity was VIII (Pereira de Sousa, 1928). In the 1909 event, Benavente village (40 km NE of Lisbon) was almost destroyed and in Lisbon a different spatial damage distribution occurred: in the East zone - the most affected - the intensities were VI and VII, and in the West

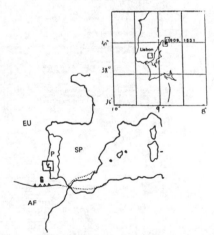

Fig. 1 - Portugal's tectonic emplacement.
Legend: EU - Eurasian plate; AF - African plate; G - Gorringe banc; P - Portugal; SP - Spain.

one the highest intensity was V (Zbyszewski, 1963).

These data lead us to conclude that it isn't the

path or the source, the responsible for these differences, but the different geotechnical site properties. We're then dealing with the so-called Site Effects.

To validate theoretic models of some Lisbon's alluvial basins and of S. Jorge Castle hill produced with a seismic microzonation experiment realized in 1981, and to evaluate the site effects on those places a new field work was performed in October 1991 by the Geophysics Center of Lisbon's University in cooperation with the Laboratoire de Geophysique Interne et Tectonophysique of Grenoble, France. Fourteen portable seismic stations were used to record the seismic impact produced by three explosions made at the Tagus river (Fig. 2). For the results here after presented only those records from shot B were used.

2 SITE EFFECTS

The main site effects are the changes produced in the seismic signals spectrum, namely amplification and resonance in surface layers. Such effects can only be considered site effects if it happens always, whatever the seismic action may be, i.e. if for each site and for different earthquakes, the spectrum and the damage distribution change in the same way.

If the observed effect is common to several sites, it will be difficult to distinguish between source and site effects. The most favourable geologic structures for producing site effects are the alluvial basins and topographical accidents.

The main methodology used to study local effects is spectral analysis of displacements, velocity or acceleration records made at rock and alluvium sites (Fig. 3) during the same seismic action. Those made in outcropping rock site are considered representative of the motion in bedrock beneath alluvial station.

Building damages are mainly produced by horizontal motions - vertical motions amplitudes are 2/3 of horizontal ones, in average. Most methods used to study the effects of local site conditions are based in the assumption of main soil motion being caused by upward propagation of shear waves from underlying rock formations laid in planner, horizontal layers with infinite lateral extension. In the far-field case, source and path effects, specially the last one, are the most important; while in the near-field site effects cannot be neglected. With these assumptions we have in the first layers, a one-dimensional vertical propagation of shear waves due to bedrock vibration.

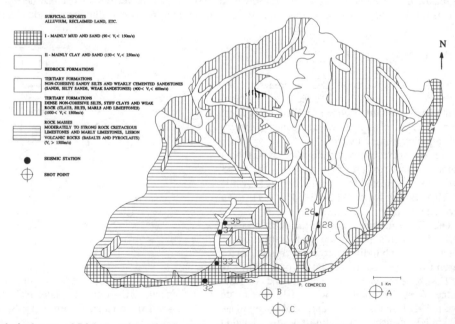

Fig. 2 Geotechnical map of Lisbon, with seismic stations location (adapted from Teves Costa and Mendes Victor, 1992)

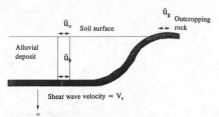

Fig. 3 - Soil and rock configuration.
Legend: $\ddot{u}_o$ - max. acel. in outcropping rock; $\ddot{u}_b$ - max. acel. in the same rock formation but under a soil column, and $\ddot{u}_g$ - max. acel. at surface (adapted from Lysmer et al., 1971).

Analytical methods used to predict the soil response are mainly based in the following points (Aki, 1988; Crespellani and Madiai, 1991; Schnabel et al., 1972):

1. Determination of the characteristics of the motion likely to develop in rock formation underlying the site, and selection of an accelerogram with these features, usually of strong motions.

2. Determination of soil deposit dynamic properties: average relationships for dynamic shear modulus (G) and damping ratios for soils (η), as functions of effective shear strain (γ).

3. Computation of soil deposit response to bedrock motions: a one-dimensional method can be used if the soil structure is essentially horizontal. This analysis is usually based on the solution to the wave equation, if the motion on rock underlying the soil column is known. However the main information available on rock motions comes from surface stations, so motions in the underlying rock will only be the same as in surface if we consider rock as a rigid layer.

Dynamic soil behaviour depends on the intensity of seismic action - since shear strains are a function of seismic action -, and it will be linear and elastic if the produced strains are less than 10^{-4}, which is the value associated with the propagation of seismic waves (Coelho, 1991; Seale and Archuleta, 1991). If this value is overcomed, which happens easily with strong motions, it's necessary to account for the nonlinear characteristics of the soil. Their effects are usually felt in the reduction noted on the maximum amplification and its associated frequency - bigger intensity gives bigger reduction -, this is caused by strong damping produced by the soils histeretic behaviour, and in the attenuation increase with the deformation increase (Bard, 1983; Joyner and Chen, 1975).

3 ALCÂNTARA AND ALMIRANTE REIS AV. BASINS MODELLING

3.1 Preliminary considerations

Having in mind the ideas so far presented, geologic and geotechnical data were collected from different previous works (Mendes Victor, 1987; Teves Costa, 1989), and from the geological map of Lisbon county (Almeida, 1986) (Tables I and II).

Table I. Geotechnical features estimated for Lisbon's geologic formations (Coelho, 1985).

Type of formation	Formation	Mass density (g/cm³)	Shear waves velocity (m/s)
Surficial deposits (Embankment, alluvium, etc.)	Mainly mud and sand	≈ 1.6	90-150
	Mainly clay and sand	≈ 1.8	150-250
Bedrock	Non-cohesive, sandy silts and weakly cemented sandstones (tertiary formations)	≈ 2.0	400-600
	Dense non-cohesive silts, stiff clays and weak rock (tertiary formations)	≈ 2.2	1000-1500
	Moderately to strong rock (Cretaceous limestones and Lisbon volcanic rock)	> 2.4	> 1500

Table II. Geographic parametres and geology in each station.

Station	Epicentral distance (km)	Altitude (m)	Geology
26	3.56	45.0	≈5 m al over M_{III}^2
28	3.16	35.0	M_{III}^2
32	2.15	3.0	≈33 m al over β
33	2.10	5.0	10-20 m al over C_c^3/β
34	2.78	12.5	10-15 m al over C_c^2
35	2.93	30.0	C_c^2

al - alluvium
M_{III}^2 - Entrecampos limestones
β - Lisbon's basaltic complex
C_c - Cenomanien carbonic complex

The strong motion accelerogram used for modelling was recorded in the North pile of Lisbon's Tagus river bridge, during the 1969/2/28 earthquake, with maximum acceleration 0.0261g (25.6 cm/s^2). Corresponding to the E-W component, it was digitized at the LNEC, and interpolated with equal intervals of time (0.02 sec) (Fig. 4).

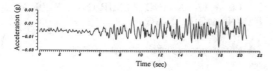

Fig. 4 - Accelerogram recorded in the North pile of Lisbon's Tagus river bridge, in 1969/2/28.

Alcântara basin has the highest alluvium thickness of Lisbon town - over 15 m -, while in Almirante Reis avenue basin thickness is hardly over 5 m. The surficial stratigraphy present in both basins is embankment and alluvium.

The geotechnical parameters that can be used to describe the soils nonlinear behaviour, namely damping and shear modulus, were estimated from standard relationships (Seed and Idriss, 1970 *in* Shannon and Wilson Inc., 1972) - those adopted for clay and sand are presented in Fig. 5-7.

For each basin the data processing started with the shot window identification in each station record, each with 10.24 sec long, where some noise was included, before and after the explosion. Than each station's spectrum was computed and, finally it was established the ratio between each alluvium station spectrum and that of the reference station in

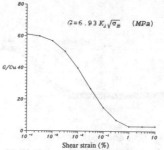

Fig. 5 - Shear modulus used for sand (D_r=75%).

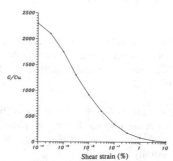

Fig. 6 - Shear modulus used for clay.

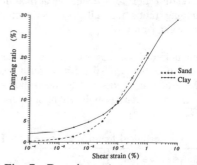

Fig. 7 - Damping ratio.

outcropping rock of the same bedrock formation beneath alluvium stations.

In alluvial deposits the lower frequencies are the most important ones. So, the aim was to reach a model that could adjust both in amplitude and in frequency the first maximum of amplitude the resulting ratio between the transverse components (TR) of Fourier spectrum, having always in mind each station's geology. The probable source of other maximums is bidimensional structures.

This analysis was always done until 8 Hz. Implicit

hypothesis present in the method used by the SHAKE program of upward incidence of shear waves lead to only use of transverse components.

The results produced by the SHAKE program were smoothed with a polynomial function of 10 degrees.

3.2 *Alcântara basin*

In Fig. 8 we can see the velocity records made in Alcântara basin seismic stations - stations 32 (Santo Amaro dock embankment), 33 (Alcântara square), 34 (Ceuta av.) and 35 (Arco do Carvalhão street). A Butterworth filter band-pass (0.2-15 Hz) with 8 poles was applied to these records.

The produced displacements were less or equal to 10^{-4}, so a linear soil behaviour can be considered.

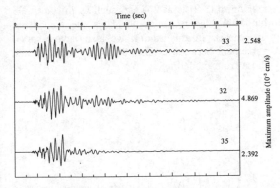

Fig. 8 - Display of velocity records made in stations 32, 33 and 35 - transverse components.

Analysing the velocity records we can see that in soil stations amplitudes are bigger and longer in time than in outcropping rock formation (station 35). In station 33 the signal length of time is approximately 8 sec, while in station 35 it is a little over 4 sec.

Signal length of time is longer in station 33 than in station 32. We would expect the opposite because this station has a bigger alluvium and embankment thickness than that one. A possible explanation to this is the compaction level present at both places. It is possible that the compaction degree produced by the construction of Santo Amaro dock embankment is greater than that one produced with filling of Alcântara streamlet.

In the low frequencies of spectral ratio between transverse components of stations 32 and 33 it can be seen two leading maximums, with amplitudes 6 and 8 at 2.4 and 5.2 Hz, respectively (Fig. 9). In the theoretic results (Fig. 10), we have a first

maximum at 2.4 Hz, with amplitude 4 and a second one at 6 Hz, but with smaller amplitude; the third one is over 8 Hz (9.2 Hz), but we can consider that it matches the maximum at 8.8 Hz in the experimental data.

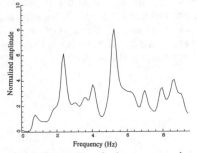

Fig. 9 - Spectral ratio between stations 32 and 35 for transverse components.

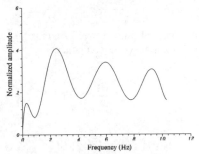

Fig. 10 - Theoretic spectral ratio

In Table III we have the theoretic model that produced these results, where η stands for the critical damping ratio and *Factor* is a multiplier applied to the shear modulus relationships (Fig. 5 and 6). The result for the total thickness of soil deposit is only 10% higher than that estimated from Lisbon's geological map (Table II).

Table III. Station 33, theoretic model parameters.

Soil type	Thikness (m)	η	Mass density (g/cm³)	V_s (m/s)	Factor
Clay	9.1	.05	1.76	230	0.9
Sand	9.1	.10	1.84	116	1.1
Sand	15.2	.10	1.92	137	1.9
Bedrock			2.72	2000	
Total	33.4				

Spectral ratio between stations 33 and 35 (components TR) shows three main maximums with amplitudes 3.0, 2.6 and 2.7, at 2.4, 4.0 and 5.6 Hz, respectively (Fig. 11). Theoretic spectral ratio (Fig.12) has a first maximum with amplitude 2.9 at 2.9 Hz, which matches both in amplitude and in frequency with the first field maximum. With the second one this doesn't happen because thickness of alluvial soil in station 33 is very small, which doesn't allows the use of more layers in the theoretic model, presented in Table IV.

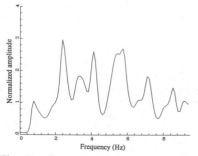

Fig. 11 - Spectral ratio between stations 33 and 35 for transverse components.

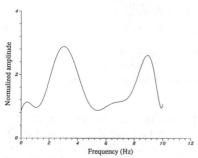

Fig. 12 - Theoretic spectral ratio

Table IV. Station 33, theoretic model parameters.

Soil type	Thickness (m)	η	Mass density (g/cm³)	V_s (m/s)	Factor
Clay	12.2	.05	1.76	213	0.9
Sand	1.5	.10	1.92	107	1.2
Bedrock			2.72	2000	
Total	13.7				

Since station 34 only had vertical component (VR), it was decide to estimate the probable model from those of stations 32 and 33, since all three stations have almost the same epicentral distance, the geology beneath them is the same and they aren't very far apart (Table II).

Analysing the vertical component of the three stations (Fig. 13) we can see that they have almost the same shape. However, spectral amplitude in station 34 is higher than in station 33. In figure 14 we have both stations 32 and 33 spectral ratios where it can be seen that station 32 has a higher amplitude ratio. So the alluvial soil thickness beneath station 34 might be higher, and the amplitude ratio in station 34 will not differ very much from that one in station 33.

In figure 15 we have the theoretic spectral ratio for station 34 and its model in Table V. The first maximum (3.2) is at 2.4 Hz, such as in station 33 but 10% higher, which validates the considered hypothesis. In this station we also have a good agreement between estimated values (Table II) and those from modelling.

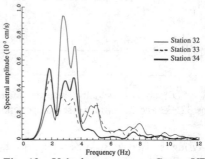

Fig. 13 - Velocity spectrums - Comp. VR

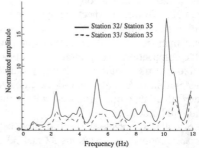

Fig. 14 - Spectral ratios for transverse components.

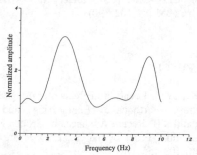

Fig. 15 - Theoretic spectral ratio

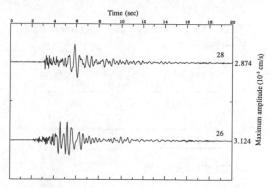

Fig. 16 - Display of velocity records made in stations 26 and 28 - transverse components.

Table V.   Station 34, theoretic model parameters.

Soil type	Thikness (m)	η	Mass density (g/cm³)	V_s (m/s)	Factor
Clay	12.5	.05	1.68	213	0.9
Sand	0.6	.10	1.84	122	1.2
Bedrock			2.72	2000	
Total	13.1				

3.3 *Almirante Reis avenue basin*

For modelling this basin we only had two stations: 26 (António Pedro street) and 28 (Anjos Churc), whose records are plotted in Fig. 16, after being filtered like in Alcântara basin.

The alluvial soil thickness in station 28 is very low (Table II), wich once again doesn't allows the use of more layers in the theoric model.

Spectral ratio between stations 26 and 28 has two main maximums, with amplitudes 3.4 and 3.2, respectively at 3.8 and 6.8 Hz (Fig. 17).

Comparing the theoretic results (Fig. 18) and the field ones we can see that they only adjust in frequency - the maximum for the theoretic spectral ratio appears at 4.2 Hz.

For a better adjust two other velocities were considered for the bedrock formation. The lower values - 1200 and 1500 - have already been used in 1D linear analysis, for the same basin (Teves Costa, 1989) and the higher value was considered from the results obtained with the last field work (MendesVictor, 1987).

The difference between the maximum in the highest velocity spectral ratio (4.9) and the lowest

(4.5) is only 0.4 ($\approx 10\%$), which isn't very significant.

The theoretic model considered for this basin is presented in Table VI (the three models used in this basin were only different in the shear velocity for the bedrock formation).

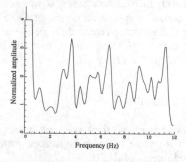

Fig. 17 - Spectral ratio between stations 26 and 28 for transverse components.

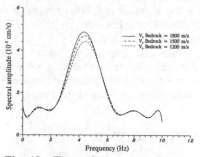

Fig. 18 - Theoretic spectral ratio

Table VI - Station 26, theoretic model parameters.

Soil type	Thickness (m)	η	Mass density (g/cm^3)	V_s (m/s)	Factor
Clay	10.0	.05	1.76	250	0.9
Bedrock			2.72	1500	
Total	10.0				

4 DISCUSSION

The estimation of some essential parameters such as shear velocity and mass density bring on some difficulties for modelling, however the achieved models are representative of the reality.

In shallow sites the results weren't very good because it was very difficult to have more layers with such low values for thickness.

The theoretic models produced allowed to verify the great importance of surficial layer, mainly clay.

Considering the existence of several stations in alluvial soil, in Alcântara basin, it was possible to note a certain amplitude function of the alluvium thickness, since the epicentric distance of these basin stations didn't differ very much, specially between stations 32 and 33 (Table II).

5 CONCLUSIONS

Theoretic values of thickness and mass density obtained for these basins have a good match with estimated ranges (Table I).

In Alcântara basin we have an alluvial deposit with thickness between 13m at the begining and 33m near Tagus river, over Lisbon's volcanic rocks (V_s=2000 m/s). Almirante Reis basin has a thinner alluvial deposit over less cohesive rock - Miocenic limestones (V_s=1200 m/s).

With this one-dimensional linear equivalent method it was possible to produce good results on the deeper sites of Alcântara alluvial basin with good fitness between theoretical and field results.

ACKNOWLEDGEMENTS

The experimental work was a component of research studies funded by Junta Nacional de Investigação Científica e Tecnológica (JNICT project number PMCT/C - N° 87484/SISM).

The author was sponsored by Junta Nacional de Investigação Científica, under the CIENCIA Program (BM N° 1025/90).

REFERENCES

Aki, K., 1988. Local Site Effects on Strong Ground Motion. Earthquake Engineering and Soil Dynamics II. Recent Advances in Ground Motion Evaluation. Proc. of the A.S.C.E. Spec. Conf., Utah, 103-155.

Almeida, F. Moitinho, 1986. Geological map of Lisbon county. Serviços Geológicos de Portugal, Lisboa.

Almeida, Isabel M.B. Moitinho, 1991. *Características Geotécnicas dos Solos de Lisboa.* PhD Thesis, Lisboa. (in portuguese)

Bard, P.-Y., 1983. *Les effets de site d'origine structurale en sismologie. Modélisation et interprétation. Application au risque sismique.* PhD Thesis, Grenoble.

Cabral, J., 1986. Neotectonics and seismicity in Portugal. Proc. 8th European Conf. Earth. Eng., Vol.1, 2.1/15-2.1/22.

Coelho, A.M.L.G., 1985. Acções Desenvolvidas no Âmbito do Programa para a Minimização do Rísco Sísmico na Área de Lisboa. Relatório 65/85. L.N.E.C., Lisboa. (in portuguese)

Coelho, A.M.L.G., 1991. *Microzonamento Sísmico,* Programa de investigação apresentado a concurso para acesso à categoria de Investigador-Coordenador. L.N.E.C., Lisboa. (in portuguese)

Crespellani, T., and C. Madiai, 1991. Seismic Response Analysis, in *Seismic Hazard and Site Effects in the Florence Area.* 10th Euro. Conf. Soil Mec. Found. Eng., Firenze, 81-89.

Fonseca, J.F.D., J.C. Nunes, P.A. Reis and V.S. Moreira, 1990. Seismicity and regional tectonics of the Portuguese Estremadura. Proc. ECE/UN Seminar on Prediction of Earthquakes - Occurence and Ground Motion, Vol.2, 857-875.

Hudson, D.E., 1979. *Reading and Interpreting Strong Motion Accelerograms.* Earth. Eng. Res. Inst., Berkeley, California.

Joyner, W.B., and Albert T.F. Chen, 1975. Calculation of Nonlinear Seismic Response in Earthquakes. *Bull. Seism. Soc. Am.,* **65,** N° 5, 1315-1336.

Lysmer, J., H. Bolton Seed, and P.B. Schnabel, 1971. Influence of Base-Rock Characteristics on Ground Response. *Bull. Seism. Soc. Am.,* **61,** N° 5, 1213-1231.

Mendes Victor, L.A., 1987. Estimation of the seismic impact in a metropolitan area based on hazard analysis and microzonation an example. The town of Lisbon. in Atti del Corso Eur. di Formazione, 183-213.

Mota, R., 1992. *Modelação unidimensional não linear - Aplicação a bacias aluvionares de Lisboa.* Master Thesis. Lisboa. (in portuguese)

Pereira de Sousa, F.L., 1928 - *O Terremoto do 1º de Novembro de 1755 em Portugal e um estudo demográfico. Vol. III - Distrito de Lisboa.* Serviços Geológicos. (in portuguese)

Schnabel, P.B., John Lysmer, and H. Bolton Seed, 1972. *SHAKE - A Computer Program for Earthquake Response Analysis of Horizontally Layered Sites.* Report nº EECR 72-12. Earthquake Engineering Research Center, University of California. Berkeley, California.

Seale, S. H., and R. J. Archuleta, 1989. Site Amplification and Attenuation of Strong Ground Motion. *Bull. Seism. Soc. Am.,* **79**, Nº 5, 1673-1696.

Shannon and Wilson Inc., 1972. *Soil Behavior Under Earthquake Loading Conditions - State of the Art Evaluation of Soil Characteristics for Seismic Response Analyses.* U.S. Atomic Energy Commissions. Los Angeles. 129-159.

Teves Costa, P., 1989. *Radiação Elástica de uma Fonte Sísmica em Meio Estratificado: aplicação à Microzonagem de Lisboa.* PhD Thesis, Lisboa. (in portuguese)

Teves Costa, P. and L.A. Mendes Victor, 1992. Site effects modelling experiment. Proc. 10th W. Conf. Earth. Eng., Madrid. 1081-1084. Rotterdam.

Zbyszewski, G., 1963. *Carta geológica da cidade de Lisboa - Notícia Explicativa da Folha 4 (Lisboa).* Serviços Geológicos de Portugal, Lisboa. (in portuguese)

Application of ground penetrating radar in engineering geology – Detection of cavities

Application du radar de pénétration à la géologie de l'ingénieur – Détection de cavités

Marília P. Oliveira & L. F. Rodrigues
Laboratório Nacional de Engenharia Civil, LNEC, Lisbon, Portugal

ABSTRACT: The use of the Ground Penetrating Radar (GPR) method of survey to solve problems of engineering geology is relatively recent. Nevertheless, it has come to be an useful tool, both in the reconnaissance of shallow geological structures and in the location of buried objects (conduits, containers, cables, etc.). This paper deals with the use of this technique in the engineering geology domain for the detection of cavities. The results presented here demonstrate the great potential of this technique as an instrument in terms of detection of cavities, mainly, in large areas along roads and highways.

RÉSUMÉ: L´application de la méthode de prospection par radar de pénétration "Ground Penetrating Radar" (GPR) à la résolution de problémes en géologie de l´ingénieur est assez récente, nonobstant, cette méthode s´est révélé trés utile soit pour la reconnaissance de structures géologiques à petite profondeur soit pour la détection d´objets enterrés (tubes, conteneurs, câbles, etc). Dans cet article on fait la description de l´application de cette technique dans le domaine de la géologie de l´ingénieur, à la prospection de cavités. Les résultats présentés démontrent le grand potentiel de l´utilisation de cette technique comme un instrument de détection de cavités, spécialement en grands tronçons au long des routes et autoroutes.

1 INTRODUCTION

Since the 1960's, the GPR method of geophysical survey has been developed and applied in different situations (Harrison, 1970; Ulrikson, 1982; Davis and Annan, 1989). Advances made, both in recent equipment and in sophisticated techniques in signal processing have permitted a significant improvement in the resolution of images and have enhanced the usefulness of the method in several domains. Possible applications in engineering range from the identification of geological structures, to the detection and delineation of cavities and buried objects, from the study of integrity and control of materials to the detection of leaks of toxic liquids.

In civil engineering and under certain circumstances, the existence of subterranean cavities, natural or even artificial, can be a source of many problems and lead to geotechnical accidents in engineering works, either before or after construction. These accidents are usually associated with the phenomena of sinking or settling, fracture and collapse of land. These types of phenomena are often observed in road and highway pavements and in the foundations of diverse structures that are located in karst zones or over fills subject to internal erosion or piping. Due to the fact that cavities can constitute major breaks in the interior of rock masses or in the interior of earth/rock fills, it becomes essential in engineering geology to locate and characterize them. In this paper we present two situations in engineering geology that are subject to this kind of problems, in which Ground Penetrating Radar is used to detect cavities.

Case history I - Along a section of highway under construction, approximately 20 Km long, located in a region of limestone formations from the Middle Jurassic period, where it is known to be some areas of marked karst formation. The objective in exploring this section was to identify anomalous zones with respect to the possible presence of natural cavities under the highway platform.

Case history II - Along a section of road approximately 3 Km long, located on a rock/earth fill on the back of a retaining wall, constructed to protect the fill of the sea action and where settling of the pavement has been verified in some areas due to the existence of voids resulting from piping in the interior of the fill. Before undertaking the repair of the retaining wall and the pavement, a survey with GPR was done to locate related anomalies, both the probable existence of empty spaces and the presence of old drainage pipes. In both cases, the main anomalies found were verified by drilling, making it possible to identify some of the existing cavities.

In addition to locating these cavities, the use of this method provided, in some areas, a better knowledge of the structure of the formations underlying the pavement of the roads.

2 BASIC PRINCIPLES OF GPR

The GPR method is based on the propagation of electromagnetic waves in geological materials. The waves are emitted at frequencies that can range from 10 MHz to 2.5 GHz, depending on the type of transmitting/receiving antennas. The system sends electromagnetic impulses into the ground through a transmitting antenna, see Fig. 1. The transmitted signal is propagated in the ground and is reflected in interfaces between different materials that have contrasting electromagnetic properties, and then captured by the receiving antenna. By moving the antennas along a profile, the successive signals obtained form a radar section. The signals are amplified and processed through the manipulation of gain levels and filtering.

Fig. 1 shows a diagram of the execution of the GPR reflection profile and the respective section of reflection relating to the model schematized.

The electrical properties of the materials determine the velocity of propagation of the electromagnetic waves, their attenuation and resulting depth attained, and the amplitude of the reflected electromagnetic waves generated by the different materials (Fenner, 1985 and Annan, 1992).

2.1 Some Weak Points of the Method

One of the major drawbacks in the application of this method is that the necessary conditions required in terms of the contrast between the electromagnetic properties of the structures to be detected and the surrounding materials are not always present. In some situations in which there is not sufficient contrast, or in areas of very high conductivity, the

detection of the target structures is often impossible.

Other limitations that affect this method, especially in the interpretation of some readings, are related to the heterogeneity of some land formations, as in the case of heterogeneous earth/rock fills. These easily generate an enormous quantity of reflections and diffractions superimposed upon one another, hiding the reflections generated by structures that we wish to detect. At times, under these circumstances, enormous ingenuity is required to extract the desired information, since it is often impossible to distinguish between the two types of reflections and diffractions generated.

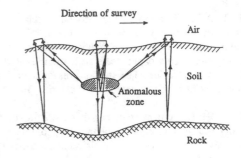

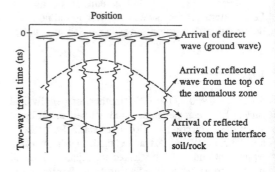

Fig. 1 - GPR reflection profiling procedure (modified from Davis and Annan, 1989).

3 METHODOLOGY, EQUIPMENT USED AND RESULTS OBTAINED

In the two cases that were studied, the continuous reflection technique was used in parallel profiles along the longitudinal axis of the roads.

For interpretation of the records and of some of the anomalies encountered, care was taken to put together a whole gamut of information, such as: information about the geology of the surface, results of boreholes previously drilled, and project elements (location of areas of earth/rock fills and excavations, location of hydraulic passageways, other elements of drainage, and cables).

3.1 Case history I

In this case our intention was to detect natural cavities along the highway platform up to a depth of approximately 8 meters, especially in the sections located in areas of excavation through the massif. The equipment used was the PulseEkko IV manufactured by Sensors & Software, Inc., Canada.

Transmitting and receiving antennas with peak frequencies of 50 MHz and 200 MHz were used with a pulser voltage of 400 V. Because of the large areas to be covered, the antennas were installed on a wooden cart and towed by a car at low speed. Fig. 2 illustrates the execution of a GPR reflection profile. The distance used between receiving and transmitting antennas was 0.55 meters and each trace was stacked 32 times.

In a first phase some preliminary studies were carried out over known cavities sites, with the objective of getting familiar with and recording the GPR response of these structures.

These studies permitted also the calibration of the readings based on a medium velocity of propagation adopted for calculations of depth.

Fig. 2 - View of the antennas arrangement for a continuous reflection profile.

The record in Fig. 3 is the image obtained over a cavity located at a depth of 1 meter. In this image, the anomalous space left by the cavity is very distinct, marked by several signal reflections in the form of multiple hyperbolas.

In the GPR image in Fig. 4, in addition to the anomaly checked and related to another cavity, we can also note a sub-vertical alignment of signal reflections that corresponds to a fault plane plainly

(a)

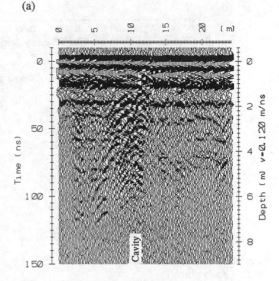

(b)

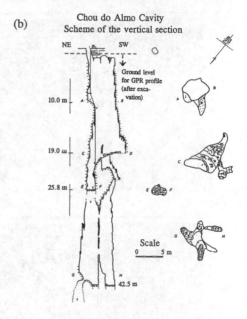

Fig. 3 - GPR section (a) of the karst cave schematized (b), (after Oliveira et al., 1992).

identifiable in the talus at the surface.

In the record of Fig. 5, equally identifiable are a series of alignments of reflections that correspond to bedding planes between successive layers of limestone.

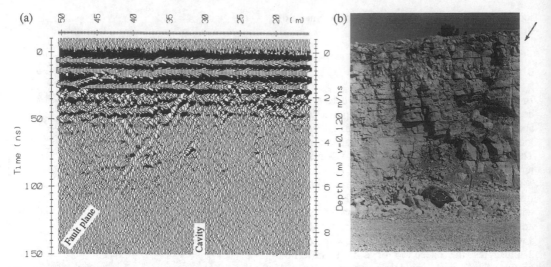

Fig. 4 - GPR section showing an anomaly related with a karst cave and an alignment of reflections (a) related with a fault plane (b)

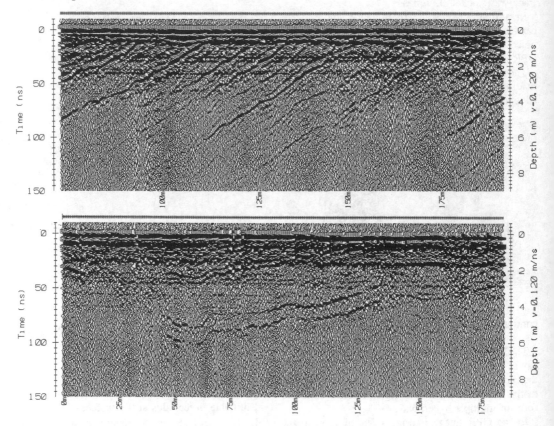

Fig. 5 - GPR sections showing bedding planes between limestone layers. Records from Torres Novas-Fátima highway, Portugal.

The anomalies clearly visible in the GPR records in Fig. 6 are some examples of anomalies that were confirmed through several boreholes. In this case, several parallel profiles reveal the existence of a continuous anomalous structure. The readings taken from boreholes drilled showed the presence of cavities from a depth of 5 meters.

3.2 Case history II

In this case we hoped to detect cavities up to a depth of around 3 meters in the earth/rock fill of the foundation of a road. The equipment used was the SIR 10 radar, made by Geophysical Survey Systems, Inc., USA, with a transmitting/receiving antenna with a frequency of 500 MHz (photograph (a), Fig. 7).

Photograph (b), Fig. 7 shows the retaining wall and the road in one of the areas where some subterranean voids and some sinking of the pavement have been verified.

In this case, in addition to the possible existence of cavities in the ground, that by themselves would generate anomalies, there was other structures, like old conduits, pipes and cables that also provoke abnormal readings. For this reason it was necessary to confirm, at the site, the different anomalies obtained.

(a)

(b)

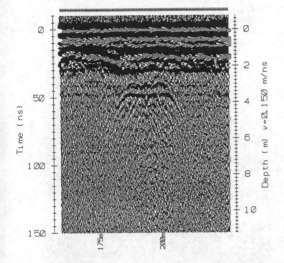

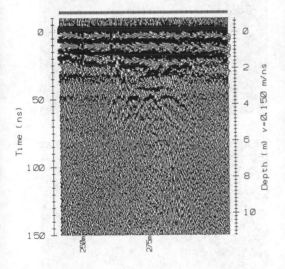

Fig. 6 - GPR sections showing anomalies.

Fig. 7 - View of the 500 Mhz antenna in a GPR profile (a). View of a part of the road where was carried out some GPR profiles (b).

As in the case described above, preliminary tests were made with the GPR over known structures, namely non-metallic conduits and cables, to facilitate interpretation of the readings.

The record (a) in Fig. 8 shows an anomaly related to an hydraulic passageway (non-metallic pipe) with a diameter of 0.6 meters. The record (b) of the same figure shows the anomalous area having essentially identical characteristics as that of record (a). After checking with a borehole, the anomaly was found to correspond to a cavity located at between 2 and 2.4 meters of depth.

In order to determine if this void was related to an old, previously unknown conduit constructed with hewn stones and mortar, or to a cavity originated by piping, the borehole was surveyed with a TV camera. These studies turned out to be extremely helpful and showed that the cavity was an empty space in the earth/rock fill with fairly wide lateral dimensions. Fig. 9 shows a series of TV camera images at various depths, respectively at 1.8, 1.9, and 2.3 meters.

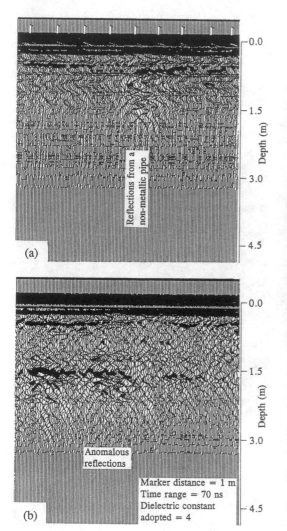

Fig. 8 - GPR sections showing a pipe anomaly pattern (a) and an anomaly caused by a cavity (b) (after Oliveira, 1994).

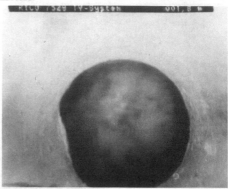

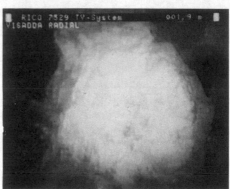

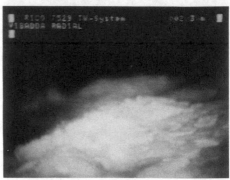

Fig. 9 - Images from a TV camera in a borehole, showing the empty space round the borehole.

4 GENERAL CONCLUSIONS

In spite of some limitations that this method presents in its application in certain situations and in the interpretation of some readings, the examples presented here show its potential as an useful tool for cavity detection.

In the examples presented here, it should be noted that the information resulting from the radar surveys was used in the planning and programming of direct exploration works, permitting the localization of some cavities, and the confirmation of many of the anomalies that were found.

5 ACKNOWLEDGEMENTS

The authors wish to thank BRISA - Auto-Estradas de Portugal, S.A. and Junta Autónoma de Estradas, for the permission to publish these results.

REFERENCES

Annan, A. P. (1992). Ground Penetrating radar. Workshop notes. Sensors & Software Inc. Mississauga, Ontario.

Davis, J. L. & Annan, A. P. (1989). Ground-Penetrating radar for high-resolution mapping of soil and rock stratigraphy. Geophysical Prospecting, 37, 531-551.

Fenner, T. J. (1985). Applications of subsurface interface radar (SIR) in limestone. Geophysical Survey Systems, Inc. Hudson, USA

Harrison, C. H. (1970). Reconstruction of subglacial relief from radio sounding echoes. Geophysics, 35, 1099-1115.

Oliveira, M. P., Rodrigues, L. F. & Coelho, M. J. (1992). Geophysical survey by ground penetrating radar at the highway Torres Novas - Fátima. LNEC report, Lisbon, Portugal.

Oliveira, M. P. (1994). Geophysical survey by ground penetrating radar at Marginal Avnue, between Alto da Boa Viagem and Paço de Arcos. LNEC report, Lisbon, Portugal, (in press).

Ulriksen, C. P. F. (1982). Application of impulse radar to civil engineering. Lund University of Technology, Doctoral Thesis, Lund, Sweden.

Investigation of the stress-strain state of the rock mass

Investigation de l'état de contrainte-déformation des massifs rocheux

J. Zalezhnev
Hydroproject, Moscow, Russia

ABSTRACT: The paper deals with the investigations of the level of field stresses and deformability of the rock mass with the help of the known method "regeneration of movements (deformations)" with the use of flat jacks. The analytic task was solved by the A. Liava, R. Mindlin-D. Chench dependences. The paper shows the results of fourteen tests of the rock mass formed by phylites.

RESUME: On étudie le problème des contraintes naturelles dans le massif rocheux. On à utilisé la méthode du vérin plat. L'étude du problème va s'appuyer sur la solution A. Liave, R. Mindlin-D. Cheng.Ce papier montre les résultats obtenus par des mesures en place dans quatorze essais en phyllade.

The ratings of underground structures are based on the knowledge of the level of field stresses in rock mass. The level of the said stresses is determined by various experimental methods. Each of them has certain advantages and drawbacks.

The method of flat jacks (Freyssinet jacks) is well known for a number of years (Talobre Rossi).

The said method implies the creation of the area of disturbances - slot in the rock mass securing the destressing around them and measurements of the movements resulting from it.

Meanwhile the primary movements take place in the process of destressing in the rock mass resulting from the field stresses.

The secondary movements take place when the inner surfaces of the slot are loaded by outer loads, as if they reestablish the field stress state around the slot.

The proposed method is used when the following activities are undertaken: preliminary activities, tests during the processing of experimental data and their analysis.

The preliminary activities began from the selection of testing site, location of each slot and places for installation of devices to measure

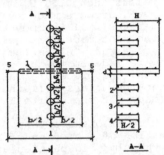

Figure 1. Pattern of marking of slots: 1 - slot; 2 - holes to install the rods of the close line; 3 - holes to install the rods of the extreme line; 4 - holes to install the unmovable rods (reference rods); 5 - points to locate the rock bolts.

up the movements as well as rock bolts (Fig. 1).

The slots were made after installation of templets and measuring devices.

The diagram of testing arrangement is indicated on Fig. 2. The slots were made by pneumatic percussion hammer by line drilling with the partial overlapping of the cross section of each previously made shot hole.

Slots were made by gradual deepening, thus by a number of runs. The

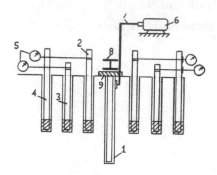

Figure 2. Diagram of the arrangement
for active loading of the slot: 1 -
hydraulic bag; 2,3 - measuring rods;
4 - unmovable rods (reference points);
5 - needle type of the device to
measure up the movements; 6 - pump
plants; 7 - hydraulic pressure line;
8 - reference line; 9 - slot sealing.

Table 1. Ultimate values of
deformation.

Orientation of a slot in the planes of coordinate axes	Deformation index	Deformation value x10³	Duration of observations (till cease of deformations),days
YOZ	ε_x	0.30 0.36 0.46 0.64 0.67	19 8 19 11 31
XOZ	ε_y	0.03 0.11 0.46 1.18	14 35 22 25
XOY	ε_z	0.33 0.40 0.78 1.00 3.41	18 51 18 13 29

depth of every run was designed to
be equal or less than 0.35 H.

The observations were made by the
measuring devices till the movements
ceased in the slot made for the entire depth of a run. Only after that
the slots were made up to the depth
of a next run. The procedure was repeated three-four times till the
slots are made for the entire H
depth.

The duration of observations over
the deformation of a section of rock
mass near the slot, in particular
during some 25-51 days in phylites
(Table 1).

Thus, the passive part of the experiment is over: the observations
over the deformation of destressing
of the rock mass around the slot
made.

Then, the inner surface of the
slot was made ready for loading with
the outer loading. The rubber hydraulic bag (with b=65 cm, length H=
=125 cm) was inserted into the slot.
The hydraulic bag was linked with
the pressure hydraulic system and
the active part of experiment began.

The active part of experiment was
performed at two stages: first
stage - loading, second stage - unloading.

The loading was performed gradually, the rate of loading did not exceed 0.05 MPa/sec.

The unloading was also performed
gradually with the decrease of

loading made at the same rate. The
movements for every stage loading -
unloading were measured with the accuracy of 0.01 mm.

The duration of every stage load-
ing-unloading at P_{max} < 5 MPa was
equal to 8-12 hours.

The diagram was made by the results of observations over the deformations of the rock mass around the
slot during cutting the slot itself
within the coordinates ε - t (Fig.3).

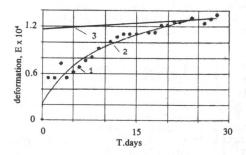

Figure 3. 1 - measure; 2 - rated;
3 - tangent line.

When analising the results of the
experiment it was assumed that the

214

a)

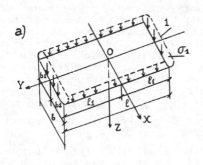

b)

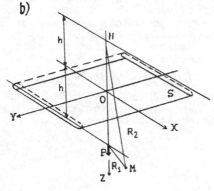

Figure 4. Rating diagram:
a) to the selection of the A.Liava
problem on stresses inside the rock
mass: σ_1 – rate of the equally distributed force on the area of semi-
space; X,Y,Z – main axes of coordinates; O – centre of gravity of the
semi-space section; b – width of the
loaded semi-space section; 1 – slot
outline.
b) for determination of movements
inside the rock mass (R.Mindlin,
D.Chench): S – semi-space area section: X,Y,Z – main axes of corrdi-
nates; P – force applied in the rock
mass at the h depth; R_1,R_2 – the distance from certain N point with coordinates (0, 0, 2h) to M point with
coordinates (X,Y,Z), where the movements are determined; 1 – slot outline.

rating diagram of the semi-space
loaded equally with the distributed
pressure σ_1 (Fig. 4a,b) applicable
to the loaded slot section.
   Then, the equation, describing the
stress-strain state of the loaded
rock mass section, was used, respre-
sented by the Burgers model [1]:

$$\varepsilon_{1(t)} = \varepsilon_b - \frac{\sigma_1}{3} \times \left(\frac{e^{-G_1 t/\eta_1}}{G_1} - \frac{T}{\eta_2} \right) \quad (1)$$

where: $\varepsilon_{1(t)}$– axial deformation by
time t in the body by Burgerars,
under the permanent axial stress σ_1

$$\varepsilon_b = \varepsilon_0 + \sigma_1/3G_1$$

and

$$\varepsilon_0 = \sigma_1 \times (2/9K + 1/3G_2) \quad (2)$$

ε_b – long, ε_0 – instant deformation, meanwhile ε_0 is determined by
the diagram $\varepsilon_1 = f(t)$ with the help
of regression plot, as a section cut
by a curve on the ε_1 axis, G_1 and

G_2 – the constants of deformability of rocks, respectively; G_1 – fea-

turing the degree of variation of
elasticity by time and G_2 – the
elasticity modulus of the second degree, thus the shear modulus; η_1 and

η_2 – the constants of viscosity,
respectively η_1 – featuring the
rate of variation of elasticity by
time and η_2 – featuring the velocity

of viscosity flour; K – modulus of
volume compression determined at the
known values of axial (ε_1) and lateral (ε_3) deformations by the ratio

$$K = \frac{G_1}{3} \times (\varepsilon_1 + 2\varepsilon_3) \quad , \text{ or at the}$$

known value of deformation modulus –
$$K = \frac{E}{3} \times (1 - 2\nu) .$$

   The values of parameters of deformability and viscosity, which are in
equations (1) and (2) are determined
by the known methods (1).
   The difference in deformations
$\varepsilon_b - \varepsilon_{1(t)}$ – is the deformation
ε_q, corresponding to the temporary compression strength happened in
the rock mass.
   On the basis of the long term observations over the destressing of
the rock mass section around the
slot the following empirical equation was obtained:
$$\varepsilon_{1(t)} = a \times \varepsilon \times 10^{-4} - b \times \varepsilon \times 10^{-4} \times$$
$$\times e^{-0.17t} + c \times \varepsilon \times 10^{-6} \times t \quad (3)$$

215

Table 2. Values of destressing deformation $\varepsilon_1(t)$ rated by the ratio (3).

Time (t), days	Deformation, $\varepsilon_1(t) \times 10^{+6}$
1	60.3
5	91.0
10	110.0
20	126.0
30	133.0
40	138.0
50	144.0
60	149.0
70	154.0
80	160.0
90	165.0
100	170.0
200	222.0
300	275.0

The assessment of development of deformation, accounting for the plastic deformation of phylites was made on the basis of this equation (Table 2).

Fourteen tests were made at the rock mass, formed by phylites, including the observations over the destressing and active loading. The results of ratings on the basis of these tests are given in Table 3.

The active part of experiment covers the loading of inner space of the slot by the equally distributed loads (σ_1), the deformation modulus was determined in the same part of experiment $E_{(x,y,z)}$.

At the same time, the values of specific force in the point on the axis of symmetry of loaded section, where the movements were measured (Fig. 4a), were calculated by the Liava ratio (2) out of the following expression:

$$\sigma_{(x,y,z)} = (2P/\pi) \times K_{(x,y,z)} \qquad (4)$$

where: $P = S \times b_1$; S - the area of the slot surface (semi-space) loaded by th specific forces σ_1.

Table 3. Values of parameters of stress-strain state.

Orientation of the slot in space (plane)	Index and values of ultimate stresses σ	Value MPa	Values of deformation $\varepsilon_q \times 10^3$ $\varepsilon_q^{max} = \varepsilon_b - \varepsilon_{1(t)}$	Characteristics of deformability, x10³ MPa E	K	G₁	G₂	Characteristics of viscosity, x10⁻⁶ MPaxmin η_1	η_2
XOZ	σ_y	1.53	1.077	1.23	1.02	1.41	1.17	15.6	258.6
		1.42	0.063	19.5	16.20	6.91	15.83	58.4	1300.1
		9.95	0.359	24.0	20.00	68.90	14.00	669.0	1460.0
XOY	σ_z	3.92	0.244	13.9	11.58	35.44	4.97	1027.2	256.6
		7.05	0.332	18.4	15.30	34.64	13.37	478.7	788.8
		4.19	0.203	17.9	14.90	24.89	5.39	249.0	402.5
		6.48	1.041	5.40	4.50	33.69	1.03	909.1	87.4
		40.51	1.003	35.0	29.20	37.73	39.87	409.4	769.9
YOZ	σ_x	1.97	1.376	1.24	1.03	1.79	0.66	26.6	12.9
		12.59	0.303	36.0	30.00	70.16	25.52	1168.3	1377.8
		2.50	0.146	14.8	12.30	26.67	7.20	1006.0	0.98
		0.61	0.045	11.7	9.80	12.73	0.34	192.0	252.0
		13.64	0.611	19.3	16.10	60.47	12.89	351.9	2110.0

$$K_{(x,y,z)} = \left[\frac{1_1 \times b_1 \times (x,y,z)}{D_{(x,y,z)}} \times \right.$$

$$\times \frac{1_1^2 + b_1^2 + 2\,(x,y,z)}{D^2_{(x,y,z)} \times (x^2,y^2,z^2) + 1_1^2 \times b_1^2} +$$

$$\left. + arc\,Sin \times \frac{1_1 \times b_1}{\sqrt{1_1^2 + (x^2,y^2,z^2)} \times \sqrt{b_1^2 + (x^2,y^2,z^2)}} \right]$$

$$D_x = 1_1 + b_1 + x^2; \quad D_y = 1_1 + b_1 + y^2;$$
$$D_z = 1_1 + b_1 + z^2.$$

21_1 - length of the loaded surface,

$2b_1$ - width of the loaded surface,
x, y, z - the shortes distance from the loaded surface to the point, where the movements are measured.

The values of the deformation modulus of the rock mass along the orientation of one of symmetry axes were rated by the following equation, obtained as the result of convergence of R.Mindlin-D.Chench ratio (2).

$$E_{(x,y,z)} = A \times \delta_{(x,y,z)} \times S/_{W_{(x,y,z)}} \times (x,y,z) \quad (5)$$

where: $\delta_{(x,y,z)}$ are determined by ratio (4), S - ditto, as in ratio (4);

$$A = \left[8 \times (1-y)^2 + 1 \right] /_{16\pi} \times \left[(1-y)/(1+y) \right]$$

y - Poisson ratio, $W_{(x,y,z)}$ - movements, measured in experiments along the orientation x, y, z,
x, y, z - the shortest distances from the surfaces of loading to the points of measurements of movements along the orientation of the coordinate axes.

The result of rating of the deformation modulus values by the orientations x, y, z are indicated in Table 3.

CONCLUSIONS

1. The assessment of stress strain state was made on the basis of the test results with the use of flat jacks in the slots, meaning the known experimental method, which was effected with implementation of serious modifications. It was related to the procedure of execution of preliminary activities, fulfilment of tests, processing the results of experiments.

2. The specific features of the preliminary activities is: the slot to be loaded was made in the rock mass by the line cutting with pneumatic percussion-rotary hammer.

3. The distinguishing feature of these tests in comparison with ordinary tests is that each of them consists of two stages.

The long term observations over the destressing of the rock mass during the slot cutting and its gradual deepening were made at the first stage. The loading of the inner surface of the slot with the use of flat jacks was made at the second stage in order to reestablish the movements which took place during the actual cutting.

4. The levels of the field stresses in the rock mass, represented by the Burgers model and the parameters of the rock mass deformability were calculated.

The rating-analytic solutions of the problems by A.Liava and R.Mindlin-D.Chench for stresses and strains inside the rock mass were jointly used for the first time.

REFERENCES

Goodman, P.E. 1980. Introduction to Rock Mechanics. University of California.
Tsytovich, N.A. 1963. Mechanics of soils. Moscow.

Subgrade strength evaluation with the extended dynamic cone penetrometer
Évaluation de la résistance par l'usage d'une extension de la tige du pénétromètre dynamique

M. Livneh
Transportation Research Institute, Technion, Haifa, Israel

Noam A. Livneh
LCI, Transportation Engineers, Consultants, Israel

ABSTRACT: Experience gained in Israel with the DCP test, led to the use of an extended penetrating rod in addition to the standard one. Tests with these rods were performed and the statistical analysis showed that the CBR values calculated from the DCP values derived from a 2.0 meters penetrating rod, should be corrected. Additionally, comparative DCP tests showed that penetration at any angle other than the perpendicular leads to significant errors in the final results, thus indicating that vertical penetration is essential.

RESUMÉ: L'expérience acquise en Israël avec l'épreuve du D.C.P. a amené l'usage d'une extension de la tige du pénétromètre en addition à la tige standard. Des essais avec cette tige ont été effectués et leur analyse statistique a montré que les valeurs CBR calculées sur les mesures du DCP résultant d'une tige de pénétration de 2.0 meters de long devraient être en outre corrigées. Des épreuves de DCP comparées ont montré que la pénétration à tout angle outre l'angle perpendiculaire introduit des erreurs significatives dans les résultats définitifs et indiquent ainsi qu'il est essentiel de se tenir à la pénétration verticale.

1 INTRODUCTION

Various penetration tests are often utilized in the course of soil investigation programs. Among these tests, the Dynamic Cone Penetrometer (DCP) test is gaining more and more acceptance, especially in pavement design or evaluation projects.

Over the last decade, the DCP test for the determination of in-situ California Bearing Ratio (CBR) and, more recently, the Elastic Modulus (E), has been increasingly used in many parts of the world because it is economical, simple and able to provide rapid measurement of in-situ strength of pavement layers and subgrades without excavating the existing pavement as in the CBR test.

The use of the DCP was initiated by Scala (1956) who developed the Scala Penetrometer for assessing the in-situ CBR of cohesive soils. This penetrometer consists of a steel rod with a cone at one end which is driven into the pavement or the subgrade by means of a sliding hammer while measuring the material's resistance to penetration in terms of millimeter per blow. The angle of the original cone is 30° and its diameter is 20mm. The hammer weighs 8 kg, and the dropping sliding height is 575 mm. (See Figure 1).

The DCP was originally designed and used for determining the strength profile of flexible pavement structures (layers and subgrades). Usually, the pavement testing at a given point involves the extrusion of a 4" circular core from

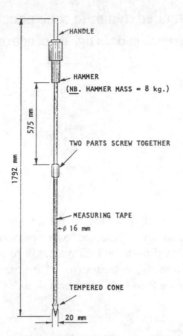

Figure 1. General details of the South African Dynamic Cone Penetrometer.

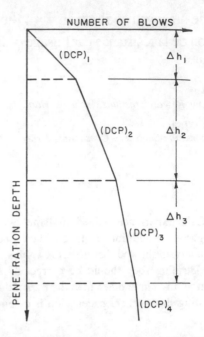

Figure 2. Average strength profile of an existing flexible pavement.

the top asphaltic layers only and inserting the DCP from the top of the base course layer down to the required pavement or subgrade layer. The properties of the asphaltic layers are directly evaluated in the laboratory by proper mechanical tests (resilient modulus, diametrical test, splitting test, Marshall test or others). The pavement parameters are continuously measured and recorded with depth by the DCP. Immediately at the end of the test, the shallow 4" hole is easily filled with either portland cement concrete (regular or fast curing), or a proper cold asphaltic mixture. When subgrade evaluation is required for design purposes, the DCP is inserted down from the top of the natural soil or compacted subgrade.

During the test, the number of blows is recorded and compared with the depth of penetration. The "DCP Value" is defined as the slope of the curve relating the number of blows to the depth of penetration (in mm per blow) at a given linear depth segment (see Figure 2).

2 DCP VERSUS CBR

In order to be able to relate the DCP value to structural pavement parameters under local pavement design technologies, extensive controlled laboratory and field tests were carried out in many parts of the world in order to correlate DCP and CBR results.

According to Livneh (1987 and 1991), the quantitative relationship between a material's CBR and DCP values is (see Figure 3):

$$\log CBR = 2.20 - 0.71 \, (\log DCP)^{1.5} \qquad (1)$$

$$R^2 = 0.85 \qquad N = 152$$

where DCP = penetration ratio in mm per blow.

This correlation was found to be valid for a wide range of pavement and subgrade materials. Also, it is worthwhile mentioning that in a recent work carried out by Webster et al. (1992) for the U.S. Corps of Engineers, the following relationship was found:

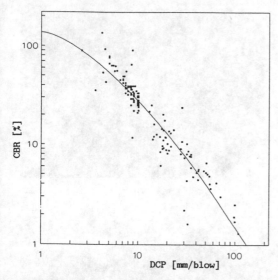

Figure 3. Relationship between CBR values and DCP values. Source: Livneh (1991).

$$\log CBR = 2.465 - 1.12\ DCP \qquad (2)$$

However, from Figure 4, it can be concluded that the relationship given in Equation 2 is very similar to the one given in Equation 1, thus constituting additional proof of the applicability and usefulness of Equation 1.

3 TESTS WITH AN EXTENDED ROD

The widespread experience gained in Israel with the DCP test, has led to the introduction of some modifications in the testing apparatus and procedure. One of these modifications is the extension of the standard penetrating rod's length to a total of about 2.0 meters, by screwing an additional 1.0 meter rod into the original one, after completing its insertion into the pavement or subgrade. Such a procedure is required when the strength profile is significant at depths exceeding 1.0 meter, or when the total thickness of the pavement layers, including the relevant subgrade layers, exceeds the length of a standard penetrating rod. The testing procedure with the extended penetrometer rod is illustrated in Figure 5 according to its stages of operation.

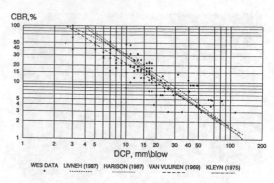

Figure 4. DCP versus CBR. Source: Webster et. al (1992)

DCP tests with standard and extended penetrating rods were performed in several clayey and sandy clay tests pits. Before carrying out a full comparative test between the results derived from the two DCP rod lengths (Figure 6 presents a sample of DCP results in one pit), the CBR values calculated from the confined DCP results should be examined to determine if they are identical with the CBR values calculated from the unconfined DCP results. In these confined and unconfined tests, all results were derived by means of the standard, 1.0 meter long penetrating rod. The confined DCP values represent values derived from a test carried out through the upper clayey layers, while the unconfined DCP values represent values derived from a test carried out after said clayey layers have been removed. Here, it should be mentioned that Equation 1 was used for calculating CBR values from DCP results obtained by both the standard and the extended penetrating rods.

Figure 7 provides a graphic illustration of the test points above and below the line of equality. The Grubbs (1973) statistical test carried out on the confined and unconfined DCP results, indicates that at a confidence level of 5% ($\alpha = 0.05$), one cannot reject the null hypothesis regarding the identity of the confined and unconfined DCP results. This applies both to the identity of the variance of errors of measurement

Figure 5. Operation of the DCP apparatus with the extended penetrating rod: (a) Photograph of the standard DCP device with the extended penetrating rod (unattached). (b) Insertion of the standard penetrating rod almost to its maximum depth. (c) Detaching the weight guide from the "buried" penetrating rod. (d) Screwing the extension rod to the "buried" standard rod. (e) Screwing the weight guide to the extension rod. (f) Operating the device after screwing the extension rod and the weight-guide.

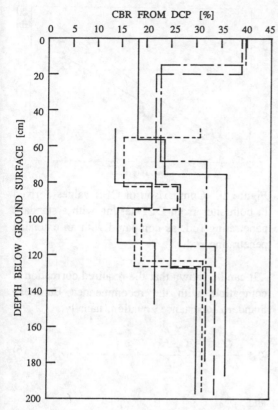

Figure 6. CBR values (derived from DCP data) versus the depth below the ground surface, in various test pits.

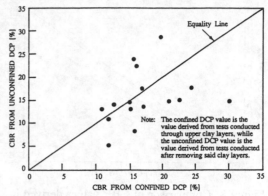

Figure 7. Comparison of CBR values derived from confined DCP testing as compared with unconfined DCP testing.

$$CBR_{L=1.0m} = 0.84\ CBR_{L=2.0m} \tag{3}$$

$$R^2 = 0.83 \qquad N = 18$$

where $CBR_{L=1.0m}$ = CBR result calculated according to Equation 1, from DCP values obtained by means of the standard 1.0 meter rod; $CBR_{L=2.0m}$ = CBR result calculated according to Equation 1, from DCP values obtained by means of the 2.0 meter rod.

Similar correlative tests were carried out in an additional clayey site in Israel. The results of this study are given in Figure 9 and yield the following regression:

$$CBR_{L=1.0m} = 0.77\ CBR_{L=2.0m} \tag{4}$$

$$R^2 = 0.83 \qquad N = 19$$

When the results of the above two sites are combined, the regression analysis leads to the following final relationship (see also Figure 10):

$$CBR_{L=1.0m} = 0.80\ CBR_{L=2.0m} \tag{5}$$

$$R^2 = 0.83 \qquad N = 37$$

The conclusion which may be drawn from the above tests is that one should deduct 20% from

and to the identity of the additive unknown bias associated with the above two testing methods.

The above statistical analysis made it possible to integrate all of the results of the confined and unconfined tests obtained by means of the standard 1.0 meter penetrating rod and to compare them with results obtained by means of the 2.0 meter penetrating rod. Figure 8, which presents such a comparison in graphic form, indicates the existence of a deviation between the results of the two testing methods. CBR values obtained by means of the long penetrating rod were greater than the CBR values obtained by means of the shorter (standard) rod. Thus, the CBR values derived from the non-standard penetrating rod must be corrected according to the following regression equation:

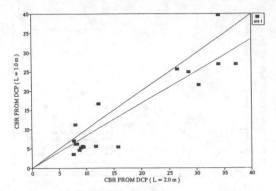

Figure 8. Comparison of CBR values derived in site 1 from DCP testing with a standard penetrating rod, as compared with an extended penetrating rod.

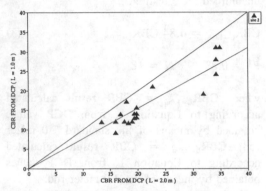

Figure 9. Comparison of CBR values derived in site 2 from DCP testing with a standard penetrating rod, as compared with an extended penetrating rod.

the CBR values derived from DCP results obtained by means of a 2.0 meter penetrating rod. If this is not done, the obtained material strength will probably represent an over-estimation.

4 SEMI-THEORETICAL EXPLANATION

The above conclusion pointing to the necessity of correcting the CBR values calculated from the DCP values derived from penetration tests with an extended rod, may be explained by other findings given in the technical literature.

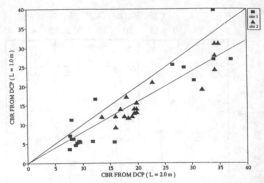

Figure 10. Comparison of CBR values derived in both sites from DCP testing with a standard penetrating rod, as compared with an extended penetrating rod.

It can be shown that the required correction is compatible with the recommended European Standard Resistance equation, namely:

$$q_d = \frac{G_f}{G_f + G_g} \times \frac{G_f \cdot H}{A \cdot L} \quad\quad (6)$$

and

$$q_d = \eta \times \frac{G_f \cdot H}{A \cdot L} \quad\quad (7)$$

thus,

$$\eta = \frac{G_f}{G_f + G_g} \quad\quad (8)$$

where q_d = soil resistance value; G_f = hammer's weight; G_g = penetrating rod's weight; L = depth of penetration for one blow; H = free hammer drop height in mm; A = cone base cross-section area and η = apparent energy transfer efficiency factor.

In view of the above findings, Equation 6, 7 and 8 may explain the experimental DCP results given in the previous section. For the standard

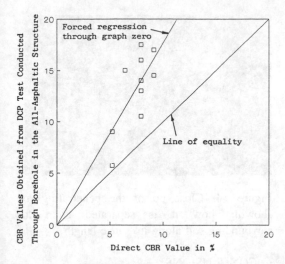

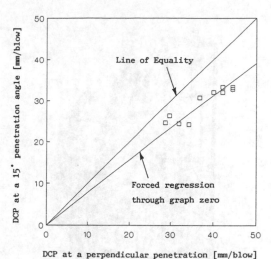

Figure 11. CBR results from the clay subgrade of a 50cm. all-asphaltic structure. Source: Livneh et al (1990).

Figure 12. DCP values obtained from vertical penetration and from a penetration deviating 15° from the perpendicular.

DCP apparatus the ratio G_f/G_g equals 2.23, while for the extended rod apparatus G_f/G_g equals 1.78. Thus the value of η for the standard rod is 0.69, while for the extended rod, η is 0.64, a reduction of about 10%. Thus, the driving energy required to insert the extended rod into soil with the same strength to the same depth, is 10% greater than that required to insert the standard rod.

This means that, for the same soil, the DCP values derived from the extended rod are likely to be lower by 10% than the values derived by means of the standard rod. This constitutes an artificially recorded increase in soil strength, solely attributable to the extension of the penetrating rod. This theoretical finding corresponds well to the experimental results presented in Figure 10.

5 FRICTION EFFECTS

Occasionally, the CBR value derived from DCP testing are higher than anticipated and a comparison with other in-situ strength testing results confirms this deviation. A good example of this phenomenon are the DCP values obtained

from subgrades confined beneath conventional or all-asphaltic pavements, which sometimes yield deviant strength results as compared with the CBR values obtained by direct CBR testing. A recent example of such a deviation is given in Figure 11, which depicts DCP tests conducted in a clayey subgrade following the drilling of cores from the entire 50 cm. depth of the all asphaltic pavement. Confinement cannot be used to explain these results, and the most appropriate explanation is the possibility that friction forces develop along the penetrating rod as a consequence of non-vertical penetration. In order to test this assumption, a series of tests were conducted in which the rod was inserted both vertically and at a deviation of 15° from the perpendicular.

The statistical analysis indicates the fact that there is a difference between the two tests. The mean CBR value for the vertical penetration was 6.6%, while the CBR value for the penetration at an angle of 15° was 8.9%. A constrained regression produces the following correlation (see also Fig. 12):

$$DCP_{15°} = 0.78 \times DCP \qquad (9)$$

225

Figure 13. Attaching the frame base to the DCP apparatus.

$$R^2 = 0.92 \qquad N = 10$$

where DCP = value obtained by means of the standard vertical penetration; $DCP_{15°}$ = value obtained by means of a penetration deviating $15°$ from the perpendicular.

The meaning of the above regression is that the DCP value decreases (i.e., the strength increases) when the penetrating rod is inserted at an angle deviating from the perpendicular. This fact can be used to explain the results presented in Figure 11.

In view of the above findings, it is essential to ensure that testing process be free of any skin friction along the penetrating rod. This can be achieved by attaching a base frame to the DCP apparatus (see Figure 13) in order to ensure vertical penetration, thus keeping the material away from the penetrating rod as shown in the close-up presented in Figure 14.

Figure 14. Close up of the penetrating rod showing how it is separated from the surrounding soil along its entire length.

6 CONCLUSIONS

The widespread experience gained in Israel with the DCP test, has led to the introduction of somemodifications in the testing apparatus and procedure. One of these modifications is the extension of the standard penetrating rod's length to a total of about 2.0 meters, by screwing an additional 1.0 meter rod into the original one, after completing its insertion into the pavement or subgrade. Such a procedure is required when the strength profile is significant at depths exceeding 1.0 meter, or when the total thickness of the pavement layers, including the relevant subgrade layers, exceeds the length of a standard penetrating rod. Comparative tests have indicated that one should deduct 20% from the CBR values derived from DCP results obtained by means of the 2.0 meter penetrating rod.

Additionally, it was found that it is important to ensure that the penetration process be free of any skin friction which may develop. This can be achieved by attaching a base frame to the DCP apparatus (see Figure 13) in order to ensure vertical penetration, thus keeping the material away from the penetrating rod.

REFERENCES

Grubbs, F.E, 1973. Errors of measurement, precision, accuracy and the statistical comparison of measuring instruments. *Technometrics*, Vol. 15, No. 1, pp. 53-66.

Harrison, J.A.1987. Correlation between California Bearing Ratio and Dynamic Cone Penetrometer strength measurements of soils. *Proceedings of Instn Civil Engineers*, Part 2, pp. 833-844.

Livneh, M. and Ishai, I. 1987. Pavement material evaluation by a Dynamic Cone Penetrometer. *Proceedings of Sixth International Conference Structural Design of Asphalt Pavements*, Vol. I, pp. 665-676, University of Michigan, Ann Arbor.

Livneh, M. 1987. The use of Dynamic Cone Penetrometer in determining the strength of existing pavements and subgrades. *Proc. 9th Southeast Asian Geotechnical Conf.*, Vol. 2, pp. 9.1-9.10, Bangkok.

Livneh, M., Ishai, I. and Livneh, N.A. 1990. Carrying capacity of unsurfaced runways for low volume aircraft traffic. *Phase II: Application of the Dynamic Cone Penetrometer*. Air Force Engineering and Services Laboratory (AFESC). Tyndall Air Force Base, Florida.

Livneh, M. 1991. Verification of CBR and Elastic Modulus values from local DCP tests. *Proc. of the 9th Asian Regional Conf. on Soil Mechanics & Foundation Engrg.*, Vol. 1, pp. 45-50, Bangkok.

Scala, A.J. 1956, Simple methods of flexible pavement design using cone penetrometer. *New Zealand Engineering* Vol. 11, No.2.

Van Vuuren, D.J. 1969. Rapid determination of CBR with the portable Dynamic Cone Penetrometer. *The Rhodesian Engineer*, Vol. 7, No. 5, pp. 852-854.

Webster, S.L. Grau, R.H. and Williams R.P. 1992. Description and application of dual mass Dynamic Cone Penetrometer. *US Army Engineer Waterways Experiment Station, Instruction Report*, No. GL-92-3.

APPENDIX

The following figures display the procedure for extracting the extended rod from the soil.

Figure 15. Extraction of the extension rod from the soil, with the help of an hydraulic jack.

Figure 16. Releasing the extension rod from the standard rod, in order to extract the standard penetrating rod from the soil.

On interrelationships among mechanical, electric, and piezoelectric cone penetrometer test results

Au sujet des relations parmi les résultats des essais avec les pénétromètres au cône mécanique, électrique et piézoélectrique

Fred H. Kulhawy
Cornell University, Ithaca, N.Y., USA

Paul W. Mayne
Georgia Institute of Technology, Atlanta, Ga., USA

ABSTRACT: For many years, the cone penetration test has been a common and valuable site investigation tool. Many stages of evolution have occurred in the development of this test type, from early mechanical cones to recent piezocones, resulting in numerous equipment variations and subsequently different test results. This paper examines some of these variabilities and presents recommended correlations to interrelate the different cone test results.

RÉSUMÉ: Depuis plusieurs années, le test du pénétromètre au cône est devenu un outil précieux et très utilisé pour la reconnaissance du sol. Différentes étapes dans l'évolution technique de ce type de test ont conduit à une grande diversité de matériel, des anciens cônes mécaniques aux récents piézocônes, et de ce fait à de grands écarts de résultats entre les différents équipements. Cet article décrit quelques-unes de ces variabilités et présente les corrélations recommandées pour faire le lien entre les résultats des différents tests.

1 BACKGROUND

The cone penetration test (CPT), once known as the Dutch cone test, is a versatile sounding procedure that can be used to classify the materials in a soil profile and to estimate their engineering properties. The CPT is becoming perhaps the most popular and versatile in-situ test in the world (DeBeer et al. 1988). In the CPT, a conical penetrometer tip is pushed slowly into the ground and monitored. The earlier versions of the CPT still are used widely and are known as mechanical friction cone (Begemann) penetrometers (Fig. 1b). Some of these penetrometers, such as the Delft mantle cone (Fig. 1a), lack a friction sleeve and measure only tip resistance. These devices obviously provide less information about the soil conditions. Modern devices, such as those shown in Fig. 2, contain electrical transducers to measure both tip and side resistances as the instrument is advanced and are known as electric friction cone penetrometers. In the United States, the electric cone in most common use is the Fugro-type cone. Unless otherwise noted, this cone is assumed. More recently, piezoelectric cone (or, simply, piezocone) penetrometers (CPTU) have been developed that

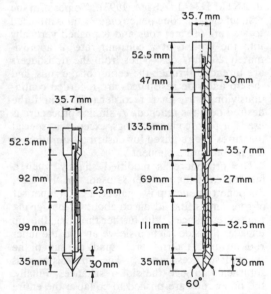

a) Delft Mantle Cone   b) Begemann Friction Cone

Fig. 1 Mechanical cone penetrometers (ASTM 1993)

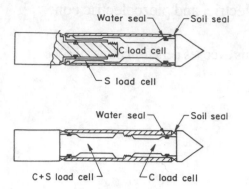

Fig. 2   Typical designs of electric cone penetrometers (Schaap & Zuidberg 1982)

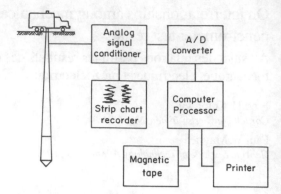

Fig. 3   Electric CPT data acquisition system
(DeRuiter 1982)

measure the pore water stresses during penetration, as well as the cone tip resistance and sleeve
side resistance.

In this paper, an overview is given of the inter-
relationships among various mechanical, electric,
and piezoelectric cone penetrometers. Where
possible, useful qualitative measures are presented to guide the test result interpretations.

2 PROCEDURE

The detailed procedure for the CPT is described
in ASTM D3441 (ASTM, 1993). To perform the
test, an electric cone penetrometer tip is attached
to a string of steel rods and is pushed vertically
into the ground at a constant rate of approximately 20 mm/sec. Wires from the transducers
are threaded through the center of the rods, and
the tip and side resistances are recorded continuously on a strip chart recorder (Fig. 3) until the
desired depth is reached. A similar procedure is
used for piezocone soundings, except that special
measures are required for ensuring saturation of
the porous stone element.

The procedure is modified slightly when a
mechanical penetrometer is used. In this case,
the penetrometer tip is connected to an inner set
of rods and is first advanced about 40 mm, giving
the tip resistance. With further thrusting, the tip
engages the side friction sleeve and, as the inner
rods advance, the rod force equals the sum of the
tip and side resistances. The tip resistance is
subtracted to give the side resistance. Finally,
the outer rods are pushed to collapse the entire
device, and the process is repeated at approximate 200 mm intervals. This mechanical process
has several important sources of error not characteristic of the electrical process and, where avail-

able, the electric penetrometer is recommended.

With standard mechanical and electric cones,
the two most useful parameters measured by the
test are the tip resistance, q_c, and the side resistance, f_s. Piezocones also provide readings of the
maximum pore water stress, u_m. The CPT can
be used where a sample is not needed and soil
conditions do not prevent its penetration. In general, the CPT is less suitable in soils containing
gravelly soils, cobbles or boulders, or cemented
seams.

3 COMPARISON OF MECHANICAL AND ELECTRIC CONES

Cone penetrometers have been in general use
since the 1930s in Europe, but only within the
past two decades have they gained wide usage in
the United States. Cone penetrometers can be
employed in a variety of soils and, especially
when performed with an electrical tip, provide a
continuous log of soil conditions. Furthermore,
when comparisons of cone soundings using both
electric and mechanical cone tips are made, the
profiles give similar trends, as shown in Fig. 4.
However, the electric cone provides more tip
detail and shows less scatter in the side resistance
profile, indicating that soil boundaries can be
located more accurately with an electric penetrometer tip. Note that in this and subsequent
figures, atmospheric pressure (p_a = 101.3 kN/m^2
= 1.058 tons/ft^2 = 1.033 kg/cm^2 = 14.7 lb/in^2)
is introduced to make all stress terms dimension-
less.

The mechanical and electric cones do not give
the same results, largely because of the different
geometry of the cones. The Delft and Begemann
cones shown in Fig. 1, as well as the Gouda cone

230

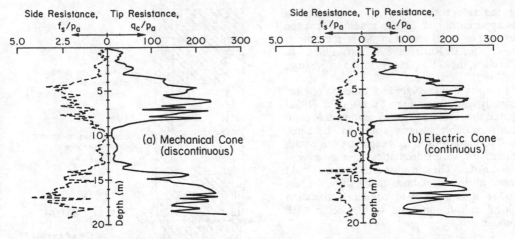

Fig. 4   Comparison of soundings using Begemann mechanical and Fugro electric cones
(DeRuiter 1971)

(similar to the Delft), all have a reduction in diameter beyond the cone tip. In contrast, the Fugro electric cone has the same sleeve and tip diameter. Approximate correlations between these mechanical and electric cones have been suggested (e.g., Schmertmann 1978, Smits 1982, Rol 1982, Amar 1974, Baligh et al. 1978). These studies generally have shown that q_c for electric cones is greater than q_c for mechanical cones in sands, while the reverse is true in clays and silts. To quantify these studies further, data were summarized from 14 sands and 10 clays and silts tested by both Fugro electric cones and several mechanical cones. The results are shown in Fig. 5 and indicate a good correlation. The regression statistics given include the number of data points (n), coefficient of determination (r^2), and standard deviation (S.D.).

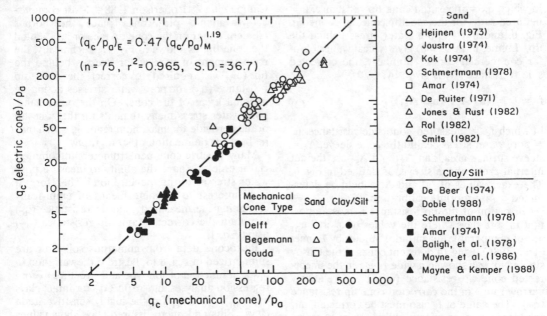

Fig. 5   Correlation of q_c between Fugro electric cone and three mechanical cone types
(Kulhawy & Mayne 1990)

231

For the side resistance (f_s), the mechanical cones apparently give higher readings than the electric cones in all soils. In sands, the ratio is about 2 (e.g., DeRuiter 1982, Smits 1982). In marine clays, the ratio varies from 2.5 to 3.5 (e.g., Dobie 1988).

CPT results also can vary as a function of electric cone type. One study by Lunne, et al. (1986) compared the results of 14 different types of commercially-available electric cones in the same sand. The variations in q_c were relatively small, but values of f_s varied dramatically, in some cases by a factor of 3. These results undoubtedly would influence all interpretations made from the test results, so it is prudent to conduct verification and local calibration tests with specific CPT equipment.

4 COMPARISON OF ELECTRIC CONES AND PIEZOCONES

The introduction of the piezocone (CPTU) and the resulting comparative studies of the CPT and CPTU have shown that all cone tip and side resistances must be corrected for pore water stress effects acting on unequal areas of the cone geometry. The corrected tip resistance (q_T) is given by:

$$q_T = q_c + (1-a)u_{bt} \qquad (1)$$

in which q_c = measured cone tip resistance, a = net area ratio for the specific cone, as defined in Fig. 6, and u_{bt} = pore water stress behind the tip. Lunne, et al. (1986) give typical values of "a" for commercial cones. Similarly, the corrected side resistance (f_t) is given by:

$$f_t = f_s = (u_s A_{s2} - u_{bt} A_{s1})/A_s \qquad (2)$$

in which f_s = measured cone side resistance, u_s = pore water stress behind the cone sleeve, A_s = sleeve surface area, and A_{s1} and A_{s2} are the net internal areas of the sleeve, as defined in Fig. 6. The exact values of "a" and A_s should be determined by appropriate calibration.

Correction of the tip resistance is most important in soft clays where the values of q_c and u_m are of comparable magnitude. Studies by Lunne, et al. (1986) using 14 different cones at the Onsoy site in Norway showed a wide range in the uncorrected cone tip resistance (q_c), but a relatively narrow range in the corrected cone tip resistance (q_T). The value of f_s also must be corrected, but this correction requires additional pore water stress measurements behind the cone sleeve.

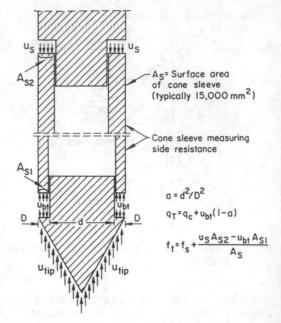

A_S = Surface area of cone sleeve (typically 15,000 mm^2)

Cone sleeve measuring side resistance

$a = d^2/D^2$

$q_T = q_c + u_{bt}(1-a)$

$f_t = f_s + \dfrac{u_s A_{S2} - u_{bt} A_{S1}}{A_S}$

Fig. 6 Unequal end areas of electric friction cone

These additional measurements are not yet routine for commercial CPTU testing.

One further complication with the piezocone is that its design has not yet been standardized (e.g., Campanella & Robertson 1988). Most commercially-available piezocones place the porous element either on the cone tip face or just behind the cone tip, as shown in Fig. 7. Technically, the measurement of pore water stresses behind the tip (u_{bt}) is required to correct the cone tip resistance (q_c) for pore water stresses acting on unequal areas of the cone. On the other hand, pore water stress measurements on the cone tip or face provide the maximum reading, which may be best for delineation of stratigraphy.

Many electric cone penetrometers in commercial use do not have the ability to measure pore water stresses during penetration. Therefore, it is of interest to examine means of empirically correcting the measured cone tip resistance (q_c) to obtain the corrected cone tip resistance (q_T) using Eq. 1.

Piezocone data from numerous soil sites are summarized in Fig. 8 to illustrate the variation in u_{bt} as a function of soil type and structure. From regression analyses, u_{bt} = 0.53 q_T for intact clays and u_{bt} = 0.58 q_T for the highly sensitive Leda clays. Silts and micro-fissured clays show values of u_{bt} that are only a small fraction of q_T. For

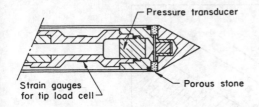

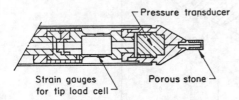

Fig. 7 Common piezocone geometries
(Campanella & Robertson 1982)

fissured clays, u_{bt} is about zero. These trends will be useful for estimating the corrected cone tip resistance on a preliminary basis.

Fig. 9 shows comparable data to illustrate the variation in behavior when pore water stress measurements are made at the tip or mid-tip (u_t), in contrast with those behind the tip (u_{bt}). It is clear that higher values are measured on the tip for all soil types and structures. The basic reason for these differences is that tip measurements (u_t)

are being made on soil undergoing compression, while measurements behind the tip (u_{bt}) are being made on soil undergoing some degree of dilation as well. It is clear from Figs. 8 and 9 that cautions must be exercised in interpreting and utilizing the pore water stress data.

5 SUMMARY

The cone penetration test is a very useful sounding procedure to evaluate soil sites and their engineering properties. However, there are mechanical cones, electric cones, and piezocones, and numerous manufacturer variations within each category. Each cone type has specific characteristics that must be addressed before any "standard" correlation relationships can be used. In this paper, a number of these characteristics have been summarized and reviewed, and recommended correlations have been given to interrelate the different cone test results.

ACKNOWLEDGMENTS

This study is part of ongoing research for the Electric Power Research Institute under RP 1493. A. Hirany is the EPRI Project manager. A. Avcisoy prepared the figures, and L. Mayes prepared the text.

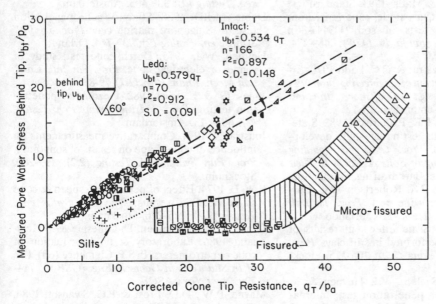

Fig. 8 Relationship between u_{bt} and q_T in CPTU tests (Mayne et al. 1990)

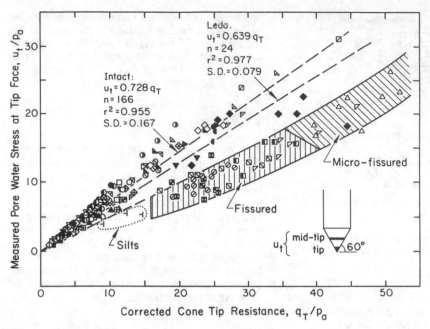

Fig. 9   Relationship between u_t and q_T in CPTU tests (Mayne et al. 1990)

REFERENCES

Amar, S. 1974. Use of static penetrometer in Laboratoires des Ponts et Chaussees. *Proc. Eur. Symp. Pen. Testing (2.2)*: 7-12. Stockholm.

Amer. Soc. Testing & Matls. 1993. Standard test method for deep, quasi-static, cone & friction-cone penetration tests of soil (D3441-86). *Annual Book of Standards (4.08)*: 469-474. Philadelphia:ASTM.

Baligh, M.M., V. Vivatrat & C.C. Ladd 1978. *Exploration & evaluation of engineering properties for foundation design of offshore structures (Report R78-40)*. Cambridge:M.I.T.

Campanella, R.G. & P.K. Robertson 1982. State-of-the-art in in-situ testing of soils: developments since 1978. *Proc. Conf. on Updating Subsurface Sampling of Soils & Rocks & Their In-situ Testing*. Santa Barbara:Eng. Foundation.

Campanella, R.G. & P.K. Robertson 1988. Current status of piezocone test. *Proc. 1st Intl. Symp. Pen. Testing (1)*: 93-116. Orlando.

DeBeer, E.E. 1974. Scale effects in results of penetration tests performed in stiff clays. *Proc. Eur. Symp. Pen. Testing (2.2)*: 105-115. Stockholm.

DeBeer, E.E., E. Goelen, W.J. Heynen & K. Joustra 1988. Cone penetration test: international reference test procedures. *Proc. 1st Intl. Symp. Pen. Testing (1)*: 27-52. Orlando.

DeRuiter, J. 1971. Electric penetrometer for site investigations. *J. Soil Mech. Fndns. Div. (ASCE)*. 97(SM2):457-472.

DeRuiter, J. 1982. Static cone penetration test: state-of-the-art report. *Proc. 2nd Eur. Symp. Pen. Testing (2)*: 389-405. Amsterdam.

Dobie, M.J.D. 1988. Study of cone penetration tests in Singapore marine clay. *Proc. 1st Intl. Symp. Pen. Testing (2)*: 737-744. Orlando.

Heijnen, W.J. 1973. Dutch cone test: study of shape of electrical cone. *Proc. 8th Intl. Conf. Soil Mech. Fndn. Eng. (1.1)* 79-83. Moscow.

Jones, G. & E. Rust 1982. Piezometer penetration testing. *Proc. 2nd Eur. Symp. Pen. Testing (2)*: 607-613. Amsterdam.

Joustra, K. 1974. Comparative measurements on influence of cone shape on results of soundings. *Proc. Eur. Symp. Pen. Testing (2.2)*: 199-200. Stockholm.

Kok, L. 1974. Effect of penetration speed & cone shape on CPT results. *Proc. Eur. Symp. Pen. Testing (2.2)*: 215-216. Stockholm.

Lunne, T., D. Eidsmoen, D. Gillespie & J. Howland 1986. Laboratory & field evaluation of cone penetrometers. In S.P. Clemence (ed), *Use of In-Situ Tests in Geotech. Eng. (GSP 6)*: 714-729. New York:ASCE.

Mayne, P.W., D.D. Frost & P.G. Swanson 1986. *Geotech. report CEBAF project, Newport News, VA*. Washington:Law Engineering.

Mayne, P.W. & J.B. Kemper 1988. Profiling OCR in stiff clays by CPT & SPT. *Geotech. Testing J. (ASTM).* 11(2):139-147.

Mayne, P.W., F.H. Kulhawy & J.N. Kay 1990. Observations on development of pore water stresses during cone penetration in clays. *Can. Geotech. J.* 27(4):418-428.

Rol, A.H. 1982. Comparative study on cone resistances measured with three types of CPT tips. *Proc. 2nd Eur. Symp. Pen. Testing (2)*: 813-819. Amsterdam.

Schaap, L.H.J. & H.M. Zuidberg 1982. Mechanical & electrical aspects of electric cone penetrometer tip. *Proc. 2nd Eur. Symp. Pen. Testing (2)*: 841-851. Amsterdam.

Schmertmann, J.H. 1978. *Guidelines for cone penetration test performance & design (Report FHWA-TS-78-209).* Washington:Federal Highway Administration.

Smits, F.P. 1982. Cone penetration tests in dry sand. *Proc. 2nd Eur. Symp. Pen. Testing (2)*: 881-887. Amsterdam.

Short note on rock quality description (RQD) functions

Remarques sur les fonctions de la 'rock quality description' (RQD)

B.H.Sadagah
King Abdulaziz University, Jeddah, Saudi Arabia

Z.Şen
Istanbul Technical University, Turkey

ABSTRACT: A simple procedure is proposed for extracting all of the possible Rock Quality Designation (RQD) values from observed scanline or boreholes discontinuity spacings. The new term RQD function is defined as the change of average RQD index with any given number of discontinuities. For very big discontinuity numbers these functions are expected to arrive asymptotically to the classical single RQD value. The methodology is applied to some field data which lead to empirical RQD functions. Comparison of empirical RQD functions reveals significant interpretations about the homogeneity or local heterogeneity of the fracture pattern existed in the fractured rock mass in concern.

RÉSUMÉ: Une procedure simple est proposé au but d'extraire toutes les valeurs du "Rock Quality Designation-RQD" tirées d'après les spaces de ligne de balayage des discontinuités et ces des forages. Le nouveau terme RQD est défini que le change d'indice moyen de RQD avec chaque nombre des discontinuités. Pour des grandes nombres des discontinuités cettes fonctions arriveront a une convergence avec les valeurs traditioneles simples de RQD. La méthodologie a été appliquée à quelques données de terrain qui conduisent à des fonctions numériques de RQD. Le comparison des fonctions numériques de RQD a révélé d'interpretations significatives en qui se concerne l'homogeneité ou l'hétérogénéité locales du système de fissures qui existe dans la roche en question.

1 INTRODUCTION

Rock masses are not continuous bodies in nature due to the presence of various types of discontinuities (joints, faults, fractures, fissures, cleavages etc.) in different directions. These features are critical defects present in every rock mass. They change to a considerable extent surface and subsurface strength, permeability and deformation characteristics of regional intact rock mass. The mechanical behavior of any rock mass can be governed by the characteristics of those discontinuities. However, a complete evaluation concerning the discontinuities can face difficulties, mainly due to the complex behavior and insufficient data.

Discontinuity pattern in any rock mass is recognized through boreholes or surface scanlines. The quantity of discontinuities in a rock mass is expressed simply as the discontinuity frequency which is defined by Hudson and Priest (1979) as the mean number of discontinuities per unit length intersected along a borehole or scanline (measuring tape) set upon a rock exposure. Numerical analysis of the joint spacing distributional forms have been studied by many researchers. They found that the distribution is either lognormal (Steffen et al., 1979; Bridges, 1975; Einstein et al., 1979 and Şen and Kazi, 1984) or negative exponential (Call et al., 1976; Priest and Hudson, 1976; Baecher et al., 1977; Einstein et al., 1979 and Wallis and King, 1980). The

RQD was first introduces by Deere (1963) and later elaborated by Hudson and Priest (1979) as the percentage of borehole or scanline intact lengths that are greater than 0.1m. Subsequently, RQD has been discussed by many researchers, (Priest and Hudson, 1976; Şen, 1984; Şen and Kazi, 1984).

2 DISCONTINUITY DISTRIBUTION

Existence of set of complex discontinuity patterns in rock mass practically affects stresses, deformation, water movement, recharge and grouting properties of the host rock. Therefore, it is preferable to employ simple but effective technique that lead to detailed quantitative assessments of RQD in spot. Various techniques describing collective effects of discontinuities in rock mass are given by different researchers, (Barton et al., 1974; Bieniawski, 1976; and Şen, 1992).

In practice, the prime task of the engineer is to make a detailed discontinuities survey with the purpose of representing the discontinuity pattern in a rock mass using a reliable model. Robertson (1970) and Piteau (1970) were among the first to describe discontinuity survey techniques. Discontinuity survey can be carried out on various types of natural (outcrops, trenches) and artificial exposures (boreholes cores, shafts, tunnels, artificial cut or rock slope faces). In case of very limited exposures, where it is, in general, physically difficult to examine a simple scanline measurement or borehole log of the discontinuities, systematic sampling techniques based on both limited areal or linear extents can be used. For this purpose, Priest and Hudson (1976) suggested that the length of the scanline should be 50 times the mean discontinuity spacing in length, in addition to the performance of three orthogonal scanlines of equal length.

For line spacing of discontinuities, the scanline survey is usually placed at the base of a given rock face. The determination of the quality of the rock mass is recognized from determination of the discontinuity frequency, λ. Hudson and Priest (1979) indicated that, λ, could be measured perpendicular to a single set of parallel and planer discontinuities, where, λ,

could be measured in a plane along a line perpendicular to the discontinuity set by counting the number of discontinuities along a certain length of scanline.

According to the present state-of-the-art, the best way to describe the dimensional intensity is through the estimation of fracture intensity for each joint set along the mean vector dimension. If one wants to obtain an estimate of fracture intensity in three dimensions, it can be obtained by combining one dimension intensity and joint size.

3 STATISTICAL TREATMENT OF THE FRACTURED ROCK MASSES

Statistical treatment of the characteristics of rock mass fracture system was discussed by many authors such as Baecher (1983 and 1984), Silva (1985), Young and Hoerger (1986 and 1988). Hoerger and Young (1987), Şen (1984, 1992), Şen and Kazi (1984) and Hudson and Priest (1979).

The numerical simulation of discontinuity distribution within the rock mass went into a recent development by Şen (1990) who used Monte Carlo method whereby it is possible to attach risk statements for maximum and minimum RQD values representing the rock mass along a scanline. He then produced a curve which represents the maximum and minimum values of RQD in a rock mass. This numerical technique enables the design engineer to estimate how much percent of low quality rock mass could be present within the whole (high quality) rock mass.

4 BASICS OF THE USED STATISTICAL TECHNIQUE

It is convenient to study the pattern of discontinuity spacing values using their probability density distribution which could follow either negative exponential, normal, lognormal or uniform distribution. One or more spacing pattern may combine to form a single probability distribution. This depends on the stress intensity in the direction of discontinuity set. Accordingly, the mean discontinuity frequency along one discontinuity set will vary due to the length of the scanline. If the length of the scanline becomes longer, it will cover a

238

larger number of discontinuities, and hence will be more precise to give reliable and stable λ values. On the other hand, the scanline gets shorter, it may give an unstable lower or higher RQD values than the actual one. Based on this through a new computer program in order to calculate the different expected (high and low) RQD values which may occur and vary along the total length of the scanline. The new introduced computer program could be applied on a field data to calculate (i) the higher and lower existing RQD (Rock Quality Designation) values, (ii) the mean value of RQD along any length within the whole scanline length. Hence to calculate the rock quality description RQD function which is the change of average RQD index with any given number of discontinuities.

5 APPLICATIONS

A number of scanlines were collected on different types of rock masses. These rock masses were selected according to the intensity of fractures along the highway in Jeddah city, Saudi Arabia (see Figure 1). A variety of rock types were chosen for the application of RQD function concept. These rock types are granite, quartz diorite and schistosed granodiorite. A representative rock mass for each rock type was carefully selected according to the intensity of fractures exhibited. In each rock mass, the measurement of the spacing between successive discontinuities intersecting the horizontal scanline was taken along the existed slope face cut of the rock mass. All the rock masses were scattered along the highway in the eastern vicinity of Jeddah city. The nine selected rock masses are low, medium and highly fractured rock masses, and their intensity category is represented by three rock types.

Each scanline was chosen to be placed in somewhere to include most of the discontinuities represent the fracture patterns in the rock mass under study. Each scanline is more than 20 meters continuous. Every rock mass was represented by one scanline. The geotechnical properties of the studied rock masses are given in Table 1.

Each rock type in each category is represented by a graph showing the relationship between the RQD values and the number of discontinuities. RQD functions for the low,

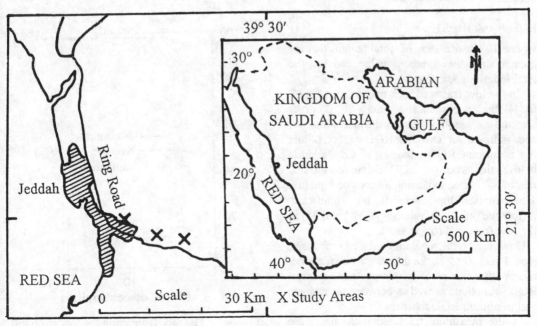

Figure 1. Location map of the studied area.

Table 1. Geotechnical properties of the studied rock masses.

Property / Rock Type	Type of Fracturing	Unconfined compressive strength, MPa	Friction angle, ϕ	Joint Roughness Coefficient, JRC
Granite	low	170	37	6.5
	medium	105	37	7
	high	77	38	6
Granodiorite	low	150	36	5.5
	medium	100	35	4.5
	high	90	37	5
Schistosed granite	low	160	36	8
	medium	95	36	5
	high	110	37	6.5

medium and highly fractured rock masses are represented in Figures 2 to 4, 5 to 7 and 8 to 10, respectively. In these Figures the RQD values are represented by maximum, average and minimum values. The three curves in each figure converge asymptotically for large scanlines to the classical single-valued RQD defined by Deere (1963) as

$$RQD = 100\,(L_s \backslash L) \qquad (1)$$

where L_s is the sum of total lengths that are greater than the threshold value and L is the total length of scanline.

In the low fractured rock masses, (see Figs. 2 to 4), the maximum and average RQD curves are almost identical for all the discontinuities and indicate an excellent RQD quality. While the minimum RQD curve is slightly a little below the average RQD curve for all the readings, hence indicating a very good quality. This implies that there is no significant difference in RQD values (78, 84 and 85) in the low fractured rock masses.

However, in the medium fractured rock mass, (see Figs. 5 to 7), the differences increase between the curves of maximum and average RQD functions as well as between the average and minimum RQD function.

In the maximum fractured rock mass, (see Figs 8 to 10), the differences between RQD

functions tends to be even larger than the medium and low fractured rock masses. While the distance between the average and minimum RQD attain its maximum value and where the curve represents the minimum, RQD tend to approach the origin of the graph in all examples of the highly fractured rock masses.

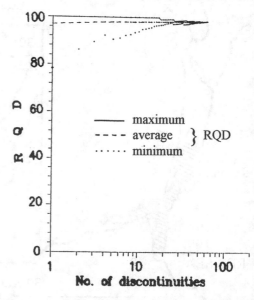

Figure 2. The RQD function for low fractured rock masses.

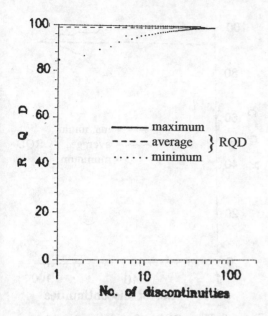

Figure 3. The RQD function for low fractured
rock masses.

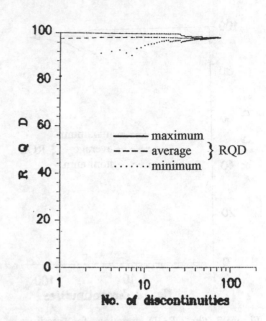

Figure 5. The RQD function for medium
fractured rock masses.

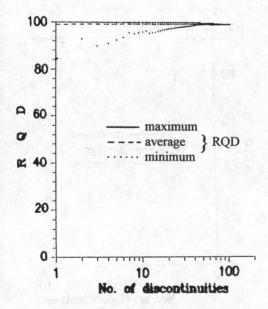

Figure 4. The RQD function for low fractured
rock masses.

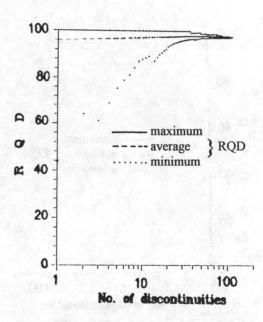

Figure 6. The RQD function for medium
fractured rock masses.

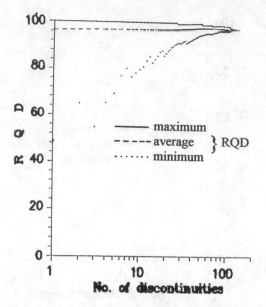

Figure 7. The RQD function for medium fractured rock masses.

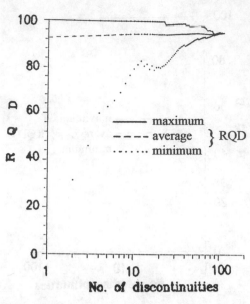

Figure 9. The RQD function for highly fractured rock masses.

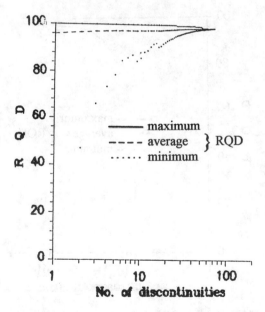

Figure 8. The RQD function for highly fractured rock masses.

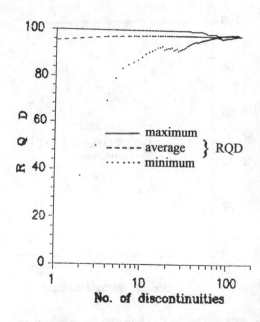

Figure 10. The RQD function for highly fractured rock masses.

6 DISCUSSION AND CONCLUSIONS

Although the mathematical concepts indicate that it is untrue to have a significant value of RQD at a zero number of discontinuities and compare the corresponding RQD value with the final RQD of the rock mass.

In the highly fractured rock masses, there is a large probability of having a low RQD rock mass, while there is a low probability of observing a low RQD rock masses within the low fractured rock masses.

Comparing the low and highly fractured rock masses at Figures 2 to 4 with Figures 8 to 10, it could be indicated that at few cases of discontinuities measured along the scanline, the shape of the minimum RQD curve indicates the quality of the rock mass.

At the low fractured rock masses, about 20 readings of discontinuities are enough to calculate the final RQD of the rock mass, while the number of required discontinuities increases in case of medium fractured rock masses and increases again to about 70 in case of highly fractured rock masses.

REFERENCES

Barton, N., Lien, R. and Lunde, J. 1974. Engineering classification of rock masses for the design of tunnel support. *Rock Mechanics* 6:189-236.

Baecher, G.B. 1983. Statistical analysis of rock mass fracturing. *Mathematical Geology* 15: 329-348.

Baecher, G.B. 1984. Geostatistics, reliability and risk assessment in geotechnical engineering. In G. Verly et al.(eds.) *Geostatistics for natural resources characteristics*: 731-744. Dordrecht: D.Reidel.

Baecher, G.B., N.A. Lanney and H.H. Einstein 1977. Statistical description of rock properties and sampling. *18th U.S. Symposium of Rock Mechanics, Keystone, Colorado*: 5c1- 1-5c1-8.

Bieniawski, Z.T. 1976. Rock mass classification in rock engineering. In Bieniawski, Z.T. (ed.), *Proceedings of the Symposium on Exploration for Rock Engineering*, Johannesburg:97-106.

Bridges, M.C. 1975. Presentation of fracture data for rock mechanics. *2nd Australia-New Zealand Conference on Geomechanics*, Brisbane: 144-148.

Call, R.B., J. Savely and D.E. Nicholas 1976. Estimation of joint set characteristics from surface mapping data. *17th U.S. Symposium on Rock Mechanics*, 2B2-1-2B2-9.

Deere, D.U. 1963. Technical description of rock cores for engineering purposes. *Rock Mechanics and Engineering Geology* 1: 18-22.

Einstein, et al. 1979. Risk analysis for rock slopes in open pit mines. *Part I and V, Contract J0275015 MIT.*

Hudson, J.A. and S.D. Priest 1979. Discontinuities and rock mass geometry. *International Journal of Rock Mechanics and Mining Sciences & Geomechanics Abstracts* 16: 339-362.

Piteau, D.R. 1970. Geological factors significant to the stability of slopes cut in rock. *Proceedings of the Symposium on Open Pit Mines*: 33-53. Cape Town: Balkema.

Priest, S.D. and J.A. Hudson 1976. Discontinuity spacing in rock. *International Journal of Rock Mechanics and Mining Sciences & Geomechanics Abstracts* 13:135-148.

Robertson, A.M. 1970. The interpretation of geological factors for use in slope theory. *Proceedings of the Symposium on Open Pit Mines*: 55-71. Cape Town: Balkema.

Şen, Z. 1984. RQD models and fracture spacing. *Journal of Geotechnical Engineering, American Society of Civil Engineering* 110: 203-216.

Şen, Z. and A. Kazi 1984. Discontinuity spacing and RQD estimates from finite length scanline. *International Journal of*

Şen, Z. and A. Kazi 1984. Discontinuity spacing and RQD estimates from finite length scanline. *International Journal of Rock Mechanics and Mining Sciences & Geomechanics Abstracts* 21: 203-212.

Şen, Z. 1990. RQP, RQR and fracture spacing. *International Journal of Rock Mechanics and Mining Sciences & Geomechanics Abstracts* 27: 135-137.

Şen, Z. 1992. Rock quality charts based on cumulative intact lengths. *Bulletin of the Association of Engineering Geologists* 24: 1-11.

Silva, A.J. 1985. A new approach to the characterization of reservoir heterogeneity based on the geomathematical model and kriging technique *SPE 14275*. Richardson, Texas: SPE.

Steffen, O., et al., 1975. Recent developments in the interpretation of data from joint surveys in rock masses. *6th Reg. Conf. for Africa on Soil Mechanics and Foundation*: 17-26.

Wallis, P.F. and M.S. King 1980. Discontinuity spacings in a crystalline rock. *International Journal of Rock Mechanics and Mining Sciences & Geomechanics Abstracts* 17: 63-66.

Young, D.S. and S.F. Hoerger 1986. Spatial analysis of fracture orientation. In K.Saari (ed.), *Large rock caverns:* 1111-1121. Oxford: Pergamon.

Optimum moisture content evaluation for dynamic compaction of collapsible soil

Évaluation de la teneur en eau pour la compactation dynamique des sols collapsables

Kyle M. Rollins
Civil Engineering Department, Brigham Young University, Provo, Utah, USA

Stan J. Jorgensen
Horrocks Engineers, American Fork, Utah, USA

Todd Ross
Inberg-Miller, Riverton, Wyo., USA

ABSTRACT: The influence of moisture content on dynamic compaction efficiency was evaluated at six field test cells, each with a progressively higher average moisture content. The soil profile consisted of sandy silt and test cell moisture contents ranged from 6% to 20%. At each cell, compaction was performed with a 4.54 metric ton weight dropped from a height of 24.3 m. Compaction efficiency was evaluated using: (1) crater depth measurements and (2) undisturbed samples before and after compaction. Crater depth increased by a factor of 4 as moisture content increased. The depth of improvement was relatively unaffected by the moisture content, however, the degree of improvement increased substantially up to a moisture content of about 17% and then decreased rapidly. This optimum moisture content is similar to that predicted by lab testing. Vibration levels also decreased significantly as moisture content increased.

RESUME: L'influence de l'humidité sur l'efficacité de la compaction dynamique a été évaluée sur six aires d'étude sur le terrain (pour test), en augmentant progressivement à chaque site la teneur moyenne en humidité. La composition du sol consistait en un dépôt sablonneux et la teneur en humidité dans les cellules s'échelonnait entre 6% et 20%. À chaque site, la compaction a été exécutée avec une charge de 4.54 tonnes que l'on laissait tomber d'une hauteur de 24,30 m. L'efficacité de la compaction a été évaluée en utilisant: (1) le calcul de la profondeur du cratère produit par la chute de la charge et (2) les échantillons non perturbés avant et après compaction. La profondeur du cratère augmentait par un coefficient 4 au fur et à mesure que la teneur en humidité augmentait. La profondeur à laquelle une amélioration était observée a été relativement inaffectée par la teneur en humidité, cependant, le degré d'amElioration a augmenté de façon significative jusqu'à atteindre une teneur en humidité d'environ 17% après quoi il a diminué rapidement. Cette teneur en humidité maximale est comparable aux prédictions faites en laboratoire. Le niveau des vibrations a également diminué de façon significative au fur et à mesure que la teneur en humidité augmentait.

1 INTRODUCTION

Dynamic compaction is a process for densifying soils by repeatedly dropping a large weight (10 to 20 metric tonnes) from heights of 20 to 30 m (80-100 ft) with a crane. This process has successfully densified a wide range of soils in-place to depths greater than 14 m (46 ft) (Lukas, 1986). Over the past 10 years, dynamic compaction has been employed at several sites in the western United States to mitigate the hazard posed by collapsible soils (Rollins and Kim, 1994). This procedure has been reasonably successful in reducing the potential for settlements associated with wetting of collapsible soils.

Although it is common practice in the construction of earth dams to pre-wet a dry borrow area prior to excavation and placement, dynamic compaction is normally performed at the natural moisture content of the soil. Limited field test data and experience with lab proctor testing indicates that compaction efficiency is related to the soil moisture content and that an optimum compaction moisture content likely exists (Rollins and Rogers, 1990). This would suggest that increased compaction efficiency could be obtained by pre-wetting the soil to the optimum moisture content prior to dynamic compaction. In addition, limited field test data suggest that vibrations associated with dynamic compaction could be reduced by increasing the natural moisture content of the soil. However, if the moisture content becomes too great, large pore water pressures would be generated upon impact and little densification would occur.

To investigate the influence of soil moisture content on dynamic compaction, a series of full-scale dynamic compaction tests were performed at six test cells with different average moisture contents. The field test procedure was similar to a large scale Proctor test. Investigations were carried out before and after the treatment at each test cell so that it would be possible to determine: (1) how the moisture content variation affected the degree of improvement (2) if an optimum moisture content could be observed for dynamic compaction, and (3) whether or not the degree of improvement observed in the field could be predicted by lab testing procedures.

2 SITE CHARACTERIZATION BEFORE DYNAMIC COMPACTION

2.1 General Site Conditions and Geology

The test site was located on the lower end of an alluvial fan at the base of the Wasatch Mountain Range in Nephi, Utah, U.S.A. Most of the upper portion of the profile was deposited by mudflows and consists of sandy silts with some thin layers of silty gravel. Annual rainfall is less than 350 mm/yr (13.8 in/yr) and except for surface layers, the soils are never saturated subsequent to deposition. Repeated deposition has produced a layer of collapsible soils which is at least 30 feet thick. Shales in the drainage basin above the site provide sufficient amounts of clay to act as a binder and create a meta-stable structure which is strong in its dry natural state but can settles significantly when wetted

under load. Accidental wetting events near the test site have caused damage to houses, roadways and streets.

2.2 Description of Soil Profile

The soil profile at each of the test cells was defined by excavating a trench adjacent to the cell to a depth of 6m (20 ft) using a trackhoe. The soil profile was quite consistent from cell to cell and consisted of a reddish brown low plasticity sandy silt with 0.3m-thick silty gravel layers at depths of 0.6 and 1.8 m (2 and 6 ft) below the ground surface. The silty soil has a liquid limit between 20 and 25 and a plastic index of about 6. The fine-grained soil generally classifies as ML or CL-ML material according to the Unified Soil Classification system. A typical gradation consists of 20% sand, 70% silt and 10% clay. The gravel layers generally contained between 10 to 15% cobble size materials and classified as GM soils.

While the trenches provided an excellent opportunity to evaluate the consistency of the soil layering, they were excavated primarily so that high quality block samples could be obtained. Block samples of the fine-grained soil were carefully cut from the sidewalls of each trench at approximately 0.6m (2 ft) intervals to a depth of 6m (20 ft). The block samples were then wrapped in plastic bags and transported to the lab for testing.

2.3 Density and Oedometer Testing

To provide a baseline for evaluating improvement at each test cell, ring samples were carefully trimmed by hand from each block sample obtained in the field. Based on the results of this testing, plots of the void ratio as a function of depth were determined for each test cell prior to treatment as shown in Figure 1. The void ratios were generally between 0.9 and 1.1 and tended to decrease somewhat with depth. Natural moisture contents prior to moisture conditioning were generally found to vary between 6 and 12%.

In addition to determining the in-place void ratio, the collapse strain upon wetting was determined for each ring sample by oedometer

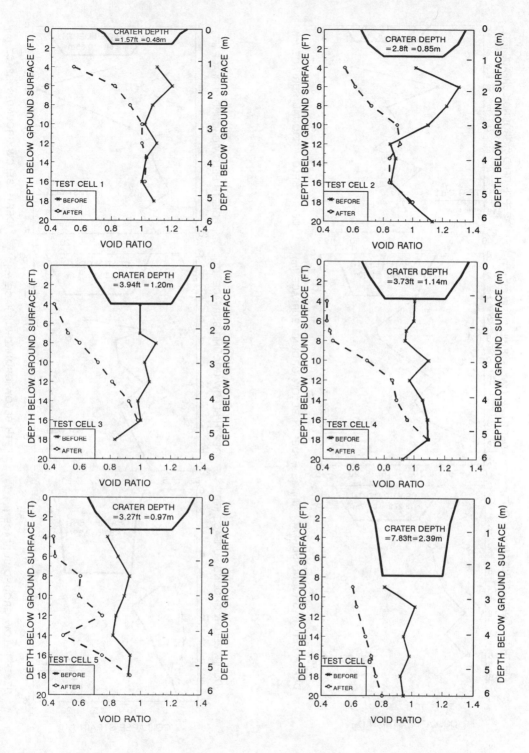

Figure 1 Void ratio profiles as a function of depth before and after dynamic compaction at test cells 1 through 6.

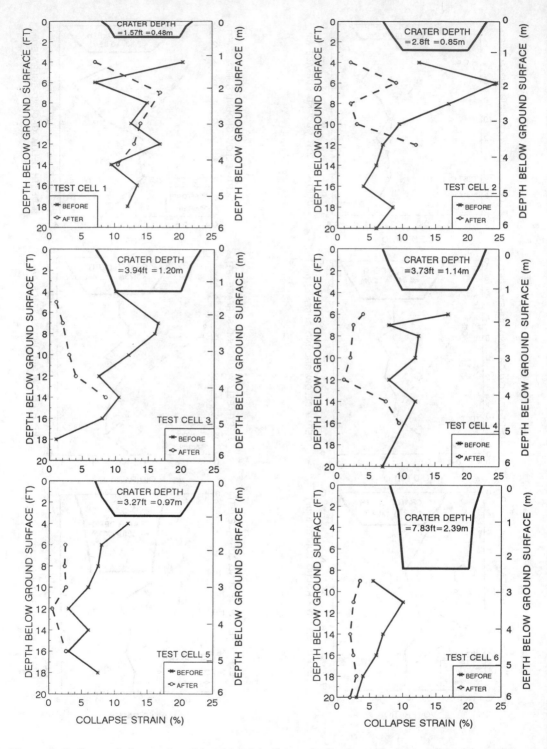

Figure 2 Collapse strain as a function of depth before and after dynamic compaction at test cells 1 through 6.

testing. Each sample was gradually loaded to a pressure of approximately 100 kPa (1 tsf) at which point the sample was saturated. The collapse strain, ϵ_c, upon wetting was determined using the equation,

$$\epsilon_c = \frac{\Delta e}{(1 + e_o)} \qquad (1)$$

where Δe is the void ratio change following wetting and e_o is the initial void ratio at the start of the test. The test was then continued to a final pressure of 1800 kPa (18 tsf). Profiles of collapse strain versus depth for each test cell are shown in Figure 2. The collapse strains shown in Figure 2 generally decrease with depth, as was the case with the void ratio plots, and typically range from 5 to 20%. The variation in the collapse strain is greater than was observed in the void ratio data as might be expected.

3 MOISTURE-DENSITY RELATIONSHIPS AND MOISTURE CONDITIONING

3.1 Moisture-density relationships

To estimate the optimum moisture content and the range of moisture contents which would be desirable to obtain in the field testing, lab Proctor testing was performed. Disturbed samples of the soil were obtained and compacted at three constant energy levels with increasing moisture contents using procedures specified by ASTM D-698 and D-1557. The relationship between dry density and moisture content for the silty soil at the test site is shown in Figure 3 for compactive energies associated with the modified proctor, the standard proctor and one-third of the standard proctor.

The results of the testing show the typical increase in dry density up to an optimum moisture content followed by a decrease in dry density beyond the optimum moisture content. As the compactive energy decreases, the maximum dry density decreases and the optimum moisture content increases. A line connecting the optimum moisture content for each compactive energy in Figure 3 would roughly parallel the zero air voids line. Typically, the energy applied to a site by

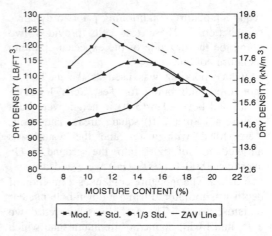

Figure 3 Relationships between dry density and moisture content for compactive energies defined by modified proctor, standard proctor and one-third of the standard proctor.

dynamic compaction is between 40 and 60% of that applied in a standard Proctor test. The applied compactive energy for this study was expected to be near the low end of this range. For this compactive energy level, the Proctor testing suggests that the optimum moisture content would be between 16 and 18%.

3.2 Moisture conditioning

Ideally, it would be desirable to produce a soil profile near optimum along with two profiles on the dry side and two profiles on the wet side of optimum for utilization in the field testing. In addition, it would be ideal to have the moisture contents at each test cell separated by 2 to 3% and for each soil profile to have a relatively constant moisture content within the anticipated depth of improvement. However, these objectives must be tempered by the difficulty of producing cells with uniform moisture contents at arbitrary values under field conditions. Even in the construction of earth dams where the compaction moisture content is particularly important, it is generally not possible to achieve variations of less than about $\pm$ 2 percentage points about the desired moisture content.

Test Cells 1 and 2 had average moisture contents of approximately 6 and 10% respectively in their natural state and did not

249

require moisture conditioning prior to dynamic compaction. These test cells provided two points on the dry side of the expected optimum moisture content.

Surface ponding was used as the method of moisture conditioning for Test Cells 4 and 5. Small soil berms 1m (3 ft) in height were built around a 3.8m (12 ft) square area. The berms were filled with water and the water was allowed to infiltrate into the ground. On average, 3800 L (1000 Gal) were added every two days. The moisture content as a function of depth was monitored using a down-hole nuclear moisture probe. The probe was lowered down a 75 mm (3 in) diameter aluminum tube which extended to a depth of 6m (20 ft). The progression of the wetting front with depth as a function of the volume of water applied is illustrated in Figure 4.

The soil behind the wetting front generally reached a final moisture content between 20 and 25% which corresponds to a degree of saturation between 50 and 70%. Following wetting, the test cells were allowed to dry out and drain until a desired moisture content was reached. The ponding operations at the two test cells were offset by a month in an effort to create a difference in the average moisture content in the two test cells. Unfortunately, the average moisture content at both cells treated by surface ponding differed by only a percentage point or two and was close to the predicted optimum moisture content of 17% at the time of compaction.

At Test Cell 3, three 300 mm (6 in) diameter holes were drilled to a depth of 6m (20 ft) using a hollow-stem auger prior to surface ponding. The vertical holes gave the water access to the deeper soil layers without causing the moisture content in the upper layers to increase to the level normally required to cause downward transmission of the water. With the addition of 7600 L (2000 Gal) of water, the soil moisture content was increased to an average of roughly 17% to a depth 4.3m (14 ft below the ground surface. At the time of compaction, the average moisture content at this test cell had decreased to approximately 15%, which was somewhat below the expected optimum moisture content.

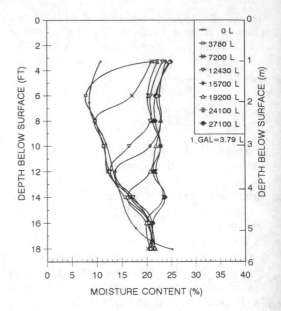

Figure 4   Increase in moisture content as a function of depth after each application of surface water and subsequent infiltration.

Therefore, this test cell provided a third point on the dry side of optimum.

Test Cell 6 was prepared with the goal of obtaining a soil profile with a uniform moisture content above 20% which would represent a point on the wet side of the expected optimum. About two weeks before compaction, a soil berm was placed around the test cell and surface ponding was performed using the same procedures employed for Test Cells 4 and 5. Approximately 30,300 L (8000 Gal) of water was applied to the test cell. This produced a soil profile with a moisture content between 18 and 22% to a depth of 5.5m (18 ft) at the time of compaction. A plot showing the moisture content as a function of depth for the six test cells immediately prior to compaction is shown in Figure 5. Although the variability of the moisture content with depth is greater than ideal, the range in average moisture contents was sufficient to allow a comparison of the moisture content effects on compaction efficiency.

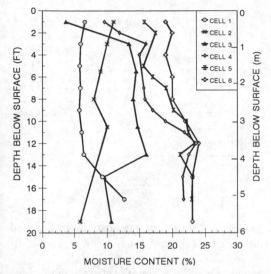

Figure 5 Moisture content as a function of depth for each test cell immediately prior to dynamic compaction operation.

4 DYNAMIC COMPACTION OPERATIONS

4.1 *Compaction Procedure*

The compaction process was carried out with 60 tonne capacity crane having a free-spool drum and a single cable. The tamper was a 4.54 metric tonne (5 ton) cylindrical reinforced concrete weight with a steel shell. The tamper was 1.3m (4.25 ft) in diameter and 1.3m (4.25 ft) high. The weight was dropped seven times at each corner of the test cell from a height of 24.3m (80 ft). The time interval between drops was typically 1.5 minutes. To compare the field compaction testing with the lab Proctor tests it is necessary to divide the applied energy by the treated volume. The center to center spacing of the drop weight was 2.4m (8 ft), which leads to a treated surface area of 5.95 m^2 (64 ft^2). The anticipated depth of improvement, D_{max}, can be estimated based on the following equation proposed by Lukas (1986) for low plasticity silts,

$$D_{max} = 0.4 \sqrt{W H} \qquad (2)$$

where W is the weight of the tamper in metric tonnes and H is the drop height in meters.

Equation (2) predicts that improvement will occur to a depth of 4.3 m (14 ft). Therefore, the energy per volume used in the field compaction testing is approximately 30.2 tonne-m/m^3 (3.12 ton-ft/ft^3), which is about half of the energy per volume used in the standard proctor test.

4.2 *Crater Depth Measurements*

During the compaction process, the crater depth following extraction of the tamper weight was measured after each drop using a level and a stadia rod. A review of this data indicates that the variation in the final crater depths at any given test cell was generally less than $\pm$ 10% of the average final crater depth at that cell. Final crater depths ranged from 0.48m (1.57 ft) at the driest cell to 2.34m (7.83 ft) at the wettest cell. Figure 6 presents the average normalized crater depth versus the number of drops at each test cell. The crater depths have been normalized by dividing by the square root of the drop energy so that they can be compared with the typical range of normalized crater depths reported by Mayne et al (1984). From these curves, it is clearly evident that crater depth is directly related to the amount of moisture present in the soil. The penetration is most influenced by the moisture content near the ground surface.

The crater depth curve for the driest cell (7%) was slightly less than the typical range, while the curve for the wettest cell (20%) was more than twice as large as the typical range observed for dynamic compaction. The curve for Test Cell 2, with a moisture content of 10% plotted almost down the center of the typical range; however, the curves for cells with a moisture content near the optimum (15 to 18%) plotted near the maximum limit of the typical range.

The observed increase in crater depth with moisture content likely relates to the fact that the shear strength and hence the penetration resistance decreases as the moisture content increases. The large penetration of the weight at the highest moisture content made it very difficult to extract the weight from the ground and significantly increased the time between drops. In fact, the force required to extract the weight eventually snapped a strap and caused

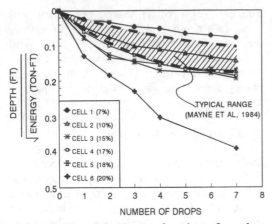

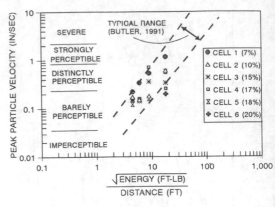

Figure 6 Crater depth as a function of number of drops for each test cell.

Figure 7 Vibration attenuation with scaled distance for each test cell.

damage to the crane which prevented compaction at all four corners of Test Cell 6.

4.3 *Vibration Attenuation*

The most common limitation of dynamic compaction is the effect of ground vibration on adjacent facilities. The high energy ground vibrations produced by dynamic compaction can be felt over significant distances. Generally, a particle velocity of 25 mm/sec has been considered the threshold of structural damage.

A portable seismograph was used to obtain vibration measurements at each cell. Throughout the compaction process, readings were taken with the seismograph at distances of 15.2, 30.5, 45.7, and 61 m (50, 100, 150 and 200 ft) from the point of impact. A plot of the peak particle velocity (PPV) as a function of scaled distance from the impact point is shown in Figure 7. The scaled distance is the square root of the energy per drop divided by the distance from the drop and is used to facilitate comparison with data from other investigators.

Figure 7 shows that at short distances from the drop point (highest scaled distances) there is a general trend for the PPV to decrease with increases in the average moisture content. The driest cell had a PPV near the upper limit of the range defined by Butler (1991) while the wettest cell had a PPV below the lower limit. As the distance from the drop point increases (scaled distance decreases) the effect of moisture content

becomes much less significant since most of the soil between the drop point and the seismograph is still relatively dry. At these higher distances the measured PPV's are generally well within the range defined by past experience.

5 SITE CHARACTERIZATION FOLLOWING DYNAMIC COMPACTION

5.1 *Post Compaction Sampling*

Approximately 2 weeks following the dynamic compaction, 6m (20 ft) deep trenches were excavated through each test cell so that high quality undisturbed block samples could be obtained directly below drop point locations. The samples were necessary to determine the improvement in density in comparison with the initial conditions defined previously. Previous experience at the site had shown that it was not possible to obtain undisturbed samples of moist collapsible soils using conventional thin-walled Shelby tube procedures. Block samples were again cut from the side walls of the trench at approximately 0.6m (2 ft) intervals below the base of the crater to a depth of approximately 6m (20 ft).

In addition, the moisture content was determined at each depth immediately after sampling. The average moisture content in the upper 4.3m (14 ft) of the profile was 7, 10, 15, 17, 18, and 20% for Test Cells 1 through 6, respectively.

Immediately below the base of the crater, the soil was extremely dense and brittle. The soil had the appearance of a weathered rock with horizontal laminations. Because of the brittle nature of the soils in this region it was very difficult to obtain a block sample less than 0.5m (1.6 ft) below the base of the impact crater.

The trenches also made it possible to view the deformation pattern created by the compaction process. At the wettest location, Test Cell 6, the weight appeared to punch through the soil and the gravel layers at 1 and 6 foot depth were abruptly displaced as if on a shear plane. However, at all the other test cells, the weight created a more concave shaped deformation pattern and the gravel layers at depth were still continuous with the greatest deformation occurring directly below the footprint of the tamping weight.

5.2 *Post Compaction Density and Oedometer Testing*

Ring samples were once again cut from the block samples in the lab so that the void ratio and density could be determined for each test cell as a function of depth. The profiles of void ratio after dynamic compaction for each test cell are shown in comparison with the profiles before treatment in Figure 1. The cells are arranged in order from driest to wettest. A visual inspection of the plots indicates that there is some decrease in the void ratio to a depth of between (4.3 and 4.9m (14 and 16 ft) below the ground surface. This is in good agreement with the depth of improvement predicted by equation (2). While the depth of improvement is not significantly affected by the moisture content, the degree of improvement does appear to increase with the moisture content. For example, the area between the before and after curves for Test Cell 1, which was the driest, is much smaller than that observed at Test Cells 4 and 5 which were near the expected optimum moisture content.

Oedometer collapse tests were also performed on each of the ring samples to determine if the dynamic compaction was effective in reducing the potential for collapse and how this reduction was influenced by the soil moisture content. However, because the moisture content of some of the samples was relatively high, these samples settled significantly prior to saturation of the sample at the 100 kPa (1 tsf) pressure and did not collapse. To allow comparison with the collapse strain measured prior to dynamic compaction, the Δe_c in equation (1) was re-defined as the difference between the initial void ratio and the void ratio at a pressure of 100 kPa (1 tsf). Profiles of the "collapse strain" as defined in this manner are shown in Figure 2 in comparison with values prior to treatment. These plots show that reductions in collapse strain are greater when dynamic compaction is performed with the water content above that in the dry natural state.

6 COMPARISON BETWEEN LAB AND FIELD COMPACTION TESTING

Plots of dry density versus moisture content following the dynamic compaction in the field are shown in Figure 8. The plots were prepared for depths of 0.6, 1.2, 1.8, 2.4, 3.0 and 3.6m (2, 4, 6, 8, 10 and 12 ft) below the base of the impact crater. The dry density and moisture content for each point represent the weighted average values above the specified depth at one test cell. The data in Figure 8 indicate that there is indeed an optimum moisture content associated with dynamic compaction. The maximum density decreases and the optimum moisture content decreases with depth below the ground surface. This likely results from the fact that the compactive energy produced by the dynamic compaction decreases with depth and lab proctor testing indicates that the optimum moisture content increases as the compactive energy decreases.

The dry density versus moisture content curves for the modified proctor, standard proctor and one-third of standard proctor are also shown with dotted lines in Figure 8. In general, the field curves have shapes similar to that predicted by the lab testing and the dry densities are typically between that for the standard proctor and the one-third of standard curves. Considering that the energy employed in the dynamic compaction work was about 50% of the standard proctor energy, the agreement is quite good.

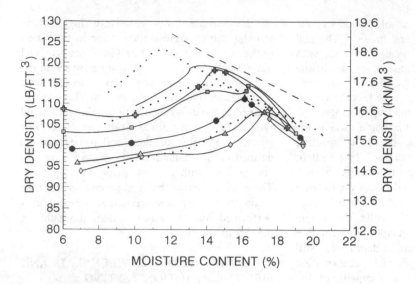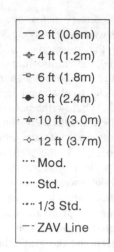

Figure 8 Comparison of lab moisture vs density relationships with relationships based on the field testing using dynamic compaction.

7 CONCLUSIONS

Based on the results of the field and lab tests conducted in this dynamic compaction testing program the following conclusions are made:

1. The concept of an optimum moisture content is valid for dynamic compaction. The optimum moisture content tends to increase with depth because the compactive energy produced by the dynamic compaction decreases.

2. Measured dry densities and optimum moisture contents in the field are in reasonable agreement with those predicted by lab testing considering the compactive energy employed in the dynamic compaction.

3. The crater depth increases as the moisture content increases. Crater depths ranged from 0.48m (1.57 ft) at 7% moisture to 2.4m (7.83 ft) at 20% moisture. At moisture contents beyond the optimum, crater depths become excessive and do not always reflect greater improvement. The deeper craters make extraction of the weight difficult and time-consuming.

4. There is a tendency for vibrations to decrease when compaction is performed at higher moisture contents. However, this reduction becomes insignificant at greater distances when the wetted area being treated is small.

8 REFERENCES

Butler, R.C. 1991. Ground improvement using dynamic compaction. *Geotechnical News*. June, pp. 21-27

Lukas, R.G. 1986. *Dynamic compaction for highway construction, Vol. 1 design and construction guidelines*. U.S. Federal Highway Administration, Report No.FHWA/RD86/133, Washington, D.C.

Mayne P.W., J.S. Jones, and J.C. Dumas 1984. Ground response to dynamic compaction. *J. Geotechnical Engineering*, ASCE, vol. 110 no. 6, pp. 757-74

Rollins, K.M. and G.W. Rogers 1991. Stabilization of collapsible alluvial soil using dynamic compaction. *Geotechnical Special Publication No. 27*, ASCE, vol. 1, pp. 322-33

Rollins, K.M. and J. Kim 1994. U.S. experience with dynamic compaction of collapsible soils, *Proc. ASCE Annual Convention*, Atlanta, GA, In-Press

Helium and Radon as gas tracers in the unsaturated zone

Hélium et Radon comme traceurs de gaz dans la zone non saturée

A. Keppler & W. Drost
GSF, Institute for Hydrology, Neuherberg, Germany

T. Fierz
Solexperts AG, Schwerzenbach, Switzerland

L. Kovac
STU, Institute for Geotechnics, Bratislava, Slovakia

ABSTRACT: Helium and Radon were used as gas tracers within the vadose zone in order to determine the travel time of the marked soil gas between an injection and an extraction borehole and to describe the dispersivity of the soil matrix. The results of the vadose zone tracer testing give valuable insight into the transport mechanism of contaminants and can be used to optimize and to reduce the costs of a soil venting system on a remediation site.

RÉSUMÉ: L' application des traceurs gazeux Helium et Radon pour la détermination du passage du gaz du sol entre un point d'injection et un point d' observation offre la description conséquente de la dispersion du gaz dans le sol. Le résultat montre et elucide le mécanisme du transport des contaminents dans le sol et est employé pour l' évaluation meilleure et la réduction du frais des traveaux pour la restauration du sol qui est contaminé par résidus gazeux d´industrie.

1 INTRODUCTION

A detailed knowledge of the vadose flow is neccessary for a better understanding of the transport of volatile pollutants like organic hydrocarbons (VOC) in the vadose zone. The aim of a field experiment at a remediation site in Germany was to demonstrate a pneumatic connection between two boreholes and to determine the flow condition within the soil matrix in order to prove a successful remediation of a contaminated area.

2 GAS TRACERS IN THE VADOSE ZONE

The new vadose tracer testing method is illustrated in figure 1. A known volume of conservative helium-4 is introduced into an injection borehole, which is drilled down to the groundwater-table and sealed with Bentonite. The tracer is mobilized by the induced flow field and is transported by the soil gas flow to the vapor extraction well. The helium tracer concentration of the extracted vapor is measured on-line with a mass-spectrometer, which allows a continuous real-time monitoring.

The gas tracer measurement result in a helium breakthrough curve, which depends on the dispersivity of the soil matrix. Interpretation of the observed breakthrough curve and the measured tracer recovery rate allows the description of the dispers behaviour of the vadose zone.

The results can be used to optimize the number, position and the distance of boreholes used for soil venting systems, which helps to minimize the costs of installation and maintenance.

In order to obtain corroborativ results the active He-tracer testing method was combined with a passive tracer experiment using in-situ radon . Radon is a radioactive noble-gas which is produced by the radioactive decay of the naturally occurring element Radium within soil, water or rock. The concentrations within the vadose zone are 30-100 kBq/m^3 and therefore several orders of magnitude higher than the concentration in the athmospheric air or in the

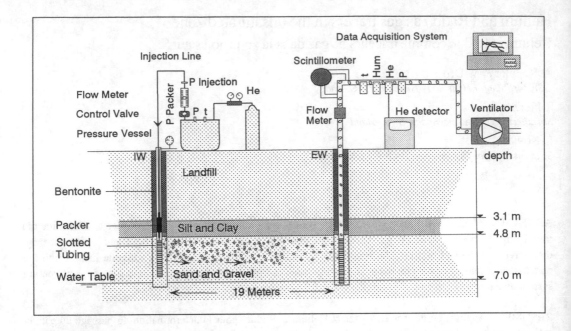

Figure 1: Configuration of the in-situ vapor tracer experiment including the geological-hydrogeological situation of the test site Heddernheim (Germany)

precipitation. The radon content in the vadose zone and the groundwater is not constant. It is influenced by the water-content of the soil and shows short-time and seasonal variations. The underground concentration is increasing with the water-saturation of the vadose zone and the surface flux depends on climatic conditions.

The Radon concentration of the soil gas from the extraction borehole represents a mixture of vadose zone levels and athmospheric levels due to the dilution effects of passive venting wells, which feed radon-poor air into the underground system. As well as the He-experiment, the Rn-tracer experiments are based on the stimulus-response principal, where a system is stimulated by an impulse at location A and the response is observed at location B.

By sealing a venting well with an air-tight cap or by removing a cap from a sealed well at a certain distance from an extraction well, a "negative" Radon step function is induced giving an indirect measure of the soil dispersivity. The result of the step-function can be monitored on-line using Scinti-

lation counters or a Lucas cell at the vapor extraction borehole.

The passive method is a simple tool to test a possible pneumatic connection between two or more boreholes.

3 THE TECHNICAL SETUP

The technical basis for the helium gas tracer testing was developed by SOLEXPERTS LTD together with the Paul-Scherrer-Institute (Villingen, CH) and the Institute for Hydrology (GSF, Munich, D) as part of the Migration Experiment at the Grimsel Test Site supported by NAGRA (Wettingen, CH).

The gas tracer methods were used as part of a remediation plan at a VOC-contaminated site at Heddernheim, Germany (management: MG GEO-CONTROL), where soil venting systems have been installed to decontaminate the unsaturated zone.

The main goal of the test was to determine the radius of influence for different vapor extraction wells. The position of the different wells is shown

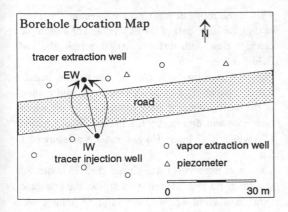

Borehole Location Map

tracer extraction well

EW

IW
tracer injection well

○ vapor extraction well

△ piezometer

road

0   30 m

Figure 2: Borehole location map of the test site
Heddernheim (Germany).

in the borehole location map (figure 2) and a
simplified geological cross-section of the remediation site is shown in figure 1.

A vacuum pump was extracting soil gas with a
constant flow rate of 900 l/min from a vapor
extraction well (EW) located 19 m from the tracer
injection well (IW). The most permeable zone is a
layer of sand and gravel with a thickness of 2.2 to
2.5 m in a depth between 4.8 and 7 m below
surface. It is enbedded by the water table below and
a relatively impermeable silt and clay layer above.

The injection borehole was installed with a 2" PVC
pipe with a slotted screen length within the sand and
gravel horizon. The upper part of the borehole was
cemented and filled with Bentonite to provide an
air-tight seal. A pneumatic packer was installed
within the impermeable zone, which allows a tracer
input without disturbing the soil gas flow system.

Helium tracer input was conducted using a minimal
overpressure of only 15 mbar in order to avoid
alteration of the flow field within the vadose zone.
A pressure vessel equipped with a gas flow meter,
needle valve and pressure transducer was used for
the tracer input. With the known volume of the
vessel, the measured gas flowrate and the the the
constant injection pressure, the helium input function and the volume of injected tracer were defined
and calculated.

The helium concentration of the extracted soil gas
was measured on-line with a helium-leak-detector.
The tracer concentration was graphically displayed
and stored by a PC-based data acquisition system.

The radon concentration was measured bypass within a lead-protected vessel by means of a Scin-
tillation-counter detecting the γ–rays. Helium
concentration measurements were made with a
frequency of 1 minute. The detection limit for the
helium leak detector is approximately 0.2 ppm, the
background concentration about 5.2 ppm. In addition, the pressure at the extraction well, the gas
flowrate, gas temperature and humidity were
measured on-line.

Passive radon experiments were conducted together
with the helium tracer testing. Prior to the installation of the helium injection equipment, the open
injection well was continuously supplying radon-
poor athmospheric-derived gas to the permeable
layer of the vadose zone, which acted as a dilution
point source within the flow field of the extraction
well. By installing the injection equipment, the well
was sealed from the athmosphere and further dilution of the underground radon level was ruled out.
This induced step function increased the radon level
after a certain time in the vapor of the extraction
well. The athmospheric pressure and the humidity
were constant during the time of the experiment and
therefore the radon content of the soil gas and the
air close to the surface showed no variations.

4 RESULTS OF THE EXPERIMENTS

As a result of the gasflow measurements the helium
and radon breakthrough curves are shown in figures
3 and 4.

First arrival of helium occurred about 4.1 hours
after the tracer input. The same result was achieved
by the passive radon experiment. The Rn-gas concentration was rising about 4 hours after starting the
pumping at the extraction well (figure 3). At this
moment the vacuum pump is only extracting soil
gas, which is not more diluted by the athmospheric
gas with lower radon content. After 20 hours, the
radon concentration reached a constant plateau
value of normal enviromental conditions.

The breakthrough peak of the helium was observed
at about 12.6 hours after tracer injection with a peak
concentration of 45 ppm. This peak exceeds 8 times
the background concentration. The large spread of
the breakthrough curve indicates a high dispersion

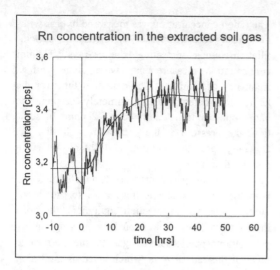

Figure 3: The radon concentration measured at the vapor extraction well (T_0= start of the vacuum pump)

in the gasflow within the sand and gravel horizon. The breakthrough data are rather smoothed during the initial test phase and start to fluctuate after 18 hours.

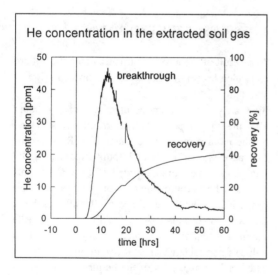

Figure 4: The helium breakthrough and recovery curve, corrected for the He-background in the soil gas (T_0= start of the He-injection).

This is the result of not constant vacuum flowrates during the latter part of the experiment or a multiple channel flow with different length within the test field.

Figure 4 is showing a plot of the helium breakthrough curve. The recovery rate is the summation of the helium concentration multiplied by the vapor flow rate and divided by the helium input volume M_0, which is 97 l pure He-gas under a pressure of 1 bar.

The tracer experiment was closed after 70 hours. At this time, 1.5 ppm helium were still in the extracted vapor and about 42 % of the injected helium was recovered at the extraction site. The fate of 58 % of the injected tracer volume may be related to several possibilities:

- The observation time was to short, a longer test period would give a higher helium recovery rate. It could be possible, that additional peaks could appear because of longer flow pathes.

- An indefinite amount of helium migrated on its pathways upward through possible discontinuities in the silt and clay layer.

- A certain percentage of helium was retarded by the soil matrix within dead end pores and static flow zones.

The tracer tests indicate that the vapor extraction well is influencing a fairly large radial area and is efficiently mobilizing vapor-phase constituents. Based on the results of the vadose zone tracer testing, a number of vapor extraction wells were deemed to be unnecessary and taken out of operation in order to save costs.

5 CONCLUSIONS

The planning of a soil venting system for remediation of VOC contaminated sites assume detailed informations about the dispersive flow system within the vadose zone and the radius of influence of individual vapor extraction wells.

Helium gas and the noble-gas radon have been proved to be a valuable tracer tool for flux measurements within the unsaturated zone:

- Both elements are not toxic and show an inert behaviour. They act as a conservative tracer and are not affected by adsorption, absorption and other chemical processes.

- They can be measured on-line at a low detection limit, which avoids cost for sampling and laboratory analyses.

- Helium has a very low and constant background concentration.

In-situ tracer investigations on a remediation site help to reduce the costs for a long-term operation of soil venting systems. By reducing the number of required boreholes, installation and operation costs can be minimized without reducing the quality of soil remediation. The use of passive Radon tests has been shown to be a valuable tool for the characterisation of the vadose zone. It is a useful tool to test a possible pneumatic connection of two boreholes on a simple way, where quantitative tracer flow and recovery rates are not required. It can also be used as preparatory experiment for helium tracer tests.

6 FUTURE PLANS

It is neccessary to improve the sensitivity of the radon measurement by using a Lucas-cell or an α-Scintillation-counter. This technique makes an on-line observation of the more frequent α-radiation possible, which results from the radioactive decay of radon-219, -220 and -222.
Further radon tracer tests in the unsaturated zone with different geological situations are neccessary to investigate the application of the method without a reference tracer.

REFERENCES

Cothern, C.R. & Smith, J.E. (1987): Enviromental Radon; Plenum Press, New York.

Fierz, T., Fisch, H., Herklotz, K., Schwab, K., Bielesch, H., Keppler, A., 1993: Durchführung eines Tracerversuchs mit Helium und Radon in der ungesättigten Zone im Rahmen einer Bodenluft-sanierung eines Altlaststandortes; In: Altlasten-spektrum, ITVA, November 1993.

Lederer, M. & Hollander, J. (1968): Table of Isotopes; Wiley & Sons, New York.

Philipsborn v., H. (1990): Radon und Radonmessungen. Teil I: Eigenschaften, Meßgrößen und Methoden; Die Geowissenschaften, 8. Jahrg. 1990, Heft Nr. 8.

Philipsborn v., H. (1990): Radon und Radonmessungen. Teil II: Geräte und Verfahren, Vorkommen und Verbreitung, Strahlenbiologie und Strahlenschutz; Die Geowissenschaften, 8. Jahrg. 1990, Heft Nr. 10.

SPT – Compressive strength correlation for some hard soils-soft rocks

Corrélation SPT – Résistance à la compression pour quelques roches tendres et sols durs

Maher Atalla
Egy-Tech Group, Consulting Engineers, Cairo, Egypt

ABSTRACT: Based on the site investigation for four sites, a trial has been made to correlate the SPT or its extrapolation with the unconfined compressive strength for five geometrials, namely: shale, soft sandstone (poorly cemented sandstone), hard clay, weak calcareous mudstone (marl) and calcareous cemented sand (Calcarenite). The trend of this correlation is compared with the findings of Stroud for insensitive clays, soft rocks and glacial materials (Stroud line), and also with the extension of Terzaghi and Peck line for cohesive soils (T & P line). Another two lines are suggested to cover the correlation between the SPT and compressive strength for the different considered geomaterials.

RÉSUMÉ: Basé sur l'investigation du sol pour quatre sites, un essai a été fait pour corrélationner le SPT ou son extrapolation avec la résistance à la compréssion de cinq géomatériaux: shale, grès tendre (faiblement cimenté), argile dure, glaise calcaire tendre (marne) et sable calcaire cimenté, (calcarenite). La tendence de cette corrélation est comparée avec les conclusions de Stroud pour les argiles inscnsibles, les roches tendres et les matériaux glacials, aussi avec l'extension de la lignc dc Terzaghi et Peck pour les sols cohésives. Deux autres lignes sont proposées pour la corrélation entre SPT et la résistance à la compréssion des différents géomatériaux considérés.

1. INTRODUCTION

It is meant by hard soils - soft rocks, the materials which lie in the transion zone between soils and rocks. Alternatively, the terms rock similar soils or weak rocks are occasionally used. However, hard soils - soft rocks are used in the context that they mean weak rock materials (whether argillaceous, arenaceous or calcareous) not including weak rock masses due to closely spaced discontinuities.

While the use of the standard penetration test (SPT) in the estimation of the allowable bearing stress and settlement prediction was originated by Terzaghi and Peck (1948) for cohesionless and cohesive soils, only from mid seventies the SPT was introduced as a tool for investigating hard soils-soft rocks. Typical criticism to the interpretation of the SPT results can be summarized as it needs an engineering judgement and subjected to several corrections. SPT is sometimes described as crude, misleading and unreliable.

However, the indication that the SPT can be used to assess the properties of weak rock materials was reported on by Stroud(1974), Stroud and Butler (1975), Tomlinson (1976), Meigh and Wolski (1979), Nixon (1982), Decourt and als (1988), Atalla (1993), Nevels and Laguros (1993), Stamatopoulos and Kotzias (1993).

From the other side, sampling and then laboratory testing of the weak rock materials would be practically difficult due to crumbling, included fragments of stones and the unavoidable disturbance of samples. Moreover, samples tested for unconfined compressive strength are usually fissured and results would be much affected by these fissures. Class of samples, got by core drilling, are considered usually as that corresponding to classification and stratification purposes.

This criticism to both the SPT and unconfined compressive strength suggest that trying to use them for rather complicated materials as weak rocks would be highly complicated task and needs a lot of evidence to be able to draw conclusions. However, due to the humple nature of both these tests, the trials to correlate between them, was discussed and reported on by the typical references for such attempts of Stroud (1974), Nixon (1982) and Meigh, Wolski (1980).

Five geomaterials which could be classified as hard soils-soft rocks are studied in the present work. Data was developed from the site investigation for four projects involving the considered geomaterials, namely: shale, soft sandstone, (Atalla, 1981),

Hard Clay (Atalla, 1989), weak calcareous mudstone or marl (Atalla, 1978) and calcareous cemented silly sand or calcarenite (Atalla and Benltayef, 1993). Table 1 describes the materials tested and the considered sites.

2 CORRELATION PARAMETERS

2.1 Penetration resistance

In the considered four sites, the penetration resistance has been measured by the SPT if the penetration of 45 cm is achieved. Where full penetration is not achieved, the test is stopped at 50 blows and the actual penetration noted in centimeters. The idea of extrapolating the number of blows to correspond to 30 cm penetration i.e N = 100 as equivalent for 50/15 cm or N = 300 for 50/5 cm is appealing. However, this would be done only as a tool for judgement about how the tested hard soil-soft rock is hard, in other words, how hard is hard.

2.2 Compressive strength

Undisturbed samples of weak rock materials are difficult to obtain and in many cases are unavailable. In the considered sites, samples for laboratory testing were those resulting from core drilling, SPT spoon (but very limited due to high penetration resistance), and by cutting blocks from test pits if the considered layer lies near the surface. In the cases where large diameter bored piles were drilled, block samples were trimmed from the product of the pile drilling.

Regarding the compressive strength, the

unconfined compressive strength, q_u, was carried out for the four sites and the undrained shear strength, c_u, was carried out for the hard clay site. q_u and c_u were found to be reasonably compatible.

Cylinders were used to carry the unconfined compressive strength. However, for the marl site, prisms were used with prism height double its width. In this case, the unconfined compressive strength was taken as equal to 3/4 the prism compressive strength.

2.3 Correlation factor f_1

Stroud (1974), Stroud and Bulter (1975) extensively compared the SPT N values with the undrained shear strength c_u for several insensitive clays (the most famous of which is London clay), soft rocks and glacial materials, using Stroud's simple correlation

$$c_u = f_1 N \quad kN/m^2$$

where f_1 is a constant essentially independent of depth and of discontinuity spacing and varried for the different tested materials form 6 to 4 kN/m^2 with plasticity indices from 10 to 65 % respectively.

In the present paper, f_1 was considered as equal to :

$$f_1 = \frac{q_u}{2N} = \frac{c_u}{N} \quad kN/m^2$$

The way used by Stroud (1974) to correlate with a very reasonable degree the N values with the mass shear strength from triaxial tests was by adjusting the horizontal scales to which they are plotted to give the best fit while the vertical axis is the depth below ground level.

3 MATERIALS TESTED

3.1 Oil storage tanks site, Sebha

For the considered shale strata, 50 blows of the SPT gave a penetration (to the nearest inch) ranging from 10 cm to 25 cm. Thus the extrapolated N would be equal to 150 to 60. The scatter of N value with depth is shown in Fig. 1. With respect to the soft Sandsone layer, Fig. 2, the N value ranged between 120 to 200.

Resutls of the unconfined compressive strength, q_u, for both soft shale and soft sandstone samples are given in Fig. 1 and Fig. 2 where q_u for the soft shale ranged between 250 to 1450 kN/m^2 and between 2500 to 7400 for soft sandstone. A statistical analysis for these results is given in Table 2.

When plotting the N and q_u values against depth, the best fit of the horizontal scales was adopted giving $f_1 = \frac{q_u}{2N}$ for the soft shale as 4 kN/m^2 and f_1 for soft sandstone as 14 kN/m^2.

A reduction of 15 to 20% was noticed in the shale N value for layers under the G.W.L. Such a reduction in the "strength" was not remarked in the unconfined compressive srength. The random scatter for both materials and both tests shows no depedence of depth. Results are obviously affected by samples fissures and discontinuity spacing in general.

Table 1. Materials tested and considered sites

Material	Description	Site	Depth of Testing
Soft shale	Hard calcareous silty clay, fissured and fissile. Natural moisture content less than the plastic limit. SPT N Values and q_u see Fig. 1.	Oil storage tanks, Sebha, Libya	20 m
Soft sandstone	Highly cemented sand interbeded in the shale layers and occasionally mixed with claystone. SPT N values, qu see Fig. 2.	Oil storage tanks, Sebha, Libya	20 m, thickness vary from 3 to 5 m
Hard clay	Overlain by 10 m of sand, organic silty clayey sand and cemented sand. Natural moisure content very close to ten platic limit. SPT, q_u see Fig. 3.	Tall Building, Bengazhi, Libya	35 m
Calcareous weak mudstone (marl)	Variety of calcareous sandy silty clay including limestone fragments and boulders and in some locations sandwiched between limestone layers SPT, q_u see Fig. 4.	El-Khoms II cement plant, El-Khoms, Libya	20 m
Calcareous cemented silty sand (calcarenite)	Overlain by about 10m of fill, marine deposits, silty sand. Considered material consists of alterations of very dense calcareous cemented silty sand marine bioclastic carbonate sand (calcarenite). Limestone layers are interbeded. SPT, q_u see Fig. 5	Adminsitrative offices Towers, Tripoli, Libya	35 m

Table 2. Measured values of SPT N and unconfined compressive strength q_u for the tested materials.

No.	Material	SPT N				q_u, kN/m2			
		No.	Min.	Max.	Mean	No.	Min.	Max.	Mean
1	Soft shale	60	60	150	100	23	250	1450	750
2	Soft sandstone	9	120	200	170	10	2500	7400	4600
3	Hard clay	135	18	250	50	20	72	1015	500
4	Cal.weak mudstone	31	36	200	100	6	930	1920	1200
5	Calcarenite	103	50	150	100	8	2500	10000	4000

Range of values* comprising mean ± one standard deviation or for 90% of resutls

Material	N	q_u, 2 cu kN/m2
1	85 - 115	400 - 1100
2	140 - 200	2800-6400
3	20 - 80	200 - 800
4	55 - 145	1000 - 1900
5	50 - 150	2500 - 10000

* Measured value in the whole tested depth.

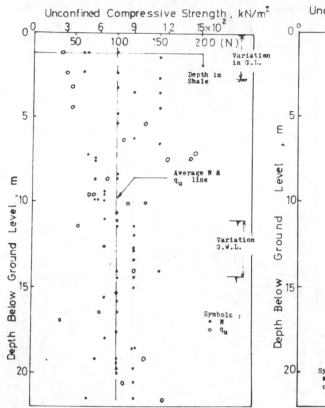

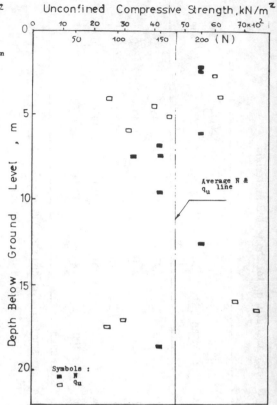

Fig.1 Correlation between unconfined
compressive strength and N for
soft shale

Fig. 2 Correlation between unconfined
compressive strength and N for
soft sandstone

3.2 Tall building site, Benghazi

Typical properties of the considered hard
clay layer which starts at about 10m depth
from the ground level are in average :
Specific gravity 2.7, water content 30% unit
weight 20 kN/m^3, liquid limit 70%, clay
content 50%, plastic limit 30%. The measured
SPT N values, unconfined compressive
strength and undrained shear strength for
the clay layer increase with depth with a
range as given in Fig. 3 and Table 2. The best
fit of the horizontal scales suggest a value of
f_1 for the considered hard clay as equal to 5
kN/m^2. f_1 in this case is independent of
depth.

3.3 Cement plant site, El-Khoms

Typical characteristics of the considered
calcareous weak mudstone (marl) are:

unit weight	18 to 25 kN/m^3
water content	7.0 to 12%
% passing sieve	
No. 200 by washing	30 to 70 %
plasticity	varies from non-plastic to highly plastic

Statistical analysis of the SPT N values and
the unconfined compressive strength are
given in Table 2. Variation of N with depth is
shown in Fig. 4. The ratio between half the
average unconfined compressive strength
and the average N for the tested mudstone
i.e. f_1 is estimated as 6.0 kN/m^2.

3.4 Administrative offices towers, Tripoli

Typical properties of the considered calcareous cemented silty sand (calcarenite) are given in Table 2 where in eight unconfined compressive tests, q_u ranged between 2500 and 10000 kN/m^2 with a mean of 4000 kN/m^2. another three values of q_u two of which are higher than 10000 kN/m^2 and

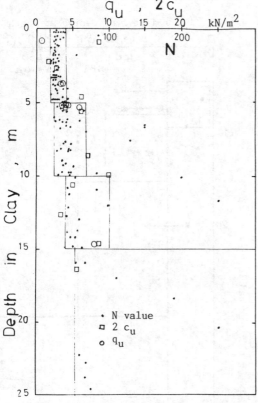

Fig. 3 Correlation between unconfined compressive strength, undrained shear strength and SPT N value, hard clay

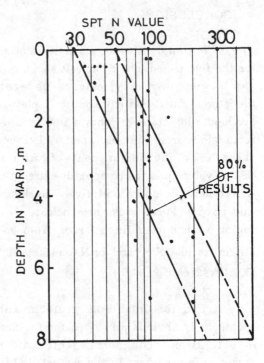

FIG. 4 SPT N VALUES OF MARL AS FUNCTION OF DEPTH IN MARL

one of 1100 kN/m^2 were excluded when estimating the mean value. The obvious scatter of q_u for the calcarenite may be explained by the wide range of carbonate-sand mixtures from calcareous sandstone to arenaceous limestone.

With respect to the N values, a simialr scatter is noticed (Fig. 5) but a range of 50 to 150 may correspond to the variation of N with depth. The best fit of the horizontal scales suggest a value of f_1 for the tested calcarenite as equal to 20 kN/m^2.

4 CONCLUSIONS

The SPT- compressive strength correlation for the five tested materials: soft shale, soft sandstone, hard clay, calcareous weak mudstone (marl) and calcarenite is plotted on double log scales as shown in Fig. 6. A rectangle representing each material covers about 90% of the measurements of both the SPT N values and the compressive strength. Accordingly, it can be tentatively concluded that hard soils-soft rocks are enclosed in a region with border lines for q_u from 0.4 MN/m^2 to 10 MN/m^2 and for N from 30 to 300 (50 blows/5 cm).

A trend fo continuous geotechnical spectrum is obvious and open for further investigation. The term geotechnical spectrum is used here in the context that it covers the range from soils to rocks with the hard soils-soft rocks as central position of the spectrum. Johnston and Novello (1993) indicated also the existance of such a spectrum.

The general trend of the proposed correlation is compared with findings of Stroud (1974, 1975) and also with the extension of the comparison between the SPT N values and unconfined compressive strength for cohesive soils proposed by Terzaghi and Peck (1948). These two lines (Fig. 6) are reasonably compatible with the proposed hard soils - soft rocks zone. Two another lines parallel to those of Stroud and Terzaghi and Peck are proposed with f1 = 10 and f_1 = 20 kN/m^2 to cover the rest of the tested materials. In other words, f_1 may vary from 4 to 20 kN/m^2 for the tested hard soil-soft rock matrials.

ACKNOWLEDGEMENTS

The material used in this paper is worked out basically from the author's technical consulting reports and the quoted in the paper published works. The author would like to thank Al-Fateh University civil engineering deparmtnet and the consulting engineering office for the kind permission to publish this work. Thanks are extended to Prof. Dr. M.A. Benltayef for many helpfull and stimulating discussions.

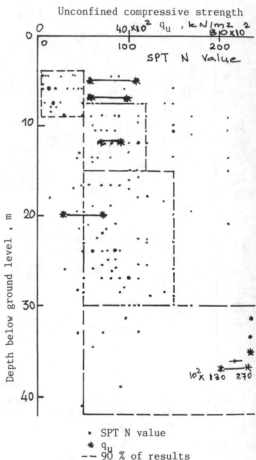

Fig. 5 Correlation between unconfined compressive strength and N for Calcarenite

268

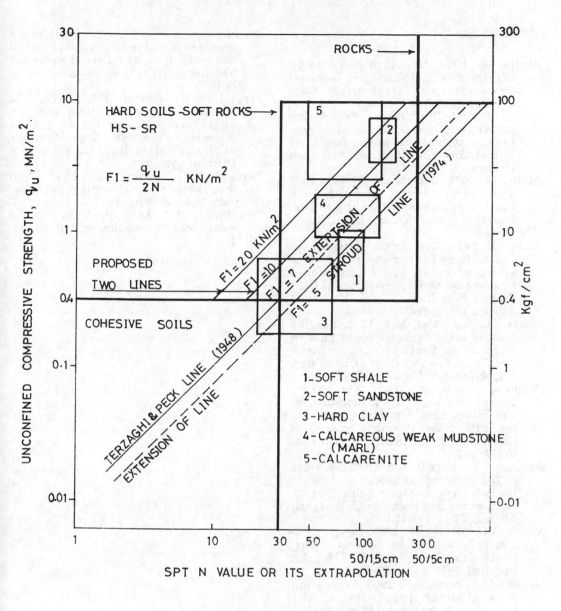

Fig 6 SPT–Compressive strength correlation for the
tested materials compared with Stroud´s line
and extension of Terzaghi and Peck line

REFERENCES

Atalla, M. 1978. Standard penetration resistance of El-Khoms marl-Technical consulting report, Consulting Eng. Office, Al-Fateh -Univeristy.

Atalla, M. 1981. Evaluation of Engineering properties of soft shale for oil storage tanks site. Bulletin of the faculty of Petroleum and Mining, Al-Fateh University, Vol.2, No. 2.

Atalla, M. 1989. Correlation between SPT and undrained shear strength for a hard clay layer. Journal of Eng. Research Center, Tripoli, Libya, Vol. 1, March.

Atalla, M. 1993. Description, classification and testing of calcareous rock-similar soils. proc. of the International Symposium on Geotech. Eng. of hard Soils - Soft Rocks, Publ. Balkema, Athens, Vol. 1; 21-27.

Atalla, M., Benltayef, M.A. 1993. Loading tests of large diameter bored piles in hard soils. Proc. of the 5th Arab Conflerence on structural Eng., Tripoli-Libya, Vol. 2: 1218-1229.

Decourt, L., Muromachi, T., Nixon, I.K., Schmertmann, J.H., Thornburn, S. and Zolkov 1988. Standard penetration test SPT: International reference test procedure, Proc. ISOPT 1, Orlando, Vol. 1: p. 3.

Jonston, I. W., Novello, E.A. 1993. Soft rocks in the geotechnical spectrum. Proc. of the Int. Sym. on Geotechnical Eng. of Hard soils - Soft Rocks, Publ. Balkema, Athens, Vol. 1: 177-183.

Meigh, A.C., Wolski, W. 1980. Design parameters for weak rocks. General Report, Proc. 7th Eur. Conf. Soil Mech., Brighton, 1979, Vol. 4: 59-79.

Nevels, J., Laguros, J.G. 1993. Correlation of enginering properties of the hennessey formation clays and shales. Proc. of the Int. Sym. on Geotecnical Eng. of hard Soils-Soft Rocks, Publ. Balkema, Athens, Vol. 1: 215-221.

Nixon, I.K. 1982. Standard penetration test: state of the art report. Proc. of the 2nd Eur. Sym. on Penetration Testing, ESOPT II, Vol. 1: 3-24, Amsterdam.

Stamatopoulos, A.C., Kotzias, P.C. 1993. Refusal to the SPT and penetrability, Proc. of the Int. Sym. On Geotechnical Eng. of Hard Soils - Soft Rocks, Publ. Balkema, Athens, Vol. 1: 301-306.

Stroud, M.A. 1974. The standard penetration test in insensitive clays and soft rocks. Proc. Eur. Sym. Penetration Testing ESOPT I, Stockholm, Vol. 2: 367-375.

Stroud, M.A., Butler, F.G. 1975. The standard penetration test and the engineering properties of glacial materials. Proc. Sym. Eng. Behaviour of Glacial Materials, University of Birmingham.

Terzaghi, K., Peck, R.B. 1948. Soil mechanics in engineering practice. John Wiley & Sons., P. 300.

Tomlinson, M.J. 1976. Preface, piles in weak rocks, "Geotechnqiue, Vol. 26, No. 1.

Direct shear tests for hydroelectrical projects in Mexico

Essais de cisaillement direct pour des projets hydroélectriques au Méxique

Roberto Uribe-Afif, Leonard Cañete-Enriquez & Sergio Herrera-Castañeda
Comision Federal de Electricidad, Mexico

ABSTRACT: Results of shear strength tests are presented, as a revisionof the experience about the evaluation of rock masses. All tests are employed in the characterization of sites for several hydroelectrical projects.

RÉSUMÉ: Les résultats des épreuves de cisaillement qu'on présent ici font partie de la révision de l'expérience d'évaluation des milieux rocheux. Tous les essais ont été conduits pour la caractérisation du lieu pour divers projets hydroélectriques.

1 INTRODUCTION

This paper presents the results of the - shear strength of rock masses of various lithologies, with discontinuities of diffferen characteristics obtained in several projects. A brief description of the methodology of essays for each project - is presented. The lithology of the place and the characteristics of the disconti- nuties that wer essayed are also. The - plots of the shear strength agains the - normal strength of each discontinuity - essayed are included. Finally, the re- sults are compared to evaluate the impor- tance of filling characteristics on the shear strength a given of rock mass.

2 ANTECEDENTS

In order to know the strength and defor- mability of the rock masses, in addition to the state of their internal stresses, in places were large projects are built, like dams, thermoelectric and nucledelec- tric plants, the Federal Comission of - Electricity (FCE) personnel have applied various tests methods in the laboratory and in the field to this purpose. These tests trat have been carried out, not very often due to its high cost. The projects in wich the tests have been per- formed are: C.H. Caracol, Gro.,C.H.

Chicoasen, Chis., C.H. La Angostura, Chis, in operation, P.H. Aguamilpa, P.H. Zima- pan, Hgo., in Construction, P.H. La Paro- ta, Gro., P.H. Itzantun, Chis., P.H. Boca del Cerro, Tab., in study, The Results of all tests have been used to Analysed the different Geotechnic Problems as: Inesta- bility of natural slopes, excavations and design and construction of open and under ground, foundations dams and structures - spillway.
The paper has also the purpose to present the experiences that had been obtained - during some years in shear strength tests that have been done in FCE.

3 IMPORTANCE OF THE TESTS FOR THE PROJECTS

As everybody knows the specific purposes of sher strength tests are to determine the peak and residual resistances in fun- ction of the normal stress acting on the shear plane or weakness plane. This pla- ne can be a fracture, a discontinuty, a plane of stratification or a contact bet- ween two kinds of rock, a foliation plane etc: The geothechnic problems may invol- vegreat economic consequences so for ins- tance, the price of treatment and support that are required to assert the safety of the works and any damages that will indu- ce failure due to a weak surface in the rock mass or the structures lodge settled on it. Therefore, the shear tests had been required to get information about of the discontinuities of the rock masses, and in this way go through with more complete studies of stability and good performance of hydroelectric plantas.

4 TEST PROCEDURE

The tests both in the field and in the laboratory, have a similar procedure (ref.1) that is described as follows: The discontinuity to test is selected and the relevant geological characteristics of it are registered, orientation, dip rugosity, filling type, etc., one block containing - the weakness plane. That will be essay is isolated. The usual dimensions of the - block oscilates between 30 x 30 and 60 x 60 cm and wide of 90 cm. The specimen is covered with concrete leaving free only n+E weakness plane. (fig. 1). Some specimens have been tested with it's natural - moisture content and other have been saturaded previously.
The test is performed in two stages, first The consolidation that consists in applying a normal stress on the specimen, measuring the displacements normal to the - weakness plane. With time until the change of displacement will be less than a - previously fixed limit or when the 100% of consolidation curve is reached. The second stage consists in applying a tangential stress on the front part of the specimen, the stress is gradually applied to get a maximum. That corresponds to failure and is defined as the maximum resistance value. Depending of the characteristics of the weakness plane, after failure a gradual or sudden increased in the rate displacement - stress on the specimen is observed. From this moment the tangential stress is applied in such way that the - displacement of the top of the specimen - is maintain to a constant speed of 0.5 mm /min. This value can be larger or sumaller but it has been observed that this speed does not have influence over the friction angle obtained in this waw (ref. 2). During this part of the test, after failure the normal and tangential displacements - are continuously registered in order to - detect any tangential charge variation. The test ends when the tangential displacement reaches 1.5 cm. After the test samples are taken from the filling of the - weakness plane to perform the conventional index tests (granulometry, consistens limits, water contined, etc.) The obtained data are plotted in the usual normal -- stress vs vertical displacement, normal

stress vs tangencial stress maximum and resiqual graphics.
These last two are described in point No. 6, in a brief form for each one of the - projects already mentioned.

5 CHARACTERISTICS OF THE DISCONTINUITIES TESTED IN EACH PROJECT

P.H. Aguamilpa, Nay., Pseudostratification plane with fill clay - sano. In the "Colorines" Lithological Unit.
Rhyodácitic lithic tuff pseudostratified P.H. Itzantun, Chis., Contact of the limestone units U2-U3, ligthly plastic clays located as stuff.
P.H. Itzantun, Chis., Brown clays of low compresibility (cl) located between limestone units 1 and, 3 to 5 mm wide combined with small fragments of limestone.
P.H. Zimapan, Hgo., Contact of doctor formation (limestone and dolomitic limestone) with the soyatal formation (carbonoseus shale) without any filling.
P.H. La Parota, Gro., Follation planes very well definite by the alternancy of micas bed (biotite) with quartz feldespatic beds.
As a complement an artificial contact was tested making a paralell cut to the foliation, as a contact between rock blocks.
P.H. Boca del Cerro, Tab., Brown clay with haloisite in medium proportion and high - compresibility. It is interstratified in the tenosique formation unit a constitute for a dolomitic limestone recristalized, coarse-medium grain and karstic structure. It was also tested a sample with a rock-rock contact in a clay limestone.
C.H. Caracol, gro., Bed planes with clean surfaces, of the mezcala formation (alternative sequence of shales and arenites.
C.H. Chicoasen, Layer of arenites of variable wide, almost 5 cm interstrafied in limestone beds.
C.H. La Angostura, Chis., Contact limesto ne-breccias calcareous, where can be observed as a clay of 5 mm of wide, but with many contacts rock to rock.
In figure No. 2 it is shown in the plasticity chart of soils the classification of the clay fillings tested.
The descriptions in this work correspond to the originals given by the different authors consulted.

6 RESULTS

In figure 3 to 4 as presented in the plottings of tangential stress vs. tangenti-

272

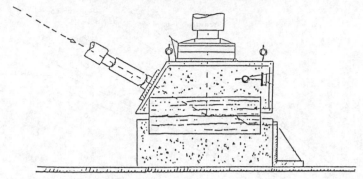

FIG. 1. DISPOSITIVE TEST

al stress vs tangential displacement, and tangential stress vs normal stress, of - the weakness surfaces of all tests perfor med.
In the graphics are included the vawes of the maximum and residual friction angle (it depends of the case) of each kind of test, thus the resistence or mohr's enve- lopments adjusted to a straight line or a curve according to the behavior of each - specimen.
The graphics presented correspond to the most representative of the samples beha- vior among the series of tests performed in each site.
The maximum resistance values obtained - are less notably minor than those of re- sidual resistence, due to the fact that only is possible to obtain one value of maximum resistence for each speciment, - and in the other hand, the residual resis tence can be determined several times, - using different normal stresses in the same specimen.

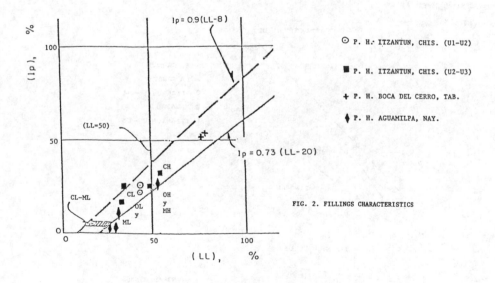

FIG. 2. FILLINGS CHARACTERISTICS

⊙ P. H. ITZANTUN, CHIS. (U1-U2)

■ P. H. ITZANTUN, CHIS. (U2-U3)

+ P. H. BOCA DEL CERRO, TAB.

♦ P. H. AGUAMILPA, NAY.

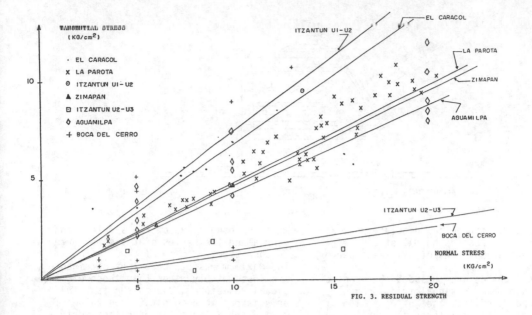

FIG. 3. RESIDUAL STRENGTH

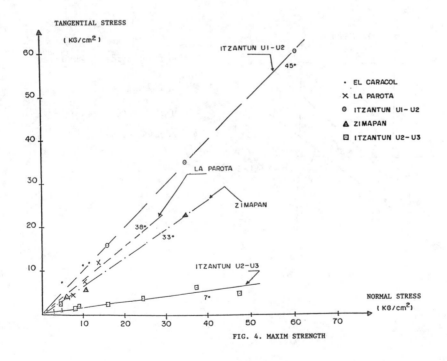

FIG. 4. MAXIM STRENGTH

The tangencial stress vs displacement behavior of the rock surfaces with weakness
planes containing filling of ine soil, -
granular or a combination of both, present a ductil failurg, in such way that -
it is not observed a maximum strength in
the moment of failure occurring this, also for the rock-rock contacts (Boca del -
Cerro) where the rock is a very soft and
easily dispersed.
In the case Itzantun project (fig.3) of
clay fill, between limestone units U1 and
U2, a peak resistence was observed, possi
ble due to very thin clay layer (3 mm).
This maximum resistence migth be controlled by the fragments of limestone or by
the rock to rock contacts that exist   -
along the surface of the specimen.
The values of the maximum friction angles,
vary between 28° and 45° for all samples
tested which have a complete rock-rock -
contact. The residual friction angles flu
ctuate between 26° and 42° correspondent
to specimens with less contacts rock-rock
or thinner clay fillings. The planes of -
close foliation or clay sand filling, - '
with gravel, are also inside this group -
in the case of weakness planes with rock
to rock contact / the residual resistence
seems to be influenced by the "Detritus"
wich is formed when the weakness plane -
fails or is crushed, this remanent material is pure friccionant, and the angles
of residual friction obtained are characteristics of this kind of materials.
For the rocks of Aguamilpa and Boca del -
Cerro projects, were applied 3 normal   -
stresses increasing as follows: 5, 10 and
20 kg/cm2 and 5, 10 and 15 kg/cm2, respectively, ending with the application of
the normal stress used at the beginning.
Observing that, the residual strength value is greater than interpreting of this
effect could be the dissipation of porous
pression provocated by the confinance given to the material by the application of
previous greater normal stresses.
The weakness surfaces that contain clay
filling exclusively, presented angles of
residual friction between 6° and 3° that
are the tipical values to the clay (ref.3)
it was not observed any relation between
the plasticity index of the clayey materials and the angles of friction obtained.

REFERENCES

"Resistencia al Corte Directo". Procedi-
    miento interno de la Oficina de Mecáni-
    ca de Rocas, Gerencia de Ingeniería Ex-
    perimental, CFE. 1992.
"Informe 88-44-GR, Resultados de las prue
    bas de laboratorio realizadas con nú-
    cleos de roca de los barrenos 1, 3, 4,
    G-1 y G-3 y muestras superficiales", Ge
    rencia de Ingeniería Experimental y Con
    trol, CFE, 1988.
"Manual de Diseño de Obras Civiles", Tomo
    B.3.3. Comisión Federal de Electricidad.
Archivo interno de la Oficina Mecánica de
    Rocas de la Gerencia de Ingeniería Ex-
    perimental y Control, CFE.

In situ tests to evaluate the influence of soil structure on deformability

Essais en place pour l'évaluation de l'influence de la structure du sol sur la déformabilité

Roberto Meriggi & Maurizio Soranzo
Department of Georesources and Territory, University of Udine, Italy

ABSTRACT: Results from in situ tests using the screw plate device and the pressumeter are compared with laboratory triaxial tests on isotropically consolidated soil samples taken to failure in drained conditions, in order to highlight the influence of cementation bonding between the grains of the solid skeleton on strain behaviour in soil.

RESUME : On compare dans cet article les résultats de tests effectués in situ en utilisant la plaque hélicoidale et le pressiomètre de Ménard, et ceux qui ont été obtenus grace à des essais de compression triaxiale à consolidation isotrope et portés à la rupture dans des conditions de drainage, afin de mettre en relief l'influence des liens de cimentation qui existent entre les grains de l'ossature solide sur le processus de déformation du sol.

1. INTRODUCTION

The structure of granular sedimentary soils depends on the environmental conditions in which sedimentation occurred and is characterized by the preferential orientation of grains in the solid skeleton. This structural anisotropy, defined as "inherent" by Ladd et al. (1977), is one of the causes of mechanical anisotropy in soils whose strain behaviour depends on the direction of stress. If cementation bonds develop among the grains of the solid skeleton, the structural arrangement is stiffer than that in which only porosity is present. Correct assessment of the soil strain moduli, which must be used when calculating settlement, must therefore include the influence of the cementation bonds of the structure on the strain behaviour of the soil. However, both the most commonly used in situ and laboratory tests cannot identify this aspect and generally underestimate the real value of the strain modulus. That is, during in situ penetrometric tests, both dynamic and static, tip penetration causes failure of the soil immediately around the tip and modulus E is therefore calculated in a field of plastic strains in which the material has lost its original structure and cementation bonds no longer exist. Instead, in the laboratory, the strain modulus may be measured by triaxial compression tests, even for very low strain values, although testing is generally carried out on reconstituted samples which lack the characteristic structure of the original material.

2. REVIEW OF STRAIN CHARACTERISTICS OF GRANULAR SOILS

The strain behaviour of a granular soil sample subjected to a variation of axial compression stress is defined in plane σ', ε (fig.1a) by the stress-strain curve, the strain modulus in drained conditions E', defined as the ratio between stress and vertical strain, and by Poisson's ratio μ between minor and major strains. Since the strain behaviour of soils is markedly non-linear, even at low values, no single value for modulus E' can be defined, and therefore tangential or secant lines are used to define tangent modulus Et and secant modulus Es for all points of the stress-strain curve. The tangent moduli at the origin of axes, ε ==> 0, are defined as initial moduli Ei. Instead, in order to define the strain behaviour of a soil subject to a stress level lower than the maximum level of stress undergone, modulus Eur is used, evaluated as slope during an unload/reload phase in compression testing (fig. 1b).

4 to 6 times higher than that of normally consolidated ones. The micro - and macrostructure of cohesionless soils appears both as preferential orientation of the grains and as the bonds which develop between the soil particles during sedimentation and give them characteristics of initial anisotropy in a transversal direction (Ladd et al., 1977; Ricceri & Soranzo, 1980). Neglecting for the moment the complexity of strain behaviour of soils, in design practice, settlement calculations are often carried out using moduli E50 and E25, evaluated respectively as 50% and 25% of the peak deviatoric stress, and using formulas taken from elasticity theory.

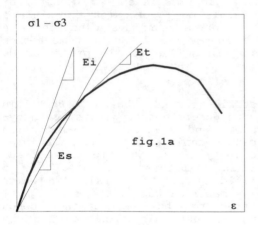

Fig. 1a - Moduli of deformation for non linear stress-strain behavior: first load.

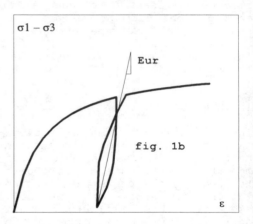

Fig. 1b - Moduli of deformation for non linear stress-strain behavior: unload-reload.

The extent of strain moduli depends on the type of material, micro- and macro-structure of the deposit, drainage conditions, and in situ stress state (Battaglio & Jamiolkowsky, 1987).

In particular, the latter, understood as the set of factors stiffening the solid skeleton: mechanical overconsolidation, viscous hardening, cementation, etc., level of strain and grain-size composition being equal greatly differentiates the magnitude of the moduli which, for overconsolidated materials, is

Instead, when undertaking analysis which considers soil strain state more correctly, calculation models which simulate the real non-linear elasto-plastic and/or elastic behaviour are used. One of the most popular models for non-linear analysis approaches the stress-strain curve with a hyperbola (Kondner, 1963; Kondner & Zelasko, 1963) (figs. 2a and 2b).

In conditions of primary load and axial symmetry at σ'_3 constant, the stress-strain curve (fig.2a), which has a maximum value for de

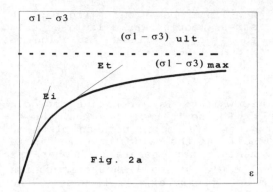

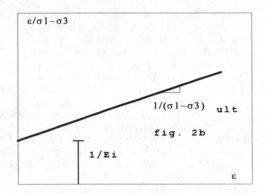

Fig. 2. Transformed hyperbolic stress-strain relationship.

viatoric stress $(\sigma'_1-\sigma'_3)_{max}$, and an asymptotic value $(\sigma'_1-\sigma'_3)_{ult}$ approximately fits the equation:

$$(\sigma'_1-\sigma'_3) = \cfrac{\varepsilon}{\cfrac{1}{Ei} + \cfrac{\varepsilon}{(\sigma'_1-\sigma'_3)}} \qquad (1)$$

where, in plane ε, $\varepsilon/(\sigma'_1-\sigma'_3)$ represents a straight line (fig. 2b). The intercept of this line on the axis of the ordinates is the reciprocal of the initial value of modulus Ei, and the reciprocal of the slope is asymptotic value $(\sigma'_1-\sigma'_3)_{ult}$. How the values of tangent modulus Et or secant modulus Es depend on the strain level reached may thus be defined by means of the following equations (Duncan et al., 1980a):

$$E_t = E_i \times (1 - Rf \times f)^2$$

$$E_s = E_i \times (1 - Rf \times f)$$

where

$$Rf = \cfrac{(\sigma'_1-\sigma'_3)_{max}}{(\sigma'_1-\sigma'_3)_{ult}}$$

is the correlation factor, and

$$f = \cfrac{(\sigma'_1-\sigma'_3)}{(\sigma'_1-\sigma'_3)_{ult}}$$

is the degree of mobilitation of the stress.

Moduli Ei and Eur increase with confining stress. As stated by Janbu (1963), the variation may be represented by the experimental equations:

$$Ei = Ke \times Pa \times (\sigma'_3/Pa)^{ne}$$

$$Eur = Kur \times Pa \times (\sigma'_3/Pa)^{nur}$$

in which K and n are experimental constants and Pa the reference atmospheric pressure. Duncan et al. (1980a), testing reconstituted samples, showed that modulus Eur is greater than initial modulus Ei according to a ratio varying between 1.2 and 3, whereas Clough et al. (1981), using triaxial tests on natural and artificially cemented sands, highlighted the influence of cementation bonding on the mechanical behaviour of sands. Cemented samples turned out to have greater initial stiffness and greater peak resistance than uncemented samples but, once peak resistance had been overcome and at strains sufficient to break all residual bonds, the stress-strain curves were similar.

3. THE INVESTIGATED SITES

Samples were taken from two sites in the Lignano peninsula, along the littoral belt separating the

279

Bassa Friulana lagoon from the northern Adriatic. The peninsula is characterized by mainly magnesian alluvial formations transported by the Tagliamento river (Brambati, 1970). In each of the examined sites, two in situ site investigations were carried out down to a depth of 6.0-7.5 m from ground level, along vertical lines at a short distance from each other. In one boring the stratigraphy was identified, and reworked and (where possible) undisturbed samples were taken and later subjected to grain-size analysis and isotropically consolidated and drained triaxial tests in the laboratory. In the other boring, immediately adjacent, samples were taken at regular intervals and at the same sampling depths with a screw plate device. The Lignano 1 samples, taken from near the mouth of the Tagliamento, were mainly composed of medium-grained sand mixed with varying percentages of silt (10-30%). Only at depths between 3.6 m and 4.3 m did the fine fraction turn out to be richer (45%) and plastic. The Lignano 2 samples, taken near the beach, were composed of medium- and fine-grained sands, mainly only slightly silty. Mineralogical analysis on some samples indicated carbonates, dolomite and calcite, generally 68-85%. The clay fraction (less than 2μ) was composed mainly of illite, with subordinate chlorite.

4. EXECUTION AND INTERPRETATION OF SCREW PLATE TESTS

These tests were carried out using the equipment sketched in fig. 3 (Kay & Parry, 1982), with a screw diameter of 162 mm. The plate was screwed into the soil down to the test depth and then, using the set of rods inside the rotating system, subjected to load increments to a maximum of $\Delta p = 500$ kPa. Settlement w on the plate was measured at each applied load and at least two unload/reload cycles were carried out during each test (fig. 4). The stress-strain curve shows marked

non-linear behaviour (fig 5). The stress modulus was calculated using the elastic linear solution of Brown (1988):

$$E = A \times B \times \Delta p/w \qquad (2)$$

in which A is a constant depending on Poisson's ratio and on the degree of adhesion between plate and soil. As the values of load increments p were generally higher than $2 \times \sigma'_v$ (σ'_v = lithostatic confining pressure), following Brown (1988), the test results were interpreted in the hypothesis of partial adhesion and thus, using Keer's (1975) solution and a mean value for Poisson's ratio of 0.25, a value of A = 0.42 was obtained. A typical stress-strain curve during the load tests is shown in fig. 4 and shows marked non-linear behaviour. In order to express the variability of the modulus according to load increment, a hyperbolic interpretation similar to Kondner's was used (see above).

The experimental data were plotted in plane W,W/(AxBxΔP) and parameters $1/E_i$ and $1/P_{ult}$ of the interpolationline were determined. Moreover, for each unload/reload cycle, moduli E_{ur} were determined following equation (2).

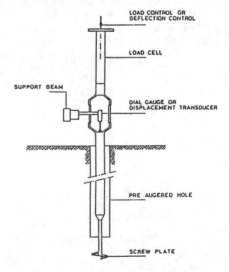

Fig. 3 Schematic view of the screw plate device

280

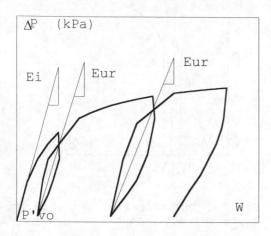

Fig. 4 - Unload-reload screw plate test results.

In order to assess the trend of moduli Ei and Eur according to the effective lithostatic pressure acting at the depth at which each test was carried out, the values measured in soils with similar grain-size characteristics were grouped together and correlated (figs. 5a, 5b). This correlation did not take into consideration the highest values measured in the shallow part of the soils, which had greater stiffness, probably due to overconsolidation.

As may be seen, the Ei values measured in the fine silty sands (fig. 5a) plot over a quite well-defined variability field (r = 0.95-0.99). Although there were not enough samples of medium grain silty sands to carry out a reliable correlation, the slope of the interpolation lines is practically the same. The same interpretative criterion was used for the Eur values (first unload/reload cycle), and regular growth in the modulus may be observed only for the fine silty sands (fig. 5b).
Fig. 6 shows the variation in the Eur/Ei ratio, measured in cohesionless soils, according to effective lithostatic pressure. Very different behaviours were clearly observed. In Lignano 2, modulus Eur always showed values higher than modulus Ei: Eur = (1.0-1.5)x Ei, substantially fitting the laboratory results of Duncan et al. (1980a). In Lignano 1, the same value, at depths greater than 2.0 m, was always lower than 1: Eur = (0.81-0.96) x Ei. This trend may be explained qualitatively by hypothesizing that the sediments in the two sites underwent cementation by the clay minerals occurring in the finest fraction (mainly illite).

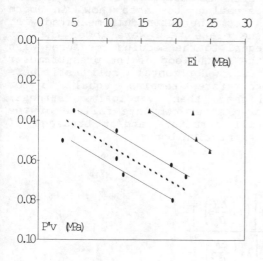

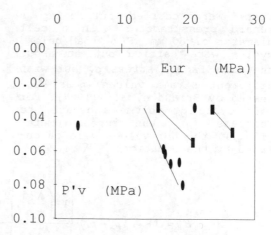

Fig. 5 - Screw plate tests: strain moduli versus effective confining pressure. ● fine silty sand; ■ medium silty sand.

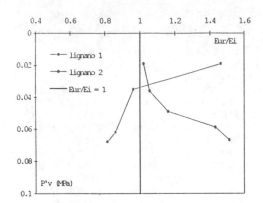

Fig. 6. Screw plate measured moduli ratio, Eur/Ei, versus effective confining pressure.

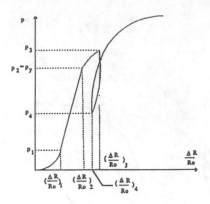

Fig. 7. Unload/reload cycle during a pressuremeter test.

Cementation bonds cause the initial stiffness of the solid ske-leton to be higher than that due to the simple arrangement of in-coherent grains in the soil. The load increments induced by the plate in the first load phase are sufficient to destroy these weak bonds, so that only the stiffness of the structural arrangement of the grains is measured during the reload phase.

5. EXECUTION AND INTERPRETATION OF PRESSUREMETER TESTS

Tests were carried out using a Menard pressuremeter with a cell diameter of 60 mm. Five load/unload cycles were carried out at about the same radial stress σ_r but with different strain values ε_θ as sug-gested by Briaud et al. (1987). For each cycle of volume increase and decrease of the pressuremeter (fig.7), the Eur values were calcu-lated by the equation:

in which p_3 and $(\Delta R/Ro)_3$ are re-spectively pressure and radial strain measured at the point where the unload/reload curves meet, and p_4 and $(\Delta R/Ro)_4$ are the same values measured at the end of the unload cycle.

Figs. 8 and 9 show the Eur values measured both with the screw plate device and with the pressuremeter as a functions of depth of testing. Since the strain conditions in the two tests were different, the first vertical and the second radial, the determined moduli cannot be com-pared directly. However, it is in-teresting to note how in both sites, the Eur moduli measured with the plate test were lower than or equal to the modulus measured with the fifth loop of the pressuremeter test. Horizontal soil stiffness therefore remains equal to or higher than vertical stiffness, even after complete failure of the soil structure and/or breakage of cementation bonds.

$$E_R = (1+v)\cdot(p_3-p_4)\cdot\frac{\left[1+\left(\frac{\Delta R}{R_o}\right)_3\right]^2+\left[1+\left(\frac{\Delta R}{R_o}\right)_4\right]^2}{\left[1+\left(\frac{\Delta R}{R_o}\right)_3\right]^2-\left[1+\left(\frac{\Delta R}{R_o}\right)_4\right]^2}$$

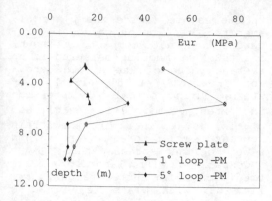

Fig. 8. Comparison between moduli measured by screw plate tests and pressuremeter tests:LIGNANO 1.

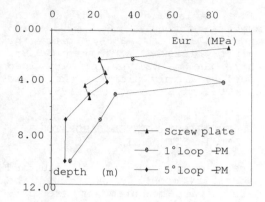

Fig.9. Comparison between moduli measured by screw plate tests and pressuremeter tests:LIGNANO 2.

6. INTERPRETATION OF TRIAXIAL TESTS

In order to compare the laboratory versus in situ strain behaviour of soils, drained consolidated triaxial compression tests were carried out on samples taken from borings. Since most samples were loose soils, they were reconstituted in the form of cylinders to a volume weight of $\gamma = 19$ kN/mc, the mean value measured in undisturbed samples. A method currently used in routine testing (Bishop & Henkel, 1962) was used: sand, mixed with water, is dropped inside a rubber membrane from a constant height. (It should be noted that this method, obviously, does not recreate either the in situ stress anisotropy conditions existing during sedimentation or the cementation bonds between the grains).

Four cylinders were prepared for each sample which, after complete saturation, were consolidated at four different cell pressures, 60, 120, 180 and 240 kPa, and then taken to failure by increasing the deviatoric stress $(\sigma_1-\sigma_3)$ in drained conditions. During the failure stage of each sample, once $\varepsilon = 1\%$ had been reached, an unload/reload cycle was carried out in order to assess modulus Eur. The stress-strain curve for each sample taken to failure was interpreted using the modified hyperbolic method (Kondner & Zelasko, 1963) to determine initial modulus Ei and each pair of values (Ei,σ_3) was plotted in a logarithmic diagram to determine K_e and n_e.

The values obtained for the two sites are shown in Table 1, together with k_{ur} and n_{ur}, which characterize modulus Eur.

SITE	SPECIMEN	DEPTH	n	k	n	k
-	n	m	e	e	ur	ur
LIGNANO S1	1	0.85	1.08	509	1.32	1149
	2	2.60	0.85	396	1.18	2086
	3	3.20	0.86	415	1.13	1862
	4	3.95	0.91	211	1.41	2567
LIGNANO S2	1	2.25	0.80	1346	0.42	3030
	2	3.75	0.87	520	0.98	1554
	3	4.25	0.96	1036	1.27	2712
	4	5.25	0.90	561	1.28	3550

Table 1 - Values of k numbers and of exponents n for moduli Ei and Eur, measured by triaxial isotropic compression tests.

The variability field of the number of moduli, K_e = 211-1346 and K_{ur} = 1149-3550, fits the values of Lambe & Whitman (1968), while the n_e values (primary load modulus) mainly plot in the field n_e = 0-1 of Byrne et al. (1987).

7. COMPARISON BETWEEN LABORATORY AND IN SITU TESTS

Direct comparison of in situ and laboratory moduli cannot really be made, since measurements are carried out in very different conditions. This is beacause, although - unlike the pressuremeter test - the plate test allows measurements of a vertical strain modulus like the triaxial test, the disturbance which the screwing of the plate produces in the tested soil may be considered minimum. The solid skeleton of the soil thus has the structural anisotropy (Ladd et al., 1977) due to the process of in situ sedimentation and possible cementation bonds which developed among the granules - characteristics which the samples subjected to triaxial compression cannot have, even though they are reconstructed with the same relative density that they originally had in situ.

Moreover, the soil volume before plate testing is characterized by initial strain anisotropy (Ladd et al., 1977) since main geostatic strains σ'_h and σ'_v are linked at rest coefficient Ko, the value of which depends on the strain history of the deposit. This is:

$$Ko(OC) = Ko(NC) \times OCR^n$$

in which OCR is the degree of overconsolidation and n the average value of the exponent (0,5). However, during consolidation in the triaxial cell, the sample is subjected to an isotropic strain state, since both vertical and radial pressure are equal to σ'_3.

The initial conditions of the test are therefore very different from the strain state existing in situ. The load applied by the plate induces in the soil elements making up the sample (twice the plate diameter) strain paths which are different and not completely known, and thus the modulus measured is an average value of the strain behavior (Bellotti et al. 1986). Instead, during triaxial compression the extent of main stresses σ'_1 and σ'_3 is easy to check and the strain modulus is therefore measured along a well-defined stress path.

Comparison with laboratory-measured moduli can therefore only be carried out with plate test results and greatly simplified hypotheses. A comparison was attempted by plotting for every boring both in situ and laboratory modulus values, according to effetive confining pressure σ'_v acting at the average depth of sampling. Since direct comparison was not possible because the laboratory values were measured according to σ'_3 and not σ'_v, the moduli were calculated by means of Janbu's equation for the value of octahedric stress σ'_m acting at the sampling depth:

$$\sigma'_m = \sigma'_v \times (1 + 2Ko)/3$$

Since no information on the stress history of the deposit were available, a reference value for at rest coefficient Ko was arbitrarly chosen at 0,5. The results obtained are shown in figs. 10 and 11. They reveal the importance of the initial state of soil in determining the extent of the moduli and their ratios. In the Lignano 1 site (fig. 10), the initial Ei values measured in situ were generally higher than the laboratory values. When Ei values were compared with in situ Eur ones, the latter were generally lower than in situ Ei values or near laboratory values. In the other site, Lignano 2 (fig. 11), the ratios between the moduli were completely different. At the lower and more significant depth, laboratory values were 2 - 4 times higher than in situ values, and were also higher than the reload modulus.

This different behavior may be interpreted by referring to the initial state of soil. In sandy sedi-

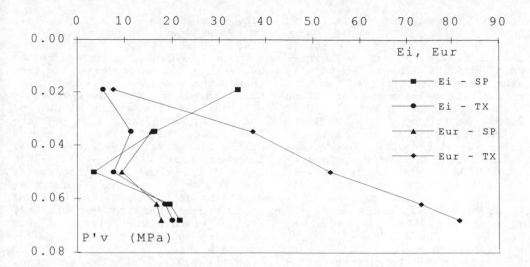

Fig. 10. Comparison between moduli measured by screw plate tests and triaxial tests: LIGNANO 1

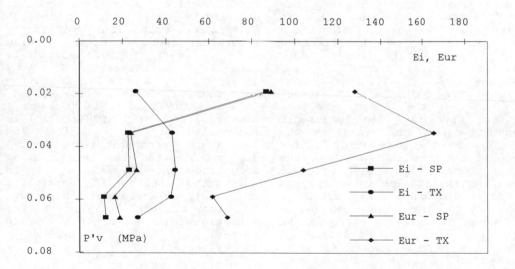

Fig. 11. Comparison between moduli measured by screw plate tests and triaxial tests: LIGNANO 2

ments with cementation bonds, as in Lignano 1 (fig. 10), where these bonds are probably the inter-grain strength of clay minerals, initial strain modulus Ei is generally higher than unload/reload modulus Eur, and the value measured during testing falls between Eur/Ei = 0,82 - 0.96. The initial moduli measured in the laboratory on reconstituted samples are generally lower than the in situ values or are comparable with them. The ratio between the laboratory-measured moduli was Eur/Ei = 2 - 8. Instead, in sites with no cementation bonds among the soil particles (fig. 11), shown by in situ ratios of Eur/Ei = 1,0-1,4, the laboratory-measured stress-strain curves led to an over-estimation both of the extent of the single moduli and of that of their ratio, which varies between 2 and 6.

8. CONCLUSIONS

Unload/reload cycles carried out during screw plate tests allow an estimate of the extent of strain moduli in relation to the structure taken on by the soil during sedimentation. Eur/Ei ratios of less than one indicate soils with cementation bonds and/or a soil structure making the solid skeleton stiffer. When the load increments transmitted by the plate to the soil exceed the mechanical resistance of the cementation bonds, the soil becomes completely destructured and the strains following the unload/reload phase are higher in relation to the lower degree of stiffness. The average ratio in the Lignano 1 site, characterized by mainly clay minerals, was Eur/Ei = 0.82 - 0.96.

In soils lacking cementation bonds or ones in which the structure is not particularly stiff, unload/reload modulus Eur is generally higher than initial modulus Ei, since the load increments undergone before the unload phase increase the density of the soil, stiffening its structure. For the Lignano2 sediments, the in situ ra-

tio was Eur/Ei = 1.0 1.4, i.e., near the lower value reported by Duncan et al. (1980).

In both sites, the unload/reload cycles of pressuremeter testing showed the progressive reduction of Eur, expressed by an average ratio of 0.33-0.77 between the modulus of the fifth loop and that of the first loop. The Eur values measured on the plate test were always lower than or equal to the fifth-loop value found with the pressuremeter test, and thus the horizontal stiffness of the soil remains equal to or higher than the vertical one, even after complete failure of soil structure and/or breakage of cementation bonds.

In the plate test, the extent of the in situ strain moduli turned out to be very different from those determined in the laboratory. The two types of test revealed that structured soils have an initial stiffness which is greater than that which may be measured with triaxial CID test and that, unlike the laboratory situation, modulus Eur may be lower than the modulus Ei. In non-structured soils, the laboratory-measured initial modulus and the unload/reload moduli are generally overstimated. In such soils it is not possible to correlate in situ and laboratory values. This can only be done if the stress history of the soil (OCR and Ko) is known with sufficient precision and laboratory reconstituted samples are consolidated in conditions of anisotropy Ko. However, even with this method, neither the conditions of structural anisotropy characterizing the original soil nor the cementation bond - factor on which initial soil stiffness depends - can be recreated.

REFERENCES

BATTAGLIO M., JAMIOLKOWSKI M. (1987). Progettazione geotecnica: metodi di calcolo e parametri. Analisi di deformazione. XIII C.G.T. Torino.

BELLOTTI R., GHIONNA N., JAMIOLKOW-
SKI M., LANCELLOTTA R., MANFREDI-
NI G.,(1986). Deformation Cha-
racteristics of Cohesionless
Soils from In-Situ Tests. In-Situ
'86 Proc. Spec. Conf. GED ASCE,
Virginia Tech., Blacksburg.

BISHOP A.W., HENKEL D.J. (1962).
The Measurement of Soil Properties in the Triaxial Test. (2nd ed-
n), Edward Arnold, London.

BYRNE P., M., CHEUNG H. AND YAN L.
(1987). Soil parameters for deformation analysis of sand masses. Can. Geoteh. J., Vol. 24,
366-376.

BRAMBATI A. (1970). Provenienza,
trasporto accumulo dei sedimenti
recenti nelle lagune di Marano e
Grado e nei litorali tra i fiumi
Isonzo e Tagliamento. Mem. Soc.
Geol. It.,Vol. 9, 281-329, 61 ff.

BRIAUD J.L., COSENTINO P.J. AND
TERRY T.A., (1987). Pressuremeter
moduli for airport pavement design and evaluation. FAA Report
DOT/FAA/PM-87-10, Civil Engineering Dept., Texas A&M Univerity,
College Station, Texas

BROWN. P.T. (1988). Performance and
interpretation of screw plate tests. Geotechnical Engineering,
Vol. 19, pp. 143-159.

CLOUGH G., SITAR N., BACHUS
R.C.,SHAFFI RAD N. (1981). Cemen-
ted Sands under static loading.
J.Geotech. Engng. Div., ASCE 107,
GT6, pp. 799-817.

DUNCAN J.M., BYRNE P., WONG K.S. e
MABRY P. (1980a) Strength,
Stress-Strain and Bulk Modulus
Parameters for Finite Element
Analyses of Stresses and Move-
ments in Soil Masses. Report No.
UCB/GT/80-01, Dept. of C.E.,
Univ. of California, Berkeley.

JANBU N. (1963). Soil Compressibility as determined by Oedometer
and Triaxial Tests. Proc. ECSMFE,
Vol. 1, pp. 191-196.

KAY J.N., PARRY R.H.G. (1982).
Screw Plate Tests in a Stiff
Clay. Ground Engineering, No. 6,
pp 22-30.

KEER L. M. (1975). Mixed Boundary
Value Problems for a Penny-Shaped
Cut. Journal of Elasticity, 5,
pp. 89-98.

KONDNER R.L (1963). Hyperbolic
Stress-Strain Response: Cohesive
Soils. JSMFE, ASCE, SM1.

KONDNER R.L., ZELASKO J.S. (1963).
Hyperbolic Stress-Strain Formula-
tion for Sands. Proc. II Paname-
rican Conf. on SMFE, Rio De Janeiro.

LADD C.C., FOOT R.,ISHIHARA K.,
SCHLOSSER F. AND POULOS
H.G.(1977).Stress-Deformation and
Stregth Characteristics, SOA Re-
port, Proc. of IX ICSMFE, Tokyo.

LAMBE T.,W. e WHITMAN R.,V. (1968).
Soil Mechanics. John Wiley &
Sons, New York, NY, p. 159.

RICCERI G., SORANZO M.(1980). Parametri elastici e resistenza a
rottura di sabbie sature in prove
triassiali drenate. RIG. Vo. XIV,
n. 2, pp. 133-143.

Classification et identification automatique des objets géomécaniques dans les diagraphies instantanées

Automatic classification and identification of geomechanical objects by means of instantaneous parameter logging

F. Cremoux & J. Malzac
LERGGA Université Bordeaux I, Talence, France

RÉSUMÉ : Depuis quelques années, les diagraphies instantanées ont connu un grand essor en raison de leur faible coût. Elles donnent rapidement des informations sur les propriétés mécaniques des terrains traversés par des forages destructifs. Pour faciliter leur interprétation, nous avons développé une méthodologie de lecture automatique des données enregistrés. Tout d'abord, les zones homogènes sont repérées sur le log de manière à situer la limite des différentes formations, puis l'analyse statistique permet de classer et d'identifier les objets géologiques rencontrés.

ABSTRACT : Over the last few years, instantaneous drilling parameter logging has known a great soaring due to its low coast. They give quickly a right estimation of mechanical properties of the ground by rotary drilling. In order to make easy its interpretation, we have developed specific tools for the automatic reading of the record data. First, the statistically homogenous zone limits where identified, this is followed by an inter zone statistic analysis which made possible the classification and characterization of the tested ground.

INTRODUCTION

Cette étude a un caractère méthodologique plutôt que d'exploration, puisque nous disposons des diagraphies instantanées d'un forage carotté. La connaissance du log géologique nous a permis d'étalonner et de vérifier la validité des procédés mis en oeuvre. Diverses mesures géophysiques ou de laboratoire (logs de calcimétrie et de diagraphies différées) complétaient notre information sur les terrains traversés. Les faciès rencontrés par ce forage peuvent être classés en deux grandes catégorie, les argiles et les calcaires, mais cette distinction reste schématique : chacun de ces types de formation est affecté par des passées appartenant à l'autre type. Par ailleurs, les calcaires sont plus ou moins compacts.

Les paramètres que nous avons utilisés sont :
- la vitesse d'avance (Va),
- la pression sur l'outil (Po),
- la vitesse de rotation (Vr)
- le couple de rotation (C_t).

Le pas d'échantillonnage relativement important (dc l'ordre de 10 cm) ne nous a pas permis d'effectuer un travail statistique fiable sur les bancs de faible épaisseur pour lesquels le nombre de mesures restait insuffisant. Mais il est évident que les mêmes méthodes appliquées à des données plus serrées (avec un pas d'échantillonnage de l'ordre du centimètre ou du demi-centimètre) permettraient une analyse plus fine.

I - LE DECOUPAGE EN ZONES STATISTIQUEMENT HOMOGENES.

Les signaux enregistrés au cours des diagraphies instantanées sont caractérisés par un niveau de bruit considérable, auquel viennent s'ajouter les artefacts liés aux interruptions du forage (changement de tige, carottage, incidents divers...) Par ailleurs, la structure même de ces signaux est complexe. En effet, examinée à grande échelle, une formation paraît homogène si les bancs qui la composent montrent une répétition approximativement régulière d'alternances, alors qu'une analyse plus fine révèle ses différentes composantes. Le même phénomène se reproduit si on focalise sur un seul de ces bancs. Ainsi, dans les terrains rencontrés lors d'un forage destructif ou carotté, les formations successives s'organisent comme un empilement emboîté de structures localement gigognes. Leurs caractéristiques géomécaniques relèvent de la même logique complexe et se reflètent dans les paramètres enregistrés lors des diagraphies instantanées. Dans cette arborescence de structures, chaque noeud correspond à une échelle d'observation spécifique de l'objet géologique ou géomécanique concerné. Une interprétation automatique du signal devra donc suivre la logique de cette complexité. L'analyse commencera par une lecture de l'ensemble du forage qui sera découpé en zones, lesquelles seront à leur tour analysées et décomposées en structures plus fines. Plusieurs niveaux d'analyse peuvent être ainsi envisagés successivement. Un tel découpage ne peut-être effectué sur tous les signaux à la fois. Il portera donc sur un paramètre composite, la forabilité (F) obtenu à partir des paramètres mesurés. Ce paramètre qui condense toute l'information disponible en un signal unique sera calculé par la relation ::

$$F = (Va/Po).(Vr/Cr)$$

L'information utile à une échelle d'observation donnée apparaît à une échelle supérieure comme un bruit blanc ou effet de pépite qui nuit à la lisibilité du signal. La première phase du traitement sera donc un filtrage, automatiquement adapté à la fenêtre d'observation. La fréquence de coupure du filtre coupe-haut est déterminée par l'analyse de la courbe représentative du logarithme du spectre d'amplitude.

La courbe de l'entropie qui traduit la variabilité spatiale est ensuite calculée à partir du signal ainsi filtré. Pour une zone statistiquement homogène à l'échelle d'observation considérée, le signal possède une stationnarité d'ordre deux et l'entropie croît linéairement (Crémoux & al 1990). La courbe de l'entropie se présente donc comme une succession

de segments approximativement rectilignes et ses points de brisure marquent les limites des zones statistiquement homogènes. On peut les localiser grâce à un réseau neuronal adapté à la reconnaissance de formes, assimilable à un opérateur de dérivation d'ordre deux.

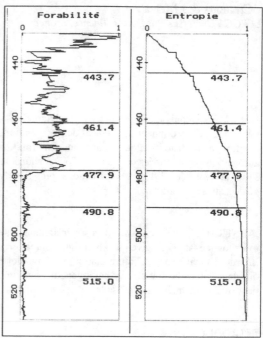

Fig 1 : Exemple de découpage de la courbe d'entropie de la forabilité.

Mais le problème reste complexe en raison du fort niveau de bruit qui subsiste. Un filtrage plus sévère ou un lissage numérique présente l'inconvénient d'émousser les points de brisure de la courbe et diminue encore sa lisibilité. Il faut donc s'accommoder du bruit et adapter les dimensions du dispositif neuronal à celles des structures recherchées de manière à négliger les ruptures de pente non significatives. Ces dimensions sont calculées en fonction de la dimension caractéristique (elle-même évaluée à partir de l'entropie et/ou de la portée du variogramme). L'ensemble de ce processus auto-adaptatif est ensuite appliqué isolément à chacune des zones homogènes obtenues par ce premier découpage. Avec une fenêtre d'observation plus étroite, l'analyse plus fine permet de détecter des zones homogènes d'un ordre supérieur (Bourget & al 1992).

La réalisation de ces étapes successives décrit l'empilement des structures imbriquées rencontrées par le forage.Cet ensemble a fait l'objet d'une maquette informatique et les résultats présentés sur deux niveaux de décision imbriqués montrent toute la pertinence de la méthode puisque les diverses homogénéités ainsi identifiées sont en bonne concordance avec les résultats fournis par les diagraphies différées effectuées sur la zone test proposée.

2 CLASSIFICATION ET CARACTERISATION DES ZONES STATISTIQUEMENT HOMOGENES.

Le log composite de la forabilité ainsi découpé en zones statistiquement homogènes, il faut maintenant essayer de déterminer les caractéristiques mécaniques des roches constituant chacune de ces zones. Ces caractéristiques mécaniques peuvent s'envisager sous trois angles qui dessinent la "signature" d'une formation :

- la dureté
- la résistance au cisaillement
- l'homogénéité de la formation

Il convient pour cela de revenir aux paramètres initiaux : Va, Po, Cr et Vr. Toutefois, ces paramètres mesurés dans des unités différentes montrent des grandeurs très disparates. On homogénéisera ces données d'amplitudes fort diverses en réduisant à 1 la valeur maximale obtenue sur l'ensemble du forage (ou de tous les forages concernés si l'on souhaite établir des corrélations entre plusieurs puits voisins ou constituer une banque de données).

2.1 *La dureté.*

Dire que la vitesse d'avance diminue lorsque la dureté de la roche augmente est une banalité. La valeur moyenne de ce paramètre lors de la traversée d'une zone homogène semble donc un bon indice de la dureté de la formation considérée. Toutefois, cette valeur demande à être corrigée en fonction de la pression sur l'outil, car il n'est pas moins évident que, dans un matériau donné, la Va croît si on augmente la Po. L'idéal serait de maintenir la Po constante, mais les contraintes de rentabilité conduisent souvent les foreurs à la modifier de manière à optimiser le processus de forage. Par ailleurs, la mesure de la Po inclut forcément celle de la réaction du terrain, laquelle est susceptible de variations : à la limite, elle chute si le forage traverse une cavité. On peut s'affranchir de ces variations en considérant non plus la Va seule mais la vitesse d'avance "réduite", notée Var = Va/Po.

2.2 *La résistance au cisaillement.*

La vitesse de rotation diminue si la résistance au cisaillement augmente, et la valeur moyenne de la vitesse de rotation dans la zone considérée sera un indice de cette résistance, mais avec les mêmes restrictions que pour la Va et la dureté : la vitesse de rotation devra être corrigée en fonction du couple imposé au train de tige. Là encore, on considérera la vitesse de rotation réduite : Vrr = Vr/Cr.

2.3 *L'homogénéité de la formation.*

Nous avons vu que nous considérions comme homogène, à une échelle d'observation donnée, une formation qui pouvait montrer une grande variabilité à un niveau supérieur d'analyse, avec une fenêtre d'observation beaucoup plus étroite. Il faudra donc estimer cette variabilité qui constituera un indice supplémentaire de caractérisation de la formation traversée. Cette variation peut concerner la dureté, la résistance au cisaillement, ou les deux paramètres en même temps.

On peut évaluer la variabilité de la formation par celle du signal dont on calculera l'écart-type. On peut affiner notre analyse en considérant non seulement la variabilité de la forabilité, mais aussi celle de tous les paramètres, y compris celle de la vitesse d'avance réduite ainsi que celle de la vitesse de rotation réduite.

Par ailleurs, il nous a paru intéressant, pour compléter la signature des terrains traversés par le forage, de corréler les paramètres Va et Po d'une part, Vr et Cr d'autre part, de manière à évaluer l'homogénéité de la roche selon d'autres critères que ceux de la variabilité du signal, laquelle peut être due en partie à un bruit aléatoire.

Considérons la corrélation entre Va et Po, (le même raisonnement pourra être conduit avec Vr et Cr) Nous avons énoncé plus haut que la vitesse d'avance croissait quand on augmentait la Po. La corrélation entre Va et Po semble devoir être toujours fortement positive. Mais ceci n'est vrai que dans la mesure où le matériau est homogène. Lorsque l'on passe d'une roche tendre à une roche dure, la vitesse d'avance diminue et la pression sur l'outil augmente avec la réaction plus grande du terrain.

Ce phénomène de variation en sens opposés de la Va et de la Po se produira à chaque changement de dureté. Le calcul de la corrélation entre la Va et la Po pourra donc constituer un indice de l'homogénéité de la dureté du matériau foré. Le signe de cette corrélation indiquera si le terrain est homogène ou non, et sa valeur absolue nous renseignera sur l'intensité de ce caractère homogène ou hétérogène.

Contrairement à l'écart-type ou à l'entropie, cette façon d'évaluer l'homogénéité est peu sensible au bruit. On peut raisonnablement supposer en effet que la corrélation entre le bruit aléatoire qui affecte la Va et celui de la Po est négligeable. Cet indice sera d'autant plus élevé que l'homogénéité sera grande du point de vue de la dureté.

Les mêmes observations peuvent être faites en ce qui concerne l'évaluation de l'homogénéité de la résistance au cisaillement par le calcul de la corrélation entre Vr et Cr. On pourra ainsi déterminer un indice qui caractérisera l'homogénéité du terrain du point de vue de la résistance au cisaillement.

Tableau I : Valeurs moyennes des paramètres Var, Vrr et F pour les six échantillons.

	Var	Vrr	F
Argiles 1	0,65	2,62	1,81
Argiles 2	0,53	2,32	1,26
Argiles 3	0,62	2,38	1,47
Calc. 1	0,08	1,25	0,10
Calc 2	0,19	0,65	0,18
Calc 3	0,13	0,90	0,12

Ces indices peuvent apporter leur contribution à la signature d'une formation, toutefois, ils demandent à être utilisés avec précaution : si le régime de la machine est tel que la Po ou la Cr peuvent être maintenus à peu près constantes, les corrélations envisagées ci-dessus perdent leur sens et le résultat devient aléatoire (divers échantillons prélevés dans la même formation montrent que ces coefficients de corrélation sont alors erratiques). Ils n'ont de signification et de stabilité que lorsque la variance des paramètres considérés dépasse un certain seuil.

3 EXEMPLE D'APPLICATION.

3.1 Description géologique des zones étudiées.

Les échantillons choisis pour notre étude sont prélevés dans une partie du forage comprise entre les côtes 430 m et 530 m. Le découpage effectué automatiquement par le processus décrit ci-dessus nous permet de délimiter six zones. Les trois premières sont caractérisées par un faciès argileux, tandis que les trois suivantes sont formées de calcaire.

Les argiles notées argiles 1, 2 et 3 pour les niveaux supérieur, moyen et inférieur respectivement, sont décrites dans le log géologique déduit de l'étude des carottes. Nous avons pour les argiles 1 : argiles silteuses à passées calcaires et argiles silteuses carbonatées. Pour le niveau moyen : argiles carbonatées à rares bioclastes et rides siltocarbonatées. Les argiles 3 sont formées d'une succession de diverses argiles carbonatées dont les natures sont assez voisines. Toute la série appartient à l'étage Callovien.

Les calcaires, pour lequel nous avons utilisé la même notation sont décrits ainsi : le niveau 1, constituant l'étage du Callovien inférieur, est formé de calcaires oolithiques et gravelo-bioclastiques à lits argilo-bioclastiques ; les calcaire 2 sont des calcaires gravelo-oolithiques bioclastiques à lits argileux ; le niveau 3 est formé de calcaires micritiques pisolithiques à passées bioclastiques. Ces deux derniers niveaux appartiennent au Bathonien supérieur.

La discrimination entre les argiles et les calcaires ne pose aucun problème : la seule forabilité moyenne suffit à les distinguer aisément : elle varie de 0,10 à 0,18 dans les calcaires alors qu'elle atteint de 1,26 à 1,81 dans les argiles, ce qui permet un seuil de décision automatique sans aucun risque d'erreur (Tableau I).

3.2 Les formations d'argiles.

Pour les trois formations, la dureté et la résistance au cisaillement varient dans le même sens, aussi pourrons nous les distinguer, du point de vue de la résistance en général, par la seule forabilité. Les valeurs moyennes relevées (1,81, 1,26 et 1,47 respectivement pour les niveaux supérieur, moyen et inférieur) présentent un écart relativement modeste, mais qui permet encore une discrimination automatique.

Tableau II : Ecarts-types des paramètres Var, Vrr et F pour les six échantillons.

	Var	Vrr	F
Argiles 1	0,37	0,55	1,20
Argiles 2	0,19	0,23	0,48
Argiles 3	0,18	0,15	0,43
Calc. 1	0,01	0,15	0,02
Calc 2	0,09	0,29	0,10
Calc 3	0,04	0,14	0,05

En ce qui concerne la variabilité, nous pouvons l'évaluer par l'écart-type des paramètres donné dans le Tableau II. Celui de la Var (liée à la dureté) est pour les trois formations : 0,37, 0,19 et 0,18, ce qui permet de discriminer aisément le premier niveau, mais non les deux autres. L'écart-type de la Vrr (liée à la résistance au cisaillement) pour lequel les valeurs observées sont : 0,55, 0,23 et 0,15 permettra une meilleure distinction. Par ailleurs, en ce qui concerne le premier niveau pour lequel la variabilité des paramètres est telle que les corrélations prennent tous leur sens, la corrélation entre Va et Po atteint -0,55, tandis que celle entre Vr et Cr est nulle, ce qui confirme bien l'hétérogénéité de la formation.

3.3 Les calcaires.

La dureté dans les trois zones retenues est estimée par la Var qui prend, pour les trois échantillons de calcaire supérieur, moyen et inférieur respectivement les valeurs suivantes : 0,08, 0,19 et 0,13. La Vrr montre une variation différente avec les valeurs 1,25, 0,95 et 0,90. L'ensemble de ces valeurs permet de discriminer les calcaires supérieurs des calcaires moyens et inférieurs d'une manière plus sûre que par la seule mesure de la forabilité moyenne, laquelle varie trop peu : 0,10, 0,18 et 0,12, ce qui rapproche le niveau supérieur du niveau inférieur d'une manière

qui rend hasardeuse une lecture automatique. Ceci est dû au fait que dans les calcaires supérieurs, bien que la vitesse d'avance soit plus faible que pour les deux autres formations, la vitesse de rotation est sensiblement plus élevée. Il semble d'ailleurs que ce phénomène doive être interprété non pas comme une diminution de la résistance au cisaillement, mais plutôt comme l'indice d'un fonctionnement moins performant de la machine opposée à un terrain très dur. La pénétration de l'outil étant moins bonne, il mord moins dans la roche et ripe davantage.

Du point de vue de la variabilité, approchée par l'écart-type de ces mêmes paramètres, on obtient, en ce qui concerne la dureté, les valeurs 0,01, 0,09 et 0,04 (écart-type de la Var). La variabilité de la résistance au cisaillement sera traduite par les valeurs : 0,15, 0,29 et 0,14 (écart-type de la Vrr). Pour les trois formations, l'écart-type de la forabilité prend respectivement les valeurs 0,02, 0,10 et 0,05. Le calcaire du niveau moyen se distingue nettement des deux autres, quel que soit le paramètre considéré. Mais pour les trois formations, la variabilité reste bien trop faible pour que le calcul des coefficients de corrélation entre les paramètres ait un sens.

CONCLUSION

L'outil d'analyse automatique des diagraphies différées que nous avons développé ne permet pas de déterminer la nature des terrains forés dans un environnement totalement inconnu, mais convenablement étalonné sur un premier forage, il peut rendre de grands services lorsqu'il s'agit d'identifier les formation traversées par des forages voisins. Il peut aussi nous permettre de constituer une banque de données comportant la signature des différents types de roches les plus fréquemment rencontrées lors des forages de reconnaissance. Adapté au travail en temps réel, il permettra de "reconnaître" les terrains traversés au fur et à mesure de l'avance du forage.

BIBLIOGRAPHIE

Crémoux F. & Morlier P. 1990 Une nouvelle méthode de lecture des diagraphies instantanées : l'entropie. *8ème congrès des journées de statisiques de Tours.*

Crémoux. F., Morlier P., Viguier C., Largillier J.P. 1990 Apport des techniques statistiques dans l'analyse des diagraphies de forage pour les grands ouvrages de Génie Civil. *Société Française de Géographie*

Crémoux F. & Malzac J. 1991 Optimisation of instantaneous parametter logging (detection of sand and clay cavities), *ISCASP 6 Mexico.*

Bourget M., Crémoux F. & Malzac J. 1992 Diagraphies instantanées et recherche automatique des zones homogènes. *Colloque International Géotechnique et informatique. Paris.*

7th International IAEG Congress / 7ème Congrès International de AIGI, © 1994 Balkema, Rotterdam, ISBN 90 5410 503 8

On some geomechanical aspects of high horizontal stresses

Au sujet des aspects géomécaniques des fortes contraintes horizontales in situ

C.F.Lee
Department of Civil & Structural Engineering, University of Hong Kong, Hong Kong

Owen L.White
Department of Earth Sciences & Quaternary Sciences Institute, University of Waterloo, Ont., Canada

ABSTRACT: Field observations and measurements have confirmed the existence of high horizontal in-situ stresses in many geological formations around the globe. Their occurrence and measurement in the Southern Ontario region of Canada are examined in detail in this paper, along with their possible origins and engineering implications.

RESUMÉ: Les observations et les measures faites aux chantiers ont confirmé la présence de fortes contraintes horizontales in situ dans beaucoup de formations geologiques autour du globe. Leur présence et mesure dans le sud de l'Ontario du Canada sont examinées en détail, et leur origines possibles et implications pratiques sont discutées dans cet article.

1 INTRODUCTION

As early as the 1920's, the eminent Chinese geoscientist J.S. Lee (Li Siguang) predicted that the horizontal stresses in many geological formations could be significantly larger than their corresponding vertical values. He attributed this phenomenon largely to the angular momentum associated with the rotation of the Earth, particularly the periodic changes in the speed of rotation (Lee, 1926).

Some half a century later, the state-of-the-art of in-situ stress measurement has advanced considerably. Field measurements from many parts of the world confirmed that horizontal stresses could indeed be relatively high, when compared to the vertical values. Such observations have been reported from Australia (e.g., Moye, 1958), Canada (e.g., Herget, 1974; Lee, 1981), Scandinavia (e.g., Hast, Southern Africa (e.g., Gay, 1974; Orr, 1975), and some parts of the United States (e.g., Obert, 1962). Various theories and hypotheses have been put forward to explain the origins of these high horizontal stresses, including continental glaciation and deglaciation, plate tectonics, erosion, etc. It is conceivable that some of these mechanisms or processes are inter-related. For example, it is possible that the Earth's rotation may have an impact on the relative movement of lithospheric plates, resulting in stress changes of significant proportions.

The present article will examine the occurrence of high horizontal stresses in the Southern Ontario of Canada. Their geological expressions and engineering implications will then be discussed on the basis of regional observations.

2 REGIONAL OCCURRENCE OF HIGH HORIZONTAL STRESSES

Geographically, the Province of Ontario is located in the central-eastern part of Canada. The southern part of the province is also the most densely populated area of Canada. Over the years, there has been a relatively high level of geological investigation activities to meet engineering requirements and industrial needs. Such geological investigation activities include mapping and drilling in most cases. For major engineering projects such as nuclear and hydroelectric power plants, in-situ stress measurements are also carried out using either the overcoring or the hydrofracturing technique, or both.

The most common form of overcoring technique used is the downhole, strain-relief method based on the U.S. Bureau of Reclamation borehole deformation gauge (Hooker and Bickel, 1974; Palmer and Lo, 1975). This technique is now being used routinely on many major site investigation projects. It has been proven to be fairly reliable. In routine applications, its main limitation is, of course, in the depth of coverage (i.e. usually not exceeding 100 m). For greater depths, the hydrofracturing method is used (Haimson and Lee, 1979; Haimson, Lee and Huang, 1986).

Figure 1 shows the results ot in-situ stress measurements obtained at the site of the 3400 MW Darlingtron Nuclear Generating Station, on the north shore of Lake Ontario and approximately 65 km to the east of Toronto. Presented in this figure are two sets of data, with the solid triangles and circles denoting overcoring results, and the open symbols denoting hydrofracturing results from greater depths. The triangle represents the maximum horizontal stress and the circle, the minimum horizontal stress. The overburden pressure is also indicated. It is obvious that the horizontal stresses are significantly larger than the overburden pressure, at least down to a depth of 300 m. The stratigraphy down to this depth is indicated in the figure, along with the orientation of the maximum horizontal stress.

The horizontal compressive stresses depicted in figure 1 generally lie within the range of 5-20 MPa, showing some trend of increase with depth. The overcoring results compare quite favourably with those from hydrofracturing. The orientation of the maximum horizontal stress appears to be quite consistent in the Paleozoic sediments, being generally in an ENE (east-northeast) direction. In the Precambrian basement rocks, its predominant orientation is in a NNE (north-northeast) direction.

The in-situ stress measurements from the Darlington Nuclear Generating Station site illustrate the magnitude of horizontal compressive stress that is fairly typical of Southern Ontario. Locally, the stress regime (including the orientation) could be affected by geological and topographic conditions, particularly the presence of fractures. Figure 2 illustrates a set of overcoring stress measurements obtained in a horizontally bedded shale formation in the Toronto area. This Upper Ordovician shale is, in general, thinly bedded and interbedded with siltstone and mudstone. Although high horizontal stresses were measured in some layers, low stresses were observed in others. There was a considerable amount of scattering in both the magnitude and orientation of horizontal stresses measured. It was believed that the low values of horizontal stress could have been related to vertical fracturing of the layers in question. As indicated by the values of Young's modulus (E) to the right of Figure 2, these low stress values were associated with relatively harder layers in the shaly rock mass. Locally, the magnitude of horizontal compressive stress may still approach 7 MPa in some other layers of rock. This variability in stress orientation and magnitude is not uncommon in thinly bedded shales, particularly near the surface.

Figure 3 illustrates the effect of topographic conditions on the in-situ stress regime. The overcoring data presented in this figure are from a vertical borehole located at 7.3 m from the edge of the Niagara Gorge, immediately downstream of the Niagara Falls. The maximum horizontal stress ranges from 10 MPa to 23 MPa approximately. Its orientation is consistently within the range of N15°E, to N30°E, which is generally parallel to the gorge. The minimum horizontal stress is significantly smaller in magnitude, in some cases approaching zero. This is, of course, in a direction normal to the gorge and a consequence of prolonged stress relief at the face of the cliff. The effect of stress relief is such that vertical or subvertical tension fractures (i.e., stress relief joints) could be detected

in the rock cliff, often to horizontal depths of at least 20-30 m from the face of the cliff.

3 GLACIATION AND GEOMECHANICAL EFFECT

The origin of the state of stress in the geologic formations of Ontario is not precisely known. A number of mechanisms have been postulated, including viscoelastic deformation under glacial loading, changes in the Earth's curvature due to continental glaciation, erosion, plate tectonics and the rotation of the Earth (White et al, 1973; Lee, 1978a; Lee and Asmis, 1979; Asmis and Lee, 1980).

Continental glaciation has been postulated to have a significant impact on the crustal stress field. It is generally known that at least four periods of continental glaciation had occurred in parts of the Northern Hemisphere over the last million years. In chronological order, these glacial periods have been named the Nebraskan, the Kansan, the Illinoian and the Wisconsinan Glaciations respectively, in North America. These names reflect the southern limits of the drift borders involved. It is also known that each glacial period was composed of numerous retreats and re-advances of the ice front. The thickness of the ice sheet could not be precisely determined, but a maximum thickness on the order of 3000 m-3750 m has been estimated for the Canadian glacial system during the Wisconsinan period, based on geotechnical (Harrison, 1958) and ice flow considerations (Paterson, 1972; Cathles, 1975). This would be equivalent to a vertical load of 21-28 MPa on bedrock covered by the Laurentide ice sheet during Wisonsinan Glaciation.

Under the action of such a heavy glacial load, the rock mass would undergo significant deformation and compression. For points located in the interior of the glaciated regions, the vertical loading would be fairly uniform, leading to a state of rigid lateral confinement (i.e., zero lateral strain) in the rock mass. The horizontal stress (σ_h) is related to the vertical stress (σ_h) as follows:

$$\sigma_h = \frac{\mu}{1-\mu} \sigma_v \qquad (1)$$

where μ = Poisson's ratio.

This lateral stress is elastic and should, therefore, be fully recoverable upon deglaciation. It is, however, known that under conditions of sustained loading, a creep or viscoelastic compression of the rock mass may also occur in addition to the initial elastic deformation. Significant amounts of viscoelastic compression have been observed on rocks as competent as granite, gabbro, granodiorite and limestone (Lomnitz,1965; Hardy et al, 1970). As in the case of the initial elastic response, creeping or viscoelastic compression of crustal rock under sustained glacial loading would have to take place in the absence of any significant lateral strain. For an incremental amount of vertical strain $\Delta\varepsilon_v$ the incremental horizontal stress generated under sustained glacial loading is given by (Lee, 1978a; Lee and Asmis, 1979):

$$\Delta\sigma_h = \frac{E}{2(1-2\mu)} \Delta\varepsilon_v \qquad (2)$$

where E = Young's modulus.

To illustrate the significance of Equation (2), let it be assumed that the viscoelastic compression amounts to 20 percent of the initial elastic compression under glacial loading. For a hard rock with a Young's modulus of 70 GPa and a Poisson's ratio of 0.2, one can readily calculate from Equations (1) and (2) that a 60 percent increase in horizontal compressive stress could result from a 20 percent increase in vertical compression, under conditions of sustained glacial loading and rigid confinement. Although the elastic stresses defined by Equation (1) could be fully recovered upon deglaciation, viscoelastic strains defined in Equation (2) could only be recovered partially upon unloading. Thus, a portion of these viscoelastic horizontal stresses could well be retained in the rock mass as residual stresses, subsequent to deglaciation and the elastic rebound of the crust. Such a process would probably occur during each glacial period, resulting in a progressive buildup of horizontal residual stresses. It should also be noted that there were many cycles of glacial advance and melting back of the ice front within each glacial period, i.e., many cycles of loading and unloading, with residual stresses being accumulated during each cycle. This applies also to each cycle of deposition and erosion that had occurred in the geologic past.

Upon deglaciation, isostatic uplift would occur at a rate commensurate with the rate of unloading. Dissipation of the recoverable portion of the viscoelastic stresses would take some time. It is, therefore, possible that a state of exceedingly high horizontal stress might have existed in the bedrock shortly after deglaciation. Described in the following section are compressional features in the form of folding, buckling and "pop-ups" in the bedrock surface, which have been traced to be post-glacial in age. It is conceivable that many of these features could have been formed as a result of the very high horizontal compressive stresses shortly after glacial retreat.

4   GEOLOGICAL EXPRESSIONS OF HIGH HORIZONTAL STRESSES

Long before the introduction of in-situ stress measurement techniques, there has been a considerable amount of geological evidence indicative of high horizontal stresses in parts of Northeastern North America. Compressive failures of surficial rocks, in the form of pop-ups (i.e., buckles) and high-angle, reverse faulting had been reported (Gilbert, 1986; White et al, 1973; Franklin and Hungr, 1976; White and Russell, 1982). They occur in a variety of bedrock formations including Devonian shales; Silurian limestones, dolostones and shales; Ordovician shales and limestones; as well as Precambrian hard rocks such as granite and quartzite. Most of these pop-ups were considered to be "old". They probably occurred shortly after the last glacial event (White and Russell, 1982). Evidence for this includes:

(a)   The mantling of pop-ups by later sediments;

(b)   The weathering of disturbed rock along the fold axis; and

(c)   The truncation of pop-ups by erosion surfaces.

Figure 4 shows the geologic section of a typical pop-up, occurring in Upper Ordovician shales along the Joshua Creek in Oakville, Ontario.

The size of pop-ups in Southern Ontario varies considerably. The maximum height is of the order of three metres, and the maximum wavelength approximately ten metres. Observed lengths of axial traces vary from a few tens of metres to over 1.5 km. They are essentially the result of buckling failure of surficial rocks due to the high horizontal compressive stresses. Such failure also occasionally occurs in quarry floors in the region. In the latter case, compressive failure is often triggered by the removal of overlying layers (or confining pressure) due to the quarrying operation, supplemented in some cases by surficial thermal stress effect.

In addition to the pop-ups, a number of high-angle, reverse faults have been mapped in Ontario, Quebec and Western New York State (Oliver et al, 1970; White et al, 1973; White and Russell, 1982). Their occurrence is consistent with the state of high horizontal stress upon deglaciation. These faults are often associated with the offset of glacial striae, and are therefore considered to be post-glacial in origin.

5   ENGINEERING AND MINING IMPLICATIONS

The existence of high horizontal stresses in rock has both positive and negative impacts on engineering and mining projects.

The most obvious negative impact is the stress concentration effect in open and underground excavations in rock. When a tunnel opening is created in a high horizontal stress field, there will be a concentration of horizontal compressive stress in the roof and floor of the opening. The stress concentration factor approaches three if the in-situ vertical stress is small compared with the horizontal value. In other words, if the in-situ horizontal stress normal to the longitudinal axis of the tunnel is 10 MPa, the maximum horizontal stress in the roof or the floor could approach 30 MPa upon excavation. This may result in some localized buckling and heaving of the floor as well as buckling and ravelling of the roof in some weaker rocks such as shale and sandstone (Lee, 1978b; 1980). Failure is often localized because of the variations of in-situ stresses and mechanical properties of rock along the tunnel opening, with failure occurring in the critical sections. In the case of an open cut excavation, such as a canal or a quarry, compressive failure is confined to the floor, resulting in localized floor buckling.

In general, local failure as described above resulted in inconvenience rather than major hazards on engineering projects. In underground excavations, the buckled and heaved layers of rock are often mucked out along with the rock fragments from drilling and blasting. In quarries, surficial buckles or pop-ups may cause some inconvenience to local traffic (i.e., trucking).

A major safety concern could result from stress-induced failure in deep underground mines. In many parts of the Canadian Shield, including Northern Ontario, there is evidence for the existence of high horizontal stresses and for such stresses to increase statistically with depth (Herget, 1980). These high stress levels coupled with the stress concentration effect created by room-and-pillar or longwall mining often result in rock failure in deep underground mines in the Canadian Shield. Because of the stress levels at great depths, the amount of strain energy released upon failure is relatively large, frequently leading to violent ruptures known as "rock bursts". Large rock bursts could release as much seismic energy as an earthquake of magnitude 5 on the Richter scale. Historically, rock bursts had resulted in some significant fatalities to underground miners in the Canadian Shield. Today, with improved mine designs and safety measures, such fatalities or casualties have been drastically reduced, although some rock bursts still occur on an occasional basis.

Another problem that has frequently been attributed to high horizontal stress is the long duration of stress relief in some excavations into shales and shaly limestones. In Southern Ontario, it is not uncommon that such a process of stress relief could take years or decades to complete, subsequent to the initial elastic response. This led to a progressive inward movement of the rock mass laterally into tunnel openings and open-cut excavations. In some cases, this progressive inward movement had resulted in structural distress such as the cracking of concrete lining and walls, as well as the buckling of concrete floor slab and support structures that spanned across the excavations (Lo et al, 975; Lee and Lo, 1976; Lo and Yuen, 1981). Figure 5 shows the long-term inward movement measured in the turbine pit of an underground hydroelectric power plant at Niagara Falls. While a seasonal fluctuation of the closure is evident, the cumulative movement was large enough to cause distress and buckling of some structures spanning across the pit. In other cases in the area, more rigid structures had led to more pronounced

stress buildup and structural failure, since they tended to suppress the inward movement of the rock mass. This effect, which is locally referred to as "rock squeeze", is often associated with shales and shaly limestones subjected to high horizontal stress. No evidence of hydration of swelling minerals was detected in these case histories. Today, analytical methods are available to take "rock squeeze" into consideration in design, based on experimental determination of time-dependent deformation properties of the rock units in question (Lee and Lo, 1976; Lee and Klym, 1978; Lo and Yuen, 1981). Where necessary, precautionary measures could be used to minimize the adverse effect of "rock squeeze" on an underground structure. Such measures include the use of a relatively flexible lining or support system, and allowing for a period of months between mining and lining, such that a large part of the time-dependent deformation could occur prior to lining installation. In many cases, it was found that the time-dependent deformation of the rock mass tended to be more pronounced initially, tapering off in the course of time.

It should be pointed out here that the effects of high horizontal stresses are not all negative. As noted earlier, such stresses may lead to some localized overstressing and buckling failure. In other words, they might enhance the occurrence of some localized stress-induced instability. This is, however, compensated by their tendency to reduce the occurrence of fracture-induced instability of joint blocks and rock wedges in the roof and haunch of an underground opening. The high horizontal stresses provide a clamping force that helps to stabilize otherwise loose blocks and wedges, making the opening essentially self-supporting. While nominal bolting and chain-link wire mesh are still required for safety reasons, rock support requirements tend to be significantly reduced.

Of greater significance is the effectiveness of high horizontal stresses in making the joints and fractures watertight, thereby reducing the amount of seepage and water control requirements. This helps to speed up the rate of excavation, reducing the cost of construction or mineral extraction. There are case histories of tunnelling and underground mining in Southern Ontario and adjacent areas, where dry rocks are encountered along the entire length of the tunnel or mine drift, with no physical evidence of seepage or groundwater inflow. Such subsurface environments would be ideal for underground space development, from a geological and hydrogeological point of view. In the subsurface storage of toxic waste, including the geological disposal of nuclear waste, the most important pathway for contaminants to travel back to the biosphere is through the groundwater system. A watertight rock mass would minimize the probability of this occurring, thereby enhancing the isolation capability and safety of the waste storage/disposal system. In this regard, the contribution of the high horizontal stresses becomes very significant. It has also been said that, for an underground waste disposal system, the weakest link could well be in the sealing of shafts leading to the underground vaults or disposal rooms. In other words, it may be difficult to guarantee the long-term effectiveness of the shaft sealants in alleviating the formation of a gap between the sealant and the surrounding rock mass. If, on the other hand, the rock mass has a tendency to progressively move inward under the influence of high horizontal stress, the probability of such a gap forming could be much reduced in the long term. Hence, with proper methods of design and construction, high horizontal stresses could be put to use by man under some specific circumstances.

6 CONCLUDING REMARKS

The above discussions summarize our current understanding of the occurrence, possible origin and engineering impacts of high horizontal stresses in Southern Ontario and vicinity. It is noted that high horizontal stresses could have both negative and positive impacts on engineering and mining projects. Their geological expressions, in the form of surficial compressive failures or pop-up, are also briefly examined.

Much interest has been generated in recent years in correlating in-situ stresses with earthquake occurrence and prediction. Obviously, the limited amount of in-situ stress data available is generally from shallow depths, which are small compared with focal depths. In-situ stress values also tend to be quite variable in magnitude and orientation, making it difficult for earthquake prediction or risk assessment to be carried out solely on the basis of stress calculations. However, with additional measurements accumulating in the years and decades ahead, it is hope that the in-situ stress regimes in seismically active zones can be better defined. This may allow some patternistic or qualitative differentiation to be made between the stress regimes in seismic and aseismic zones, thereby contributing to seismic risk assessment and mitigation. Also required is an instrumentation capability to minitor transient stress changes at depth, which may trigger the occurrence of seismic events.

The study of geomechanics has contributed significantly to man's endeavours in energy and resource development, and in the mitigation of natural hazards. It still has a long way to go and many more contributions to make in the decades ahead.

REFERENCES

Asmis, H.W. & C.F. Lee 1980. Mechanistic modes of stress accumulation and relief in Ontario rocks. *Proc. 13th Canadian Rock Mechanics Symposium*: 51-55. Toronto.

Cathles, L.M. 1975. *The viscocity of the Earth's mantle*. Princeton University Press, p. 38.

Franklin, J.A. & O. Hungr 1976. Rock stresses in Canada. *Proc. 25th Geomechanical Colloquy*: Salzburg, Austria.

Gay, N.C. 1974. In-situ stress measurements in Southern Africa, *Proc., International Symposium on Recent Crustal Movements*, Zurich.

Gilbert, G.K. 1886. Some new geologic wrinkles, *Amer. J. Sci.* 3rd series 32: 324.

Haimson, B.C. & C.F. Lee 1980. Hydrofracturing stress determinations at Darlington, Ontario. *Proc. 13th Canadian Rock Mechanics Symposium*: 42-50. Toronto.

Haimson, B.C., C.F. Lee & J.H.S. Huang 1986. High horizontal stresses at Niagara Falls, their measurement and the design of a new hyroelectric plant, *Proc., International Symposium on Rock Stress and Rock Stress Measurements*, Stockholm.

Hardy, H.R., Jr., R.Y. Kim, R. Stefanko & Y.J. Wang 1970. Creep and microseismic activity in geological materials, *Proc. 11th U.S. Symposium on Rock Mechanics*: 377-413. Berkeley.

Harrison, W. 1958. Marginal zones of vanished glaciers reconstructed from the pre-consolidation-pressure values of overridden silts, *J. Geol* 66: 1:72-95.

Hast, N. 1969. The state of stress in the upper part of the Earth's crust, *Tectonophysics* 8:189-211.

Herget, G. 1974. Ground stress determinations in Canada. *Rock Mechanics* 6:53-64.

Herget, G. 1980. Regional stresses in the Canadian Shield. *Proc. 13th Canadian Rock Mechanics Symposium*: 9-16. Toronto.

Hooker, V.E. & D.L. Bickel 1974. Overcoring equipment and techniques used in rock stress determination. *U.S. Bureau of Mines Information Circular 8618.*

Lee, J.S. 1926. The fundamental cause of evolution of the Earth's surface - features, *Bulletin, Geological Society of China* 5:3/4.

Lee, C.F. 1978a. A rock mechanics approach to seismic risk evaluation. *Proc. 19th U.S. Symposium on Rock Mechanics*: 1:77-88.

Lee, C.F. 1978b. Stress-induced instability in underground excavations. *Proc. 19th U.S. Symposium on Rock Mechanics*: 1:165-173.

Lee, C.F. 1980. A classification of rock mass instability underground. *Proc. International symposium on the Safety of Underground Works*: 14-20. Brussels.

Lee, C.F. 1981. In-situ stress measurements in Southern Ontario. *Proc. 21st U.S. Symposium on Rock Mechanics*: 1:435-442.

Lee, C.F. & H.W. Asmis 1979. An interpretation of the crustal stress field in northeast North America, *Proc. 20th U.S. Symposium on Rock Mechanics*: 1:655-622. Austin.

Lee, C.F. & T.W. Klym 1978. Determination of rock squeeze potential for underground power projects. *Engineering Geology* 12: 181-192.

Lee, C.F. & K.Y. Lo 1976. Rock squeeze study of two deep excavations at Niagra Falls. *Proc. ASCE Specialty Conference on Rock Engineering* 1:116-140. Boulder.

Lo, K.Y. 1978. Regional distribution of in-situ horizontal stresses in rocks of Southern Ontario. *Canada Geotechnical Journal* 15:371-381.

Lo, K.Y. & C.M.K. Yuen 1981. Design of tunnel lining in rock for long term time effects. *Canadian Geotechnical Journal* 18:24-39.

Lo, K.Y., C.F. Lee, J.H.L. Palmer & R.M. Quigley 1975. Stress relief and time-dependent deformation of rocks. Final Report to National Research Council of Canada, Special Project S-7307.

Lomnitz, C. 1965. Creep measurements in igneous rocks. *Journal of Geology* 64:473-479.

Moye, D.G. 1958. Rock mechanics in the investigation and construction of T1 underground power station, Snowy Mountains, Australia, *Proc. ASCE-GSA Symposium on Engineering Geology*, St. Louis, Missouri.

Obert, L. 1962. In-situ determination of stress in rock. *Mining Engineer*, 1983.

Oliver, J., T. Johnson & J. Dorman 1970. Postglacial faulting and seismicity in New York and Quebec. *Canadian Journal of Earth Science* 7:579-590.

Orr, C.M. 1975. High horizontal stresses in near-surface rock masses. *Proc. 6th African Conference on Soil Mechanics and Foundation Engineering*: 201-206. Durban, South Africa.

Palmer, J.H.L. & K.Y. Lo 1975. In-situ stress measurements in some near surface rock formations. *Canadian Geotechnical Journal* 13:1-7. Thorold, Ontario.

Raterson, W.S.B. 1972. Laurentide ice sheet: estimated volumes during late Wisconsin. *Reviews in Geophysics* 10:885-917.

White, O.L., P.F. Karrow & J.R. MacDonald 1973. Residual stress relief phenomena in Southern Ontario. *Proc. 9th Canadian Rock Mechanics Symposium*: 323-348. Montreal.

White, O.L. & D.J. Russell 1982. High horizontal stresses in Southern Ontario - their orientation and their origin. *Proc. 4th Congress of International Association of Engineering Geology*, New Delhi: 5:39-54.

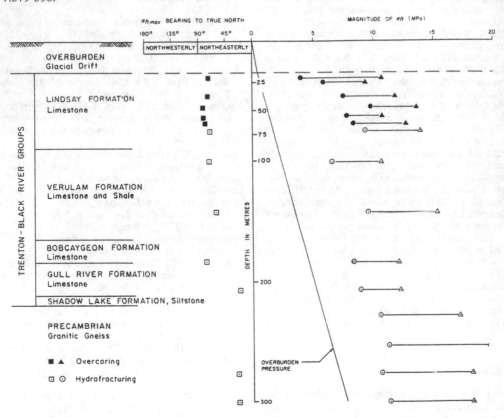

FIGURE 1: HYDROFRACTURING AND OVERCORING STRESS DETERMINATIONS AT DARLINGTON SITE

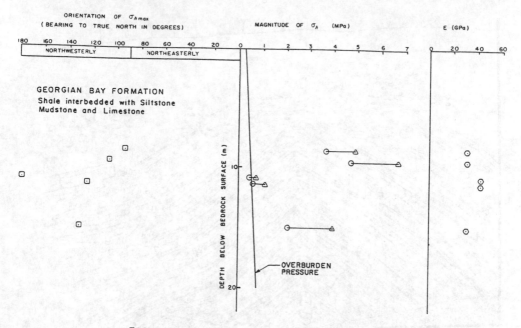

FIGURE 2: OVERCORING STRESS MEASUREMENTS
AT LAKEVIEW PLANT, MISSISSAUGA

ONTARIO POWER GENERATING STATION , NIAGARA FALLS

DOWNHOLE OVERCORING AT 7.3 m FROM EDGE OF CLIFF

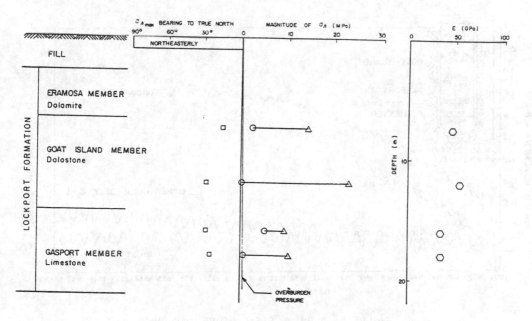

FIGURE 3: OVERCORING STRESS MEASUREMENTS AT
ONTARIO POWER GENERATING STATION,
NIAGARA FALLS (7.3m from edge of gorge)

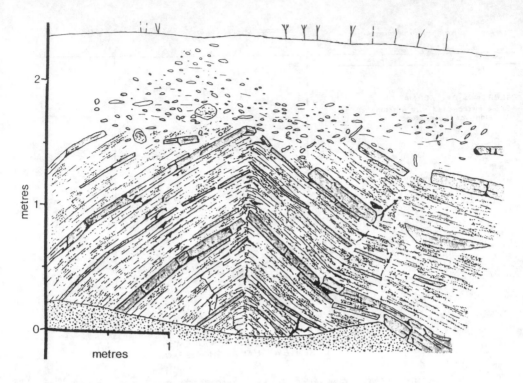

FIGURE 4 : NATURAL "POP-UP", JOSHUA CREEK, OAKVILLE, ONTARIO

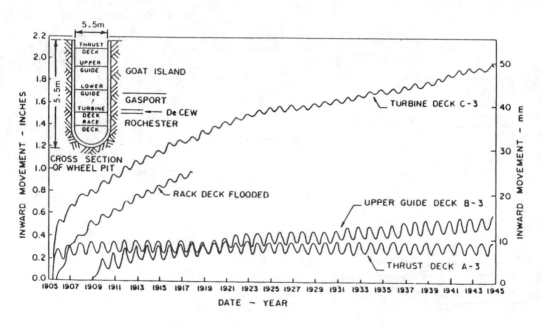

FIGURE 5: TURBINE PIT CLOSURE 1905-1945

In situ experiment on the fracture creation and propagation over a rock slope made of tertiary mudstone

Des expériences in situ sur la création et propagation d'une fracture dans une pente de roche argileuse d'âge tertiaire

Isao Suga
CTI New Technology, Fukuoka, Japan

Kunio Watanabe
Saitama University, Faculty of Engineering, Urawa, Japan

Shinji Yamawaki
Construction Technology Institute, Tokyo, Japan

ABSTRACT: Fracture creation and propagation phenomena occurring over a rock slope made of a tertiary soft rock must be studied in detail for protecting the slope from erosion. Many small fractures are newly created and developed by the repetition of the drying and the wetting processes. This phenomena were studied by a large scale in-situ test carried out at a test site in the Okinawa Prefecture that is the southern-most part of Japan from August 4 to August 10, 1992. Evaporation rates were continuously measured at 4 different points on the rock surface by 4 evaporation sensors. It was clearly found that the fracture creation and propagation phenomena occurred during a single drying process can be devided into four stages; (1) creation of many small "vertical fractures", (2) enlargement of some of those vertical fractures, (3) creation of many "parallel fractures" and (4) development of the parallel fractures.

RESUME: Les phénomènes d'initiation et de propagation d'une fracture se produisant sur une surface pentée dans des roches peu indurées d'age tertiaire doivent être étudiés en détail pour protéger la pente de l'érosion. De nombreuses petites fractures sont nouvellement créées par mouillages et séchages répétés. Ce phénomène a été étudié du 4 au 10 aout 1992 sur un site dans la préfecture d'Okinawa dans le Japon sud ouest. Les taux d'évaporation ont été mesurés en continu en quatre points de la surface affleurante par quatre sondes d'évaporation. On montre clairement que les phénomènes d'initiation et propagation de fractures se produisant durant une seule phase de séchage se font en quatre étapes: (1) apparition de nombreuses petites "fractures verticales", (2) élargissement de certaines petites fractures verticales, (3) création de nombreuses "fractures parallèles" et (4) propagation des fractures parallèles.

1 INTRODUCTION

Many small fractures are created on and beneath a slope made of tertiary mud stone by the repetition of the drying and the wetting processes. Some of these fractures are continuously developing toward the deeper part and finally cause the small scale rock falls. Small fractures created below the cutface can store and transport a large amount of rainwater and then drastically decreases the strength of the shallow part of the slope with the slaking phenomena. Surface erosion may thus occurres when an intence rain is given. The occurrences of the rock fall and the surface erosion must be carefully avoided for the long term maintainance of the rock slope (Yamaguchi 1993). The basic study on the fracture creation and propagation phenomena is quite important for making a technique by which the severe damage of the rock slope can be avoided.

Many investigations have been performed to make clear the property changes of mud stone, that are mainly due to the repetition of the drying and the wetting processes. Most of those investigations are based on the results of laboratory tests (Nakano, 1967, Franklin & Chandra 1972, Franklin 1984, Pasamentoglu & Bozdag 1993). The phenomena occurring within a drying process can be easily simulated by changing the temperature and the humidity of air surrounding a rock specimen in a laboratory. The rock specimen is gradually drying by the evaporation of water contained in this specimen. Wetting process is also easily simulated by giving water from the surface of the specimen. However, evaporation rate given in the laboratory test differs from the actual ones. There are also some differences in the wetting process when the laboratory test conditions are compared with the actual ones. The laboratory test can not completely simulate the actual processes. There is another disadvantage in the laboratory test. Many large fractures are initially existing in an actual rock mass and these fractures may much influence on the fracture creation and propagation phenomena. However, the effects of those initial fractures are very difficult to be studied because the size and of the rock used for a laboratory test is small.

A large scale in-situ test is thus needed to make clear the mechanism of fracture creation and propagation occuring in an actual environment. The present authors carried out an in-situ test to study well the phenomena.

2 IN-SITU TEST

An in-situ test was carried out from August 4 to August 10, 1992 at the Okinawa Prefecture that is the southern-most part of Japan. Figure 1 shows the location of this test site. The rock of the site is almost composed of the Shimajiri mud stone of the Miocene age and contains many micro-fossiles such as foraminifera living in seawater. A slope was selected for this in-situ test. Figure 2 presents the cross-section of the selected slope. The bottom part of this slope was covered with a concrete wall. The surface gradient of the test site shown in Figure 2 was 35°, the slope length was 6.5m and the width of the slope was about 30m. This slope was excavated at about 1 year before the start of this test.

The test site of 3m wide and 2.5m high was selected on the slope. This site was carefully excavated by hands to 30cm deep for removing the weathered rock. Figure 3 is the photograph of this test site. Figure 4(a) shows the initial fractures observed just after the excavations. The dotted part in the center of this area implys a thin tuff layer. Four solid circles are the points at which the evaporation

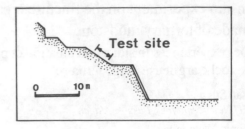

Figure 2. Cross-section of the slope used in this test

Figure 3. Test site

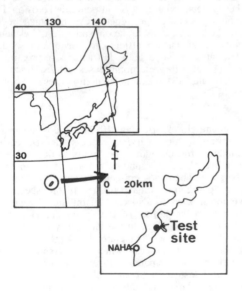

Figure 1. Location of the test site

rates were continuously measured. Evaporation rates were measured using four evaporation sensors (Watanabe, 1991, Watanabe et al., 1992, Watanabe and Bossart, 1992) that is briefly explained in the

next chapter. Figure 4(b) presents the devision of this site. The fracture creation and propagation were carefully observed and mapped over the shaded sub-area in Figure 4(b).

Figure 5 shows two vertical profiles of saturation distribution measured before the surface excavation at two different points. It is found that the saturation is very low in the shallow part from the surface to the depth of 3.0-8.0 cm. The saturation was larger than 75% and almost uniform in the part deeper than 10cm. The saturation of the new slope surface excavated was about 90%.

Rain were observed twice during this test period, in the early morning of August 6 and August 7. The total height of the former rain was 1-2mm and was not so intense. The duration time was about 2 hours. The latter rain was accompanied with a large typhoon and continued about 20 hours. The total height of the latter rain was 50mm.

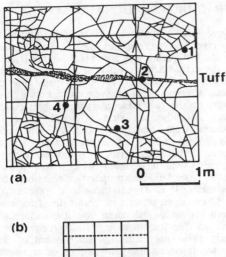

(a)

Tuff

0   1m

(b)

Area studied

Figure 4. Initial fractures observed over the test site (a). (b) presents the area studied.

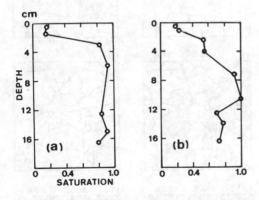

Figure 5. Initial saturation distribution

3 EVAPORATION MEASUREMENT

Figure 6 schematically shows the fundamental idea of the evaporation measurement. When the wind velocity is low, a thin layer in which vapor in air is transported by moleculor diffusion in the normal direction to the rock surface is formed. The

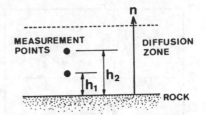

Figure 6. Fundamental idea of the evaporation measurement

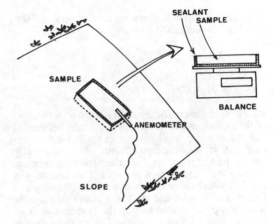

Figure 7. Calibration technique for estimating the effect of the turbulent motion of air

evaporation rate, in the other words, the vapor flux transported in this normal direction within this thin layer can be estimated from the absolute humidity gradient formed in this layer and the diffusion coefficient of vapor in air by the following equation:

$$E_V = -D_m \frac{\partial \theta}{\partial n}$$

where, Ev is the evaporation rate, Dm is the moleculor diffusion coefficient, n is the co-ordinate axis in the normal direction to the surface and 0 is the absolute humidity. The absolute humidity gradient can be measured by the sensor with measuring two absolute humidities at different two heights. The diffusion coefficient Dm can be estimated from the temperature measured. However, the thickness of this layer formed above the rock surface becomes too much small under the usual wind condition that we are facing in an actual field. The vapor is transported not only by the molecullor diffusion but also the turbulent motion of air. The evaporation rate Et under this wind condition was approximated by the following equations:

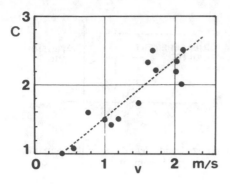

Figure 8. The relation between coefficient C and the average wind velocity v

$$E_t = C \cdot E_v$$

where, C is the coefficient and it may be a function of the average wind velocity. The relation between C and the average wind velocity is studied before the evaporation measurement. Figure 7 schematically shows the technique used for estimating the relation. A platy rock sample was prepared. This sample was fixed in a depression on the surface as shown by this figure. The bottom and the side walls of this sample were completly sealed to stop the evaporation from these surfaces. Evaporation could occur from the top surface. The weight of this sample was measured many times. The weight of this sample gradually decreased due to the evaporation. The true evaporation rate could be estimated from the weight change of this sample. At the same time, evaporation rate was directly measured using the sensor. Wind velocity was also measured at 5mm above the top surface of this sample using an anemometer. The constant C could be estimated by comparing the true and the measured evaporation rates. Figure 8 shows the relation between C and the average wind velocity v measured at 5mm above the surface. The wind velocity was averaged in 10 minutes. The relation between C and the average velocity v measured at 5mm above the surface was roughly approximated by the following equation:

$$C = 1.7v + 0.7 \qquad (v \geq 0.2\,{}^m/_s)$$

However, the maximum error included in this approximation equation is large and it reached more than 50 % in some cases. The major reason why such large error was observed is mainly due to the difficulty in the quantitative estimation of the effect of the turbulent motion of air. The error becomes extremely large when the average wind velocity is stepwisely changed. Although the error was not small, this sensor was used for getting the general tendency of the evaporation rate change.

4 FRACTURE CREATION AND PROPAGATION

Figure 9 shows the fracture creation and propagation observed over the rock surface of shaded area in Figure 4(b). Figure 9(a) presents the initial fractures. (b) shows the fractures pattern at 2 hours after the in-situ test start. It is clearly found that many new small fractures (vertical fracture) were created during this 2 hours. (c) shows the fractures pattern observed at 32 hours after the test start. Although some new fractures were created between the time period between (b) and (c), the number of the fractures created during this period was not so many as compared with it in the first 2 hours. However, another typecal fracture propagation was observed in this period. It is the enlargement of some selected fractures. (d) in this figure shows the fractures of which gap width and depth from the surface were enlarged. The maximum gap width reached to 1.5 mm. Following with this enlargement of these selected fractures, the gapwidth of other fractures decreased. Figure 9(d) shows the fractures observed at the noon of August 6. It was rain in the early morning of this day. The surface was at once wetted

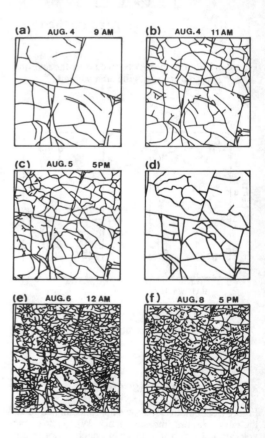

Figure 9. Fracture creation and propagation over the shaded area in figure 4(b)

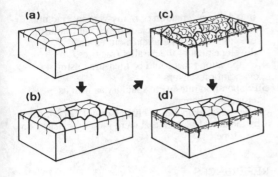

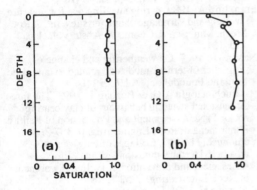

thin rock pieces are detached from the surface by the wetting and the drying of the surface. As shown by figure 10(d), the zone of parallel fractures are developed toward the deeper part.

Figure 11 shows the change of the saturation profile after the intense rain fall. Figure 11(a) shows the vertical saturation distribution measured at am 11:00 in August 7. The saturation was almost uniform and about 90%. Figure 11(b) presents the distribution measured at pm 5:00 in this day. It is clearly found that the saturation just beneath the surface much decreases during this period of 6 hours and large saturation gap was formed at 2.0-3.0 cm depth beneath the surface.

5 EVAPORATION MEASURED

As mentioned above, the accuracy of the evaporation measurement technique used in this study was not so high because of the influence of the turbulent motion of air. However, because the wind condition was almost uniform over the small test area, the spatial variability of the evaporation could be studied.

Figure 12 shows the evaporation changes measured at 4 points shown in figure 4. As the general tendency, the evaporation rate was large in day time and decreased toward evening. During night time, the rate became negative. The negative value implys the occurrence of dew drop. The evaporation rate was spatialy changed. The rate measured at No. 2 point shows the minimum value. The reason why the rate from this point is the minimum can be explained from the geologic feature of this point. As shown in Figure 4, this point was located on a thin tuff layer. As the hydraulic conductivity of this layer is less than the

Figure 10. Four stages of the fracture creation and propagation

Figure 11. Saturation distribution change after the heavy rain

and then dried. Many small thin pieces of rock were detached from the surface and these pieces were bended by the drying. Many new fractures developing in the parallel direction to the slope surface (parallel fractures) were also found beneath the surface. The depth of the zone in which these parallel fractures were concentrated was about 5 mm. The number of these parallel fractures was increased and this depth reached to 1.0 cm in some places until the evening of this day. (f) presents the rock surface after the intence rain fall that was accompanied with the typhoon. The surface of this slope was eroded and the enlarged fractures were observed on the surface.

Figure 10 schematically shows the process of the fracture creation and propagation observed. This process seems to be devided into 4 stages. At first, as shown in Figure 10(a), many small vertical fractures of which depth are less than 1.0 cm are created. Then, some of these fractures are enlarged and the gap width and the depth of these selected fractures becomes 1.0mm and 3.0-4.0 cm, respectively. (c) presents the creation of many small parallel fractures beneath the surface. When rainwater is given, many

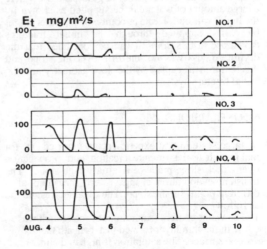

Figure 12. Evaporation changes measured at 4 different points

mudstone, so that water could not be easily transported towards the surface. On the other hand, the maximum rate was observed at point No. 4. This point was located in-between two initial fractures. It seems that a relatively large amount of water might be supplied through those fractures. The evaporation rate was also small at point No. 1 except for August 9. Because August 9 was after the typhoon, the slope surface conditions might be changed. One possible reason of this small evaporation may be due to the height of this point. This point was located at the highest level. In usual case, the higher part tends to be in less saturation condition because the water flows downwardly. The evaporation rate from the lower saturation part is less than the high saturation part.

As shown the evaporation rate changes measured in August 5, the time at which the evaporation rates reached their peaks were about 1:00 pm for points No.3 and No.4. On the other hands, the time was about 10:30 am for point No. 1 and No. 2. The potential evaporation rate is essentially decided by the radiation and the wind condition. Because the radiation and the wind condition was thought to be uniform over the small site, the time of the maximum evaporation had to be the same. The difference observed in those times must be controlled by another effect. It may be the difference of water content beneath the surface. The evaporation rate becomes large when the water content of rock is high. This water content is essentially governed by the balance between the evaporation and the water supply from deeper part of rock. Even though the potential evaporation becomes large, the evaporation rate must decrease when the water content decreased. The difference found in those times when the evaporation rates reached their peaks might present the spatial variability of the water supply condition. The peak evaporation rate also becomes large at the place where a large amount of water can be supplied as shown by the high evaporation rate at point No. 4. Water might be easily supplied through the initial fractures developed at this point. It can be concluded that the drying process was not uniform over the slope and this feature was caused by the slope geometry and the initial fracture system.

6 CONCLUSIONS

A large scale in-situ experiment was carried out for making clear the fracture creation and propagation over a rock slope made of a tertiary mud stone. The transient evaporation changes were measured at 4 different points on the slope. The obtained results are as follows:
1) The water content (saturation) of rock much was low in the thin layer formed just beneath the slope surface. The saturation is high and almost uniform under this layer.
2) The thickness of this thin layer is about 3.0-8.0cm.
3) The fracture creation and propagation phenomena occurring in the first drying period can be devided

into 4 stages; creation of many vertical fractures, enlargement of some selected vertical fractures, creation of many parallel fractures and developement of the parallel fractures.
4) Evaporation rates could be measured using 4 evaporation sensors.
5) Evaporation rates were spatially as well as temporaly changed.
6) The spatial evaporation change might be controlled by the large fractures that were initially existing on the slope.

REFERENCES

Franklin J.A. & Chandra R. 1972. The slake-durability test. Int. J. Rock Mech. Min. Sci. & Geomech. Abstr. vol.9, p. 325-341.
Franklin J.A. 1984. A ring swelling test for measure swelling and shrinkage characteristics. Int. J. Rock Mech. Min. Sci. & Geomech. Abstr. vol.21 p. 113-121.
Nakano R. 1967. On weathering and change of properties of tertiary mudstone related to land slide. Soil and Foundation, Vol.7. p.1-14.
Pasamehmetoglu A.G. & Bozdag T. 1993. The three dimensional swelling behaviour of clay bearing rocks. Proc. Assessment and Prevention of Failure Phenomena in Rock Engineering. p. 357-360.
Yamaguchi H. 1993. Collapse of cut slope of weathered tertiary mudstone ground. Proc. Assessment and prevention of Failure Phenomena in Rock Engineering. p. 521-528.
Watanabe K. 1991. Evaporation measurement in the validation drift- Part 1. Stripa Project TR 91-06. p. 1-131.
Watanabe K., Noguchi Y., Sakuma H. & Sutani Y. 1991. Transient change and spatial variability of the evaporation on a tunnel wall. Proc. 7th Congress of ISRM. p. 647-650.
Watanabe K. & Bossart P. 1991 Complementary methods for measuring water inflow into the ventilation drift, TR 91-34, NAGRA. pp. 19-34

A new equipment used for measuring evaporation in a field

Un nouvel équipement utilisé dans la mesure de l'évaporation sur le terrain

Kunio Watanabe
Saitama University, Faculty of Engineering, Urawa, Japan

Yoshihito Tsutsui
Yokohama City, Japan

ABSTRACT: The direct evaporation measurement in a field is very important and indispensable for making clear the moisture movement in the shallow part of underground. A new equipment that is useful for the field measurement of the evaporation was developed. The basic idea of this measurement technique is identical to it of the ventilation test that has been used in some tunnels for measuring the small inflow rate of groundwater. A small part of ground surface is covered with a box made of transparent plates. Air is continuously injected into this box and exhausted. The evaporation rate from the ground surface covered with this box can be calculated from the air flow discharge rate and the absolute humidity difference between injected and exhausted air. The accuracy and the applicability of this equipment have been successfully examined by some field and laboratory tests.

RESUME: La mesure directe de l'évaporation sur le terrain est très importante, en particulier pour clarifier les mouvements d'humidité dans la partie superficielle du sous-sol. Un nouvel équipement pratique pour mesurer l'humidité sur le terrain a été développé. L'idée de base est identique a celle des essais de ventilation utilisés dans certains tunnels pour mesurer les faibles taux de venues d'eaux interstcielles. Une partie de la surface du sol est recouverte par une boite faite d'un matériau transparent. L'air est continuellement injecté et extrait de la boite. Le taux d'évaporation de la surface couverte par la boite est calculé a partir du taux de décharge du flux d'air et de la différence absolue d'humidité entre l'air entrant et sortant. La mise en oeuvre et la précision de ce systerne sont examinées par quelques essais en laboratoire et sur le terrain.

1 INTRODUCTION

Evaporation is one of the most important phenomena closely relating to many engineering geologic problems such as irrigation, cultivation, weathering, and so on. Many previous authors have been proposing some techniques for estimating the precise evaporation rate. However, most of those techniques are essentialy based on some meteorological theories and the heat balance (Penman 1948). Although the average evaporation rate from a wide area can be estimated by those meteorological techniques, the evaporation from an arbitrary point on ground surface has been very difficult. The evaporation rate may differ from place to place because the vegetation, moisture content and geologic features of ground are not uniformly distributed. For the reason, the direct measurement of the evaporation from an arbitrary point on a ground surface has been needed.

The evaporation from water surface can be measured using pan-type evaporation-meter. The evaporation from a surface of a small size specimen is also estimated with measuring the weight change of the specimen because evaporation decrease the total weight of the specimen. Although these technique have been widely used, the evaporation rate from an arbitrary point on ground has been very difficult.

The present authors have developed a evaporation sensor for measuring the evaporation rate from the low permeable tunnel wall (Watanabe 1991, Watanabe et al 1991. Watanabe and Bossart 1991). The absolute humidity gardient in a thin boundary layer formed just above the rock wall can be measured using this sensor. The vapor transportation in the normal direction of the wall is mainly caused by the molecular diffusion in this thin layer. For the reason, the evaporation rate, in the other words, the vapor flux in the normal direction can be estimated with maltiplying the absolute humidity gradient measured in this layer by the diffusion coefficient of vapor. Under the wind condition observed in many tunnels, the thickness of this layer is about 1.0cm except for some special places auch as the area in the vicinity of the exit of a ventilation pipe. The absolute humidity gradient can be measured when this thickness is 1.0cm and more. However, the thickness become very small under the usual wind velocity condition that we are facing in a field and it is almost inpossible for measuring precisely the absolute humidity gradient. The vapor transportation by the turbulent motion of air must be also considered (Suga, Watanabe & Yamawaki 1994). Because the wind velocity on an actual ground surface is high and not steady, there is a large difficulty to measure the evaporation rate using this sensor.

With the aim to measure the evaporation in a field, the present authors newly developed a equipment. In this paper, the fundamental idea of the measurement technique and the structure of this equipment are shown with some data obtained in a grass field and a laboratory.

2 MEASUREMENT EQUIPMENT

The basic idea of this equipment is essentialy identical to the ventilation test that have been used in tunnel for measuring the small inflow rate of groundwater. Figure 1 schematically shows the idea. A part of the ground surface is covered with a box made of transparent plates. The size of this box is 30cm high, 50cm wide and 60 cm long. Air is continuously injected into this box and then exhaust from this box through two pipes attached on both side walls. When vapor is coming out from the ground surface by the evaporation and/or the transpiration, the absolute humidity of the exhaust air must be larger than it of the injected air. Under this condition, the evaporation

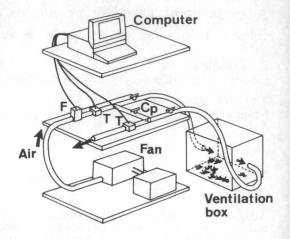

Figure 2. Equipment. F: Flowmeter, T: A couple of the thermistor and a humidity sensor and Cp: short-cut pipe.

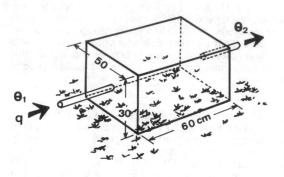

Figure 1. Idea of the evaporation measurement

from the ground surface can be estimated by the following equation:

$$Ev = q(\theta_2 - \theta_1)$$

where, θ_1 and θ_2 are the absolute humidites of the injected and exhaust air respectively and q is the air flow dischage. The absolute humidity can be calculated from the relative humidity and the temperature measured by a couple of a thermistor and a humidity sensor. The fundamental idea is very simple. Figure 2 shows the total system of this equipment. Air is injected into the ventilation box

using a fan. The flow rate is measured by a flowmeter (F). T in this figure implys a couple of a thermistor and a humidity sensor. All data are sent to a micro-computer and absolute humidities and evaporation rate are calculated. All data and the calculated results are stored in a floppy disk. The minimum time interval for taking data and calculating the evaporation rate is 2 minute.

Cp in this figure is the short-cut pipe used for the calibration of the thermistor and the humidity sensor. When air is flowing only through this short-cut pipe, the absolute humidities measured in both injected and exhaust air must be same. The accuracy of these sensors can be calibrated by comparing these absolute humidity values.

The interstices between the ground surface and the side walls of this box must be completely sealed to avoid the leakage of air through these interstices. Many interstices are formed because the ground surface is rough. This leakage was stopped using a frame made of steel as shown by Figure 3. The height of the steel frame is about 20 cm and this frame penetrates into ground. The transparent ventilation box is installed on this plate and the interstice between the ventilation box and the frame is completely sealed.

This equipment can be also used for the evaporation acceleration test with installing a heater behind the fan and injecting warm air into the box. The accelaration of the evaporation is important for drying up the ground surface within a short time and to simulate the arid condition. The moisture movement in a shallow part will be easily studied using a evaporation accelation equipment.

It must be carefully tested whether or not the absolute humidity difference can be precisely measured using two couples of a thermistor and a

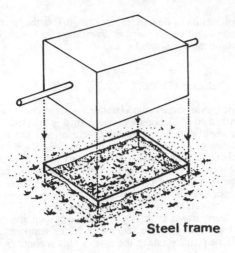

Figure 3. Installation of the ventilation box on a steel frame

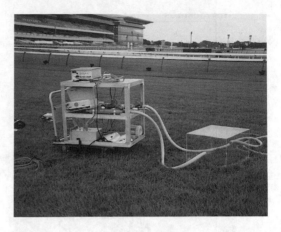

Figure 4. Evaporation measurement at a grass field

humidity sensor. Some field and laboratory tests were performed to examine the applicability and accuracy of this measurement technique.

3 APPLICABILITY OF THE EQUIPMENT

The accuracy and the applicability of this equipment were examined by a field and some laboratory tests. The transient change of the evaporation from a grass field of the Tokyo Horse Racing Course was measured. This grass field has been well maintained and the grass are iniformly growing over a wide area. The evaporation rate from this field was thought almost uniform and the evaporation rate is mostly determined by the net-radiation (Fitzpatrick and Stern 1965, Chang 1970). The measurement was carried out in August 25, 1993. Figure 4 shows the photograph of this measurement. Air flow discharge can be arbitrary controlled. The discharge was fixed as 75l/min in this test. A solid line in figure 5 presents the transient evaporation rate change measured from 11:00 am to 5:00 pm. The transient change of the net-radiation is also presented by a broken line. The net-radiation rate was low in the period from about 1:00 pm to 2:00 pm. A cloud hided sun in this time period, so that the net radiation became low. The evaporation rate also decreased during this period. It was thought from this fact that the general tendency of the evaporation change can be measured using this equipment because the evaporation rate essentially governed by the net radiation. The transient evaporation change, however, is not completely parallel to the radiation change. This feature may present the effect that the grass itself controlls the evaporation from its leaves. Although the

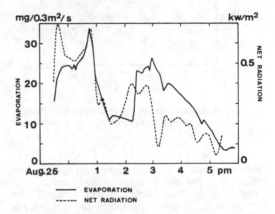

Figure 5. Evaporation and net radiation measured at a grass field

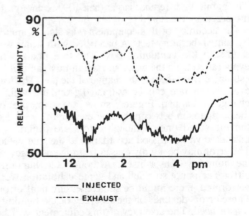

Figure 6. Relative humidity difference between the injected and exhaust air

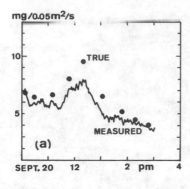

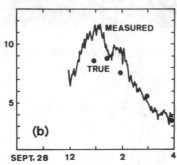

Figure 7. Two examples of the comparisons between the true and the measured evaporation

evapotranspiration is much influenced by the net radiation, it can not be estimated only from some meteorological data because plants controll the transpiration. Figure 6 presents the comparison between the relative humidities of the injected and the exhaust air. It is found that the relative humidity differences between those air were larger than 10%. The humidity difference larger than 10% is enough to me precisely measured.

The accuracy of this equipment was also examined by some laboratory tests. A pan filled with water was settled in the ventilation box. The true rate of the evaporation from the water surface in this pan can be estimated from the weight change of the pan. The true evaporation rate can be estimated from the weight change of this pan. Figure 7 shows two examples of the comparison between the true evaporation rate and the measured one. Although errors are found, the measured value is good accordance to the true value. The error may be mainly caused from the accuracy of the humidity sensor itself. Although the humidity difference is not so small and some calibration were performed, there might be some measurement error. These errors depends on the accuracy of the humidity sensor used. The accuracy of this equipment will be improved by replacing the sensor with the better one. Although those problems are remained yet, it was found that the evaporation can be measured and the

applicability of this equipment is high. It is the largest advantage that the evapotranspiration rate can be continuously measured in a field.

4 ACCELERATION TEST

The evaporation accelaration tests were also performed in a laboratory. Warm air was injected into the ventilation box. Five tests were carried out with changing air temperature. The air discharge was $75l/min$. A pan filled with water was settled within the box for measuring the true evaporation rate. Figure 8 shows three examples of the transient evaporation changes measured. Air temperatures in the ventilation box (a) 15°C, (b) 35°C and (c) 57°C. It is clearly found by this figure that the evaporation rate is increased with the increment of air temperature. The evaporation rate in the case of (c) is about 5 - 6

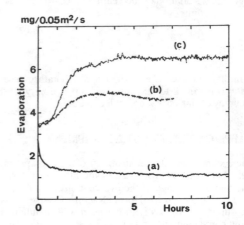

Figure 8. Results of the evaporation acceleration tests

Table 1. Comparison between the true and the measured evaporation rates

Experiment number	Temperature in box °C	Measured rate	True rate
Case 1	15	1.41	1.32
Case 2	35	3.77	3.40
Case 3	37	4.89	4.36
Case 4	57	6.68	6.16
Case 5	57	6.56	7.40

*Unit of the evaporation rate is mg/s
*Size of the pan used in this experiments was 25cm long and 20cm wide

312

times larger than the case of (a). The evaporation rates reached to their steady value at about 1 hour after the test start.

Table 1 presents the comparison between the measured and the true evaporation rates. The measured value shows the good accordance with the true value. The error was about 10%. This error may be also due to the accuracy of the humidity sensor. Although this problem is remained yet, It is concluded that the evaporation can be increased using this equipment.

5 CONCLUSIONS

A new equipment for measuring the evaporation rate in the actual field was developed. The applicability of this equipment and the accuracy of the measurement were examined by some field and laboratory tests. Although some problems on the accuracy of the humidity sensor is remained yet, the applicability of this equipment was made clear with following results:
(1)The absolute humidity difference between the injected and the exhaused air can be measured.
(2)The transient change of the evaporation rate measured on a grass field was almost parallel to the change of the net radiation as the general tendency.
(3)The error contained in this evaporation measurement is thought to be caused by the accuracy of the humidity sensor.
(4)The accuracy of this measurement will be improved by replacing the sensor with a better one.
(5) This equipment can be applied to the field measurement of the evaporation.

REFERENCES

Chang, Jen-hu 1970. Global distribution of net radiation according to a new formula, Ann. Assoc. Amer. Geoge., 62, p.340-351.
Fitzpatrick, E.A., Stern, W.R. 1965. Components of radiation balance of irrigated plots in a dry monsoonal environment, J, Appl. Meteor. 649-660.
Penman, H.L. 1948. Natural evaporation from open water, bare soil and grass. Proc. Roy. Soc., London, A-193, p.129-145.
Watanabe, K. 1991. Evaporation measurement in the validation drift- Part1. Stripa Project TR 91-06. p.1-131.
Watanabe, K., Noguchi, Y., Sakuma, H. & Sutani, Y. 1991. Transient change and spatial variability of the evaporation on a tunnel wall. Proc. 7th Congress ISRM, p.647-650.
Watanabe, K. & Bossart, P. 1991. Complementary methods for measuring water inflow into the ventilation drift, TR 91-34 NAGRA, p.19-34.
Suga, I., Watanabe, K. & Yamawaki, S. 1994. In-situ experiment on the fracture creation and propagation over a rock slope made of tertiary mud stone, Proc. 7th Congress IAEG.

Mechanical behaviour of a residual soil from Botucatu sandstone

Comportement mécanique d'un sol résiduel du grès de Botucatu

L.A. Bressani, F.B. Martins & A.V.D. Bica

Civil Engineering Department, UFRGS, Porto Alegre, Brazil

ABSTRACT: Special triaxial tests were carried out on fully saturated specimens of residual soil from Botucatu sandstone of Southern Brazil. The results were indicative of the existence of an yield surface associated with the soil structure. The Young modulus was calculated considering the stress paths followed during the tests. A servo controlled triaxial machine which incorporates instrumentation to measure locally the specimen strains is also described.

RESUMÉ: On a fait des essais triaxiaux spéciaux en éprouvettes complétement saturées de sol résiduel de grés Botucatu, occurent au Sud du Brésil. Les résultats ont indiqué l'existence d'une superficie de glissement associée à la structure du sol. Le module d'Young a été calculé en considérant le chemin de tensions suivi pendant l'essai. On a aussi décrit un appareil triaxial contrôlé à l'aide d'un ordinateur, qu'incorpore d'instrumentation pour mesurer localement les déformations des éprouvettes.

1 INTRODUCTION

In the Southern region of Brazil there is a characteristic sandstone of eolic origin deposited under desertic conditions, the Botucatu Formation, which occupies around 1.3 million square kilometres. Most of it has been covered by volcanic rock (basaltic flows of the Parana Basin). It also occurs between the basaltic layers in some places.

The sandstone has colours varying from yellowish to red and a characteristic cross stratification due to its eolic origin.

The sandstone has considerable variation on its unconfined compressive strength and has been subjected to different degrees of weathering according to covering, topography, drainage and climate. At Rio Grande do Sul State the Botucatu formation appears in the centre of the State as a band of varying width with a general east-west direction. One important area in which it appears is to the north of Porto Alegre, where several towns are gradually occupying areas with soils derived from the formation. As a consequence there is an increasing number of problems related to superficial erosion, piping and slope stability.

2 MAIN SOIL CHARACTERISTICS

The Botucatu sandstone has a grading varying between fine to medium sand (average grain diameter is 0,25mm) with the majority of the grains being of quartz and feldspar (IBGE,1986).

The sandstone gives origin to residual profiles with thickness varying greatly depending of the particular conditions at the site. The topsoil (A horizon) is generally thin with thickness less than 0,5m. The B horizon varies from zero to 3m or 5m depth. This horizon has a greater proportion of clay size particles and it does not show the stratification of the original rock. This soil presents a mosaic of variations in colour and the mass is strongly heterogeneous in strength. There are parts with clear clay concentrations and parts with material very similar to the original sandstone.

The C horizon is the most important having up to 25m thickness. Its strength is a function of the weathering and of the original rock cement, generally iron oxides. The stratification and the layering is still perfectly visible and there is a great homogeneity in colour and field strength with depth.

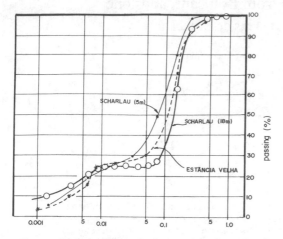

Figure 1 - Grading of the residual soil of Botucatu sandstone from two sites.

One of the main problems of this horizon is related to erosion of exposed soil on slopes of small inclination. It is common the formation of gullies in such cases.

The soil has an uniform grading as shown in Figure 1. In the figure are also shown the grading of C and B horizons of another location 10 km North of the borrow pit site (Estância Velha site) and it can be seen the small variation between the two sites at the same horizon.

The Atterberg limits of samples obtained from the two sites are given on Table 1. Note that the natural voids ratios are quite high compared with the equivalent voids ratio at the Liquid Limit (e_L - Vaughan, 1988).

Table 1 - Main physical characteristics of Botucatu Sandstone residual soil (C horizon).

	Scharlau	Est. Velha
LL	21	21
LP	17	-
e_L	0.56	0.56
e_{nat}	0.73	0.60-0.68
w_{nat}	13	13 - 14
G	2.67	2.69

3 TRIAXIAL EQUIPMENT

The tests referred here were carried out in a triaxial equipment with a self control of the stress paths developed at UFRGS by Martins (1994). A software was also developed for test control which is able to maintain linear stress paths or to keep the radial strain null during axial loading tests (k_0).

The equipment has a triaxial cell similar to that described by Bishop and Wesley (1975), servo controllers of axial and confining pressures, transducers for local measurements of axial and radial strains, internal axial load cell and pressure transducers. All transducers are connected to a microcomputer through an interface (CIL 6580).

The local transducers of axial strain follow the mechanical arrangement proposed by Clayton and Khatrush (1986). The radial strain transducer followed the design developed by Bressani (1990). The axial strain devices have a linear range of 1.5mm corresponding to around 2% axial strain of the specimen. The radial devices are linear over 1.6% strain of the specimen. The general characteristics of the Hall transducers are described by Clayton and Khatrush (1986) and Soares (1992).

The design of the servo controllers used for confining pressure and the lower chamber pressure (axial load) was based on a direct current motor acting on a screw control piston. A gear box was used with a 1:50 reduction in speed. The motor can be switched on or off with clockwise or anti-clockwise direction by relays from the CIL interface. Acting through these relays the software can control very precisely the stress paths followed during the test (Martins, 1994).

4 SAMPLING AND SPECIMEN PREPARATION

The samples used in this study have been taken from an old borrow pit (Vila Scharlau site) with a slope of around 20m height. Cubic blocks 30cmx30cmx30cm have been taken from a steep slope with great care due to the material brittleness. The depth was 4.20m from the original surface. The samples were covered with paraffin wax and transported inside wooden boxes with soft material to protect them. At the laboratory they were kept inside the controlled humidity room.

The specimen preparation followed the usual procedure employed with soil samples. The material is very easy to shape by wearing it using a nylon string. The final dimensions were 50mm diameter and 100mm high.

The specimens were assembled in the triaxial machine immediately after the moulding process described. The local instrumentation was then glued in previously marked positions at the rubber membrane. The application of vacuum of 20kPa inside the specimen was necessary to keep the instrumentation in position. The same pressure was used when the triaxial cell was filled with water (confining pressure). Deaired water was then percolated through the specimen from base to top. The saturation was obtained with back-pressure. To avoid the cycles of loading/unloading which are common in the normal back-pressure procedure and its deleterious effects on the soil structure (Bressani and Vaughan, 1989), small increments of cell pressure and back-pressure were made simultaneously. A constant effective confining pressure of 20kPa was maintained throughout the process. The pressures were typically increased to 170 and 150 kPa. Having in mind that the equipment used had a maximum total confining pressure of 700 kPa, the saturation with back-pressure was not used in the tests of null radial strain so that the maximum levels of pressure could be used.

5 DIRECT SHEAR AND TRIAXIAL TEST RESULTS

5.1 Direct shear tests

Direct shear tests were carried out to identify the strength envelope at large strains of the soil, as the local strain measurements used at the triaxial tests had a limitation of 1.5 mm displacement. The direct shear results are presented on Figure 2. Four normal stress were used: 25, 50, 100 and 200 Kpa. All four tests had a clear peak of strength. Two strength envelopes were so derived, one for peak strength and the other for values of strength corresponding to the maximum deformation of the test. The peak strength envelope has a curved format specially at low stress levels. The envelope derived from strength at 10% strains has a cohesive intercept of 7.5kPa and an angle of internal friction of 35^0.

5.2 Triaxial tests

Four linear stress path tests and one test with radial null strain were chosen to verify the soil behaviour. Figure 3 shows the stress paths followed by the tests. The results in terms of deviatoric stress versus axial strain and volumetric versus axial strain are presented on Figure 4.

The test in which the stress path follows a constant angle of 135^0 on the p'x q diagram presented a clear peak. The stress path of this test was able to

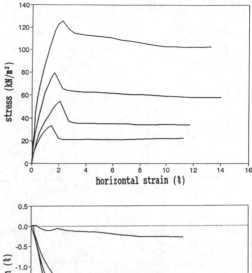

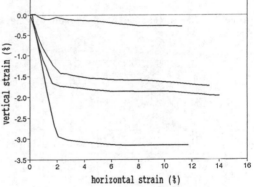

Figure 2. Results of direct shear tests.

surpass the strength envelope obtained from direct shear tests at large strains. Note that an over-consolidated soil presents peak at some stress levels but the point of peak strength of this test do not correspond to the maximum volume deformation rate which would be normal for an over-consolidated soil. Such behaviour of peak strength non-coincident with volume variation agrees very well with results described by Vaughan (1988).

For the test with vertical stress path (90^0) the maximum stress happened without a peak, with an axial strain of the order of 0.5%. The maximum stress which the specimen withstand was coincident with the strength envelope.

In the tests with stress paths corresponding to 45^0 and k = 1.2, the specimens did not reach the strength envelope. However, these two stress paths showed a clear change in behaviour with significant reduction of stiffness at low strain levels.

Some workers have observed the same

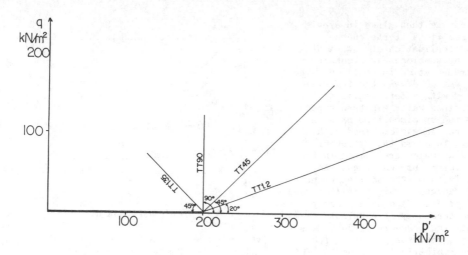

Figure 3. The stress paths followed by the tests.

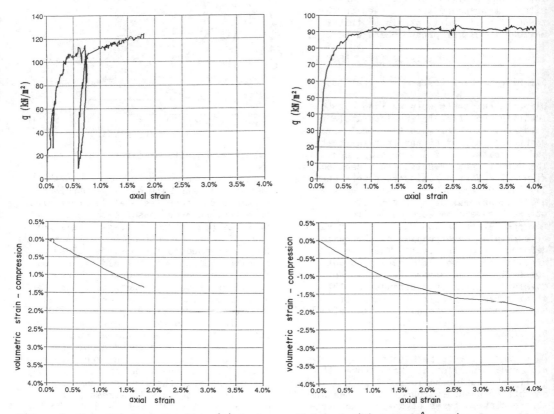

Figure 4. Triaxial test results. (a) test 45⁰ on p' x q;

Figure 4. (b) test 90⁰ on p' x q;

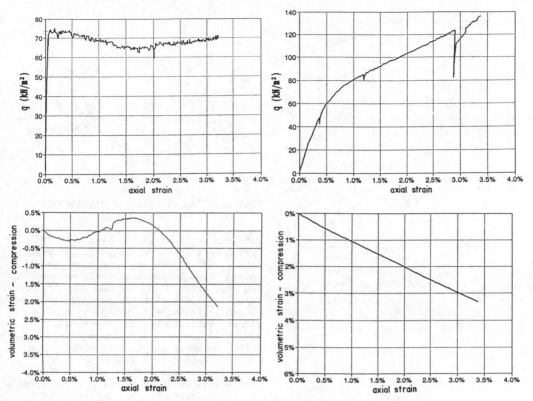

Figure 4. (c) test 135^0 on p' x q;

Figure 4. (d) test $\sigma_1/\sigma_3 = 1.2$;

general behaviour in different natural materials. Leroueil and Vaughan (1990) related this behaviour to the presence of a soil structure. This structure allows the soil to maintain a much higher voids ratio than it could have if the classical concepts of stress history and effective stress were applied and gives to the soil the brittle characteristics shown at some stress levels. According to those authors the structure can have different origins such as silica deposition, carbonates, hydroxides or organic matter on the particle contacts, cold welding, residual structure of original rock, etc.

The effect of structure can also be seen on one dimensional test results of residual soils such as those presented by Leroueil and Vaughan (1990) (Figure 5). Note that in these results the curve of the natural soil surpass the curve of the remoulded soil. The relation between stress and voids ratio cannot be reproduced after the structure is destroyed.

5.2.1. Yield curve

The change in behaviour observed in the stress-voids ratio curve of some natural soils was called yield point by Vaughan (1988). The yield point is determined through the extension of the straight lines which represent the stress-strain curve before and after the change in stiffness.

When the yield points obtained from different tests are plotted in the stress space they define a surface in the p' x q diagram which is called the yield surface (Leroueil & Vaughan, 1990). According to those authors the surface can be divided in three zones: compression, shear and swelling (Figure 6). The yield surface presents in general an asymmetric format. Some yield surfaces obtained from different natural materials such as clays, residual soils and soft rocks have been presented by Leroueil & Vaughan (1990). The authors observed that for residual soils the yield surface is centered around the isotropic axis. The sedimentary soils

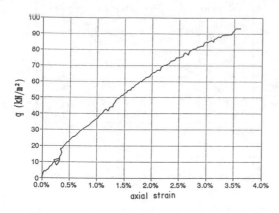

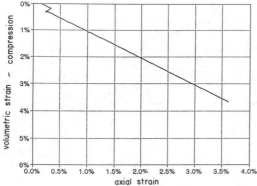

Figure 4. (e) test with null radial strain.

have the yield surface centered around the K_0 axis. The results obtained on the Botucatu sandstone residual soil are presented on Figure 7. The yield surface cross the strength envelope at low levels of stress. This can explain the peak

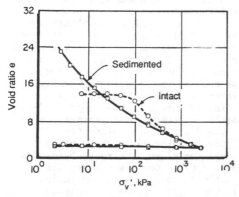

Figure 5. The results of one-dimensional tests on structured and remoulded soil.

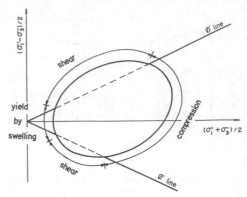

Figure 6. The yield curve division.

strength obtained in the test with 135^0 stress path as the yield point was reached with a stress combination higher than the normal strength envelope would allow. In the test with 90^0 stress path the failure point coincides with the yield point and the soil structure is stable at that stress combination. The tests with 45^0 and k = 1.2 showed a yield point well below the shear strength envelope which was not reached during the tests.

It is worthwhile to examine the stress path followed by the test with null radial deformation. At low levels of stress, the stress path followed just bellow the yield surface and above the shear strength envelope of large strains obtained from direct shear tests. Similar behaviour has been observed by other workers in structured soils (Leroueil and Vaughan, 1990, Bressani, 1990).

5.2.2. Young modulus

The determination of the Young modulus of the specimens following different stress paths requires some special calculations. The calculations take into account the variation of the two effective principal stresses, the variation of axial and radial strains and the Poisson's ratio. The material was supposed to be isotropic in these calculations as there is no data available to calculate otherwise. This hypothesis needs to be reviewed later on as the material has a clear horizontal stratification. The Young modulus obtained from these calculations are presented in Figure 8.

The results show a clear difference between the tests according to the stress paths followed by the specimens. The test with the stiffer behaviour was the unloading compression (135^0), followed by

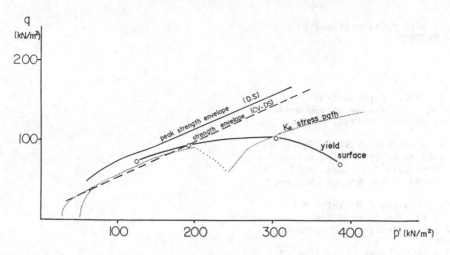

Figure 7. The yield curve obtained from Botucatu sandstone residual soil tests.

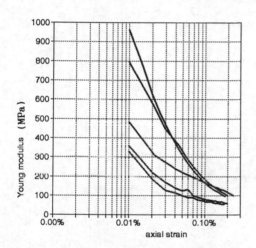

Figure 8. The Young modulus obtained from the tests. The curves are from tests (top to bottom) 135^0, 90^0, 45^0, k_0 and $k = 1.2$.

the vertical stress path test and the compression loading.

There is a clear correspondence between the results of the null radial strain and the test with $k = 1.2$ constant stress ratio in which very small radial strains were observed.

These observations seem to indicate that the apparent Young modulus (as calculated) is directly affected by the stress path inclination.

6. CONCLUSIONS

A servo controlled triaxial equipment with local instrumentation was built to perform tests with different stress paths. The equipment was used to test a residual soil from Botucatu sandstone (C horizon). The soil has an in situ voids ratio (0.73) which is higher than the voids ratio at the Liquid Limit (0.56).

Five triaxial drained tests were carried out with specimens fully saturated. The results allowed the determination of a yield curve similar to those found in other natural materials. There is a clear change in behaviour after the yield point.

The strength envelopes of peak and large strains were obtained from direct shear tests. The peak envelope showed a clear curvature.

The Young modulus was calculated for each test taking into consideration the stress paths followed. The initial stiffness measured at 0.01% axial strain had values of up to 960MPa. The soil was considered to be isotropic for the calculation. The results indicate that the stiffness is influenced directly by the direction of the stress path. The two tests with lower stress path inclination also showed the less stiff behaviour ($k=1.2$ and k_0).

ACKNOWLEDGEMENTS

The second author was supported by the federal agency CNPq. Some of the equip-

ment was built with funds provided by the
state agency FAPERGS.

REFERENCES

Bressani, L.A. 1990. *Experimental
studies of an artificially bonded soil*.
PhD Thesis, Imperial College, London.
530p.
Bressani, L.A. & Vaughan, P.R. 1989.
Damage to soil structure during triaxial
testing. *Proc. 12th Int. Conf. Soil
Mech. and Foundn Engng*, Rio de Janeiro,
1:17-20.
Bishop, A. W. & Wesley, L.D. 1975. A
hydraulic triaxial apparatus for
controlled stress testing. *Géotechnique*,
25,4:657-670.
Clayton, C.R.I. & Khatrush, S.A. 1986. A
new device for measuring local axial
strain on triaxial specimens.
Géotechnique, 36, 4:593-597.
IBGE, 1986. *Levantamento de recursos
naturais, Folha SH.22, Porto Alegre*.
Fundação Instituto Brasileiro de
Geografia e Estatística, Rio de Janeiro.
Leroueil, S. & Vaughan, P.R. 1990. The
general and congruent effects of
structure in natural soils and weak
rocks. *Géotechnique*, 40, 3:467-488.
Martins, F.B. 1994. *Triaxial tests on the
Botucatu sandstone residual soil*. CPGEC,
UFRGS, MSc dissertation (in portuguese).
Soares, J.M.D. 1992. *Local measurement of
strain with transducers based on Hall
effect principle*. CPGEC, UFRGS. 31p.
(in portuguese).
Vaughan, P.R. 1988. Characterising the
mechanical properties of in-situ
residual soil. *Proc. 2nd Int. Conf.
Geomechanics in Tropical Soils*,
Singapore, 2:469-487.

Engineering properties of soft marine clay as studied by seismic cone testing

Étude des propriétés géotechniques d'une argile molle marine par l'essai sismique de cône

Yasuo Tanaka & Daizo Karube
Department of Civil Engineering, Kobe University, Nada, Japan

Kiichi Tanimoto
Construction Engineering Research Institute, Nada, Kobe, Japan

ABSTRACT: The seismic cone testing was carried out in a soft marine clay which is still in the middle of consolidation process under the load of man-made island constructed above it. The results of seismic testing shows a variation of seismic velocity with depth which is in accordance with the consolidation state of the clay. By comparing the piezocone test data with the result of seismic cone testing, it was also found that and the dynamic shear modulus increases non-linearly with the increase in undrained shear strength of clay.

RÉSUMÉ: Nous avons fait l'essai sismique de cône dans l'argile fragile marine. Le resultat de l'essai sismique de cône indique bien que l'argile a été soumise a une sous-consolidation. Le module dynamique de cisaillement augmente d'une façon non-linéaire avec la resistance au cisaillement non drainé.

1 INTRODUCTION

In Osaka Bay of Japan, a thick deposit of soft marine clay covers the surface of seabed and many man-made islands are being constructed along the coast of the Bay. For example, the city of Kobe has been reclaiming three large man-made islands at the near-shore of the city and the total area gained from the reclamation works amounts to be nearly 1500 ha. For the construction of these islands, the geotechnical properties of the soft clay is a major importance as it governs the settlement, the stability and the seismic response of man-made island and furthermore the complication arises as these geotechnical properties changes with time and in spaces as the consolidation of soft clay progresses.

The consolidation and strength properties of the soft marine clay in Kobe area have been studied previously in order to assess the settlement and stability of soft clay, for example Tanimoto et al. (1982) and Tanimoto and Tanaka (1984). On the other hand, the study on dynamic properties of the soft marine clay is very limited. However, the dynamic property of clay is very important for the seismic analysis of the man-made islands on soft clay as it is known that the seismic wave from basal rock is usually much amplified near the ground surface especially when there is a soft clay layer beneath the structure.

In order to investigate the dynamic properties of soft marine clay in-situ, the piezocone test apparatus equipped with seismic sensor (so called Seismic-Cone) was used and the testing was performed at one of the man-made islands in Kobe. The soft clay layer at the investigation site is still consolidating and thus a particular interest was placed on examining the depth variation of the dynamic properties and the penetration resistances. The results of seismic testing shows a variation of seismic velocity with depth which is in accordance with the consolidation state of the clay and the result of seismic cone testing was analysed by comparing with the data from the piezocone testing.

2 GENERAL DESCRIPTION OF TEST SITE

The test was performed at one of the man-made islands in Kobe. The water depth in the area is about 15 m and the seabed consists of very soft marine clay of post-glacial origin at the surface which is underlain by a thick layer of sand and gravel of Pleistocene deposit. The thickness of the soft clay varies from 15 to 20 m typically in the area and its consistency shows a very highly plastic nature of the clay. The thickness of the sand and gravel layer varies from 50 to 70 m in the area, and below this layer there are multiple layers of Pleistocene clay and sand. The geological study of the seabed deposit in Osaka Bay shows that there are altogether fourteen layers of marine clay which are inter-bedded with the layers of sand and/or gravel, and these formation are the results of sea level changes that occurred during the glacial advances and retreats. The top most marine clay is locally known as alluvial clay, and its deposition has started about 10000 years ago.

The man-made island is reclaimed by placing granular fill materials, mainly decomposed granite or crushed rocks, on the top of the soft marine clay. Usually the fill is placed rather rapidly without installing vertical drains and after the fill elevation reaches a certain level the vertical drains, usually sand drains, are installed in the areas where the settlement control is necessary. Thus the consolidation of soft clay occurs under one-dimensional condition for the early stage of reclamation work, and during this stage a large variation of soil stiffness with depth is expected.

3 SEISMIC CONE TESTING

Figure 1 shows a typical arrangement of the seismic cone testing used in this study. As noted earlier the fill thickness above the soft clay layer is large, about 15 to 20 m, and therefore the casing has to be installed near to the top of clay surface. The testing procedure employs a down-hole method of seismic logging and the cone penetration is stopped at certain depth intervals, in this study every 1 m intervals, to initiate the recording of the seismic waves which is generated at the surface. The seismic shot was generated by hitting the end of wooden plank placed on the surface using a sledge hammer. Campanella et al. (1986) have reported first the result of seismic cone testing they have developed and showed the usefulness and effectiveness of determining the shear wave velocity in the ground. It is also shown by many researchers that the result from the seismic cone test agrees very well with the one obtained from the cross-hole method for example Campanella et al. (1986), Larsson and Mulabdic (1991) and Baldi et al. (1988).

As to the seismic cone apparatus, Fig. 2 shows the details of the seismic cone which is commercially available from Hogentogler of US. It consists of cone tip with its angle of apex 60 degrees having a cross-sectional area of 10 cm2. Behind the cone tip, a porous element to monitor the pore water pressure is mounted and then a cylindrical sleeve for measuring the frictional resistance. The maximum capacities of the measurements are 100 MPa for q_c, 1 MPa for f and 3.5 MPa for u. The seismic sensor is mounted horizontally behind the friction sleeve. During the cone penetration with penetration rate of 1 cm/sec. which is a Japanese standard for the cone penetration, the measurements are taken of the cone tip resistance, q_c, and the pore water pressure, u, and the frictional resistance, f. It may be noted that the cone tip resistance reading needs to be corrected to account for the pore water pressure acting behind the tip

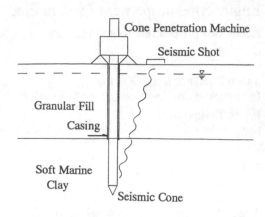

Fig. 1 Schematic of Seismic Cone Testing

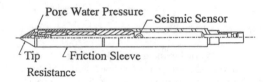

Fig. 2 Seismic Cone Apparatus

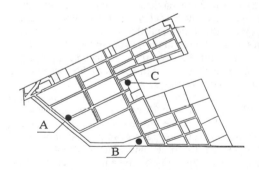

Fig. 3 Location of Test Sites

(Campanella & Robertson 1988) and the corrected value of tip resistance, q_t, is used in this paper.

The seismic cone testing was performed at two different foundation conditions; the one without the installation of vertical drains and the other with the treatment of vertical drains. Figure 3 shows the location of test sites on the island, and the sites A and B are located in the areas without the drains and the site C is located in an area with the drains. It may be also noted that in each area a bore hole for taking undisturbed soil samples is drilled in adjacent to the hole for the seismic cone testing, and the undisturbed samples are used to determine the consistency indices as well as the undrained shear strength based on unconfined compression test.

3.1 *Tests in areas without the drains*

The test result at the sites A and B are shown in Figs. 4 and 5 respectively. In each figure, depicted are the depth profiles of the consistency indices (w_P, w_n, w_L), the unconfined compression strength (q_u), the cone tip resistance (q_t) together with the frictional resistance (f), the pore water pressure (u), and the shear wave velocity (V_s). The data of frictional

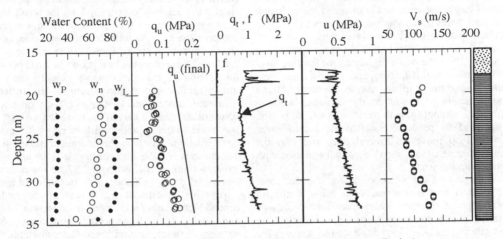

Fig. 4 Depth Profiles of Test Results at Site A (Area Without Drains)

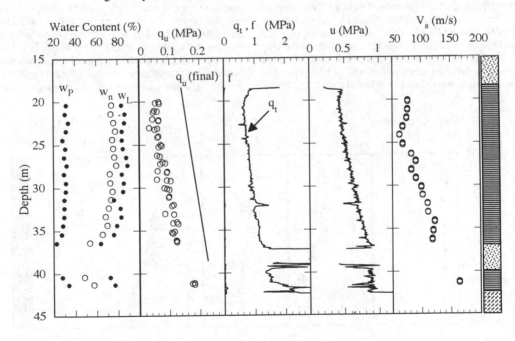

Fig. 5 Depth Profiles of Test Results at Site B (Area Without Drains)

325

resistance (*f*) is almost invisible because of the negligible amount of measured value. It may be noted that the shear wave velocity was determined from the travel time versus depth curve of the seismic records and the difference in travel time between the two measurement depths are used to compute the velocity by assuming the travel path length is approximated by the vertical distance between the measurement depths in the clay. Since the seismic shot at the surface was given fairy close to the hole for the seismic cone, about 3 m away, and also the depth of fill above the clay is large, the error in the assumed travel length was less than 0.2 cm that was later concluded from the back analysis of measured velocities.

As can be seen from Fig. 4, the penetration at site A started from the depth slightly above the clay layer and some scattering of q_t and f indicates some mixture of granular fill and clay at that depth. After a couple meters of penetration, the resistance becomes smoothly changing with depth. At site B, on the other hand, the penetration started after the cone was pushed into the clay layer slightly and thus the penetration resistance shows a smooth variation with depth from the start.

The trends of the piezocone data as obtained from the sites A and B are nearly the same as the tip resistance and pore water pressure decreases somewhat from the top towards the mid level of the clay stratum and then increasing slightly towards the bottom in both figures. This is a typical feature of penetration data as often observed in a clay stratum of under-consolidated state. Tanaka and Sakagami (1989) have already described the interpretation method of the excess pore water pressure in the clay from such piezocone data, and therefore the detailed accounts for this observation will not be presented here. It may only noted here that the amount of pore water pressure measured during the penetration in clay depends on both the undrained shear strength of clay and the excess pore water pressure due to consolidation. The pore water pressure which depends on the undrained shear strength of clay decreases towards the middle of stratum because the clay stratum which is undergoing one-dimensional consolidation shows the weakest zone at the middle of its stratum.

The under-consolidated state of the clay at these two sites are clearly seen from the depth profile of shear wave velocity, especially from the result at site A, and also from the profiles of unconfined compression strength with depth. Low values of the shear wave velocity are found at the middle of the layer, and there seems to be a good similarity between the trends of the shear wave velocity and the unconfined compression strength. In order to illustrate more clearly the evidence of under-consolidated state of clay at the two sites, an estimate was made on the strength profile of clay at the end of consolidation. Based on the undrained triaxial tests, Tanimoto et al. (1982) have suggested the undrained strength ratio, c_u/σ'_v, is about 0.37 for the normally consolidated clay in Kobe. Using this value of c_u/σ'_v, a possible profile of q_u values at the end of consolidation is obtained based on the present fill load and assuming that the unconfined compression strength is twice the undrained shear strength. The estimated profiles are shown in the q_u versus depth graphs in Figs. 4 and 5. It is clear from these figures that the clay at the two sites are still consolidating under the fill load. It is also evident from the figures that the consolidation state of clay at site B is lower than that at site A by exhibiting much larger difference between the present and the final strength profile with depth. The measured values of shear wave velocities

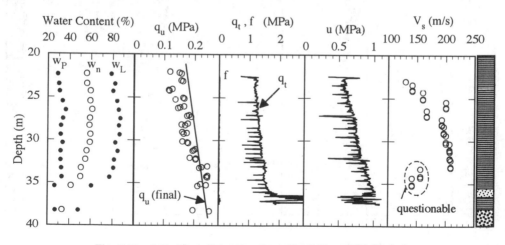

Fig. 6 Depth Profiles of Test Results at Site C (Area With Drains)

between the two sites are also different and it also indicates the consolidation state at site B is lower than that at site A.

3.2 *Tests in an area with the drains*

At site C the sand drains were installed in the soft clay layer few years before the investigation, and therefore the consolidation of clay at this site should have been nearly completed. Figure 6 shows the test results at this site which is presented in an identical manner to those given in Figs. 4 and 5. It may seen that there are slightly erratic readings of q_t and u as compared with those for sites A and B, but this is believed to be caused by more frequent interruptions of cone penetration as the penetration becomes harder for this site.

It is evident from the comparison of these figures that the clay at site C exhibits higher unconfined compression strength and higher values of the penetration resistance, the pore water pressure and the shear wave velocity in comparison with those of sites A and B. The measured value of q_u is very close to those estimated at the end of consolidation, and therefore the consolidation state at this site is much advanced. The seismic cone testing shows the velocity value ranging from 130 to 200 m/sec.which is significantly higher than the areas without the drains. The velocity profile with depth seems to indicate that the velocity increases towards the middle of stratum and decreases towards the top and bottom ends. However, the lower velocities measured near the base of clay need to be carefully examined because of much weak seismic signals at these depth which was coupled with some noise problem encountered at this site. Although the velocity shown in the figure is based on the best estimate of travel

time, some caution is needed to interpret the velocity data near the base of clay.

4 CORRELATIONS AMONG TEST DATA

In the following, the correlations among the test data as obtained from the laboratory test, the piezocone test and the seismic cone test were examined. As our main interest is on the dynamic property of the clay, the correlation study is first focused on the relationship between the shear wave velocity and the undrained strength of the clay. The unconfined compression strength represents the undrained shear strength from the laboratory test, and the value of $q_t - \sigma_{vo}$ is usually considered as a parameter which represents the undrained shear strength of piezocone data (Campanella & Robertson 1988). Figures 7 and 8 present the relationships between V_S and q_u and between V_S and $q_t - \sigma_{vo}$ respectively. It can be seen from these figures that the shear wave velocity increases almost linearly as both the values of q_u and $q_t - \sigma_{vo}$ increase based on the data from sites A and B, although the data from site C shows some scattering. As noted earlier, the V_S data of site C at lower depth is somewhat less reliable and therefore the data points in Fig. 7 corresponding to these depths need more scrutiny. When these data points are disregarded, the indication of Figs. 7 and 8 would be that the increase of V_S will be larger as the undrained shear strength of clay increases. Although more test data are needed, there may be a non-linear relationship between the shear wave velocity and the undrained strength of the clay.

The relationship between the $q_u/2$ and $q_t - \sigma_{vo}$ is presented in Fig. 9. The data points show much less scattering as compared with those of Figs. 7 and 8, and there seems to be a linear relationship between the

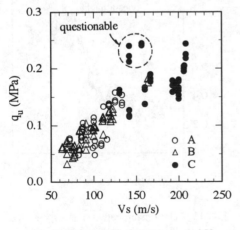

Fig. 7 Relationship between q_u and V_s

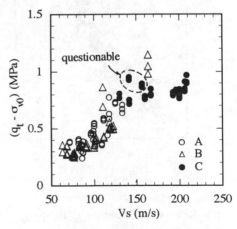

Fig. 8 Relationship between $q_t - \sigma_{vo}$ and V_s

two parameters. The use of piezocone data to evaluate the undrained shear strength is usually done by introducing an empirical parameter N_{kt} in the following equation; $c_u = (q_t - \sigma_{vo}) / N_{kt}$. When a straight line passing through the origin of graph is drawn, the line implies N_{kt} value ranging from 8 to 10. This value is in a good agreement with the experimental data available for the clays in Kobe, for example Tanaka et al. (1989).

The dynamic shear modulus of clay, G_o, is evaluated next from the seismic cone data using the $G_o = \rho V_s^2$ relationship. Figures 10 and 11 present the relationships between G_o and $q_u/2$ and between G_o and $q_t - \sigma_{vo}$ respectively. As discussed above, the shear wave velocity may increase more rapidly with the increase of undrained shear strength and therefore the logarithmic increases of G_o was plotted against the changes in $q_u/2$ and $q_t - \sigma_{vo}$ values. Although there is still some scatter in data points, for example the scattering in Fig. 10 is the exactly same data as discussed in Fig. 7, the relationships depicted in these figures clearly show that the dynamic shear modulus increases almost exponentially with the increase of undrained shear strength of clay. The undrained shear strength of clay is controlled mainly by the effective stress and the stress history, and for normally consolidated clay the undrained shear strength should increase linearly with the effective stress. Therefore, the results shown in Fig. 10 and 11 should mean that the relationship between the dynamic shear modulus and the effective stress is non-linear though the exact form of the relationship is yet to be determined.

5 CONCLUSIONS

The seismic cone test was used to evaluate the in-situ dynamic properties of soft marine clay which is in the process of consolidation under the load of man-made island. The following conclusions may be drawn from the study performed herein;

1) The seismic cone test may be effectively used to determine the variation of dynamic properties with depth within the soft clay stratum which is undergoing one-dimensional consolidation. The depth profile of V_S obtained from a clay layer in

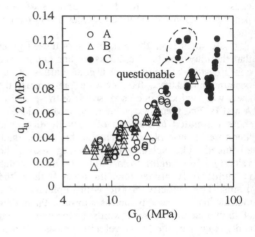

Fig. 10 Relationship between $q_u/2$ and G_o

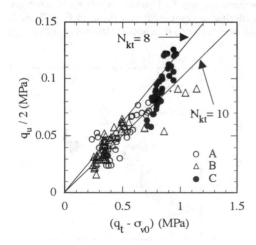

Fig. 9 Relationship between $q_u/2$ and $q_t - \sigma_{vo}$

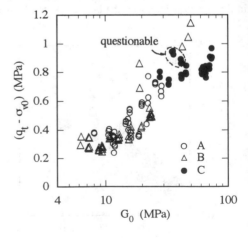

Fig. 11 Relationship between $q_t - \sigma_{vo}$ and G_o

328

under-consolidated state clearly shows a zone of very low stiffness in the mid level of stratum.

2) There is a good linear relationship between the undrained shear strength and the $q_t - \sigma_{vo}$ parameter of the piezocone test, and the value of empirical parameter Nkt to correlate these two is found to be 8 to 10 approximately for the soft clay in Kobe. Thus the piezocone test can be effectively used to evaluate the undrained shear strength of clay.

3) The relationship between the dynamic shear modulus and the undrained shear strength is studied using the experimental data obtained herein, and the study strongly indicates that the dynamic shear modulus increases almost exponentially with the increase of undrained shear strength of clay.

ACKNOWLEDGEMENT

This research project was supported by Development Bureau of Kobe City government, and special thanks are due to Messrs. Takeyama, Nobukawa, Yoshii. A generous support for the test equipment is given from Hogentogler Co. of US and their technical assistance is much appreciated. Our appreciation is also extended to those student, Messrs Shirakawa, Pudyo and Yoshikawa for their assistance in preparing this manuscript.

REFERENCES

Baldi, G., D.Bruzzi, S.Superbo, M.Battaglio & M.Jamiolkowski 1988. Seismic cone in Po River sand. *Proc. of 1st Int. Symp. on Penetration Testing, ISOPT-1* 2:643-650

Campanella, R.G., P.K.Robertson & D.Gillespie 1986. Seismic cone penetration test. *Use of in situ tests in geotechnical engineering, ASCE Geotech. Spec. Publ.* 6:116-130

Campanella, R.G. & P.K.Robertson 1988. Current status of the piezocone test. *Proc. of 1st Int. Symp. on Penetration Testing, ISOPT-1.*1:93-116

Larsson, R. & M.Mulabdic 1991. Shear moduli in Scandinavian clays. *Swedish Geotechnical Institute, Report* 40.

Tanaka, Y. & T.Sakagami 1989. Piezocone testing in under-consolidated clay. *Canadian Geotechnical Journal* 26:563-567

Tanaka, Y., T.Sakagami & M.Furutani 1989. The use of piezocone (CPTU) in laand reclamation work. *Tsuchi to Kiso (Jornal of Japan Society of SMFE)* 37:41-46 (in Japanese)

Tanimoto, K., Y.Tanaka & M.Nishi 1982. An investigation of marine clay in Kobe area. *Proc. of 7th Southeast Asian Geotech. Conference* 1:865-878

Tanimoto, K. & Y.Tanaka 1984. Setlement of reclaimed land on thick deposit of marine clay. *Proc. 4th Australia-N.Zealand Conf. on Geomechanics* 1:341-345

7th International IAEG Congress / 7ème Congrès International de AIGI, © 1994 Balkema, Rotterdam, ISBN 90 5410 503 8

Les propriétés géomécaniques, un facteur de contrôle de l'érosion: Cas d'une zone aride du sudest espagnol

Geomechanical properties, an erosion controlling factor: A study case of an arid zone located in the Southeast of Spain

F. Berrad, L. García-Rossell & M. Martín-Vallejo
IAGM-CSIC Universidad de Granada, Espagne

ABSTRACT: Some geomechanical properties were studied for soils of the Almanzora basin badlands (Almeria, Spain). Both laboratory and field works display some connection between the soils nature and that of the modelling processes of the landscape. Thus, it is possible to differentiate between areas with superficial mass movements and others where runoff is the unique agent of erosion. The two formes can be met however in a same area as a consequence of the coincidence of different lithological formations or because of a favourishing topographic situation of the material. To understand these differences, we had recourse to the study of particle size distribution, swelling, methylen blue value and Atterberg limits.

RÉSUMÉ: Quelques propriétés ont été étudiées dans le sens d'une caractérisation des matériaux, pour des sols de badlands du bassin versant de l'Almanzora (Almeria, Espagne). Une différenciation entre zones à mouvement de versant superficiels et zones où le ruissellement est l'unique agent érosif, est possible, montrant une certaine relation entre la nature des sols et celle des processus modelant le paysage. Les paramétres retenus sont: la granulométrie, la consistance, le gonflement et l'adsorption de bleu de méthylène.

1 INTRODUCTION

La dégradation des sols définie par Duchaufour comme un changement par rapport à un état antérieur ou par rapport au climax, et définie par la F.A.O comme une diminution quantitative et qualitative de la productivité (F.A.O 1980 et F.A.O et P.N.U.M.A 1984) affecte selon les dernières cartes réalisées de vastes extensions de part le monde.

Au sein de l'environnement méditerranéen et surtout des zones à activité économique fondamentalement agricole, comme le cas du sud-est espagnol indiqué zone dont 40 à 59.9 % des terres ont un risque actuel d'érosion , et 60 à 79.9 % des terres ont un risque potentiel d'érosion élevé (CORINE 1985-1990), l'étude quantitative et qualitative de l'érosion est d'une grande importance pour tenter de mieux comprendre ses processus et par là, faciliter la recherche des méthodes de prévention et de lutte.

L'érosion hydrique est considérée comme etant le processus le plus déterminant de la dégradation physique des sols dans ces régions où se rassemblent les causes naturelles décisives et qui sont: une aridité et une irrégularité des précipitations, un relief contrasté avec présence de hauts versants à fortes pentes, un substratum imperméable et riche en sels, et un couvert végétal faible.

Pour ce travail on a retenu l'évolution en badland, ceci pour leur grande extension et aussi pour être considérés comme l'expression la plus claire d'une érosion extrême. Les travaux touchants ce sujet ont presque toujours donné plus d'importance aux facteurs anthropiques et climatiques qu'aux propriétés des matériaux surtout du point de vue caractéristiques géotechniques (Pérez Soba.A 1980).

En effet, l'idée de l'application des méthodes de la mécanique des sols à l'étude des versants remonte à 1948 (V.R.Leuzinger) mais sa mise en pratique ne date que du début des années soixantes (Avenard 1961). Ce travail a donc pour objectif d'aborder en plus des pas classiques dans ce genre d'études (géologie, pédologie, géomorphologie...) le côté des essais pour la détermination des propriétés des sols (consistance, gonflement...) dans le cadre de la mécanique des sols.

2 Zone d'étude et description des sites

2.1 Milieu naturel

Le bassin versant de l'Almanzora se situe dans le secteur oriental de l'Andalousie (sud-est d'Espagne) et s'étend des hauts plateaux de la dépression de Baza jusqu'à l'embouchure de la rivière de même nom dans la mediterranée près de la ville balnéaire de Vera dans une direction fondamentalement Est-Ouest (Fig 1).

De point de vue topographique, le bassin offre une nette dissymétrie du relief: plus accidenté au sud représenté par les "Fila

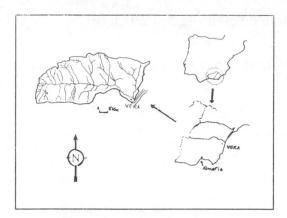

Fig. 1 Localisation géographique

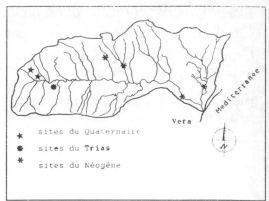

Fig. 2 Localisation de sites

bres" et moins abrupte avec des altitudes
modérées dans la "sierra de las Estancias"
au nord. Ces reliefs encerclent une cuvette
intramontagneuse néogène et quaternaire.

Géologiquement parlant, le bassin appar
tient à la cordillère Bétique et est situé
sur ses zones internes: Complexe Nevado-
Filabrides au sud et Alpujarrides au nord.

La lithologie est représentée par deux
grands groupes de matériaux: les roches
métamorphiques(schistes, micaschistes, cal-
coschistes..du Paléozoïque et du Trias) et
carbonatées des zones internes et les matériaux de remplissage du bassin néogène essentiellement marneux et conglomératiques en
plus des sédiments quaternaires colluviaux,
alluviaux et de glacis.

2.2 Distribution des badlands dans le bassin versant

La figure 2, montre les différents sites de
badlands du bassin étudiés, c'est dans les
matériaux marneux du Néogène que le
phénomène de ravinement intensif occupe de
vastes étendues très souvent limitées par
des glacis. Ces badlands se regroupent souvent autour d'une rambla qui va augmenter de
nombre et de longueur des émissaires vu la
faible résistance que les marnes opposent
aux pluies torrentielles et au ruissellement.

Les badlands du Quaternaire occupent la
partie ouest du bassin et n'ont pas la même
importance que les précédents,bien qu'ils
soient mieux représentés que ceux des schistes du Trias des zones internes.

2.3 Description des sites

Pour une comparaison entre ces différents
groupes, six sites ont été choisis, un autre
site noté "H" est étudié pour plus de détail
dans le cas du Néogène.

Le Trias: site "64", situé à 700m d'altitude, le sol affécté par le ravinement est
un sol d'altération de schistes et ardoises
dont la répartition spatiale est peu homogène.

Les badlands se developpent sur le flanc
nord d'un ravin, le microbassin duquel pro-
vient l'échantillon est en forme d'éventail

allongé de périmètre égal à 65m et de 60° de
pente moyenne. La surface est dépourvue de
végétation ce qui crée un contraste avec la
forêt de Pin qui l'entoure. L'aspect général
des autres badlands du Trias est très simi-
laire avec présence d'un grand glissement
sur le site "63" situé à 1000m d'altitude.

Dans ces matériaux d'altération le réseau
de "rills" ou rigoles est assez net malgré
une érosion laminaire avec lavage des fines
et déplacement des gallets de forme applatie
vers le bas de la pente où ils s'accumulent
en éboulis propres.

Le Quaternaire:Les sites "60" et "61":
situés à 920m au dessus du niveau de la mer.
ce sont des sols essentiellement conglomera-
tiques avec présence de sols au sens édapho-
logique du terme. Pour le couvert végétal il
y'a dominance des champs de cultures souvent
abondonnés. Les badlands sont du type "linear gully" ou ravin linéaire, créant des
ravins abruptes séparés par des interfluves
relativement plats inclinés de quelques 8°,
en continuité avec le glacis. On n'y observe
pas de phénomène de reptation "creep" ni de
solifluxion, les formes d'érosion sont quelques "piping" ou érosion en tunnel peu fré-
quente et des éboulements sur des plans
verticaux laissant des stries qui seront
élimiées par l'érosion laminaire. Pour chacun des points on a fait la distinction
entre la partie édaphique notée "A" et le
matériel (sol dans le sens de la mécanique
des sols) sousjacent noté"B".

Le microbassin du site "60" a un périmètre
de 28m et une pente de 75° pour "A" et 43°
pour "B", celui du point "61" a 39m de périmètre et une pente qui varie entre 70°
dans la partie édaphique et 65° dans le
sable au dessous.

Le Néogène: site "65": à 550 m d'altitude
des marnes grises et kakis devellopent un
modelé en badland le long d'un talus sous
le glacis quaternaire, avec des chutes de
blocs du glacis qui arrivent jusqu'au niveau
de la "rambla". les marnes sont dépourvues
de végétation.

Les microbassins sont de dimensions assez
importantes 65m de circonférence, séparés
par des interfluves de forme arrondie. les
pentes sont en moyenne de 48°. L'érosion en
tunnels est evidente ainsi que des manifestations de la présence d'argiles gonflantes
comme en témoignent les nombreuses fractures

332

dans les constructions. Les mouvements de surface sont présents mais non généralisés affectant surtout les points où le sol gris vient se superposer au sol kaki.

Site"66": présente des pentes de 40 à 60°, le périmètre est de 33m. Le régolithe noté "66S" est trés mince ne dépassant pas les 20cm. L'érosion en tunnels est trés développée au sein du régolithe et paraît profiter des nombreux systémes de diaclases remplies de gypse. Dans le voisinage du point d'échantillonage il n'y a pas de mouvements en masse, pas plus qu'on observe de manifestation claire du gonflement, le réseau de" rills" est net mais souvent dévié par les tunnels.

Site "LO": Les annotations "LOJ" et "LOG" correspondent respectivement aux matériaux de couleur jaune et gris. Ce site a la même situation que le précédent c'est à dire à moins de 200m d'altitude.Les matériaux jaunes se présentent comme des veines à l'intérieu du gris, souvent verticales mais toujours en communication avec la surface .

Site "H": choisi pour la diversité de ses formations et des processus érosifs qui s'y observent. L'altitude est de 400m, les pentes variables, les versants de longueurs variables aussi avec présence au sein de quelques uns de plusieurs types de lithologies. L'érosion adapte dans ce secteur différents processus: tunnels de types et de diamètres variés, ruissellement, et divers types de mouvements de versant.

3 PROPRIÉTÉS DES MATÉRIAUX

3.1 Propriétés chimiques

- Le pH : la connaissance de ce potentiel du sol permet une première approximation à son chimisme, surtout à son degré de saturation en bases, duquel peuvent se déduire les propriétés éléctriques du complexe coloïdal.

les valeurs obtenues indiquent des sols basiques saturés en bases et des sols salino-alcalins trés saturés en bases avec possibilité de la présence de sodium. Le pH des échantillons a été mesuré por une suspension 1:5 à 20 °c

-La conductivité éléctrique de l'extrait de saturation varie entre 1.07 mS du "60B" et 103 mS du "66S".

-L'analyse chimique: les resultats des bases d'échange et des bases de saturation figurent sur le tableau nº1, ils illustrent la richesse en bases de la majorité des échantillons.

La figure (3) montre la relation entre pH, conductivité éléctrique(CE) et le sodium d'échange.

3.2 Minéralogie

Une analyse des diffractomètres (tableau nº2) montre qu'il n'y a pas de différences significatives dans le pourcentage en minéraux argileux entre la partie édaphique et le sol sousjacent dans le cas du Quaternaire, la différence en argiles pour un même niveau n'est pas grande entre les deux types de sols quaternaires. Les matériaux du Néogéne ont un porcentage d'argiles entre 30 et 46%.

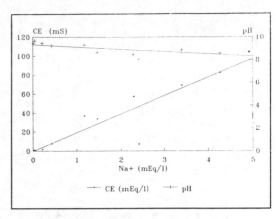

Fig. 3 Relations entre sodium d'échange, pH et conductivité éléctrique

La calcite est abondante sauf dans le cas de "61B" où prédomine le Quartz, et "65G" quartzeux aussi. Il faut signaler l'absence du gypse qui remplie pourtant en abondance les diaclases et fractures des sols néogénes.

En ce qui concerne le type d'argiles, on note une faible représentation de la kaolinite, et des valeurs importantes des Smectites sauf pour le "64" qui posséde 0% de Smectites et 87% d'Illite.

3.3 Essais d'identification

Les résultats des essais viennent regroupés dans le tableau nº3. Les limites d'Atterberg (ASTM D4318/84) donnent pour le Néogéne des sols de la classe Ml c.à.d des limons peu plastiques, exception faite de "65K" classé "MH" ou limon trés plastique. Le Trias et le Quaternaire sont représentés par la classe "SM", ce sont des sables argilo-graveleux, bien que le "60B" soit un "Cl" ou argile peu plastique.

Pour la mesure du potentiel de gonflement on a utilisé la méthode de Lambe, méthode rapide et suffisante pour une comparaison qualitative entre les différents sols étudiés. L'essai ne donne un potentiel critique(selon Lambe) que pour le "65K", les autres sols ayant des indices de gonflement de Lambe entre 0.44 et 1.17 Kp/cm².

L'essai au bleu de méthylène (NF P18-592 (1980)) devrait servir de complément pour la minéralogie, la valeur de bleu varie entre 0.4 pour le "64" et 5.4 (g/100 g de sol) pour le "65K", valeurs bien concordantes avec la minéralogie des sols.

4 DISCUSSION DES RÉSULTATS:

Globalement il y'a une assez bonne relation pour l'ensemble des échantillons entre le pourcentage en fines (argiles et limons) et les limites de liquidité et de plasticité , (Fig 4), la corrélation etant de 0.97 dans le cas de la limite de liquidité.

C'est surtout le Quaternaire et le Trias qui favorisent cette bonne corélation,les matériaux du Néogéne forment un autre groupe offrant pour des contenus en fines trés

	Ca++ meq/l		Mg++ meq/l		Na+ meq/l		K+ meq/l		SAR
	*	▲	*	▲	*	▲	*	▲	
60A	7.55	21.47	1.00	1.15	2.13	0.66	----	0.22	1.46
60B	3.78	23.76	1.46	2.82	1.4	0.53	0.09	0.16	1.22
61A	3.24	18.79	3.11	2.99	0.46	0.40	----	0.21	0.37
61B	10.11	16.67	7.33	1.91	42.84	1.93	0.24	0.2	20.5
64	34.49	10.23	54.40	7.92	11.72	6.62	0.31	0.2	2.49
65K	29.12	40.53	14.13	5.07	23.95	2.03	2.64	0.59	7.28
65G	37.65	35.43	26.98	5.27	146.7	6.76	0.91	0.32	36.5
LOG	72.00	30.80	52.64	7.36	339.8	20.6	4.62	0.53	60.8
LOJ	52.80	24.28	49.67	9.83	228.2	18.3	1.79	0.51	45.1
66R	47.95	0.193	55.10	0.13	426.7	0.09	25.0	0.04	84
66S	42.80	27.23	63.02	17.12	492.6	55.43	4.52	0.91	95.8

* Solution de saturation
▲ Bases d'echange

Tableau 1 Bases d'échange et d'extrait de saturation

	A(%)	Q(%)	F(%)	C(%)	D(%)	G(%)	S(%)	I(%)	K(%)
60A	22	22	0	53	2	0	55	38	7
60B	21	23	2	52	2	0	38	60	2
61A	16	28	3	37	16	0	59	34	7
61B	17	58	17	5	3	0	34	62	4
64	-	-	-	-	-	-	0	87	13
65K	34	22	6	34	3	1	66	30	4
65G	46	31	9	11	3	0	28	56	17
66	30	20	7	4	3	0	56	37	7

A: argiles   C: calcite   S: smectite
Q: quartz   D: dolomite   I: illite
F: feldspaths   G: gypse   K: kaolinite

Tableau 2 Minéralogie de quelques échatillons

	Fines (%)	Wl(%)	Wp(%)	Ip(%)	Ir(%)	Ih (Kp/cm^2)	Vb(g/100g de sol)
60A	42	32.5	21.5	11	0.2	0,55	1.25
60B	74	45.8	26.8	19	7.5	1	1.72
61A	40	28.5	21.1	7.3	6	0.77	1.09
61B	29	-	-	-	-	0.44	0.7
64	35	24.5	15.9	8.6	3.7	0.6	0.4
LOJ	98	49.5	31.9	17.6	5.5	0.97	3.58
LOG	98	46.4	30.9	15.6	4.4	0.97	3.69
65K	96	52	29	23		1.84	5.3
65G	96	43.3	29.3	14	0.2	1.17	2.02
66R	95	45	30	15	5.5	1.08	4.21
66S	96	48	30.6	17.4	2	0.6	4.32
HAR	85	48	28.6	19.4	-	1.15	-
HAS	91	45	29	16	-	1.15	-
HBR	98	53.5	30.5	23	-	1.38	-
HBS	96	46	29	17	-	0.81	-

(-) valeur non mesurée.

Tableau 3 Résultats des essais d'identification

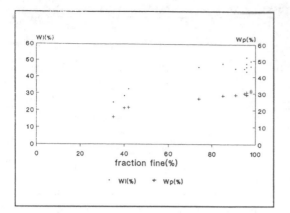

Fig. 4 Relation entre la teneur en fines et les limites de plasticité et de liquidité

similaires des valeurs de la limite de liquidité et d'indice de plasticité assez variables. Il en est de même de la relation pourcentage en fines-gonflement (r=0.65), (Fig 5).

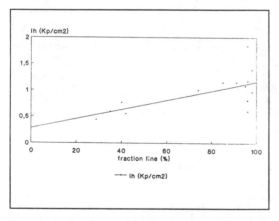

Fig. 5 Relation entre teneur en fines et gonflement

La minéralogie à elle seule n'explique pas ces différences alors qu'elle permet une interpretation satisfaisante de la valeur de bleu adsorbé, ce qui confirme le bon usage qui peut se faire de cette mesure pour détecter la présence d'argiles actives, Lautrin.D(1987), chose confirmée aussi en comparant le gonflement et l'adsorbtion de bleu, (Fig 6).
Les limites de liquidité et l'indice de gonflement donnent de curieuses corrélations avec le sodium d'échange: (Fig 7), en effet la limite de liquidité qui devrait avoir une corrélation négative avec le sodium par un effet de défloculation, parait être contrairement affectée (r=0.35). D'autre part une concentration importante du sodium d'échange conduit généralement à un gonflement plus important (Pousada Presa. 1984), cependant la corrélation observée dans ce cas est (r=-0.31).
De tout ce qui précéde on peut déduire que

les relations ne peuvent être établies en traitant les facteurs de façon individuelle et encore moins quand il s'agit de comparer des sols d'âges et de compositions differentes.
Le calcul de l'indice de consistance de De Ploey (1980) à partir de la différence entre les teneurs en eau à 5 et à 10 coups de la coupelle de Casagrande donne des valeurs allant de 3 à 7%, il n'y a donc pas de valeurs inférieures à 3, valeur au dessous de laquelle les sols sont susceptibles à l'encroûtement dans les zones tempérées, cet indice semble ne pas être valable pour des sols salins des zones semi-arides à arides, vu que la majorité des sols étudiés sont afféctéspar l'encroûtement.

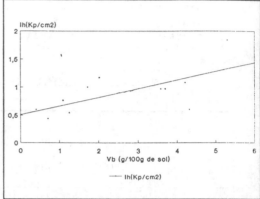

Fig. 6 Relation entre l'indice de gonflement de Lambe et la valeur de bleu de méhtylène

Une comparaison entre les phénoménes obsérvés sur le terrain et les propriétés des matériaux, montre les points suivants:
-Les manifestations du gonflement commencent avec des indices de Lambe de gonflement de l'ordre de 0.8Kp/cm2 donc à la limite d'un potentiel de gonflement marginal, cette maniféstation se refléte surtout en une structure en "popcorn", Hodges (1982), avec une chute de la densité sous une mince croûte de quelque deux millimétres et une grande densité de fissures de dissécation qui ne semble pas avoir de relation claire avec l'indice de retrait Ir= Wp - Wr.
- Le réseau de "rills" est d'autant plus net que l'indice de gonflement est faible, car l'obstruction des "rills" initiés par une forte pluie ne permet pas au prochain événement pluvieux de continuer d'éroder dans le même lieu.
-L'érosion en tunnel ou en "piping" est très favorisée par la richesse en sels des matériaux, ces tunnels peuvent servir de réseau pour le cheminement de l'écoulement sous-cutané montrant dans quelques points surtout dans le site noté "H" des coulées boueuses à leur émergence au niveau du versant et aussi au niveau du talweg. Les langues des coulées sont de petite dimension avec quelques mètres de longueur et moins de 0.6 métres de largeur, la formation de ces coulées s'explique par une augmentation de la teneur en eau du sol, il faut noter cependant qu'une mesure de la teneur en eau d'une coulée a donné 40% donc il ne serait

pas forcement necessaire de dépasser la
limite de liquidité pour obtenir une coulée,
les limites de liquidité les plus basses du
secteur etant de 42%.
- les observations de terrain rendent de
la grande complexité des phénoménes des
mouvements en masse : ils peuvent être iden-
tiques pour des matériaux d'aspect différent
"HA"et"HB" par exemple, comme ils peuvent
être distincts dans des formations identi-
ques. L'explication resideraît dans le comportement à l'eau des formations:

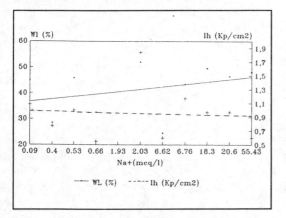

Fig.7 Relations entre sodium d'échange,
gonflement et limite de liquidité

Les matériaux marneux du Néogéne ne sont
pas totalement impérméables, le réseau de
fissures permet l'infiltration d'un minimum
d'eau permetant de dépasser la limite de
plasticité aprés une averse importante comme
il a put être constaté dans un suivi des
variations de l'humidité du sol dans les 20
premiers centimétres du régolithe du site
"H".
Dans une étude de détail du site "H",
l'ouverture de plusieurs profiles , met en
evidence la présence de plusieurs tranches
au sein des régolithes ce qui pourrait enle-
ver l'ambiguité du comportement qu'entraine
l'étude des seuls régolithe et roche mére.
Une comparaison entre la teneur en eau à
différentes hauteurs des versants montre
aussi une plus grande concentration de l'eau
à la base de ces derniers d'où une plus
grande fréquence des mouvements, une autre
observation à faire est que plus le talus
est haut plus il y'a de possibilité d'instabilité supérficielle, les mouvements etant
de plus grande fréquence pour des pentes
supérieures à 45° dans le site "H" et aussi
là où il y'a eu sapement de la base, souvent
par effondrement du toit d'un tunnel au
contact talus-terrasse de culture.
- Dans le Quaternaire on n'observe point de
mouvement de type "solifluxion" et "coulée
boueuse", ceci est evident pour le site "61"
dont le niveau supérieur présente des limites plus importantes que l'inférieur. Mais
le site "60" est aussi préservé de ce type
d'instabilité, cela doît être surement dû au
pendage casi nulle du plan séparant les
formations de consistances différentes.
-Le Trias malgrés les fortes pentes et la
difference de nature entre la roche mére et
les matériaux d'altération ne presente pas
aucun des types de mouvement en masse cités.
Cependant dans le corps même des schistes du
sîte "63", on note la présence d'un grand
glissement de terrain.

5 CONCLUSIONS GENÉRALES

- La comparaison entre des sites où se developpent différents types de badlands, montre
que la game des sols où apparaît le ravinement est très ample, est que les différences
sont essentiellement dû aux propriétés des
matériaux, qui entrainent un comportement ou
un autre.
-la superposition de matériaux dont la consistance est différente, surtout quand le
niveau supérieur possede des limites d'Atterberg inférieures à celles du niveau sousjacent, favorise les mouvements en masse à
la superficie.
- le gonflement joue un rôle important dans
l'évolution du reseau de concentration de
l'écoulement superficiel, permetant dans le
cas des sols à faible potentiel de gonflement l'enfoncement du réseau.
- la formation d'une croûte à la superficie
diminue l'infiltration mais la présence de
fissures de dissecation (trés favorisée par
le type de climat) permet un minimum d'infiltration.

RÉFÉRENCES

Alexandre, D. 1982. difference between "Ca
lanchi" and "Biancane" badlands in Italy:
71-87. Badland geomorphology and piping.
Avenard, J.M. 1989-1990. Sensibilité aux
mouvements en masse (solifluxion): 119-
129. Cah.ORSTOM, sér. pédol, vol 25, 1-2.
Avenard, J.M & Tricart,J. 1961. Téchniques
de travail et idées de recherches. Appli
cation de la mécanique des sols à l'étude
des versants: 146-156. Revue de géomor
phologie dynamique. CNRS.
Cobertura, A.1993. Edafología aplicada. ed,
Catedra: 326p. ISBN 84-376-1108-3.
De Ploey, J.1989. La conservation des sols:
38-41. La recherche Agronomie. supl. rech.
227.
Hodges, W.K & Bryan, R.B.1982. The influence
of material behaviour on runoff initiation
in the Dinosaur Badlands, Canada: 13-46.
Badland geomorphology and piping.
Lautrin, D. 1987. Une procédure rapide d'i
dentification des argiles: 75-84. Bull.
liaison lab P et ch.-152. réf. 3184.
López Bermúdez, F & Romero-Diaz, M.A. 1989.
Piping erosion and badland development in
South-east Spain: 59-73. Catena suppl 14.
López Bermúdez, F & Albaladejo, J. 1990. Fa
ctores ambiantales de de degradacíon en el
area mediterranea: 15-42. Soil degradation
and rehabilitation in mediterranean envi
ronmental conditions. ed, CSIC. ISBN 84-
00-07045-3.
Martin-Penela, A.J. 1993. Pipe and gully sy
stems development in the Almanzora basin
(south Spain). Z.sch.für Geomorphologie.
Pérez Soba, A. 1980. politica mundial ante
el problema de la erosiòn de suelos en
Almeria.ISBN-84-500-7572-6, p.57-65.
Pousada Presa, E. 1984. Deformabilidad de
las arcillas expansivas bajo succión con-
trolada. tesis doct. MOPU.

REMERCIEMENTS

Ce travail a été financé par les projets de
la CICYT NAT-0.622.0/90; AMB-93-0844-C06-02
et par le projet LUCDEME.
Nos remérciements vont aussi au Dr.Valverde
Espinosa pour sa précieuse aide.

A reversible dilatometer for measuring swelling and deformation characteristics of weak rock

Un dilatomètre réversible pour mesurer les caractéristiques de gonflement et de déformation des roches tendres

C. Jing, G. Lu & Q. Wei
Hohai University, Nanjing, People's Republic of China

P. Grasso, A. Mahtab & S. Xu
GEODATA SpA, Torino, Italy

ABSTRACT: The novel reversible dilatometer was developed in response to the need for an in situ, bore-hole device for measuring the swelling index of weak ground. The instrument has two working principles: (1) when functioning as a dilatometer, it applies a uniform pressure to the borehole and measures the radial deformation and (2) when functioning as a reversed dilatometer (or a contractometer), it responds to the swelling of the ground around the borehole. In either case, the pressure-deformation relationship is recorded during the test and is then interpreted to obtain the deformation or swelling character of the ground. The present design has the following specifications: borehole diameter=86mm; maximum depth of measurement=50m; maximum pressure on the probe=2.5MPa. The instrument has been used for field tests. The results are discussed in this paper, together with the description of the design and calibration of the instrument.

RESUME: Ce dilatomètre réversible est une innovation qui a été développé pour répondre à la nécessité qui se présentait d'avoir un instrument, in situ, capable de mesurer l'index de gonflement du sol tendre. Cet appareil présente deux fonctions différentes: (1) quand il est utilisé comme dilatomètre, il applique une pression uniforme sur le trou de forage et mesure les déformations radiales, (2) quand il fonctionne comme un dilatomètre renversé (contractomètre), il répond au gonflement du sol autour du trou de forage. Dans chaque cas, la relation pression-déformation est enregistrée durant le test et donc interprétée pour comprendre les caractéristiques de déformation ou de gonflement du sol. Le prototype actuel a les spécifications suivantes: diamètre du trou=86mm; profondeur maximale pour les mesures=50m; pression maximale sur la sonde=2.5MPa. L'instrument a été utilisé pour des tests réels. Les résultats sont décrits dans cet article ainsi que la description du projet et le calibrage de l'instrument.

1. INTRODUCTION

On exposure to water, the resulting swelling of certain rocks (e.g. anhydrite, marl) creates several problems in design, construction and operation of tunnels. The available laboratory techniques do not represent the in situ swelling behavior. Therefore, a technique needs to be developed for routine, in situ measurement of swelling. The swelling coefficient thus measured can be used for classifying weak rock for empirical design of tunnels.

The reversible dilatometer (RD) described here was developed through a joint research program between Hohai University, Nanjing, China, and Geodata SpA, Torino, Italy. The prototype RD was tested in the laboratory and then further developed for in situ application. The preliminary results are reported in this paper. The RD is being patented in both China and Italy.

2. LITERATURE REVIEW

The paper by Duncan et al. (1968) is of historic interest for the basic concepts about swelling rocks. They note that originally hard clay, clay-shales, shales, marls, and mudstone may progressively swell in the presence of water and become plastic (causing problems of stability & support).

Duncan et al. (1968) suggest a simple apparatus for measuring the axial strain, ε_a of a rock specimen,

$$\epsilon_a = \frac{\Delta\ell}{\ell} \qquad (1)$$

where:

$\Delta\ell$ = change in speciment length resulting from saturation

ℓ = original length prior to saturation

The saturation swelling pressure, σ_a, can be determined in the laboratory by measuring the dynamic modulus $E_{seismic}$, thus

$$\sigma_a = E_{seismic} \cdot \varepsilon_a \qquad (2)$$

Shales and marls containing the expansive clays (montmorillonite and illite) are common subjects of swelling, together with anhydrite which forms $C_aSO_4 \cdot 2\ H_2O$ with up to 60% expansion on hydration (Einstein and Bischoff, 1975).

The ISRM (1989) suggests laboratory tests for swelling in three configurations: (1) maximum axial swelling stress, (2) axial and radial free swelling strain, and (3) axial swelling stress as a function of axial swelling strain. At least three undisturbed samples are required for each test. The specimens are right circular disks of thickness=20-30mm for diameter of 50-100mm.

ISRM (1981) cautions against assuming the swelling strain index obtained in the laboratory "as the actual swelling strains that would develop in situ, even under similar conditions of loading and of water content." Gysel (1987) advises "to check or correct the laboratory-based swelling parameters by means of deformations or pressure measurements" performed in situ. In situ measurements of displacement of the tunnel periphery and tunnel convergence can be made by extensometers and distance measuring devices, respectively (Einstein and Bischoff, 1975; Kovari and Amstad, 1979).

The above review of the available literature suggests that the need for an in-situ swelling tester and indicates that such an instrument is not available.

3. CONCEPTUAL DESIGN OF A REVERSIBLE DILATOMETER

3.1 The Requirements

The basic requirements for the in-situ technique for measuring the swelling index are that:

1) It should be simple in terms of time and ease of use.

2) It should be relevant to the boundary conditions encountered when driving a tunnel through swelling rock.

3) It should lend itself to being standardized, based on the principles of the existing and proven methodology. As part of the standardization, the conceptual model for the technique should be verifiable using either analytical or numerical analysis procedures.

3.2 Working principle

The essential concept for the proposed technique is to simply use the borehole dilatometer for deformability measurement in a reversed sense: in other words, as a contractometer. The following are the basic elements of the reversible dilatometer, RD, (or contractometer) test:

• The RD is a deformable probe which is placed in a section of a borehole that is isolated by means of two packers.

• Water is introduced separately to the two packers.

• The pressure in the RD can be pre-set or changed to different selected levels, including that of the prevailing in-situ stress.

• Both the pressure and the volume change in the RD will be monitored as a function of time.

It is noted that the instrument whose working principle is described above will function both as a contractometer and as a conventional dilatometer.

340

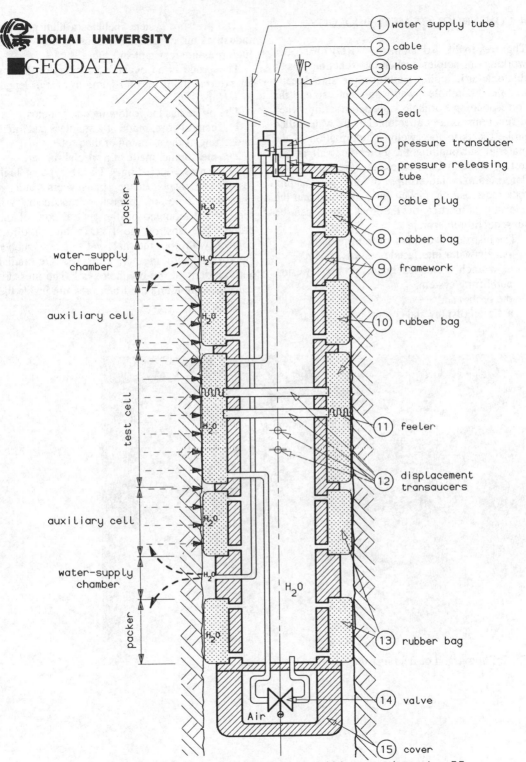

Fig.1- Schematic cross-section illustrating the RD
probe structure and its working principle

341

4. CONSTRUCTION METHODOLOGY

The reversible dilatometer (RD) has two working principles: when it functions as a dilatometer it applies a uniform pressure directly to the borehole wall and measures the corresponding borehole wall deformation; when it functions as a contractometer, its water filled probe reacts to the swelling deformation of the material surrounding the borehole wall of the test section. In both cases, the pressure versus deformation relationships of the surrounding rock mass can be conveniently recorded and the relevant properties of the rock mass can be subsequently interpreted.

The instrument consists of four modules:
- An air-water interfaced pressure system,
- a wrench-wheel operated, probe lowering and lifting system,
- the probe, and
- a digital display and control unit.

The pressure system includes a high pressure industrial air cylinder, pressure control valves, high pressure resistant soft tubes.

The probe is an expandable cylinder, having an external diameter of 70mm, an overall length of 1.18m.

The probe has the following components:
1) central core, made of a stainless steel tube and acting as the skeleton of the probe,
2) membrane, made of a special rubber,
3) sensors, including 4 LVDT type of high resolution displacement transducers and a resistance type of pressure transducer. All sensors are mounted on the central core of the probe. The tips of the LVDTs are in direct contact with the internal wall of the rubber membrane and measure directly the radial deformations of the borehole wall. The pressure transducer monitors the water pressure inside the membrane.

Fig. 2: Photograph of field instrument, RD

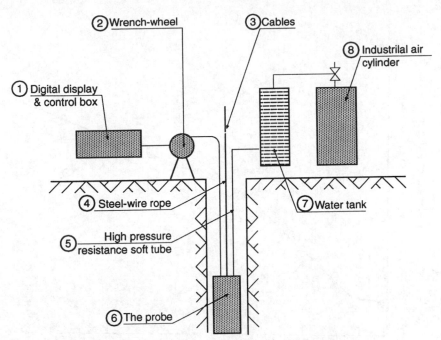

Fig. 3: Skematic illustration of the Test Setup of the Reversible Dilatometer

Table 1. Results of field test using the RD as a dilatometer

Depth interval of test section (m)	Lithological Descriptions	Maximum test pressure (kg/cm²)	Calculated E (kg/cm²)
VI 11.90-12.90	Cobbles and gravels containing sand and mud. cobbles & gravels are sub-anglular with dia=2-50mm	4	56
V 14.74-15.74	Silty clay with cobbles and gravels (dia=10-20mm). The bottom part of this section is a layer of medium-to-coarse sand with some clay	5	169
IV 16.74-17.74	Silty clay with cobbles and gravels (dia=10-25mm)	1.5	It was not possible to further increase pressure
III 18.96-19.63	Boulders, cobbles and gravels containing yellow silty clay (dia=10-35mm)	1.5	Cell pressure could not be raised
II 21.39-22.39	Silty clay containing small amount of cobbles and gravels	4	62.5
I 22.09-23.09	Boulders containing limited amount of silty clay. The top 10cm of test section is clay containing cobbles	10	1129

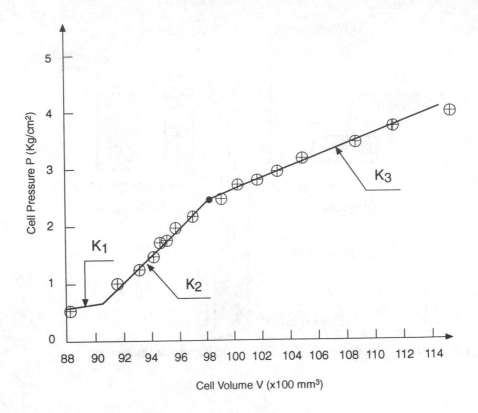

Fig. 4: Test results showing Pressure - Volume relationship

The digital display and control unit is a single-board based, microcomputer controlled, intelegent type of instrument. It can record automatically the output of the sensors and can print the results at any time since it also has a printer interface.

The probe lowering and lifting system includes steel-wire ropes and one motor-operated wheel.

The main technical specifications of the newly developed RD borehole device are:

• The boreholes to be measured should have a diameter ranging from 76-86mm.

• The measuring depth is theoretically unlimited, but was limited to 50m in the experiments.

• The maximum pressure that can be developed inside the membrane is 2.5MPa. The resolution of the pressure transducer is 10kPa.

• Test cell length is 250mm;

• The length of the two auxiliary cell is 250mm

• Total length of the probe is 1.18m

• The maximum diametral displacement of the

borehole wall (under any membrane pressure) should not exceed 20mm. The displacement transducers have a resolution of 0.05mm.

The details of the RD probe are given in Figs. 1 and 2.

4.1 Field tests

In all. 6 tests were conducted at six different depths in the borehole, BH S111, at Shantan dam site. The test setup is illustrated in Fig. 3. Table 1 summarizes the lithologic and test details of each test section. The pressure-volume relationship for test section II is illustrated in Fig. 4.

4.2 Interpretation of test results

During each test, the cell pressure (P) -vs- cell volume data were recorded as the main result of a test and their interpretation generally follows the procedure outlined below.

Step 1: Plot the cell pressure -vs- cell volume data in a scatter diagram as shown in Fig. 4.

Step 2: Using the least squares method to do piece-wise linear curve fitting and find the slopes (K1, K2 and K3) of each straight line segment. There are in theory 3 linear segments for each scatter diagram (or each test).

Step 3: Calculate the rock mass deformation modulus of the test section using the following relationship:

$$E = 2(1+\upsilon) \cdot Vy \cdot \frac{1}{K_2} \qquad (3)$$

where,

• υ is Poisson's ratio and a value of 0.2 is assumed for the alluvium,

• Vy is the cell volume corresponding to point Py (the intersection of the second and the third line segment),

• K_2 is the slope of the second straight line segment.

An example of the test results (Sec. II, Table 1) from Shantan dam site is shown in Figs. 4. Note that the two tests at Sections III and IV were not interpreted because only a very low pressure was applied to the borehole wall.

4.3 Critical review of RD

• Deformation of the hose: The hoses used in RD are plastic products which can endure a pressure up to 15 Mpa, but they are not stiff. An experiment was made to determine the deformation of a 100m long hose of 4mm dia and 1mm wall thickness. The hose was filled with water, sealed at one end, and put under a pressure of 2 MPa at the other end. The length of the hose decreased by 150 mm. In order to avoid the effect of hose deformation, an electro-magnetic valve is installed. In this way the test cell can be isolated such that it has no direct relation with the hose when the RD works as a contractometer.

• Rubber bag thickness: The deformation transducers measure the deformation of the inner wall of the test cell. In fact, it is the deformation of the outer wall of the test cell (i.e, the deformation of the borehole wall) that should be measured. The thickness of the rubber bag changes with its diameter, the bigger the diameter, the smaller the thickness. The

variation of the thickness has a direct influence on the accuracy of measurement of the borehole wall deformaion. This problem needs to be resolved.

Suppose D is the borehole diameter, d is the inner diameter of the test cell (i.e, the output of the deformation transducer), w is the thickness of the test cell rubber bag. At the condition of contact, we have:

$$D = d + 2w \qquad (4)$$

After careful analyses, it was found that w is a function of the variables: d, the test cell diameter, and p, the cell pressure. Experiments were made to define the two fonctions as follows:

$$w = f_1(p)\big|_{d = \text{constant}} \qquad (5)$$

and

$$w = f_2(p)\big|_{p = \text{constant}} \qquad (6)$$

The results showed that w is basically not influenced by p but mainly by d, and the function f_2 can be expressed as follows:

$$2w = \sqrt{780 + d^2} - d \qquad (7)$$

From Eqs. (4) and (7), we have:

$$D = \sqrt{780 + d^2} \qquad (8)$$

Thus according to Eqs. (8), D can be calculated from d. Of course, the relation shown by Eqs. (8) is obtained under the assunption of isotropic extension of the rubber bag.

• Test cell fringe: The test cell fringe refers to the two ends of the test cell. Because the two ends are fixed on the framework, there must exist a part of the rubber bag that can not contact the borehole wall when the test cell expands. The smaller the pressure within the test cell, the bigger the area which does not contact the borehole wall. While RD functions as a contractometer, it is ideal to keep the test cell volume constant. In fact, this is to keep the borehole diameter constant. With the increment of pressure within the test cell, the fringe expands progressively. Thus the borehole diameter will decrease. Simulation and calculation show that the rubber bag diameter will be reduced by about 0.37% when the fringe angle is a sharp angle. However, this variation is not considered to be significant for the performance of the RD.

• Drift (Time drift+temperature drift) and medium tests: All the outputs of the transducers were monitored in 58 hours with the environmental temperature changing from -1°C to +6°C. The results showed that they remained basically constant. Deformation transducers, which are inductance elements, were exposed to different environments, including air, distilled water, tap water, water containing rust, and magnetic fields. The results showed that the transducer performance was not affected by this exposure.

5. CONCLUSION

The RD can function both as a dilatometer, for measurement of in-situ modulus of deformation, and as a contractometer, for measuring in-situ swelling properties of the ground. The cost of the RD, however, is expected to be comparable to that of a conventional dilatometer. Geophysical logging of the borehole, prior to using the RD is recommended.

6. REFERENCES

Duncan, N, M.H Dunne, & S. Petty 1968. Swelling characteristics of rock. Water Power, May, 185-192.

Einstein, H.H. & N. Bishoff 1975. Design of tunnels in swelling rock. Proc. 16th U.S. Symp. on Rock Mechanics, Minneapolis, 185-195.

Gysel, M. 1987. Design methods for structures in swelling rock. Proc. 6th ISRM Cong., Montreal, Vol. 1: 377-381.

ISRM 1981. Suggested methods for determining swelling and slake-durability index properties. In: Rock Characterizations Testing and Monitoring - ISRM Suggested Methods Brown, E. T. (ed.), Pergamon Press, 89-94.

ISRM 1989. Suggested methods for laboratory testing of argillaceous swelling rocks. Int. J. Rock Mech. Min. Sci. & Geomech. Abstracts, V.26, N.5, 419-426.

Kovari, K. & Ch. Amstad 1979. Field instrumentation as a practical design aid. Proc. Fourth Congr. of ISRM, Montreux, Vol. 2, 311-318.

Engineering character of the fossil-weathered granite
Caractéristiques géotechniques des granites paléo-altérés

Peng Sheqin, Nie Dexin & Zhao Qihua
Chengdu Institute of Technology, People's Republic of China

ABSTRACT: By field investigation and a lots of tests of physics mechanics parameters, the engineering geological character of the fossil-weathered granite in Gongboxia hydropower station of China is studied in detail. The basic reason of the completely fossil-weathered granite having well engineering characters is revealed.It is that buried compaction make the completely weathered granite's structure and strength restore in the environmental evolution.

RÉSUMÉ: Nous présentons dans cet article des études sur les caracteristiques géotechniques de la masse granitique paléo-érodée de la central hydraulique de Gongboxia,située à la rive droite du fleuve Jaune(hanghe); les résultats obtenus par differentes méthodes montrent que la masse granitique,malgré sa paléo-érosion subie,possède de bonnes propriétés géotechniques.Celles-ci dérivent de la reproduction de la solidité et de la structure formée pendant les temps géologique,issue du compactage par la contrainte structural régionale et le poids de la couverture.

1 INTRUDUCTION

Gongboxia hydroelectric station is the forth step station in the middle-upper reaches of Yellow River. It is in the telophase of preliminary design.The main hydraulic engineering works include sloping core rockfill dam in valley and added concrete pier dam in the right bank. The foundation of the concrete dam is right on the fossil weathered granite mass. How about their engineering characters and whether it can be used as the foundation is closely related to the hydraulic arrangement and economic benefit.If this scheme is abolitioned and the scheme of channel power in the left bank is used, the engineering cost would rise a lot.So the engineering character of fossil-weathered granite is a great engineering geological problem all attentioned.

2 GENERAL GEOLOGICAL SITUATION IN DAMSITE AND MACRO-FEATURE OF THE FOSSIL-WEATHERED GRANITE

2.1 General geological situation

The strata appearing in damsite are mainly Presinian schist gneiss, lower Paleozoic intrusion granite-fossil-weathered granite(r3), Neogene Period gravel-sand(N2-Cgs), Quaternary Period sandy pebble gravel(Fig. 1).

It is known from the regional geology that the crust rised and was denuded after Caledonian orogeny, and formed the thicker weathered granites. Neogene Period the crust of this region went down, up above the weathered crust there was depth of 600 metre Linxia stratum group deposit.The weathered granite is buried by great thickness overrock and formed fossil-weathered rock.Untill lower Pleistocene Epoch,the crust of this

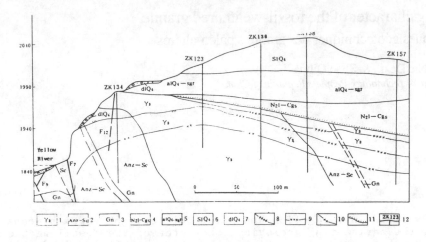

Fig 1. Geologic section at damsite

1.granite, 2.schist, 3.gneiss, 4.gravel-sand, 5.alluvial gravel layer,
6.loessal clay, 7.drift bed, 8.boundary line between completely and highly
weathered granite, 9.boundary line between highly and moderately
weathered, 10. boundary line between moderately and slightly weathered,
11.discordant line, 12.borehole and its number.

region rised again and denuded up
to now.

2.2 Geological feature of the fossil-weathered granite

The fossil-weathered granite is
under the red gravelly sandrock at
the right bank of the dam site.It
possesses integrated weathered profile section from completely weathered to highly weathered and to
moderately weathered.The completely
weathered rock has few closed
structure planes and the perfection
of rockmass is well.When it is just
exposed, the samples can be shaped
easily and the rock is difficult to
dig.But later and later its stength
decrease and the samples cann't be
shaped at all. Impaction of exposed
time for highly and moderately
fossil-weathered rock is less obvious to completely fossil-weathered
rock. There are more opened structure planes in highly fossil-weathered rock and the rockmass is blocked
by structure planes. But single
block possesses high strength.There
are few structure planes in
the moderately fossil-weathered

granite and the perfection of
rockmass is well.

3 PHYSICAL AND MECHANICAL PROPER-TIES OF FOSSIL-WEATHERED GRANITE

3.1 Physical properties of fossil-weathered granite

Weathering causes the rock structure flabby,secondary change of the
mineral,density decrease and pore
increase of the rock.From completely weathered - highly weathered -
moderately weathered - slightly
weathered is a continuous process
from quantitative change to qualitative change. Therefore a serious
studies of physical parameter of a
whole weathered section can give a
reasonable density parameter for
the engineering design.
1. Grain composition
The result of size analysis of two
sets samples is shown in table 1.
From table 1, we know that in
completely weathered granite the
percentage is more than 50 in
2-0.5mm grade. So it is called
grit according to the soil
classification. The percentage of

348

Table 1. Result of size analysis of completely fossil-weathered granite

grain size(mm)		>5	2-5	0.5-2	0.25-0.5	0.1-0.25	<0.1	<0.005
percentage	(1#)	0.27	11.23	56.06	18.12	9.74	4.49	0.47
	(2#)		2.9	63.6	13.78	15.32	4.4	

clay is less than 1 and it indicates that the physical weathering is mainly and the chemical weathering which forming the secondary clay minerals is very weak.

2. Density

The fossil-weathered granite samples are tested for their density in the site laboratory. It shows that the density of completely weathered granite is 2.3g/cm^3 or so(table 2). It needs to mention that although the samples are from a definite radial depth of investigation cave,because of the hardness of the granite and the condition restriction of getting samples, the disturbing of samples cann't be avoided.Moreover, there is a period for the cave excavation and the flaccid ring formed,so the result only can be the physical parameter of rock with some disturbing. The influence of this flabby for the hard rock such as moderately weathered, highly weathered is far inferior to the completely fossil-weathered granite.

Table 2. Density index of fossil-weathered granite in cave 77-2-2

No. of sample point	distance to cave opening (m)	density (g/cm^3)	degree of weathering
15-1	5.8	2.558	moderately
17-1	8.0	2.542	moderately
14	11.7	2.550	moderately
13	18	2.534	moderately
12	20	2.571	moderately
11	24	2.569	moderately
10	27	2.550	moderately
9	30	2.542	highly
8	32	2.520	highly
7	34	2.530	highly
6	37	2.490	highly
5	40	2.380	completely
4	42	2.420	completely
3	44	2.393	completely

In order to gain the unralexible physical parameters of the completely fossil-weathered granite, the method of preventing from relaxation is used to get samples from burrow just excavated.Their physical index of the completely fossil-weathered granite is shown as follow (table 3).

As table 3 we note that the density of the fossil completely weathered granite is 2.46-2.48g/cm, this value is just the real density parameter of this rock.

3.2 The mechanical property of the fossil-weathered granite

1. Uniaxial compressive strength

Using point loading apparatus, experiments of the compressive strength of fossil-weathered rock under different conditions are carried out. The results are listed in table 4.

The range of strength value is very large. To get representively compressive strength of different weathered zone, relative equation between compressive strength and density of every weathered zone is bulit (Fig 2,3,4). From these figures we can see that natural compressive strength and natural density are closely related.According to the relative equation and the standard value of natural density,the representive compressive strength of every weathered zone can be caculated. They are 10.7MPa, 46.4MPa, 86.9MPa respectively.

2. Shear strength of the completely fossil-weathered granite

The highly and moderately fossil weathered granite are more hard relatively.Their mechanical properties are mainly controlled by structure planes, but for the fossil completely weathered granite, the strength of the rock is the main controlling factor. Because the structure joint of the completely weathered granite has been destr-

Table 3. Physical index of unralaexible completely fossil-weathered granite

sample number	natural density (g/cm^3)	water content (%)	grain density (g/cm^3)	distance to face
B-1	2.45	4.15	2.65	farther
B-2	2.453	3.3	2.65	farther
B-2-1	2.457	3.5	2.65	farther
B-3	2.47	3.17	2.65	farther
B-4	2.46	3.51	2.65	far
B-5	2.45	3.97	2.65	far
B-6	2.452	3.95	2.65	far
B-7	2.46	3.3	2.65	nearly close
B-8	2.453	3.3	2.65	nearly close
B-9	2.49	2.7	2.65	nearly close
B-10	2.51	2.69	2.65	close
B-11	2.462	2.95	2.65	close
B-12	2.474	2.75	2.65	close

Table 4. Natural compressive strength of fossil-weathered granite

No. point	natural density (g/cm^3)	compressive strength (MPa)	degree of weathering
13-2	2.51	53.6	
13-4	2.55	49.8	
17-3	2.524	45.7	
11-2	2.545	55.1	
10-2	2.553	59.3	
11-3	2.572	112.9	moderately
11-1	2.558	49.6	weathered
10-1	2.539	38.1	
13-3	2.529	64.3	
10-4	2.546	64.2	
14-1	2.57	102.3	
12-2	2.581	143.7	
6-1	2.461	25.88	
15-1	2.493	38.86	
15-3	2.52	39.9	
9-2	2.55	62.7	
9-3	2.562	83.7	highly
8-2	2.476	25.2	weathered
8-1	2.491	32.4	
6-2	2.536	50.02	
7-1	2.56	76.3	
6	2.488	11.4	
5-1	2.365	3.877	
5-2	2.308	1.903	completely
5-3	2.384	4.36	weathered
4	2.419	7.28	
3	2.248	1.058	

oied completely. Except fracture, other crack disappeared basically, and it begins to transform into homogeneity medium.The big fracture is filled with the upward deposit and compacted by later gravity and tectonic stress.Therefore the shear strength of the rock can reveal or represent the shear strength of the completely weathered rockmass.The experiments of completely fossil weathered shear strength of several sets are performed in this research (see in table 5).

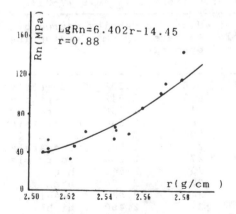

Fig 2. Relationship between density and uniaxial compressive strength of moderately weathered granite

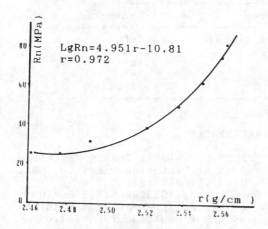

$LgRn=4.951r-10.81$
$r=0.972$

Fig 3. Relationship between density and uniaxial compressive strength of the highly weathered granite

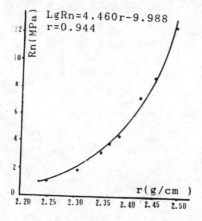

$LgRn=4.460r-9.988$
$r=0.944$

Fig 4. Relationship between density and uniaxial compressive strength of the completely weathered granite

Table 5. Shear strength of completely fossil-weathered granite weathered

number point	density (g/cm³)	water content (%)	shear strength f	c(MPa)
1-4	2.47	4.5	0.95	0.39
5-8	2.46	4.1	1.3	0.299
3-8	2.565	3.5	0.91	0.255

3. Sound wave speed in the fossil weathered granite
Sound wave speed in the rock is a parameter which can evaluate the rockmass quality.Sound wave speed is inspected in cave 77-2-2#.From the result (table 6),we find that the sound wave speed in the fossil completely weathered granite is 2800m/s and it is nearly close to the speed value in the fossil highly weathered granite.

Table 6. Sound wave speed in fossil-weathered granit

weathering dgree	sound wave speed(m/s)
moderately weathered	3500
highly weathered	3000
completely weathered	2800

4 CONTRAST BETWEEN THE WEATHERED ROCK AND THE FOSSIL-WEATHERED ROCK

4.1 Contrast of the physical mechanical character

From above we notice that fossil weathered rock have better engineering characters than the modern weathered rock's. The difference between them is listed in table 7.
    1. The mineral grains of the decomposed granite is compated and holden up again by the gravity of overcover deposit and tectonic stress.The flaccid structure will be changed toward the primary structure of the granite,in addation, the mud materal in the fossil-weathered granite play a cementing effect to the mineral grains except themselves' compaction.
    2. From completely weathered rock to the buried completely rock, it is essentially the process of changeing in structure. As the changing in structure,highly compacted and a definite cementing, completely fossil-weathered granite has cohesion,that is structure strength. And the coefficient of intrenal friction of highly compacted buried weathered granite have been greatly increased too.

Table 7. Fossil and modern completely weathered granite contrast
```
---------------------------------------------------------------
      index              modern weathered        fossil-weathered
---------------------------------------------------------------
sound wave speed(m/s)    700(right bank slope)        2800
density(g/cm³)           1.5-1.8(dry)                 2.48
compressive                    0                      3.78-10.7
strength(MPa)
macro-feature            loose                        perfectly
---------------------------------------------------------------
```

4.2 Cause analysis

Cause analysis of the different engineering character between fossil-weathered rock and the general weathered rock may begin with the environmental development which mould the feature and shape of the fossil-weathered granite. The fossil-weathered and the modern weathered rock have both gone through highly weathered effect.For completely fossil-weathered rock,if there is not the later buried processing ,their engineering character of both kinds are similar . The obviously different between the fossil weathered granitein Gongboxia and the general weathered granite is that the former is not only weathered highly but also compacted by the later gravity of overrock deposit and tectonic stress,then formed the new engineering character corresponding with the new environmental condition.It is the buried compaction that makes the completely weathered granite's structure and strength restore.

5 CONCLUSION

1. As the later environmental development, fossil-weathered rock has a well engineering character than modern weathered rock.
2. The forming of its new feature has its basical mechanism. It is the great compaction of upper gravity and tectonic stress make the fossil-weathered granite's structure and strength restore.
3. Whether the fossil-weathered rock can be used as a foundation is not in what it is called but in its real engineering character.

REFERENCES

CURRER,D.T. 1977. Deeply Weathered Rock at Victorian Damsites, Eng. Geo. 11 341-363

Sheqin,Peng.1993.Possible Evaluation of the Fossil-Weathered Granite as the Foundation for Middle-lower Concrete Dam in Gongboxia Hydroelectric Station, Thesis for academic degree of Master.

Features of watertighting grouting in low metamorphic crystalline schists

Les caractéristiques des injections d'étanchéité en roches schisteuses cristallines peu métamorphisées

S.Cvetković-Mrkić
Faculty of Mining and Geology, Beograd, Yugoslavia

ABSTRACT: Rocks of low to medium metamorphic grade, complex geology,specific fracturing, and commonly weathering to considerable depths, involve a numerosity of problems when grouted under a dam foundation. The permeability of these rocks is sufficiently high to render watertighting grouting desirable and necessary, but also sufficiently low to provide for adequate grouting effects. The difficulties in dimensioning grout curtains and selecting optimum grouting elements ensue from the very purpose of watertighting grouting, on one hand, and relevant properties of these rocks, on the other. Permeability and low groutability require specific criteria for an estimate of the grouting requirement and identification of the permissible permeability. This paper is an attempt to elucidate more closely the principial engineering-geological problems in identifying the main grout curtain parameters on a study of grouting conditions and effects for several completed projects in low metamorphosed crystalline schists.

RESUMÉ: Les roches schisteuses, peu métamorphiques, souvent décomposées profondement, apportent beaucoup de problèmes à l'injection d'étanchéité des fondation des barrages. Leur perméabilité est faible et c'est pourquoi elles sont souhaitables pour les accumulationsartificielles, mais en meme temps suffisamment grande que l'injection est obligatoire. Toutes les propriétés de cettes roches, importantes pour le projet et l'éxecution d'un voile d'étanchéité, sont très specifiques: composition géologique, fissuration, perméabilité, inconstance hydraulique, injectabilité. Ce sont les raisons pour changement les critères traditionnels à l'estimation des possibilités réelles d'imperméabilisation. C'est le sujet de cet article qui se base à la pratique d'étanchéité pour les barrages à Serbie.

1. GENERAL

Crystalline schists of low to medium metamorphic grade are traditionally considered aquifuges, consequently attractive for dam foundation and formation of artificial lakes. In these cases, however, grout curtains should be built to reduce water losses through the dam bottom and abutments, and to secure seepage stability under the new, less favourable conditions of backwater flow.

The problem of reducing water losses from a reservoir is not acute, because permeability of these rocks is generally low. The question is rather, isn't it too readily resorted to excessively deep grout curtains while hydraulic inconstancy and fractures should be adequately studied.

Another, equally important task of grout curtains is to provide conditions for an as homogenous and favourable as possible distribution of hydraulic head, gradients and velocities for a greater stability of the stucture and a protection against seepage failure. The lower the groutability of rock, the more difficult and uncertain the task.

The problems of selecting the optimum solution for a grout curtain and of inpredictability of the desired effects result from the specific manifestations relative to rock properties, which can be practically estimated on several parameters.

The terrains built up of crystalline schists of low to medium metamorphic grade have a complex geology, both in lithology and structure.

Fractures in rocks are unified in origin. They are characterized by schistosity and cleavage, resulting in close-spaced narrow cracks and a marked anisotropy.

Owing to the general geologic make-up and fractures, these rocks are a medium of low water permeability. Slightly more permeable are subsurface rocks and structurally deformed zones. Moreover, these rocks manifest certain hydraulic inconstancy caused by their mechanical properties and the character of fractures.

Such terrains are susceptible to weathering, which may locally be intensive and several metres deep. It largely controls the mechanical behaviour of rocks.

With all the above in view, the need is evident for a careful adjustment of grout curtain elements to the rock properties, with the particular emphasis on a substrantial estimate of the necessary curtain size and the permissible permeability.

2. GENERAL GEOLOGIC CHARACTER AND FRACTURES

The basis of the permeability consideration for grout curtain designs, either in Hm (hectometric), i.e. the entire dam size, or m (metric) area, i.e. grout hole sites, is the knowledge of local geology and rock fracturing.

From the knowledge of local geologies of several constructed dams in Serbia and the general picture of similar terrains, a few major characteristics can be deduced. These are primarily the variety of petrogenic elements of more or less different mineral compositions and metamorphic grades, marked traces of repeated folding and faulting, and local magmatism.

Thus, for example, quartz-muscovite--chlorite schists with occurrences of chloritic gneiss prevail in the area of Bovan Dam on the Moravica, eastern Serbia. The schists are a product of metamorphosed sediments, including sporadically preserved old sedimentary textures. Recrystallization has progressed along the cleavage of the axial fold plane. The rocks are much deformed, generally forming a steep monoclinal feature.

Fracturing of rocks in such terrains is a natural consequence of the geologic make-up, i.e. their origin, composition and fabric. The picture of fractures is formed mainly by schistosity and cleavage in addition to other structural breaks responsible for a merked anisotropy and heterogeneity. Spatial position and relationship of systems of fractures control the local structural Dm-m features. Fractures mostly have very small openings, except where in more rigid lithologic bodies and in deformed rocks. A characteristic of these rocks is the presence of

slightly consolidated, highly mobile clay fillings in fractures, particularly in the zone of weathering.

3. PERMEABILITY

The geologic features and fractures completely control the permeability and groutability of these rocks. The general hydrogeologic properties and many water pressure tests (WPT) indicated media of generally low permeability. Scrutinized on the dam site scale, WPT results indicated that small openings, although dense, restrict the permeability to 1-3 Lu. Higher permeabilities have been noted only in features of a greater structural deformation, within rigid lithologic bodies, and in shallow, weathering zones (Fig. 1).

In addition to low permeability, these media are characterized by a hydraulic inconstancy. This property expresses the rock relation to injected water of WPT, which is a rough model of the future conditions under a dam lake. A growing injection pressure increases permeability of these media highly dissected by narrow fractures, and even changes the flow pattern. The changes can be continuous of limited only to an area of hydraulic gradients; either result in a local redistribution of openings and their expansion. The conditions for classical hydraulic breaks are much fewer, due both to the typical texture of rocks and the considerable density of fractures. The behaviour of crack fillings contributes to the hydraulic inconstancy in terms of the possible local transport and compression rather than wasing out, allowed by the prevailing small openings and the marked heterogeneity. Besides the fracturing, mechanical properties and the state of rock strain are dominant in the mode and intensity of hydraulic inconstancy manifestation. That is why it is higher in shallow zones, where rocks are exposed to weathering.

The importance of a reliable estimate of the hydraulic inconstancy is multiple.First of all, water pressure permeability values which include the hydraulic inconstancy give a picture of substantially higher permeability than it actually was for the area of both natural gradients and that to be formed by the future storage lake. The difference is important in that the mechanism of fracturing change by water injection from wells can neither be equal to the mode of the future backwater effect. The neglect of hydraulic inconstancy leads to an unnecessarily large volume of grouting work and a wrong estimate of groutability.

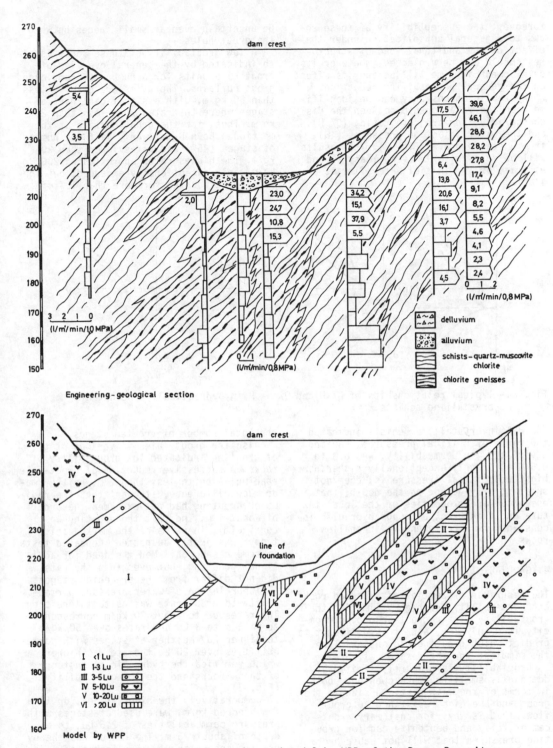

Fig. 1.- Engineering-geological section and model by WPP of the Bovan Dam site, eastern Serbia.

Moreover, the susceptibility of these rocks to a general consolidation under the effect of the additonal load of structures should not be neglected, whatever its development in time will be and its effect on the general permeability reduction.

Hydraulic inconstancy can be identified from the relationship between the discharge/injection pressure ratio, (Q/ H), and the water injection pressure, (H), from a standard WTP (Fig. 2). The results will be more reliable if at least a good estimate of the initial water flow is available.

be uncertain even at small speces between the grout holes.

A low groutability of these rocks is also indicated by the general overview of grout rate units for a number of completed grout curtains. The rates are chiefly less than 50 kg/m, with a significant part of stages where the rates were less than 20 kg/m. Thus, for the grout curtain at the mentioned Bovn Dam, nearly half the number of stages (45%) had rates lower than 20 kg/m. The highes rates approximated 200 kg/m, only exceptionally exceeded, but these never accounted for more than 10% of

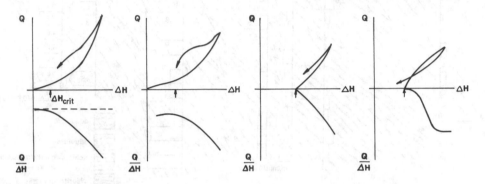

Fig. 2.- Typical relationships of Q/ H and Q/ H- H in hydraulically inconstant crystalline schists.

The WPT in crystalline schists indicated the range of critical pressures, causing a variation in permeability, was 0.3 to 0.5 MPa, or only exceptionally higher.How high the critical pressure is, does not substantially depend on the depth, that is, essential are properties of the local texture and fractures, although hydraulic inconstancy is more intensive in shallow rocks.

4. GROUTABILITY

The described properties of fissured rocks and permeability are highly limiting their groutability. Their grout-absorbing capacity is mainly low, and it is difficult to predict which grouting elements will be the simplest for the desired effects.

Groutability is primarily limited by dominantly small fracture openings, both in terms of the main particle size in the grout and the high resistance to grout flow. That is why, for customary grouts of cement, clay and bentonite and for grouting pressures that will not cause hydraulic breaks, the radius of grout spreading is not large. The continuity of grouting in the grout curtain domain will therefore

the total number of grouted stages.

Also the grout rates in various stages of grouting indicated low groutability of rocks. A successive reduction of grout-hole spacing, even to less than 2 m, neither made much difference. It meant that the radius od grouting had not exceeded the order of about 1 m, and that the grouting-up continuity, in terms of the marked anisotropy, was justly suspected. Repeated tests in some cases confirmed the need for additional grouting, but even then the rates did not differ from the preceding stages.

The relation of water pressure permeability to grout rate was quite interesting. The rates up to about 20 kg/m corresponded to values to 3 Lu and rates over 50 kg/m to higher Lu. Neither at stages with permeabilities over 20 Lu therate was higher, which verified the hydraulic inconstancy of these rocks and the low groutability (Fig. 3).

Comparatively the higest rates, up to a few hundred kg/m, were used at stages with pressure permeabilities of 2.5 Lu, to a depth of about 25 m. The reason was the hydraulic inconstancy, where increased injection pressure forced open some fissures and caused the los of grout beyond the grouting perimeter.

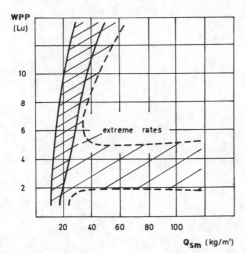

Fig.3.- Water pressure permeability,WPP,
to grout rate, Q$_{sm}$, relation in
hydraulically inconstant crystalline schists.

5. GROUT CURTAIN ELEMENTS

Bearing in mind specific properties of fractured rocks, permeability and groutability, one may raise the question of revalua-
tion of the standard criteria observed in
dimensioning grout curtains and the permissible water permeability. Insisting on
the known customary criterion of 1 Lu is
not justified in terms of either large water losses from the storage under dam or
real reduction of permeability to that value as the permissible one. A reasonable
inference is that this conventional criterion should be replaced by a value of 3 Lu,
or exceptionally slightly higher where the
hydraulic inconstancy has been proved
higher.

Water losses under dams are not high
owing to the overall low permeability.The
losses under one of the considered dams of
hydraulic head about 50 m are calculated
to less than 10 l/s without a grout curtain or less than 2 l/s with a grout curtain
of the depth equal to the hydraulic head
and 3 m spacing between grout holes. Obviously, the order of losses cannot be a justification for deep grout curtains. A
greater depth may have only parts of grout
curtains which reduce the losses through
actually permeable zones or more heavily
deformed rocks in which increased hydraulic gradients may lead to a seepage failure.

As for the grouting elements, primarily
injection pressure and rate, relative to
the groutability of these rocks, also im-

portant is the purpose of grouting. High
injection pressures are not a guarantee of
homogeneous and isotropic expansion of the
grouting radius, i.e. the assured continuity of grouting up in the curtain. The
grouting involving hydraulic breaks, for
watertighting, cannot be considered neither necessary nor productive.

6. SUMMARY

Terrains of low metamorphosed rocks are
suitable, by their generally small permeability, for formation of artificial storage lakes. However, grout curtains which
are constructed to limit water losses through the dam section and protect rocks from
seepage failure, involve many incertainti-
es and questions.

Owing to the characteristic lithogenetic
composition and fabric, and the unified
fracturing, these crystalline schists commonly have a low water permeability and a
hydraulic inconstancy. Critical pressures
which cause redistribution and expansion
of openings vary within the range od 0.3
to 0.5 MPa, and the phenomenom itself is
primarily associated with shallow-lying
rocks, exposed to weathering.

The low groutability of these crystalline schists is a consequence of the prevailing narrow fractures, and the expected
grouting effect is often unattainable even
at small distances between grout holes.

All properties of these rocks essential
for an estimate of the need and conditions
of watertighting grouting indubitably suggest the necessity of a revision of customary criteria for dimensioning grout
curtains and selecting grouting elements.

REFERENCE

Cvetković-Mrkić, S. 1992. Inženjerskogeo-
loški uslovi projektovanja i izvodjenja
injekcionih zavesa za izgradjene brane
u Srbiji, Studija (Engineering-Geological Conditions for Design and Execution
of Grout Curtains in Dam Projects of
Serbija, A Study); fondovska dokumenta-
cija, Rudarsko-geološki fakultet (docu-
mentation fund, Faculty of Mining and
Geology), Beograd.

Dilatometer tests in Brasília porous clay
Essais au dilatomètre dans l'argile poreuse de Brasília

J.A.R.Ortigao
Federal University of Rio de Janeiro, Brazil

Abstract This paper presents results of Marchetti dilatometer tests (DMT) in a soft unsaturated and collapsible porous clay of Brasília, Brazil. Soil parameters were obtained from correlations developed from saturated and sedimentary soils in Europe. DMT data for the porous clay were compared with other in situ and laboratory tests and the results presented a good agreement and repeatibility.

Introduction

The dilatometer was developed in Italy by Marchetti (1980) [9] and consists of a 14 mm thick 95 mm wide and 220 mm long blade (Figure 1) which is driven or pushed into the soil. On one face there is a 60 mm diameter steel diaphragm capable of a lateral expansion of 1 mm under gas pressure. Beneath the membrane there is an electric switch that turns on a buzzer and warns the operator when the expansion reaches 1 mm. The gas pressure is brought from the surface by means of plastic leads. At the start of the membrane expansion the operator records the initial lift-off pressure that displaces the membrane (pressure A) and at the end of the 1 mm inflation, the final pressure (B). The same procedure is repeated each 20 cm during penetration of the instrument.

The following p_0 and p_1 values are obtained:

$$p_0 = A + \Delta A \qquad (1)$$

$$p_1 = B - \Delta B \qquad (2)$$

where ΔA and ΔB correspond to membrane stiffness corrections.

Then, the following index parameters are obtained:

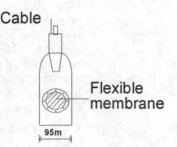
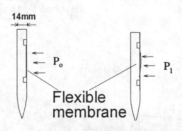

Figure 1 The Marchetti dilatometer

$$I_D = \frac{p_1 - p_0}{p_0 - u_0} \qquad \text{Material index (3)}$$

$$K_D = \frac{p_0 - u_0}{\sigma'_{vo}} \qquad \text{Horizontal stress index (4)}$$

$$E_D = 34.6(p_1 - p_0) \qquad \text{Dilatometer modulus (5)}$$

where u_0 is the in situ porepressure and σ'_{vo} the in situ overburden effective stress.

Marchetti (*op cit*) proposed a series of correlations based on Italian soils for estimating the soil type, unit weight, the in situ stress ratio K_0, the overconsolidation ratio *OCR*, the undrained strength for clays c_u, the friction angle for sands and the one-dimensional compression modulus *M*. These correlations are based on a limited number of Italian soils, being eight sand deposits and only two clay deposits. Several researchers [2, 3, 5, 7, 8, 10, 13] revalidated these correlations for other sand and clay deposits and concluded that the original Marchetti proposals led to reasonably accurate results. Lunne et al (1989) [7] proposed slight modifications to a few correlations.

Campanella and Robertson [4] interpreted the reasons why the DMT yielded so good results. They employed a special research DMT that has tiny device below the membrane that measures horizontal displacements during inflation and the whole inflation curve can be constructed, as shown in Figure 2. This research apparatus was used in clays and sands and results were compared with selfboring pressuremeter tests at the same depth, with a good agreement. The main conclusions were: the DMT is equivalent to a *poor-boy* selfboring pressuremeter test, with only two points along the expansion curve: the first one, corresponding to p_0, and the final one is p_1, corresponding to a large radial strain of 14%. The dilatometer modulus E_D is calculated as the average slope of the inflation curve between p_0 and p_1.

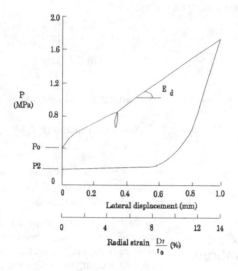

Figure 2 Interpretation of the UBC research DMT (adapted from Campanella & Robertson, 1989, [6])

South Wing
Underground line

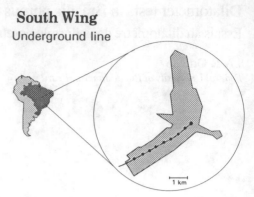

Figure 3 Site location

Site investigation programme

The work described herein is related to the construction of the Brasília Underground transportation system. It encompasses 6.5 km of tunneling employing the *NATM* (New Austrian Tunneling Method) method and over 4 km of cut-and-cover slurry-walls false tunnels. Details about the project and tunneling were published elsewhere (Ortigao and Macedo, 1993 [12])

A field investigation programme was carried out to obtain design parameters. The following in situ tests were used: boreholes and standard penetration tests (SPT), Marchetti dilatometer (DMT), Ménard pressuremeter (PMT), piezocone (CPTU) and horizontal plate loading tests (PLH). Block samples were obtained from test pits for laboratory testing. A series of triaxial and oedometer tests was carried out. This paper presents results of in situ stress ratio K_0, deformation moduli and strength parameters and compares in situ and laboratory test data.

Geology and site conditions

Regional geology and the geomorphology have been described by Ortigao and Macedo (1993) [12]. The

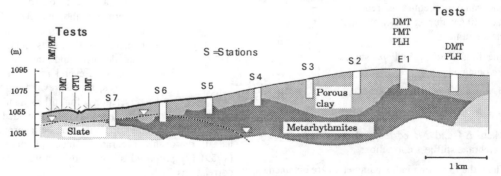

Figure 4 Soil profile along the tunnel, South Wing, Brasília

region is flat, as a characteristic of the central plateau highlands. It is covered by a layer of latosoils and lateritic soils designated here by *porous clay*. It overlies residual soils from slate or a sequence of interlayered siltstones and sandstones, that geologists call *metarhythmites*. These soils are part of the *Paranoah* formation.

Climate alternates from a 6 month rainy season to a very dry Winter, leading to laterization process. Leaching of soluble salts occurs at the top of the porous clay are deposited below. This is responsible for a large amount of pores at the top of the clay layer resulting in high void ratios, low unit weights and high permeability.

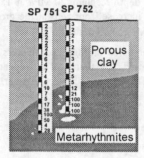

SP 751 SP 752

Porous clay

Metarhythmites

Figure 5 SPT results

This study concentrates at the south wing of Brasília (Figure 3) where the tunnel is now being excavated. Soil profile was initially investigated by 65 mm diameter 20 to 30 m spaced boreholes in which SPT's were carried out at every metre along depth. The porous red clay is 8 to 30 m thick, the SPT blowcount is low, varying from 2 to 3 blows/30 cm.

The water level is generally very deep, except at tip of the south wing (Figure 4), where it can be found at 8 to 10 m depth only. Groundwater observations show high seasonal variation of the water level due to the high permeability of the porous clay.

The bottom of the porous clay is clearly indicated by a sudden rise in the SPT values (Figure 5), as boreholes strike the residual soils below. These residual soils were investigated during excavations and show an inherent anisotropy as a dominant feature. Bedrock characteristics such as bedding and shear planes, are present in the residual soils, therefore they are *structured*. Strength and deformation properties in structured soils vary with direction in relation of bedding and shear planes. Therefore, it is unlikely that in situ tests such as those used for the investigation of the porous clay could be useful in these structured soils.

Characterization of the porous clay

A summary of the laboratory test results on the porous clay is presented in Figure 6. They were carried out on undisturbed block samples.

Atterberg limits are: liquid limit LL =50-80%, plastic limit PL =35-50% and water content w =35-55%. The clay fraction, *i.e.*, the percentage of soil particles less than 2 μm lies between 70 and 55%. The percentage of fines, *i.e.*, less than 60 μm in diameter, varies from 70 and 80%.

The average unit weight of the porous clay γ is 15 kN/m³, except at the top where severe leaching took place even lower values can be found. The void ratio is close to e =1.7 and the degree of saturation is low above the water level, about 60-80%.

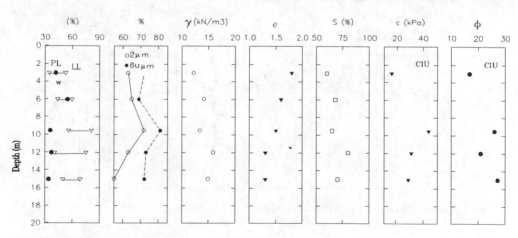

Figure 6 Summary of laboratory test data

Undrained triaxial tests on unsaturated and saturated samples were carried out. Strength of the unsaturated clay can be represented by the following Mohr-Coulomb's parameters: cohesion c from 20 to 40 kPa and friction angle ϕ in the 25-28° range. These parameters can be regarded as effective strength parameters because measured porepressures were very low in CIU tests on unsaturated samples.

Local experience with construction on this clay has proved that it is collapsible. Low rise buildings on shallow foundations tend to crack one to two years after construction. Good practice is to adopt deep foundations consisting of small diameter bored piles, even for just one story building.

DMT testing programme and results

Five DMT boreholes were carried out in the south wing of Brasília, with some 20 to 30 tests per working day. The small size and portability of the equipment and the use of a local drilling rig led to high productivity and low cost per test. Data were processed by a computer program and plotted as

indicated in figure 7 to figure 8. The results indicate a good repeatability in the porous clay. K_0 value is close to 0.6, the friction angle varies in the range of 20 to 27 degrees and the cohesion is around 20 kPa. It is remarkable that the DMT correlations yield both cohesion and friction angle for this clay, which presents a drained behaviour due to its high permeability.

Figure 9 presents the results of several boreholes together. Several and allow the following conclusions. K_0 is high at the first two metres of depth and then decreases and lies in the 0.5-0.7 range, slightly decreasing with depth. The one-dimensional compression modulus M varies from a value around 5 MPa at the ground level and increases with depth reaching 20 MPa at 15 m depth. Values of E_D increase from zero at the ground level and reach 15 MPa at 15 m depth.

Figure 10 compares strength parameters from all boreholes plotted together. The repeatability is remarkable. The friction angle lies in a narrow range

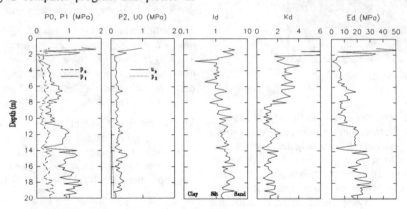

Figure 7 Typical results from DMT test (SP608)

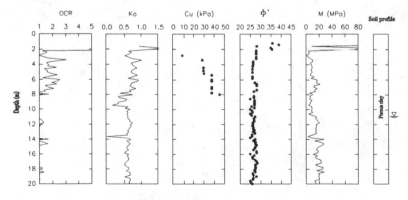

Figure 8 Typical results from DMT test (SP608)

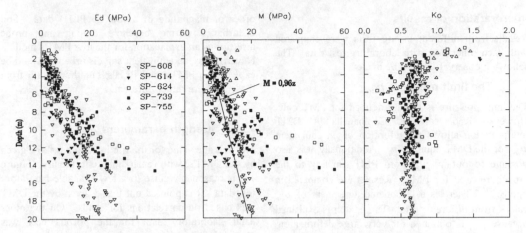

Figure 9 *Deformation moduli and* K₀ *from several DMT's in the porous clay*

from 25 to 28 degrees and does not vary with depth. Cohesion data is scattered but seems to increase with depth , as shown in this figure.

Spurious results were obtained for *OCR* and will not be presented here.

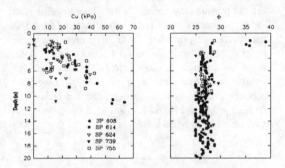

Figure 10 *Strength parameters from DMT in the porous clay*

Additional in situ tests

The in situ testing programme included horizontal plate loading tests (PLH), Ménard Pressuremeter tests (PMT) and piezocone testing (CPTU). A brief description is given below.

Horizontal plate loading tests were carried out in 1.5 m by 1.5 m wide test pits used for block sampling. A 300 mm square steel plate was jacked on the pit walls. Details of the testing procedure and analysis of results is given by Ortigao (1993) [11]. These tests yielded Young's moduli E_i at the beginning of loading and E_{ur} at an unload-reload cycle.

Ménard pressuremeter testing was carried out in two boreholes. Testing techniques conformed with standard procedures (*eg*, Baguelin al [1]). The following soil parameter were obtained: the pressuremeter modulus E_m and the limit pressure p_{lim}. In addition, an unload-reload cycle in the beginning of the test allows the determination of the unload-reload modulus E_{ur}. The corresponding shear moduli G_m and G_{ur} were obtained assuming elastic conditions.

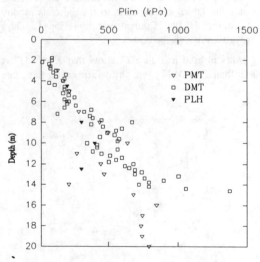

Figure 11 *Limit pressure* p_{lim} *from PMT, DMT and PLH*

Interpretation of results

Results from different in situ tests were compared with each other and with laboratory test data. The following observations can be made:

- ### The limit pressure p_{lim}

The limit pressure is usually defined for PMT only. However, Robertson and Campanella [4] DMT interpretation allows the adoption of an equivalent p_{lim} for the DMT equal to p_1. Additionally, it is also possible to obtain p_{lim} from PLH similarly to the procedure used for PMT. Results are compared in figure 11. There is a reasonable agreement of p_{lim} values from different tests. This match is explained because p_{lim} is obtained for very large deformation, or *macrodeformation* range, and beyond any influence of soil disturbance by any insertion method.

- ### Young's modulus E

Young's moduli from all in situ tests are plotted in figure 12. An arbitrary value of 0.33 for Poisson's ratio was employed to obtain E from PMT's and PLH's. Due to non-linearity in soil behaviour, secant modulus depends of the strain or stress level. This should be considered when comparing moduli from different tests and a reference strain level should be obtained for each test. As an example, Jamiolkowski et al [5] have shown that in normally consolidated sands the DMT corresponds to triaxial tests moduli at a strain of 0.1%. Equivalent comparisons for clays are not known to the author.

Results plotted in figure 12 show that E_m (PMT) is less than E_D (DMT), which in turn are in the same order of magnitude of a few E_i (PLH) data. Soil disturbance during borehole drilling and probe insertion is an explanation for the low PMT's moduli. Nevertheless, an equivalent soil disturbance would be expected in PMT's and PLH's, but the moduli from these tests are rather different.

- ### Strength parameters

The average value for the friction angle $\phi=26^o$ given by the DMT's, with remarkable repeatability, (Figure 13) is reasonably close to a few available laboratory test data. It is pointed out that ϕ given by the DMT was based on correlations for sands. On the other hand, laboratory data for the porous clay was obtained from *CIU* triaxial tests in total stresses. However, due to the low level of saturation, these values seem to agree. Field behaviour of the porous clay can be regarded as *drained* due to its high amount of pores and high permeability.

Cohesion given by the DMT correlations present a lot of scatter. Only a few laboratory reference data for the porous clay are available. Therefore, it is at least premature to state conclusions for this parameter, despite that DMT indicates the same order of magnitude of triaxial test data. Notwithstanding, at low stress levels and for unsaturated conditions it is possible to take into account a cohesion value to express a c-ϕ behaviour. At high stress levels and for saturated conditions it is advisable to disregard cohesion.

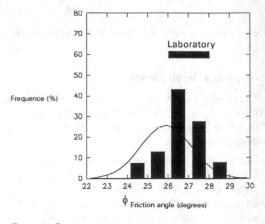

Figure 13 Gauss

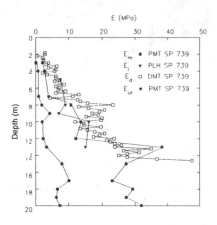

Figure 12 Young's moduli E *from PMT, PLH e DMT*

Conclusions

Main conclusions from in situ tests in Brasília porous clay are: there is no useful correlation between SPT's and deformation moduli. The SPT index is not at all sensitive to the low stiffness of the porous clay. The time consuming PLH tests, yielded results that agree with other tests. PMT's have shown very low moduli that seem to be affected by soil disturbance.

The DMT led to excellent results: very good repeatability, reasonable agreement with laboratory test data, very easy deployment and operation, low cost and high productivity.

Acknowledgments

The in situ testing programme was sponsored by the Brasília Metro Company and was carried out by Bureau Ltd, São Paulo, (PLH), Geomecânica SA, Rio de Janeiro (CPTU and DMT), Geotécnica SA, Rio de Janeiro, (PMT) and the laboratory tests by the University of Brasília. Support from Lucia Alves and Alejandro Far during the preparation of this article is appreciated.

References

1. Baguelin F, Jézéquel J F & Shields D H (1978) The pressuremeter and foundation engineering, Trans Tech Publications, 617 p.

2. Baldi G, Bellotti R, Ghionna V N, Jamiolkowski M, Marchetti S & Pasqualini E (1986) Flat dilatometer tests in calibration chambers, Proc ASCE Conf In Situ 86, Blacksburg, VA, pp 431-446

3. Campanella R G & Robertson P K (1983) Flat dilatometer testing: research and development, 1st Int Conf on the Flat Plate Dilatometer, Edmonton, Alberta, Canada

4. Campanella R G & Robertson P K (1989) Use and interpretation of a research DMT, Soil Mechanics series no. 127, Department of Civil Engineering, UBC, Vancouver, BC, V6T 1Z4, Canada

5. Jamiolkowski M, Gionna V N, Lancellota R & Pasqualini E (1988) New correlations of penetration tests in design practice, Proc 1st Int Conf on Penetration Testing, ISOPT, Florida

6. Lacasse S & Lunne T (1988) Calibration of dilatometer correlations, Proc 1st Int Conf on Penetration Testing, ISOPT, Florida

7. Lunne T, Lacasse S & Rad N S (1989) SPT, CPT, Pressumeter testing and recent developments in in situ testing, Part 1: all tests except SPT, General Report, 12th ICSMFE, Rio de Janeiro, vol 4, pp 2339-2403

8. Lutenegger A J (1988) Current status of the Marchetti Dilatometer test, Proc 1st Int Conf on Penetration Testing, ISOPT, Florida

9. Marchetti S (1980) In situ tests by flat dilatometer, ASCE JGED Journal of Geotechnical Engineering Division, vol 106:GT3, pp 299-321

10. Marchetti S (1985) Field determination of K_0 in sands, panel presentation session: In Situ Testing Techniques, Proc 11th ICSMFE, San Francisco, vol 5 pp 2667-2672

11. Ortigao J A R (1993) Dilatômetro em argila porosa, 7th Congresso Brasileiro de Geologia de Engenharia, Poços de Caldas, MG, vol 1, Setembro, vol 1, pp 309-320

12. Ortigao J A R & Macedo P (1993) Large settlements due to tunneling in porous clay, presented at the International Tunneling Conference, Toulon, Oct, published in: Tunnels et Ouvrages Souterrains, no. 119, Sept-Oct 93, pp 245-250, Paris

13. Schmertmann J H (1982) A method for determining the friction angle in sands from Marchetti dilatometer test, Proc 2nd European Symp on Penetration Testing, ESOPT, vol II, pp 853-861

Accelerated methods for field evaluation of soil properties in road survey and construction

Des méthodes rapides pour l'évaluation sur place des propriétés des sols dans la construction des routes

V.D. Kazarnovsky
Soyuzdornii, Moscow, Russia

ABSTRACT: Substantiating an optimum design solution for the road subgrade as a linear structure on the basis of the conventional approach to soil testing requires extremely much time and entails great difficulties. This problem may be solved by the development and implementation of various kinds of quick methods. Some decrease in accuracy of results obtained in single tests may be compensated by an increased number of concurrent tests. In Soyuzdornii for a number of years works have been carried out in the field of quick methods for soil testing at the stages of engineering-geological survey and construction of roads. Among solutions proposed are a method for accelerated evaluation of the peat soil compressibility by means of a portable relaxation device, a method for accelerated evaluation of optimum moisture content and maximum density of clay soils as well as a method for checking the degree of soil compaction based on testing by a dynamic device. The report contains main principles of the above methods and schematic diagrams of the devices.

RÉSUMÉ: L'argumentation d'une solution de projet optimale d'une plate-forme routière comme un ouvrage linéaire en cas de l'approche traditionnelle aux essais des sols exige trop du temps et liée aux difficultés peu surmontables. La résolution du problème est possible grâce à l'élaboration et à l'application de différentes méthodes rapides. La diminution de la précision des résultats obtenus aux essais particuliers peut être compensée dans ce cas par l'augmentation de la quantité des essais parallèles. Au Soyuzdornii on effectuait au cours de quelques années les travaux dans le domaine des méthodes d'essais des sols rapides au stade de la reconnaisance des sols de fondation et de la construction des routes. Ces solutions proposées comprennent la méthodologie d'évaluation rapide de la compressibilité des sols tourbeux à la base de l'appareil de relaxation portatif, la méthode de la détermination rapide de l'humidité optimale et de la densité maximale des sols argileux ainsi que la méthode du contrôle du niveau du compactage du sol à la base des essais par l'appareil dynamique. Les principes essentielles des méthodes citées et les schémas de principe sont exposées dans ce rapport.

A specific feature of highways from the point of view of engineering-geological provision of their design and construction is a line character of the structure. Due to this factor, the greatest difficulties happen to occur on complicated sections, on which individual designing the subgrade is required.

The most rational and economical design solution needs rather comprehensive engineering-geological data as a basis. However, most of soil test methods, widely used in practice, are initially aimed at the provision of designing engineering structures located on the restricted grounds. These methods, providing results of higher accuracies, require much time to obtain the results and are labour-consuming, which does not agree with the character of line structures and the dates for their design and construction.

In this connection for recent years the problem of developing some accelerated methods for soil tests meeting the requirements of fast obtaining single results and assessing,for a short period of time, long route sections has been considered as an urgent task in road construction of Russia.

In the first place, to solve this problem, some portable test apparatus, which could be used under field conditions, without spending time for the preservation and transportation of soil samples to a stationary laboratory is required.

Some other factor, taken into consideration in developing the above said apparatus and accelerated methods, consists in that some possible lowering of the accuracy of results obtained in standard tests can be compensated by an increase in the amount of tests, i.e. by increasing the amount of soil stratum points, for which tests are carried out, and owing to this it is possible to take into consideration more thoroughly statistics of properties of the stratum studied.

The development of methods and a portable apparatus for the accelerated determination of indices of peaty soil compressibility in using these soils for road embankment foundations is one of the examples of such works fulfilled (V.D.Kazarnovsky and Ju.A.Agapkin). The invention of those methods was due to the construction of roads in the development of very boggy, oil-gas-bearing territories of West Siberia and European North of Russia.

Standard methods come to loading a sample by some load stages untill the deformation is stabilized under each stage, the methods proposed envisage a process of successive forced deformation of the sample tested to some or other degree with recording load values, corresponding to these deformations, after the completion of the load relaxation process. The me-

thods are based on the assumption that each load has a corresponding value of soil porosity (equivalent porosity) on the straight line of the compression curve.

The above methods are realized using the so-called relaxation compression apparatus, the scheme of which is presented in Fig.1.

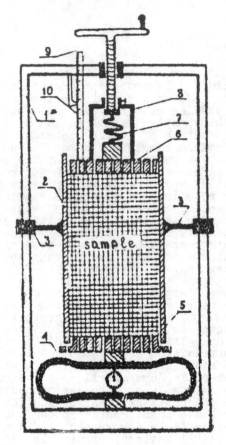

Fig.1. Scheme of the portable compression apparatus for testing increased samples of peaty soils: 1-frame with a support platform; 2-working cylinder; 3-device for fastening the cylinder to the frame; 4-bottom spring of the dynamometric device; 5 and 6-bottom and top plates; 7-top spring of the dynamometer; 8-arrester; 9-screw press; 10-strain gauge

The sample is compressed by means of a screw press to some relative deformation, then a decrease in loading is kept under observation on the dynamometer. At the moment of the end of load lowering its final value is registrated and the correction to be made in the sample deformation is determined (at the expense of dynamometer deformability). Then, the procedure is repeated, i.e. the deformation is increased by some value, and again a stabilizing value of loading is recorded. On the basis of test results the compression curve in the form of $e_p = f(P)$ or $\varepsilon = f(P)$

(where e_p = settlement modulus, ε = porosity coefficient) is obtained.

An acceleration of the test, in using these methods, is achieved owing to that the load is stabilized considerably faster than the deformation is stabilized under standard test conditions. In addition, the tests may be carried out not in the stationary laboratory, but under field conditions by means of a portable apparatus.

For the practical use of relaxation method a specific portable compression relaxation apparatus has been developed. In order to improve a reliability of test results, specimens of increased heights (to 20 cm) at the conventional sectional areas (60 sq.cm) are tested. The procedures of registrating a sample friction on the working cylinder walls using the dynamometer have been developed.

Special investigations showed that compression curves obtained by means of standard and relaxation methods are rather close to each other (Fig.2). The relaxation method can be used for soils of high compressibility, i.e. such as peaty soils.

On a level with the problems of accelerated tests of soils at the stage of survey there are some similar problems occuring at the stage of subgrade construction, in particular, evaluating a degree of soil compaction in constructing a subgrade.

Control methods for soil density usually used have some well-known disadvantages. In road building practice of Russia, express methods of control based on using apparatus and devices of mechani

cal type (penetration, etc) are preferred. For a further development of this direction the method for dynamic loading the soil layer by a falling weght acting on the soil through the spring and plate, using correlations between the soil density and the height of the weight rebound due to the action of the spring (V.D.Kazarnovsky and S.A.Avakyan) was proposed. At the same time there is a poor moment in existing practice of compaction control, which consists in that for obtaining reliable data on the degree of compaction, it is necessary to determine not only a density at every point controlled but also a maximum density by standard compaction methods. However, using standard compaction methods (method of Soyuzdornii or Proctor one) the above procedure cannot provide an express control. To overcome this difficulty, the accelerated method for determining an optimum moisture content and a maximum density of the soil (V.D.Kazarnovsky and N.Yu.Shitskaya) has been proposed. According to this method a maximum standard density ($\rho_{d\,max}$) can be calculated on the known theoretical formula, showing the relationship between the solid, liquid and gaseous phases of the soil.

The relationship can be written as follows:

$$\rho_{d\,max} = \cfrac{1}{\cfrac{W_0}{G_0 \Delta_W} - \cfrac{\lambda}{\rho_s}},$$

where ρ_s = optimum density of soil grains; W_0 = optimum moisture content (unit fractions); G_0 = residual coefficient of water saturation of the soil, determining the volume of the entrapped air at a maximum density and an optimum moisture content (unit fractions); Δ_W = water density.

As a result of the appropriate theoretical analysis the value of G_0 was proposed to be determined on the basis of a correlation relationship. In order to obtain such a relationship, many parallel tests on standard compaction of different clay soils with the plasticity index from 4 to 26

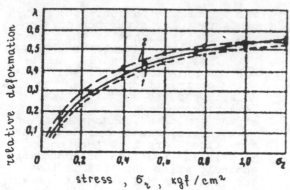

Fig.2. Comparison of compression curves
of the peaty soil obtained using standard (1) and express (2) methods

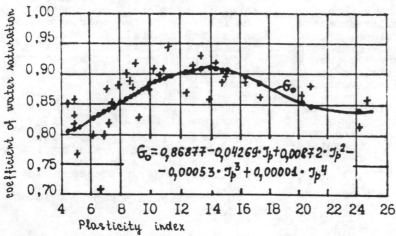

$$G_o = 0{,}86877 - 0{,}04269 \cdot J_p + 0{,}00872 \cdot J_p^2 - 0{,}00053 \cdot J_p^3 + 0{,}00001 \cdot J_p^4$$

+ - experimental points ●-points according to the equation

Fig.3. Relationship between the coefficient of water
saturation at the maximum density and optimum moisture content and the plasticity index of the soil

(mainly, cover clay soils of Central Russia) were carried out. On
the basis of statistic processing
of the results a correlation relationship has been obtained (Fig.3).
   Specific methods for an accelerated evaluation of optimum moisture
content, W_o, have been developed
on the basis of Prof. A.F.Lebedev's
methods for the evaluation of maximum molecular moisture capacities of soils. These methods proceed from the experimentally stated factor that a moisture content,
which is close to the optimum one
obtained according to Soyuzdornii
method, is gained by a thin specimen (2mm) prepared from the soil
paste (soil is on the boundary of
fluidity) due to pressing it in the
filter paper under some loading P_o
for 5 minutes. The value of P_o depends on the plasticity index of
the soil and can be calculated on
the plot of Fig.4.

370

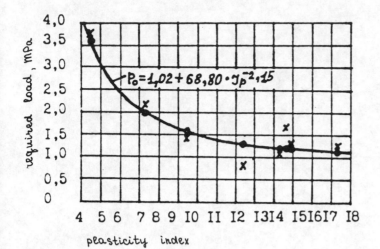

Fig.4. Compacting load required at the compression of the sample tested depending on the plasticity index of the soil

To realize the above methods, the proposed portable screw press and, also, any accelerated methods for evaluating a soil moisture content may be used.

Experiments carried out showed that there was a good conformity between the results obtained using accelerated methods and those obtained by standard methods. Thus, the optimum moisture contents differ by their absolute values not more than by $\pm$0.5% and the maximum densities – by $\pm$0.02g/cm^3. It is found that, on the whole, an evaluation of a compaction degree according to the compaction coefficient using express methods provides an accuracy which is not lower than that achieved by standard methods.

The proposed method makes it possible, even using conventional thermostatting for evaluating a soil moisture content, to accelerate determining $\rho_{d\,max}$ and w_0 by 3 hours, to reduce the required weight of the soil from 2.5-3 kg to 200 g and to improve the reliability of results owing to the possibility of determining w_0 and $\rho_{d\,max}$ practically at every point, where the density is under control,

and, also, owing to an increase in the number of tests.

Methods for the accelerated determination of an optimum moisture content can be also used at the stage of survey when evaluating a moistening degree of soils which are supposed to be used for filling an embankment.

REFERENCES

Agapkin, Yu.M., Kazarnovsky, V.D. & Kuzakhmetova, E.K. 1976. About determination of characteristics of the peaty soil compressibility on the samples of increased heights by relaxation methods. Proc. of Soyuzdornii 90, Moscow.

Avakyan, S.A., Kazarnovsky, V.D. 1983. Determination of a compaction degree of subgrade soils by dynamic compaction methods. Avtomobilny dorogi 8:628.

Methodic recommendations on the accelerated evaluation of optimum moisture contents and maximum densities of clay soils in constructing highway subgrades. 1990. Soyuzdornii, Moscow: 8.

Deformability of rock masses – Case histories and statistical analyses

La déformabilité des massifs rocheux – Études des cas et analyses statistiques

J. He
Department of Geotechnical Engineering, Chalmers University of Technology, Göteborg, Sweden

ABSTRACT: The deformability of rock masses are of vital importance in many major rock engineering projects. Field tests play an essential role in the assessment of the deformation modulus of a rock mass. A comprehensive case review of the rock mass deformability is performed, based on information in the literature. 53 reference entries and 258 deformation modulus entries were collected into a database -**DEFOR**. Important parameters, such as the deformation modulus, the rock mass factor and the plastic degree of rock masses, are discussed. General statistical features of these parameters are presented. Based on statistical analyses, some interesting issues are investigated, such as inherent scatter, the background distributions of the parameters and the relationships among the parameters.

RESUMÉ: La déformabilité des massifs rocheux est d'une importance vitale dans plusieurs projets majeurs d'ingenierie de la roche. Les essais sur le terrain jouent un rôle essentiel dans l'évaluation du module de déformation d'un massif rocheux. Une revue de cas explicite est mise au point, basée sur des informations tirées de la literature. 53 entrées publiées et 258 modules de déformation ont été introduits dans la base de données - **DEFOR**. Des paramètres importants, tel que le module de déformation, le facteur des massif rocheux et le degré de plasticité des massifs rocheux, sont discutés. Les characteristiques statistiques générales de ces paramètres sont présentées. Basé sur analyse statistique, quelques issues interessantes sont étudiées, telque le comportement propre, les distributions de base des paramètres et les relations entre les paramètres

1 INTRODUCTION

The deformation features of rock masses are of vital importance in many major rock engineering projects such as dam foundations, pressure tunnels and large rock caverns. Field testing is an essential part in the assessment of the deformation modulus of a rock mass, since Young's modulus obtained from samples in the laboratory is seldom relevant to that of rock masses.

Deformability of a rock mass concerned in this paper is characterised by the equivalent properties, which represent combining effects of intact rock blocks and discontinuities. These properties, together with the strength of a rock mass, are essential input parameters in most stability analyses of rock engineering. Adoption of such equivalent parameters implies a distinct choice of mathematical model, the continuous approach. This approach, which is still the dominant one, provides deterministic and clear predictions compared with a discontinuous model. The latter is, for example, described in the Distinct Element Method (DEM) (Cundall 1980; Lemos et al. 1985) and Discontinuous Deformation Analysis (DDA) (Shi and Goodman 1989).

A comprehensive case review of the deformability of rock masses was originally performed as a part of a research work on lined underground storage of compressive gas (He 1992). The gas pressure as high as 15 MPa was suggested stored in a cavern only 100 to 200 meters below the ground level. It was identified that the deformation of the rock mass around the gas cavern is essential for the integrity of the gas-tight liner.

The information on deformability of rock masses was collected from the literature into a data base (**DEFOR**) which arranged according to references and sites. Part of the data base was taken from that of Heuze (1980) and Bieniawski (1978). In total, 53 reference entries and 258 deformation modulus entries were obtained. The geological conditions described in available references were registered in another file for further study. The main part of the database can be found in He (1992).

The general description on deformability of rock masses as well as aspects associated with current field test techniques are described in Sect. 2. The statistics of the deformation modulus collected are presented in Sect. 3. The rock mass factor - deformation modulus normalised by Young's modulus of intact rock, is examined in detail in Sect. 4. Non-elastic deformation of rock masses is discussed in Sect. 5. And some discussions are given in Sect. 6.

2 DEFORMABILITY OF ROCK MASS

The experience on rock mass deformations accumulated mainly originates from in-situ tests, deformation monitoring and correlation with rock mass classification. In most cases, a rock mass is simplified to an elastic material, in spite of the fact that it is an anisotropic, non-linear material, often with hystereses and creep. Poisson's ratio of a rock mass can be taken either the same as the intact rock or 0.2 as suggested by He and Lindblom (1993). This is due to two reasons: i) Poisson's ratio is a relatively non-sensitive parameter in the analysis and ii) the knowledge about it is limited. The research was thus focused on the deformation modulus.

The deformation modulus (E_d) of a rock mass is a complex parameter. The name itself is distinguished from Young's modulus, which is related only to the elastic deformation. Definitions of the moduli are shown in Fig. 1, modified from Timur and Ada (1970),

$$E_i = \frac{\sigma}{\varepsilon_i} \tag{1}$$

where $E_i = E_d$, E_e and E_{st}; E_d = deformation modulus; E_e = modulus of elasticity; E_{st} = total deformation modulus; $\varepsilon_i = \varepsilon_d$, ε_e and ε_{st}, strains corresponding to E_d, E_e and E_{st}, respectively, as shown in Fig. 1.

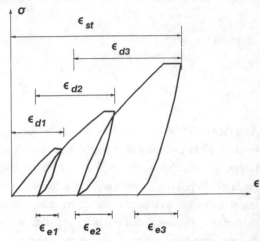

Fig. 1 Definitions of rock mass deformation modulus.

In many cases, E_d is not distinguished from E_{st}; this is only done based on the engineering judgment of factors such as durability, number of cycles and load level. However, E_d must be distinguished from E_e in field tests.

The plastic degree is defined as follows (Seeber, 1970):

$$\alpha = \frac{\varepsilon_d - \varepsilon_e}{\varepsilon_e} = \frac{E_e}{E_d} - 1 \tag{2}$$

The plastic degree is an index of plastic deformation. A larger the α value the larger plastic portion of the deformation for the rock mass. Full elastic deformation leads to a plastic degree of zero.

A rock mass factor is defined as the ratio of the deformation modulus of a rock mass to Young's modulus of intact rock (Pande et al. 1990),

$$\beta = \frac{E_d}{E_l} \tag{3}$$

where E_l = Young's modulus of intact rock.

Since the presence of discontinuities in a rock mass reduces the modulus, compared with the intact rock, Young's modulus of intact rock is considered as the upper limit of the deformation modulus. This implies an advantage, since the rock mass factor should not exceed one.

Field tests to determine the deformation modulus are regularly used for the control of dam foundations, pressure tunnels and major underground excavations. Many kinds of field tests have been developed and may be categorised as:
a. plate bearing;
b. radial jack and pressure tunnel;
c. dilatometer and bore hole jack;
d. flat jack;
e. tunnel relaxation and back analysis;
f. geophysical methods.

Field tests are often employed in spite of their high cost and time consumption. Bieniawski (1978) and Heuze (1980) recommended two different kinds of field tests to be used for a major project.

There are three preconditions for any field test:
a. proper load level is exerted on a representative rock volume;
b. displacements under the loading are recorded;
c. analytical solutions are available under the given load condition.

The third condition implies that a rock mass is assumed ideally elastic. The Boussinesq solution is applied to the plate bearing; the Kirsch solution to the radial jack and dilatometer; and the solution for elliptical holes to the flat jack. General guidelines were given by Lemcoe et al. (1980) and Obert and Duvall (1967).

Significant simplifications have been made in using the above solutions. Numerical methods were later used to eliminate these simplifications; Borsetto et al. (1983) and Faiella et al. (1983) for the flat jack; Rotonda (1987) for the plate bearing test.

The geophysical methods used for determining the deformation modulus of rock masses are primarily seismic velocity. A correlation based on an adequate database is a precondition for these methods. The correlation of seismic velocity with the deformation modulus were studied by Coon and Merritt (1970) and Ribacchi (1987). The results given by Ribacchi (1987) showed the seismic method to be very promising.

Interpretations of field tests play an important role in a deep understanding of the response of a rock mass to loading. Interesting studies were made by Dodds (1974) and Ribacchi (1987) regarding plate bearing, and Misterek (1970) regarding radial jack.

3 DEFORMATION MODULUS

Case studies form an essential part of rock mechanics. In the last three decades, a large amount of in-situ deformation modulus data have been accumulated, and reviewed by several investigators (Imazu 1986; Heuze 1980).

The main purpose of the case study presented here is to assess the inherent scatter and tendency of the deformation modulus (E_d) quantitatively. No attempt is made to correlate the deformation modulus obtained for a rock mass with other properties such as Young's modulus of an intact rock and the size of the laboratory sample.

The distributions of deformation moduli for all, metamorphic and igneous, as well as sedimentary rocks, are shown in Figures. 2, 3 and 4, respectively. The reasons for grouping igneous together with metamorphic rocks are, i) they are both crystalline rocks usually with high moduli and ii) igneous rocks often appear intimately with metamorphic rocks. The number of data, the mean and standard deviation are attached to the figures. It is noted that the standard deviation of the deformation modulus is about the same order as the mean.

The mean values and the standard deviations presented in the above figures are obtained as the arithmetic means. It is a controversial issue in the literature. It may be questioned whether the mean of the deformation moduli should be calculated as the arithmetic mean. Peres-Rodriguos (1990) found, based on a series of block tests in the laboratory, that the strain of the rock mass obeyed a normal distribution. The mean value of the deformation modulus is therefore equal to the harmonic mean. Cunha and Muralha (1990) also interpreted the

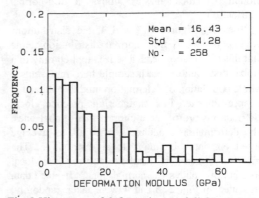

Fig. 2 Histogram of deformation moduli for all data.

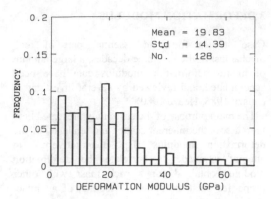

Fig. 3 Histogram of deformation moduli for igneous and metamorphic rocks.

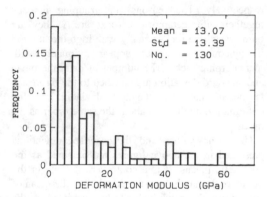

Fig. 4 Histogram of deformation moduli for sedimentary rocks.

deformation moduli of rock masses with a harmonic mean. Wittki (1990) and Charrua-Graca (1979), on the other hand, employed a geometrical mean. This implies that the distribution of the deformation moduli of rock masses obeys a log-normal distribution.

Observing Figures 2, 3 and 4, the distributions obviously deviate from normal distribution. Note that this observation is, in a sense, applied only to a global distribution - each sample here represents a whole population of deformation moduli obtained at a specific site. For engineering practice, local distributions are of more interest than global ones. The deformation modulus is, after all, an average based on data obeying a local distribution. The arithmetic mean is justified only if the local distribution obeys a normal one. If the data presented in Fig. 2, 3 and 4 were taken randomly,

the global distribution would represent the local one.

Employing the transformation

$$y = \ln E_d$$

we have

$$E_d = \begin{cases} e^{\mu_y + \sigma_y} & arithmetic \ mean \\ e^{\mu_y} & geometric \ mean \\ e^{\mu_y - \sigma_y} & harmonic \ mean \end{cases} \quad (4)$$

where μ_y = the mean value of y; σ_y = the standard deviation of y.

The arithmetic mean, as can be seen from the equations, takes the highest value, while the harmonic mean takes the lowest.

4 ROCK MASS FACTOR

Distributions of the rock mass factor (β) are shown in Figs. 5, 6 and 7, for all, metamorphic and igneous as well as sedimentary rocks, respectively. The mean value of the rock mass factor is between 0.5 and 0.6. The standard deviations of the rock mass factor are about the same as the means. The rock mass factor for sedimentary rocks is larger than that for crystalline rocks.

It is interesting to note the tendency of the rock mass factor to vary with Young's modulus of intact rock (Figs. 8, 9 and 10). The rock mass factor decreases with increases in Young's modulus for all data, as well as for igneous and metamorphic rocks. This is quite reasonable, considering that the

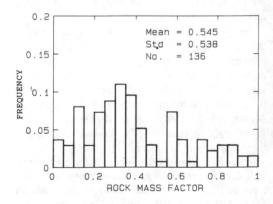

Fig. 5 Histogram of the rock mass factor for all data.

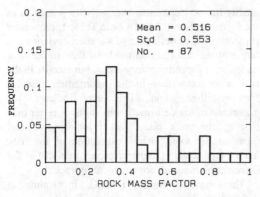

Fig. 6 Histogram of the rock mass factor for igneous and metamorphic rocks.

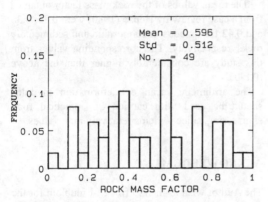

Fig. 7 Histogram of the rock mass factor for sedimentary rocks.

contribution of rock joints may be much the same both for hard and weak rocks. This means that the modulus of a intact rock does not play an important

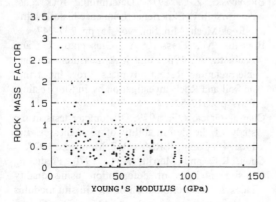

Fig. 8 Rock mass factor vs. E_i for all data.

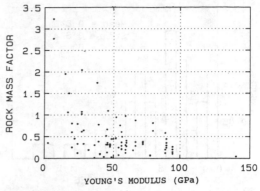

Fig. 9 Rock mass factor vs. E_i for igneous and metamorphic rocks.

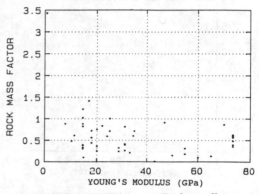

Fig. 10 Rock mass factor vs. E_i for sedimentary rocks.

role in the deformation modulus of a rock mass. The rock mass factor for sedimentary rocks, however, holds fairly constantly.

5 PLASTIC DEGREE

The distribution of the plastic degree (α) is shown in Fig. 11. Fig. 12 shows the tendency of the plastic degree versus the deformation modulus. It is noted that the data available on the plastic degree contains only sedimentary and metamorphic rocks. This might be an indication that the plastic degree is negligible for hard rocks. The plastic degree decreases with the increase of deformation modulus as can be seen in Fig. 12.

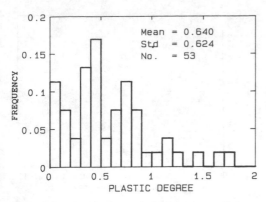

Fig. 11 Histogram of the plastic degree.

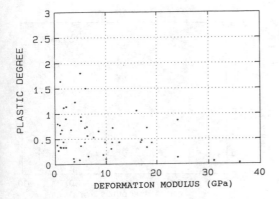

Fig. 12 Plastic degree vs. deformation modulus.

6 DISCUSSIONS AND CONCLUSIONS

As a summary, Table 1 lists the statistics of the three parameters associated with deformability of rock masses.

Table 1 Statistics of E_d, β and α.

ITEM		No.	Mean	St.d
ALL DATA	E_d	258	16.4 GPa	14.3 GPa
	β	136	.55	.54
	α	53	.64	.63
METAMORPHIC IGNEOUS ROCK	E_d	128	19.8 GPa	14.4 GPa
	β	87	.52	.55
SEDIMENTARY ROCK	E_d	130	13.1 GPa	13.4 GPa
	β	49	.60	.51

There is a discussion in the literature on which parameter, the deformation modulus or the rock mass factor, should be correlated with the rock mass quality index. Examining the statistics presented in the figures in Sects. 3 and 4 or in Table 1, the scatter (in terms of the coefficient of standard deviation), is larger in the deformation modulus than in the rock mass factor for sedimentary rocks; but smaller in the rock mass factor than in the deformation modulus for crystalline rocks. This may indicate that estimation of the deformation modulus is better than that of the rock mass factor for igneous and metamorphic rocks, while estimation of the rock mass factor leads to higher accuracy than that of the deformation modulus for sedimentary rocks.

The most common method in estimating deformation modulus is the empirical one by Bieniawski (1978). The deformation modulus is estimated base on a RMR value.

The mean values of the rock mass factor obtained from a case review by Heuze (1980) were 0.35, 0.36 and 0.42 for igneous, metamorphic and sedimentary rocks, respectively. The corresponding values from this study are considerably higher than the above (0.5-0.6).

The arithmetic mean of deformation modulus should be used with caution in practical rock engineering, since it yields relatively high values.

ACKNOWLEDGMENT

The Author wish to thank Prof. U. Lindblom for the support during the study and critical examination the manuscript. This work was financed with grants from the Swedish Government through NUTEK.

REFERENCES

Bieniawski, Z.T., 1978. Determining rock mass deformability: Experience from case histories. Int. J. Rock Mech. Min. Sci., vol. 15, pp. 237-247.

Borsetto, M., Frasson, A., Giusepptti, G. and Zaninetti, A., 1983. Flat jaok measurements and interpretation: Recent advances. Proc. Int. symp. on Soil and Rock Investigation by In Situ Testing. Paris, pp. 517-522.

Charrua-Graca, J.G., 1979. Dilatometer tests in the study of the deformability of rock masses. Int. Proc. 4th Congr. ISRM, vol. 2, Montreux, 1979.

Coon, R.F. and Merritt, A.H., 1970. Predicting in-situ modulus of deformation using quality indexes. In Determination of the in situ modulus of deformation of rock. ASTM STP 477, pp. 154-73.

Cundall, P., 1980. UDEC - a generalized distinct element program for modelling jointed rock. PCAR-1-80, Peter Cundall Associates.

Cunha, A.P. and Muralha, J., 1990. About LNEC experience on scale effect in the deformabulity of rock masses. Proc. 1st Int. Workshop on Scale Effects in Rock Masses, Loen, Norway, June 1990, pp219-230.

Dodds, D.J., 1974. Interpretation of plate loading test results. In Field testing and Instrumentation of rock. ASTM STP 554, pp. 20-34.

Faiella, D., Manfredini, G. and Roosi, P.P., 1983. In situ flat jack test: Analysis of results and critical assessment. Proc. Int. symp. on Soil and Rock Investigation by In Situ Testing. Paris, pp. 507-12.

He, J., 1992. Lined underground openings - Rock mechanical effects of high internal pressure. Doctoral thesis, Dept. of Geotech. Engr. Chalmers U. of Technology, Gothenburg.

He, J and Lindblom, U., 1993. Estimation of rock mass properties. Proc. Int. Symp. on Application of Computer Methods in Rock Mechanics. Xian, China, Vol. 2, pp. 695-702.

Heuze, F.E., 1980. Scale effects in the determination of rock mass strength and deformability. Rock Mech. and Rock Engr. vol. 12, pp. 167-192.

Imazu, M., 1986. Data base system and evaluation of mechanical properties of rock. Proc. Int. Symp. on Engr. in Complex Rock Formation. Peking, pp. 111-120.

Lemcoe, M.M., Pratt, H. and Grams, W., 1980. State-of-the-art review of rock mechanics techniques for measurement of stress, displacement and strain. Rock Store 80, Stockholm, vol. 2, pp. 927-42.

Lemos, J.V., Hart, R.D. and Cundall, A.P., 1985. A generalized distinct element program for medolling jointed rock mass - A keynote lecture. Proc. Int. Symp. on Fundamentals of Rock Riont, Björkliden, Sweden, pp. 335-44.

Misterek, D.L., 1970. Analysis of data from radial jack tests. In Determination of the in situ modulus of deformation of rock. ASTM STP 477, pp. 27-38.

Obert, L. and Duvall, W.I., 1967. Rock mechanics and the design of structures in rock. J. Wiley & Sons, Inc.

Pande, G.N., Beer, G. and Williams, J.R., 1990. Numerical methods in rock mechanics. John Wiley & Sons Ltd. West Sussex, England, 327p.

Peres-Rodrigues, F., 1990. About LNEC experience on scale effect in the deformabulity of rock. Proc. 1st Int. Workshop on Scale Effects in Rock Masses, Loen, Norway, June 1990, pp155-164.

Ribacchi, R., 1987. Rock mass deformability: In situ test, their interpretation and typical results in Italy. Proc. of 2nd Int. Symp. on Field Measurement in Geomech., Sakurai Ed., Kobe, pp. 171-92.

Rotonda, T., 1987. Interpretation of plate-loading test in tunnels. Proc. of 2nd Int. Symp. on Field Measurement in Geomech., Sakurai Ed., Kobe, vol. 2, pp. 1049-1058.

Seeber, G., 1970. Ten years use of TIWAG radial jack. Proc. 2nd Congr. ISRM, Belgrade, vol. 1, pa. 2-22.

Shi, G. and Goodman, R.E., 1989. Generalization of two-dimensional discontinuous deformation analysis for forward modelling. Int. J. Num. and Analyt. Methods in Geomach. vol. 13, pp. 359-380.

Timur, A.E. and Ada, E., 1970. A study of correlation of dynamic and static elasticity modulus determined by in-situ and laboratory tests. Proc. 2nd Congr. ISRM, Belgrade, vol. 1, pa. 2-37.

Wittke, W., 1990. Rock mechanics - Theory and applicatins with case histories. Springer-Verlag, Berlin, 1075p.

Piezocone and shear vane in situ testing of soft soils at three sites in Ireland

Essais in situ au piezocône et au scissomètre sur des sols mous en trois places en Irlande

M. Rodgers & D. Joyce
Civil Engineering Department, University College Galway, Ireland

ABSTRACT: *In situ* vane shear and piezocone tests on estuarine, lacustrine and alluvial soft soil deposits in Ireland are detailed in this paper.

RÉSUMÉ: Cet article presente les résultats des essais in situ de piezocone et au scissomètre sur des dépôts de sols mous provenant des berges des estuaires, des lacs et des fleuves en Ireland.

1 INTRODUCTION

In recent years, many new road projects have been completed in Ireland to improve its infrastructure. On some of these projects it was necessary to construct road embankments in areas of deep soft soils. This paper presents results from piston sampling, *in situ* vane shear and piezocone tests at Letterkenny, Galway City and Cork City (Figure 1). Comparisons are made between the piezocone and vane shear tests.

2. GEOLOGY OF THE THREE SITES

The soils investigated at Letterkenny, County Donegal were estuarine deposits laid down at Lough Swilly and consist of silts with a desiccated crust that is about 1 m thick. There are occasional sand partings in the deposits and the water table is less than 1 m below ground surface. The soil is slightly overconsolidated. County Donegal is part of a northeast-southwest trending Caledonian fold belt of lower Palaeozoic age. The rocks of the region are mostly schists and gneisses of Lewisian, Moinian and Dalradian age.

The soils at the Galway site consist of a sequence of soft lacustrine deposits, that have been laid down since glacial times, overlying coarse grained glacial deposits. The soft soils comprise a very soft lake mud with occasional silt and sand partings. Above the lake mud is a layer of organic clay, overlain sequentially by beds of calcareous silt (calc tufa) and a thin layer of peat. The calc tufa is formed by the precipitation of insoluble calcium carbonate from the lime rich groundwater. The water table is at or near ground surface. The site is located close to the rock interface of Galway granite and the Carboniferous limestone of the central plain of Ireland and is in the flood plain of Lough Corrib. The catchment of Lough Corrib is geologically extremely complex. The following rocks are found in this catchment: limestone, granite, schist, shale, quartzite and silurian.

The site in Cork City is situated at the Tramore River. The water table is at ground surface. The stratification of the soft soils at this site consists of peat overlying silt on organic clay. The organic clay is underlain by a soft pink silt that has sand partings which increase in frequency with depth. The site is situated in the well developed Armorican fold system of southern Ireland. This system consists mainly of Old Red sandstone with limestones and lower limestone shale outliers in some valleys.

3 EQUIPMENT

The Geonor piston sampler took 101 mm diameter and 0.9 m long good quality samples. When the samples had entered the sample tube, it was necessary to leave the sampler at the

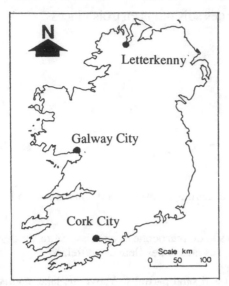

Figure 1. Site locations in Ireland

sample depth for 15 minutes to allow the excess pore water pressure, that had been generated during the taking of the sample at the inside wall of the tube, to dissipate away. If this was not done, the sample could have been lost down the sample hole. The samples were used in classification, consolidation and shear strength tests.

The Geonor penetration shear vane apparatus is very suitable for obtaining reliable undrained shear strength values. The vane used in all of the work had square ends with a height of 130 mm, a diameter of rotation of 65 mm and a vane blade thickness of 2 mm. The vane apparatus was driven down into the soil in depth intervals of 0.5 m with the vane withdrawn into a protective housing. When the apparatus had been driven to the required depth, the vane was then pushed 0.5 m below the housing to the depth where natural, rapid and remoulded shear tests were carried out. When the tests were completed the vane was returned to its housing. The values of natural vane shear strength were corrected for soil anisotropy and strain rate before they were used for stability analyses.

The University College Galway (UCG) GMF electrical piezocones have a 60^0 apex angle and a cross sectional area of 1000 mm^2 and can measure cone end resistance, q, soil-sleeve

friction, f, and pore water pressure, u. During the tests, the piezocone penetrated the soil at a rate of 20 mm per second and data were recorded every 100 mm. The filter for measuring the pore water pressure was located immediately above the cone tip. The piezocone was rated for a maximum load of 20 kN. The cone was calibrated at regular intervals and thorough deairing of the filter and cavities leading to the pore water pressure transducer took place.

4 RESULTS

The site investigation results for the Letterkenny estuarine silts are given in Figure 2. There is a good correlation between the vane shear strength, Su, and the effective end resistance, qt-u, where qt is defined as:

$$qt = q + 0.2u.$$

There is a high end resistance at 6 m depth where there is a sand parting. This sand parting is also confirmed by the drop in the pore pressure ratio, Bq, which is defined as:

$$Bq = (u - uo)/(qt-p)$$

where uo is the equilibrium pore water pressure and p is the total vertical stress. The values obtained for Bq are in accordance with data presented by Senneset et alia (1982) for silts and fine sand. The moisture content plots also indicate the sand parting at 6.0m depth. The proposed embankment loading, including surcharge, was of the order of 3m. Since the soil was overconsolidated (Figure 2) and had good drainage characteristics at low stresses, it was decided that there was no need for vertical drainage to achieve full primary consolidation within the project time period. During construction the pore water pressure was monitored and dissipated fully to equilibrium by end of the construction period (Joyce, 1988).

The results from the Galway City site are presented in Figure 3. Again, there is a good correlation between the vane shear strength and the effective end resistance. The inorganic lake muds had strengths down to 2 kPa. This was due to the light overburden of high moisture content soils. The position of the relatively

Figure 2. Vane, piezocone and moisture content data from Letterkenny

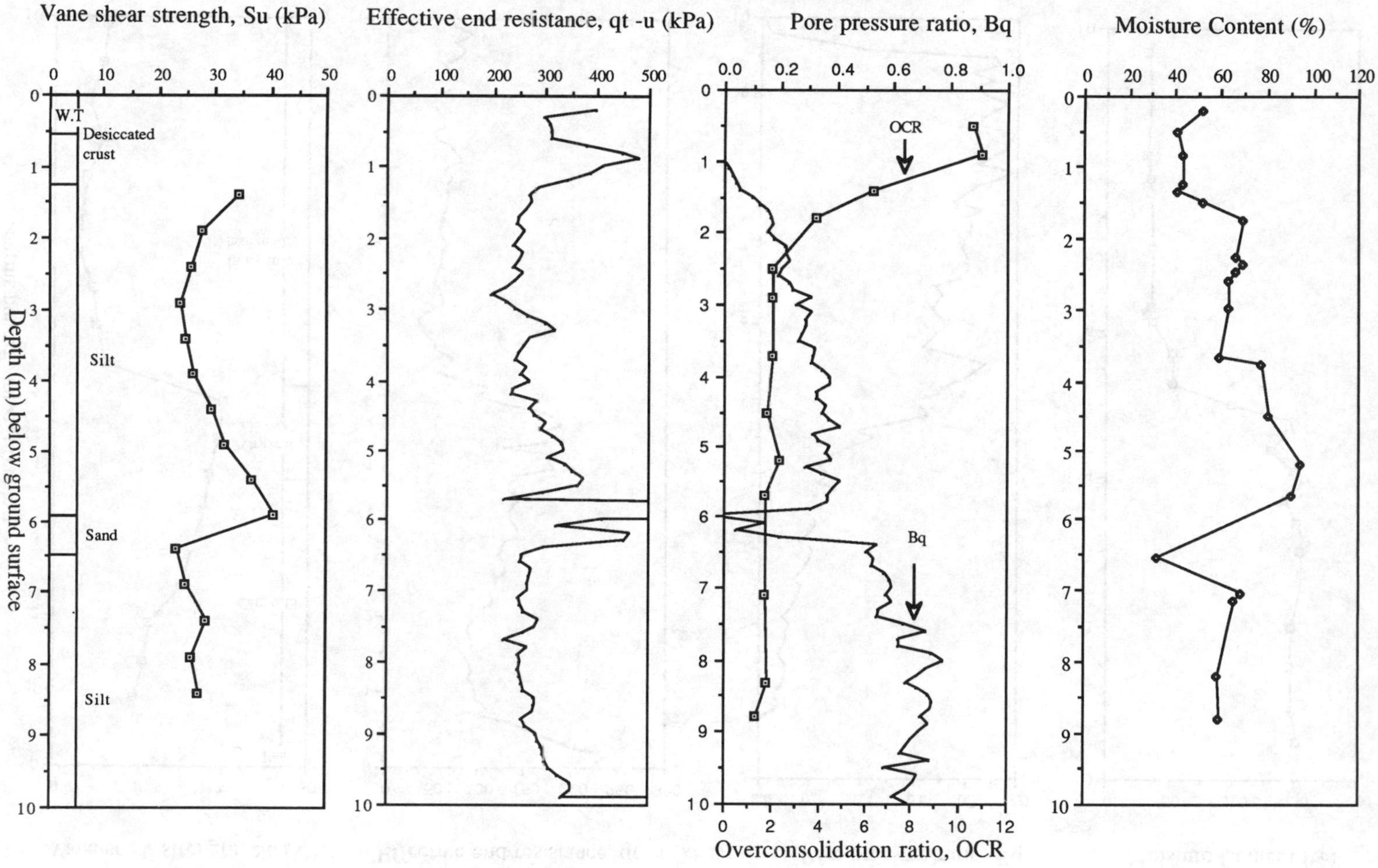

Figure 3. Vane, piezocone and moisture content data from Galway City

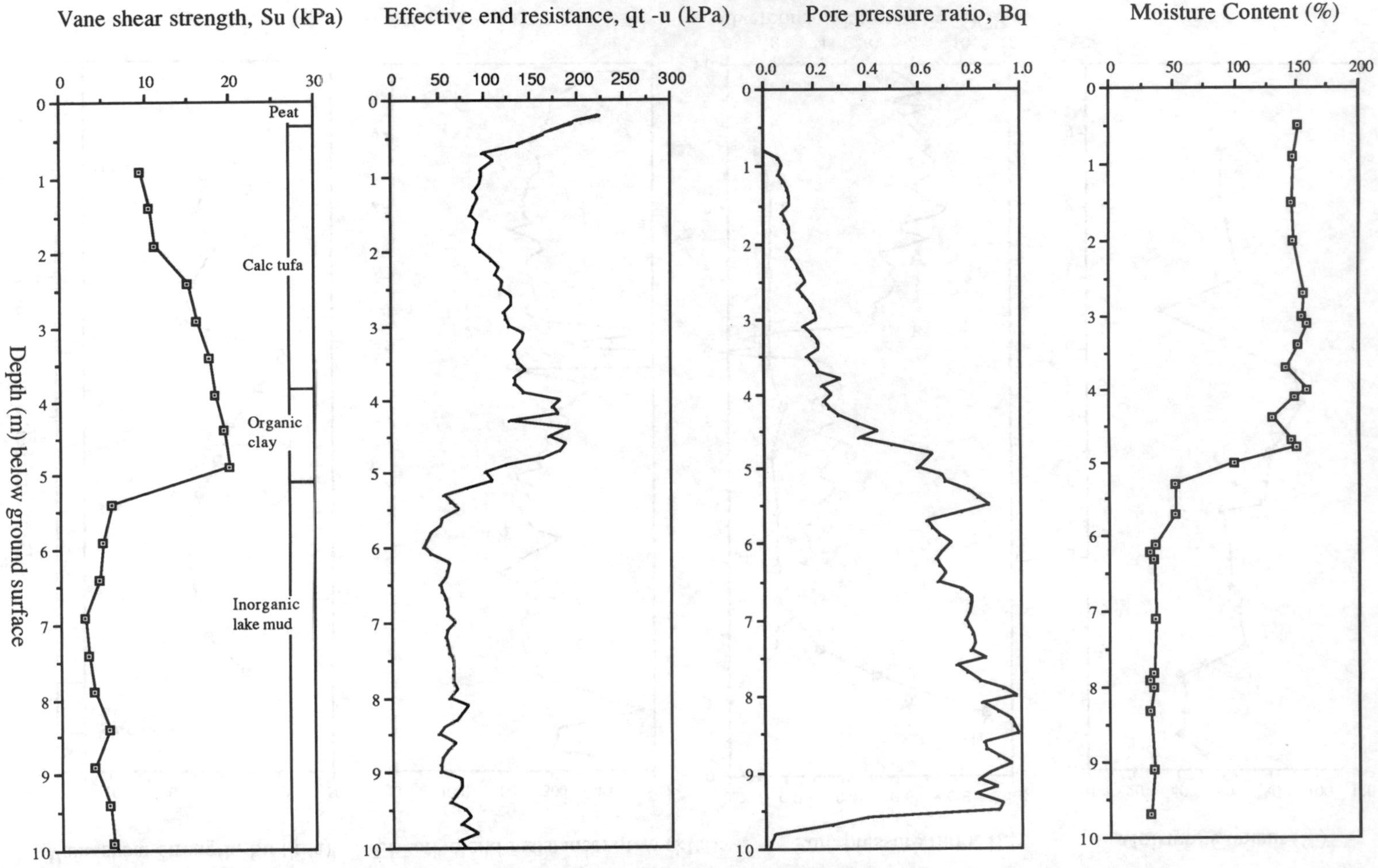

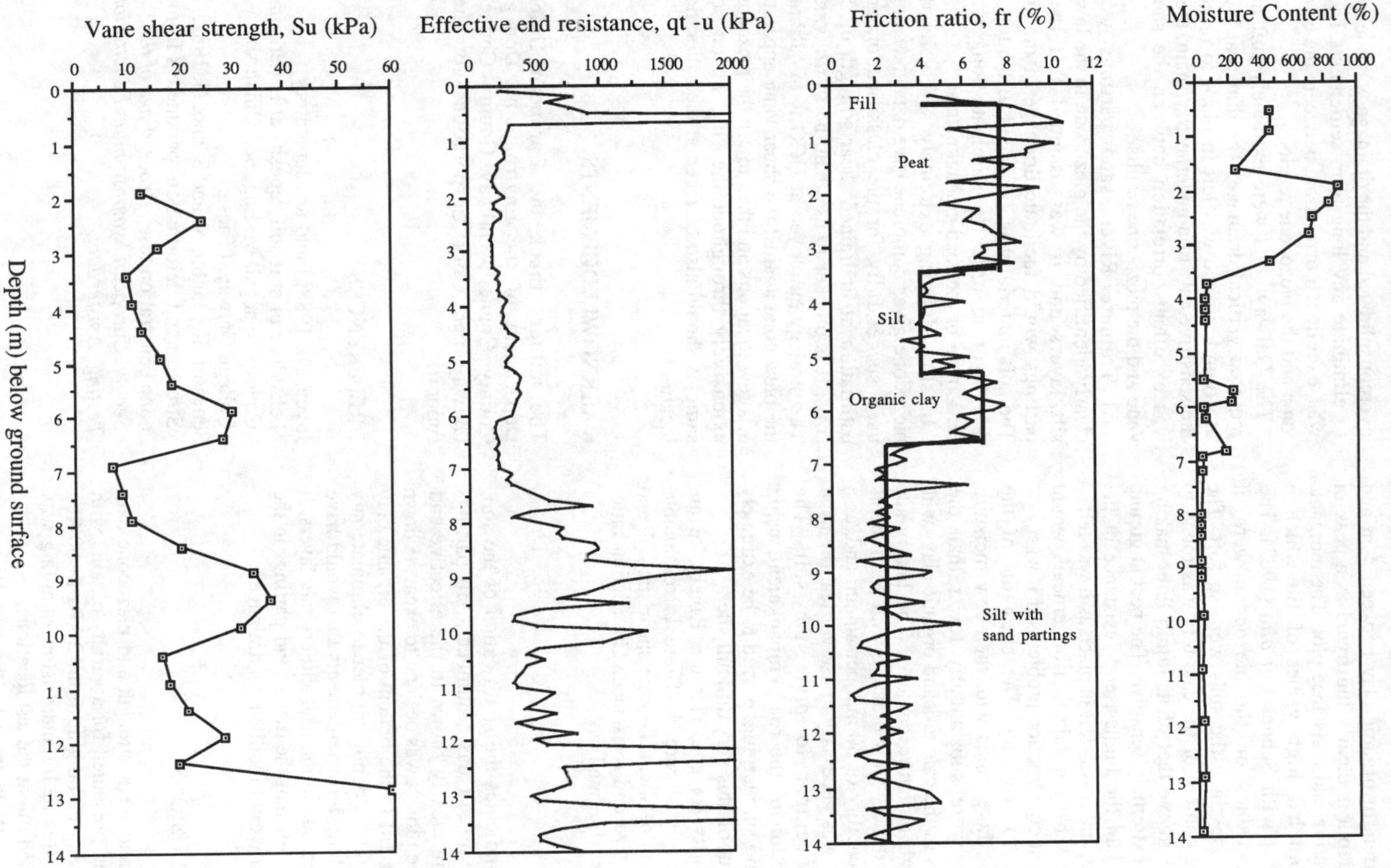

Figure 4. Vane, piezocone and moisture content data from Tramore River, Cork City

high strength organic clay between 4 m and 5.3 m depth is clearly illustrated as a peak in the effective end resistance plot. The inorganic lake mud has high values of Bq which is consistent with Senneset *et alia* (1982). The friction values on the piezocone were of limited use since the soils were so soft. The moisture content of the calc tufa was found to be 20 % above that of its liquid limit making the soil extremely sensitive. The peat at ground surface had the highest end resistance of the soft soils and it was left in place as it was felt that the sensitive calc tufa underneath would pose great construction problems if it was exposed to working. The strengths of the inorganic lake muds with their low moisture contents were substantially lower than the strengths of the calc tufa and organic clays with their high moisture contents. The lake muds were extremely slow draining with coefficients of permeability of 30 mm per annum and, as a result, it was necessary to use vertical drains at 1.0 m centres to drain the soil. The construction of the road embankment, using berms and surcharging, proved to be extremely successful on this very difficult site.

The results for the Tramore River site are presented in Figure 4. There is reasonable correlation between the vane shear strength and the effective end resistance. The friction ratio, fr, which is defined as:

$$fr = f/q$$

clearly indicates the stratification of the soil. The bottom silt has a number a silt partings which are clearly shown in the effective end resistance plot. It was decided to excavate down and back fill to the bottom of the organic clay layer at 6.5 m depth in order to eliminate any large or long term settlements that would have developed due to the peat and organic layers. It was also felt that the many sand partings in the lower silt would facilitate its drainage.

5 CONCLUSIONS

In situ vane shear strength and piezocone data have been presented for a variety of soils with shear strengths ranging from 2 kPa to 60 kPa at three sites in Ireland. Piston sampling was also carried out at the test locations.

At the Letterkenny site, the soft soils were overconsolidated and the drainage was adequate to eliminate any long term settlements. There was excellent correlation between the shear vane and the piezocone results.

The Galway soils were extremely soft and required vertical drainage. The thin peat layer at ground surface was left in place to protect the sensitive calc tufa underneath. Again, there was excellent correlation between the shear vane and the piezocone results.

At Tramore River, sand partings were clearly identified by the piezocone in the lower silt layers and it was assumed that these partings would facilitate drainage in the silt. The soft material was excavated to the top of this layer in order to eliminate long term settlement of the road embankments.

In slow draining extremely soft soils the piezocone gives comprehensive data and can be used successfully for the classification and estimation of undrained shear strength of the soils. It is quicker to use than the *in situ* vane. The strategy employed at UCG is to calibrate the piezocone against the shear vane apparatus at a few locations and then to use the piezocone extensively throughout the site. Piston soil samples should also be taken at the calibration locations.

6 ACKNOWLEDGEMENTS

The authors thank the European Union, Department of the Environment, Donegal County Council, Galway County Council, Galway and Cork Corporations for their support.

7 REFERENCES

Joyce, D., 1988. Field and laboratory behaviour of soft soil deposits at Letterkenny, County Donegal. M. Eng. Sc., University College Galway, (Thesis).

Senesset, K., Janbu, N. and Svano, G. 1982. Strength and deformation parameters from cone penetration tests. *Proceedings of the Second European Symposium on Penetration Testing, Amsterdam.*

Geological mapping of soft rock by pressuremeter test using strength criteria

Cartographie géologique de roches tendres par préssiomètre en usant des critères de résistance

B.J.Mehta

SV Regional College of Engineering & Technology, Surat, India

ABSTRACT : Pressuremeter test is an advance and more reliable insitu test for evaluating stress strain behaviour of earth deposits at considerable depth. The strata of soft or hard rock can also be investigated upto depth of capability of the instruments. The parameters obtained by pressuremeters deformation are modulus and limit pressure and they describe the strength characteristics of the strata. Also the ratio E/pl is characterized as rock strength and it is related to RQD.

Based on these characteristics the zoning and mapping can also be prepared by making grids and cadrans in the area showing the zones of various ranges of both the properties. This will help the mining engineers unable them to choose the range of mining equipments, capability of blasting, stability problem and assessability of ground water flow etc. The author have vast experience of pressuremeter testing. The interpretation of the test results are also established for computation of other behaviour.

RÉSUMÉ: L'essai préssiométrique est un essai adéquat pour l'évaluation du comportement contrainte-déformation des terrains à grande profondité. L'évaluation peut être conduite jusqu'à la profondeur à laquelle l'équipement fonctionne bien. Les paramètres obtenus sont les modules de déformation et la préssion limite qui décrivent les caractéristiques de résistance des roches. La relation module/préssion limite peut avoir une corrélation avec le RQD. Cette technique peut être utilisée pour le zonage et la cartographie des massifs a fin d'aider des ingénieurs de mines à choisir les équipements miniers plus adéquats, à décider sur l'utilisation d'explosifs, à définir la stabilité des massifs, etc.

1. INTRODUCTION

Man has conquered millions kilometers distance in the space but he has limited knowledge about the strata of the earth. He has explored not more than 12 km depth of earth crust. It is extremely difficult to explore subsoil as too much and intensive operation is necessary. However the resources from earth crust play very important role in development of civilization and advancement of science and technology. Man has developed latest methods to explore subsoil so as more and reliable information he gets. Geologist and geotechnical engineers are using advance techniques for exploration methods for assessing geological composition, availability of minerals, feasibility of irrigation projects the effort required to exavate the minerals etc. All investigation methods have certain drawbacks or limitations, at the same time, one method may be additive to other.

Pressuremeter test method, is an advance quick and more reliable insitu test which describes strength, stiffness and denseness of the strata of soil or rock at required depth. The test is based on expansion of cylindrical cavity in a predrilled borehole and at predecided depth. The latest version of pressuremeter is also introduced as self-boring pressuremeter in which possibility of disturbance to strata is minimized due to manual or machine boring. The test describes stress - strain characteristics of soil or rock strata.

2. PRESSUREMETER TEST

As the bed of soil or rock is represented by a linear elastic isotropic homogeneous mass, it is also characterized by both fundamental parameters, elastic constant and poisson's ratio. The rheological properties of elasto-plastic rock and soil by considering stress-strain behaviour of strata can be studied, and theoretical pressuremeter curve for incompressible stratum can be attempted.

The basic principle behind the pressuremeter test is the expansion of cylindrical cavity formed in subsoil stratum in order to measure the relationship between pressure and deformation for the strata. The deformation by applying known magnitude of pressure of infinitely long cylinder will take place in a pseudo-elastic stage and the analysis to arrive to obtain two fundamental parameters, the deformation modulus E_m and limit pressure pl.

A probe of standard size AX or BX

with 44 and 55 mm diameters respectively are lowered in predrilled borehole at required depth. The probe is made up of measuring cell pressurised by water and guard cell pressurised by carbon dioxide gas. The membranes of probe is exerting pressure on the wall of the borehole, thus creating more volume. Which is filled up by extra injection of fluid. Thus increase in volume for known magnitude of pressure is recorded. The pressure is increased in a proper segment and volume increased upto the failure, i.e. when the volume of the cavity double the initial, is recorded. The corrections for pizometric head elasticity of membranes and temperature applied to the observations. Based on pressure volume relationship, a pressuremeter curve is plotted, and deformation modulus and limit pressure is computed, that describe the strength characteristics of strata.

E_m, the deformation modulus can be computed as

$$E_m = (1 + \mu) \, \Gamma \, \frac{/\backslash P}{/\backslash \Gamma}$$

where μ = Poisson's ratio

Γ = radius of cavity at instance when increase of pressure occurs.

The limit pressure is the maximum pressure applied to wall of borehole when failure occur.

2.1 Interpretations of pressure-meter test results

Basically pressuremeter test describe the strength characteristic of subsoil of softrock. In geotechnical engineering practice the fundamental strength parameters are useful for design of foundation, method of excavation, and for predicting behaviour of foundation element.

For mining engineer the parameters are useful for solving problems of excavation, stability of cuts and slops, dewatering of mines, and selection and sequence of for tool and method, for excavation of bad of strata. Geologist can also predict the type of strata based on range of parameters obtained by pressuremeter tests. Discontinuity in strata can also be detected by observing continues record of the test. Folds and faults, change of geological strata, profiling, inclination of bads, soft pockets - all such phenomena can be observed by the tests, provided interpreter must have proper knowhow and he must understand the behaviour based on the past experience and research, the normal strata type are classified on the basis of range of limit pressure or on E/pl ratio.

3. PRESSUREMETER METHODS FOR GEOLOGICAL AND GEOTECHNICAL MAPS

Planners and developers when choosing locations of major irrigation projects, new towns or even community place require information on what is called the urban geology of an area. In reality most of urban geology is geotechnical information stability of the strata drainage condition characteristics of rockbads faults and folds, location of unstable slopes etc. all available in the form of maps. Such map should show location of unstable slopes, loose pockets, soil or rock that is difficult to exavate and major geological faults. Besides these the hardness, denseness and stiffness of the substrata is very much important as this will extremely essential for mining engineers to decide the method of excavation and selection of tool to exavate the same.

A simple way to represents pressuremeter information on a geotechnical or geological map is shown in fig.1. It is overlayed on the base map of the area which should be of convenient scale and can be geological map, topographical map or a city plan. If an overlay is used as transparent it is essential that some means be provided to reference

the overlay exactly to the base map so that one cannot turned around or displaced with respect to the other. The recommended form of the information should take is the cadran as shown in fig. 1. Supplemented by other markings.

General information like geological faults, unstable areas caverns or cavities, areas subjected to flooding are indicated separately on the map. The cadrans are positioned on the overlay so that their centresmark the location on the map that each cadran refers to, in other word where the pressuremeter test is carried out the cadrans are divided into eight equal segments each segment represents a certain thickness of stratum. The ground surface is at topmost position on the cadran, and the depth increases in clockwise direction. The thickness of the strata which is represented by each segment of the cadran is given by a number which is located just to the left of the top of the outside circle that is at 11 o'clock position. The small arrow which is located outside the cadran and which points towards the centre marks the highest known position of the water table. The outside annulus is reserved for symbols which characterise the type of strata : cross hatching for soft adhesive soils, dots for granular soils; completely shaded area for compact beds or rocks, material that would be hard to exavate. The cross marks the depth at which boring was stopped. This additional information is speculative but is based on geological indications at the site.

The centre of the cadran is reserved for limit pressure values. Limit pressure is given for each thickness of layer of strata which is represented by a segment of a circle.

There may not be room on the map to draw a cadran for each boring which is put down in this case a cadran might have to represent the average strata condition at a site. In fact most often a cadran will be synthesis of a number of boring and pressuremeter profiles.

389

4. CONCLUSION

The cadran map may not go for some geologist or geotechnical engineers in the sense that it does not recommend the perfect decision. This may be taken as additive to the decision making but not totally independent. The interpretation should be made by the expert personnels in the field of geology and geotechnical engineering and who is familiar with the area.

5. REFERENCES

Baguelin, F. and others. The Pressuremeter and Foundation Engineering - Trans Tech Publication 1978.

Gambin, M. - Application of Menard Pressuremeter Test Results to Foundation Design, Oxford IBH Publishing Co. New Delhi.

Ladanyi, B. - Insitu Determination of Undrained Stress Behaviour of Sensitive Clays With Pressuremeter, Canadian Geotechical Journal Vol.9.

Morrisson, N. A. and Einstein, Z. Prediction of Foundation Department in Edmoton Using an Insitu Pressure Probe, Vol.10.

Bowles Hoseph, E. - Foundation Analysis and Design, McGraw Hill Kogakusha Ltd. Publication.

Mehta B. J. - Subsoil Investigation by Pressuremeter Test, Project Report by BE IV(C) submitted to Applied Mech. Deptt. S.V.R. College, Surat-7.

Mehta B. J. - Evaluation of Subsoil Properties by Pressuremeter Test : XII ISSMFE Rio-de-Janero-Brazil 13-18 Aug. 1989.

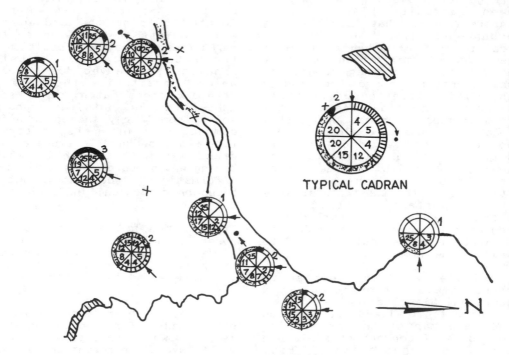

Fig.1 OVERLAY WITH SOIL CONDITION CADRANS

In situ measurement of hydraulic conductivity by borehole infiltration tests

Détermination de la conductivité hydraulique par essais ponctuels en forages

A.J. Roque & A. Gomes Coelho
Laboratório Nacional de Engenharia Civil, Site Investigation Division, Lisbon, Portugal

ABSTRACT: This paper presents the theories of punctual tests carried out in boreholes and discusses the results of tests performed in medium to low permeability soils ($10^{-3} < k < 10^{-7}$ m/s), in saturated and unsaturated zones. The comparative study of the results of constant-head tests carried out in the saturated zone, interpreted on the basis of the Hvorslev's theory (steady state flow) and of variable-head tests, interpreted on the basis of the Hvorslev's theory and Rat's theory (unsteady state flow), have shown that for the type of tested soils the most accurate interpretations for the variable-head tests is based on the unsteady state theory. The interpretation of the tests carried out above the water table was based on analytical or electrical analogue solutions of Laplace's equation, which disregard the fact of a part of the total flow occurring in soil under unsaturated conditions. This interpretation is justified on a theoretical and experimental basis, which consider that the hydraulic conductivity obtained from approximate solutions is not affected by important errors when the capillary effects are reduced.

RESUMÉ: Cet article présente les théories des essais ponctuels effectués dans des forages et discute les résultats concernant des sols de moyenne à faible perméabilité ($10^{-3} < k < 10^{-7}$ m/s), dans les zones saturée et non saturée. L'étude comparative des résultats des essais effectués dans la zone saturée à niveau constant, interprétés à partir de la théorie de Hvorslev (régime permanent) et des essais à niveau variable, interprétés à partir de la théorie de Hvorslev et de Rat (régime transitoire), montre que pour le cas des sols du type de ceux essaiés l'interprétation la plus correcte pour les essais à niveau variable est celle qui a été basée sur la théorie du régime transitoire. L'interprétation des essais effectués au-dessus de la surface phréatique a été basée sur des solutions analitiques ou d'analogies electriques de l'équation de Laplace, qui méprisent le fait d'une part de l'écoulement total arriver dans le sol dans des conditions non saturées. La justification de cette interprétation est basée sur les études théoriques et expérimentales que considérent qui la conductivité hydraulique déterminée à partir de solutions approchées n'est pas affectée d'un grand erreur dans les cas oú les effets capillaires sont réduits.

1 INTRODUCTION

The assessment of the hydraulic conductivity of soils is necessary for the resolution of numerous geotechnical problems, such as seepage through the body or the foundation of dams, design of pumping systems in excavations below the water level and characterization of geological barriers in sites for urban and industrial waste fills.

The hydraulic conductivity of soils may be determined in laboratory or in situ. Both procedures present pros and cons. In site investigation programmes, the selection and correct application of the several methods of testing presuppose a geological knowledge as accurate as possible of the area to be studied. The number, location, type and size of tests to be made within the scope of a site investigation programme must be defined in order to allow to ascertain the representativeness and the significance of the tests, involving different volumes of soil. In situ tests are generally preferable since the volume tested is higher and the disturbance of the soil is lower, in comparison with the samples tested in laboratory.

In the case of gravels, clayey silts, clean, silty or fine sands, the determination of the hydraulic conductivity by the punctual tests carried out in boreholes or piezometers is particularly indicated due to its simplicity, reliability and low cost. Since these methods can be repeated many times, they allow to characterize the distribution of the values of the hydraulic conductivity. In the case of soils of medium to low permeability, the punctual tests are sufficiently illustrative and do not present either the problems of gathering undisturbed samples used for laboratory tests, or the difficulty (and sometimes even the impossibility) of carrying out pumping tests, as a result of the insufficient flow into the borehole that does not allow a continuous pumping and a constant flow rate.

The application of the punctual tests to the assessment of the hydraulic conductivity is widely known, even though its use is often considered as a mere indicator of permeability. This general opinion is largely the result of difficulties during its execution and interpretation.

The present study is based on a research programme (Roque, 1990) carried out at the LNEC's Site Investigation Division, with the support of the Sondagens e Fundações, A. Cavaco company. The main object of the experimental work was to contribute to improvement of the design, of the execution techniques and of the theories of interpretation of punctual tests in soils with permeabilities between 10^{-3} and 10^{-7} m/s. The tests included soils composed of fine to medium sands, in the saturated zone, and medium to coarse sands, in the unsaturated zone. This paper presents and validates the theories of interpretation more adequate for the determination of the hydraulic conductivity of soils of medium to low permeability.

2 TESTS IN THE SATURATED ZONE

2.1 Theory of the variable-head tests

In the punctual variable-head tests a certain amount of water in the borehole is instantaneously extracted or added and the variation of the water level is measured until equilibrium is reached.

In the bibliography various theories of interpretation of variable-head tests are known. The most widely known theories and those most applied by the geotechnical community were developed by Hvorslev (1951) and by Rat (1968).

The "classical" theory developed by Hvorslev assumes that a steady flow is established during the execution of the test. The interpretation of the test in these conditions is made from the following general equation, similar to Darcy's law:

$$Q = khF \tag{1}$$

where Q is the flow rate, k is the hydraulic conductivity, h is the applied hydraulic potential difference and F is the shape factor that depends on the geometry of the test cavity. The units to be used for these parameters are: $Q(L^3/T)$, $k(L/T)$, $h(L)$ and $F(L)$. The general equation (1) can be used in this form in the case of a constant-head test, where the flow rate is directly measured. In the case of variable-head tests carried out in a pipe with a cross-section $A(L^2)$, the equation (1) is applied to each instant of time $t(T)$. In the instant t the flow rate Q is equal to:

$$Q(t) = A\frac{dh}{dt} \tag{2}$$

By replacing the equation (2) for the assessment of the flow rate in the equation (1) and by integrating the resulting differential equation there results:

$$\ln(\frac{h_0}{h_t}) = -k\frac{A}{F}(t_0-t) \tag{3}$$

where h_0 and h_t are the applied hydraulic potential differences in the instants t_0 and t, respectively. A second possibility is the one resulting from consideration and replacement of the variation of the flow rate between the instants t_i and t_{i+1}:

$$Q_i = A\frac{h_{i+1}-h_i}{t_{i+1}-t_i} \tag{4.a}$$

and the mean applied hydraulic potential difference:

$$h_i = \frac{h_{i+1}+h_i}{2} \tag{4.b}$$

in the equation (1).

The theory developed by Rat (1968) considers that in soils with a medium to low permeability it is hardly probable that a steady flow will be established in a variable-head test and its interpretation must therefore be made on the basis of the unsteady flow theory. In research works carried out in soils with a medium to low permeability Rat (1968, 1971 and 1974) verified that (i) in the constant-head tests the flow rate, Q, versus an applied hydraulic potential difference, h, obtained in the end of each step defined a linear relation, with the pairs of values (Q,h) aligned with the origin, which is in accordance with the equation (1); (ii) in the variable-head tests the record in a semi-logarithmic scale of t versus h (in a logarithmic scale), which was supposed to be defined in a linear relation, did not present an alignment of every test point. As the result of the tests, Rat considered that interpretation of the variable-head test by the Hvorslev theory would not be the most correct in the case of medium to low permeability soils and that the test should be interpreted by assuming that the regimen established during the test is of the unsteady type.

From the basic equation of hydrogeology:

$$\Delta h = \frac{\partial^2 h}{\partial r^2}+\frac{1}{r}\frac{\partial h}{\partial r}=\frac{S}{T}\frac{\partial h}{\partial t} \tag{5}$$

and by considering that the conditions in the limits of the equation (5) are (i) the applied hydraulic potential difference is 0 in the instant 0; (ii) the overload inside the cavity is h_0 in the instant 0; and (iii) the reduction of the head $(\partial h/\partial t)$ in the instant t is proportional to the flow rate that is percolated, Rat (1968) derived the following equation that allows to interpret the punctual variable-head tests with basis on the theory of the unsteady flow:

$$p = 0.25kFV \tag{6}$$

In the equations (5) and (6) $T(L^2/T)$ is the transmissibility, S(non-dimensional) is the porosity of the saturated aquifer material if the water table is equal to the atmospheric pressure or the storage coefficient if the piezometric surface is at a pressure higher than the atmospheric pressure, p is the declivity of the straight line tangent at the point of inflection of the curve defined by the pairs of values (h,1/t) and V (L^3) is the volume added (extracted) in the borehole.

The shape factor, F, of the test cavity is a physical magnitude that controls the flow model in the soil in the vicinity of the injection zone. In the beginning of the century the equations for the calculation of the shape factor were presented by several authors as mathematical solutions for Laplace's equation. The equations were later verified by other authors resorting for the purpose to the electrical analogs, to analytical methods and to numerical analysis methods. The F values calculated by the numerical analysis methods and by Hvorslev's formula differ by about 40% when elongated cylindrical cavities are used and by more than 100% in the cases in which L/D is small, D(L) being the diameter of the cavity. These different values of the shape factor lead to uncertainty in calculation of k that may exceed 100%. Under these circumstances the selection of the expression for calculation of the shape factor must be made on the basis of accurate criteria in order that the value of F used in the expressions (1), (3) and (6) will not exert a "negative" influence on the value of k.

The validity of the expressions presented depends on verification, with a sufficient approximation, of the following hypotheses and conditions: (a) the permeability does not depend on the hydraulic gradient; (b) the water and the soil are incompressible; (c) the soil is homogeneous and its hydraulic conductivity is constant (in space and in time) and isotropic; (d) there are no impermeable or recharge boundaries near the test; (e) there are no water losses in the junctions of the test pipes; (f) the geometry of the test zone is known and remains constant during the test; (g) the soil in the test zone is not disturbed by execution of the cavity or of the test; (h) the hydraulic head losses in the test cavity are negligible; (i) the height and radius of the cavity are sufficiently small for it to be assumed that the applied hydraulic potential difference is constant at any point of the cavity; (j) the addition (extraction) of water in the borehole causes the instantaneous variation of the water level in the borehole and the position of the water table does not suffer any alteration (item valid for the steady state); (k) the addition (extraction) of a volume of water in the borehole causes an instantaneous variation of the water level in the borehole which is propagated to the aquifer and which decreases with time, up to a sufficiently long moment where it is assumed that the disturbance has passed (item valid for the unsteady state).

2.2 Experimental programme

The variable-head tests were carried out at Foros da Lameira (Salvaterra de Magos), which is located on the left bank of the river Tagus about 60 km northeast of Lisboa. As regards the regional geology the area of study belongs to the Tertiary basin of the river Tagus. At the site, there are a well developed cover of Holocene sandy superficial deposits overlying Plio-pleistocene terraces. The thickness of the Holocene cover is variable from point to point, reference having been made to a thickness of 11 m in a borehole drilled in the zone.

At the site of the study two boreholes with 0.318 m diameter were made, about 100 m apart, with depths of 2.99 m (SA1) and 3.06 m (SA2). At the time of the execution of the tests the water table was at a depth of 1.6 m. The soil sampled in the two boreholes was a fine to medium sand after 3 m and a clay with some gravel and sand from 3 m deep. In table 1 the physical characteristics of the soil are shown.

In each borehole a tube with the following characteristics was introduced: interior diameter $\phi_i = 0.158$ m; length of the screen zone L=0.770 m (SA1) and 0.730 m (SA2). The position of the mean point of the screen zone in relation to the surface of the soil was 2.36 m in the SA1 and 2.44 m in the SA2. The screen zone con-

TABLE 1 - Physical characteristics of the soil

HOLE Ref.	SAMPLE Depth (m)	d_{10} (mm)	d_{50} (mm)	C_u	γ_t (kN/m³)	γ_d (kN/m³)	n (%)	e (%)
SA1	1.96-2.56	0.077	0.22	4.2	21.3	18.5	29	41
SA2	2.01-2.61	0.074	0.25	4.1	---	---	---	---

sisted of an helicoidal uninterrupted opening, made by Johnson, with an open area of about 33%.

Before the introduction of the tube in the borehole the screen zone was wrapped with Polyfelt geotextile TS500, since the openings were larger than a fraction of the grain size of the filter. The material of the filter was a sandy gravel and was used to fill the ring space, in the screen zone, between the tube and the soil involving the borehole. The sealing of the test cavity was accomplished by adding bentonite spheres between the tube and the soil, consolidated in layers with about 0.20 m.

In the variable-head tests the water was instantaneously added in the borehole and the variation of the level was measured by means of an electric probe as quickly as possible until equilibrium was reached. The variation of the water level was recorded with an electric probe as follows: every 15 seconds in the first minute, minute and a half, two minutes, every minute up to 5, 7 and 10 minutes. After that, the readings were made in intervals of 5 minutes until the end of the test. In SA1 and SA2 the volumes of water added were 25, 35 and 50 l, each of these volumes having been added at least three times.

2.3 Results and discussion

Figures 1 and 2 show the curves of the test carried out in the borehole SA2, corresponding to the addition of 50 l of water. The record of the variation of the water level in the borehole versus time agree with the interpretation of the punctual test according to Hvorslev's theory. In figure 2 the record of the variation of the water level in the borehole versus the inverse of time was dealt with according to Rat's theory.

In tables 2 and 3 the results of every tests carried out in SA1 and SA2 are presented, and they are interpreted by the Hvorslev and Rat theories. In order to interpret the results, an interactive computer programme was developed, which permits to pause, to display the distribution of the in situ data and to select the values (t,h) to be used in the linear regression. The programme also allows to select the appropriate equation for the calculation of the hydraulic conductivity, which may consider (or not) the influence of the upper boundary (water table) and/or the lower boundary (impermeable level or contrasting permeability level).

The interpretation of the results from each theory allows the following conclusions:

(a) The values t versus h in figure 1 do not fit a linear relation during all the test, the points corresponding to the final phase being deviated in relation to the alignment defined by the initial values, contrary to what was supposed to happen according to the Hvorslev's theory. The path followed by the experimental curves shows that the deviation of the points (t,h) occurs when the difference between the water level in the borehole and the water table becomes very small.

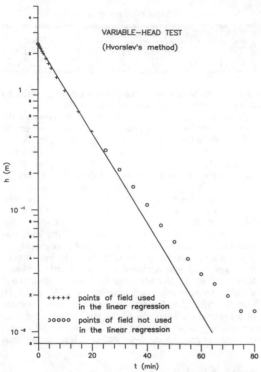

Fig. 1 - Variable-head test carried out in SA2 with addition of 50 l of water. Interpretation according to Hvorslev's theory

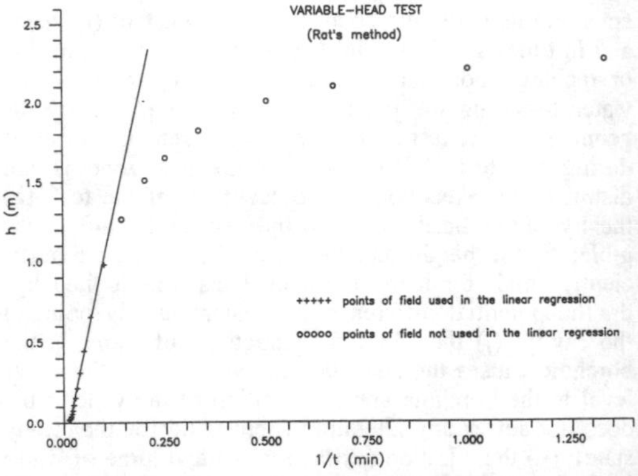

Fig. 2 - Variable-head test carried out in SA2 with addition of 50 l of water. Interpretation according to Rat's theory

TABLE 2 - Summary of variable-head test results carried out in SA1

HOLE Ref.	TEST Ref.	VOLUME $(x10^{-3} m^3)$	HYDRAULIC CONDUCT. BY HVORSLEV'S METHOD $(x10^{-4} m/s)$			HYDRAULIC CONDUCT. BY RAT'S METHOD $(x10^{-4} m/s)$		
			with infl. of the bound.	without infl. of the bound.	reg. coef.	with infl. of the bound.	without infl. of the bound.	reg. coef.
SA1	1	25.0	1.26	1.12	0.998	0.23	0.20	0.999
	2	25.0	1.23	1.11	0.999	0.21	0.18	0.999
	3	25.0	1.18	1.04	0.998	0.21	0.19	0.998
	4	35.0	1.22	1.11	0.999	0.22	0.19	0.999
	5	35.0	1.11	0.97	0.999	0.19	0.17	0.997
	6	35.0	1.06	0.93	0.999	0.19	0.17	0.997
	7	50.0	1.21	1.11	0.999	0.21	0.18	0.999
	8	50.0	1.27	1.12	0.999	0.22	0.19	0.998
	9	50.0	1.24	1.11	0.999	0.21	0.18	0.998

TABLE 3 - Summary of variable-head test results carried out in SA2

HOLE Ref.	TEST Ref.	VOLUME $(x10^{-3} m^3)$	HYDRAULIC CONDUCT. BY HVORSLEV'S METHOD $(x10^{-4} m/s)$			HYDRAULIC CONDUCT. BY RAT'S METHOD $(x10^{-4} m/s)$		
			with infl. of the bound.	without infl. of the bound.	reg. coef.	with infl. of the bound.	without infl. of the bound.	reg. coef.
SA2	1	25.0	0.52	0.46	0.998	0.08	0.07	0.998
	2	25.0	0.54	0.48	1.000	0.07	0.06	1.000
	3	25.0	0.53	0.47	1.000	0.08	0.07	0.996
	4	35.0	0.62	0.55	1.000	0.10	0.09	1.000
	5	35.0	0.85	0.76	0.999	0.15	0.13	0.998
	6	35.0	0.83	0.74	0.997	0.11	0.10	0.998
	7	35.0	0.54	0.48	1.000	0.08	0.07	0.999
	8	50.0	0.59	0.53	0.999	0.10	0.09	0.998
	9	50.0	0.54	0.48	0.999	0.10	0.09	0.998
	10	50.0	0.56	0.49	0.999	0.09	0.08	0.997

394

(b) The same pairs of values (t,h) were plotted in accordance with Rat's theory (fig. 2). These experimental curves show that the initial readings are disregarded up to 4 minutes in SA1 and up to 7 minutes in SA2. Custódio (1976) considers that this initial deviation occurs whenever the height of the water in the borehole is important in relation to the thickness of the aquifer. The experimental curve in figure 2, as well as the other curves obtained in the same programme test, are in accordance with the theoretical curves obtained by Rat. Just as the theoretical curves admit a tangent at the inflection point near the origin, or a tangent at the origin, if L is very high, also the experimental curve shown in figure 2, and the others, admit tangents at the origin, as a result of the length of the test cavity being more or less equal to the thickness of the aquifer.

(c) The hydraulic conductivity of the soil in SA1 and SA2 determined by Rat's theory is about 80% smaller than that determined by the Hvroslev's theory.

(d) Concerning the geometric conditions in which the tests were carried out, the influence of the boundaries on the hydraulic conductivity is not particularly relevant. The value of k is about 15% smaller when the influence of the boundaries is disregarded.

(e) The results of the SA1 and SA2 tests with different volumes (25, 35 and 50l) have shown that the volume of water added had no influence on the value of k and that the punctual tests are reproducible.

2.4 A comparative analysis with the results of the constant-head tests and the studies found in the bibliography

The experimental work also included some constant-head tests. In SA1 and SA2, 4 and 6 tests were carried out with different and increasing flow rates, respectively. Figure 3 shows a plot of Q versus h for the tests carried out in SA2. Table 4 presents the results of the tests carried out in SA1 and SA2. These experimental data define a linear relation passing at the origin. The graph also shows that during execution of the tests the flow in the soil was laminar, that there were no siltation phenomena of the cavity and that a steady flow has been established.

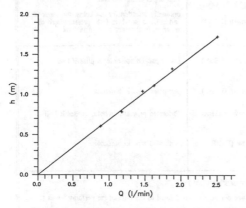

Fig. 3 - Constant-head tests carried out in SA2

TABLE 4 - Summary of constant-head test results carried out in SA1 and SA2

HOLE Ref.	NUMB. OF TESTS	HYDRAULIC CONDUCT. (x10⁻⁴ m/s)		REG. COEF.
		with infl. of the bound.	without infl. of the bound.	
SA1	4	0.32	0.28	0.998
SA2	6	0.13	0.12	1.000

The interpretation of the punctual constant-head tests carried out in this study, as well as in other studies consulted in the bibliography, did not present any special difficulties. Therefore their interpretation based on the theory developed by Hvroslev has not been refuted.

Nowadays, it seems to be of general acceptance that the permeability of a soil may be more accurately determined by the punctual constant-head tests. On the basis of this assumption, the main purpose of these tests in our study was to conduct a test which would allow to calibrate the results of the variable-head tests, interpreted by Hvroslev's and Rat's theories. Values of k from our constant-head tests are approximately 80% smaller than those from the variable-head tests interpreted by Hvorslev's theory; and they are approximately 50% and 35% higher than those from the variable-head tests, interpreted by Rat's theory, in SA1 and SA2, respectively. According to the experience of other researchers (Tremblay and Erikson, 1987) the value of k determined from the constant-head tests is bigger than when determined from a variable-head test. It can be therefore deduced that interpretation by Hvroslev's theory leads to values of k by excess. According to Rat (1968) the interpretation of the variable-head tests with basis on Hvroslev's theory is feasible only in highly permeable soils (k > 10⁻³ m/s), in which the steady flow may be reached during the test.

3 TESTS IN THE UNSATURATED ZONE

3.1 Theory of tests

The saturated hydraulic conductivity of a soil in the unsaturated zone may be determined by a set of various methods. The classical infiltrometers with a double-ring or simple-ring, sealed or not, the pressure infiltrometer, the Guelph's permeameter and the infiltrations from boreholes, are examples of some of the methods used in the determination of the hydraulic conductivity, k_s. This notation was adopted for the hydraulic conductivity in order to differentiate the permeability measured in saturated zones and in unsaturated zones. Of the various methods, only the tests carried out in boreholes allow to determine k_s for depths greater than some meters.

The first equations for the calculation of k_s, from infiltration tests carried out in boreholes, were nearly all developed about 40 years ago, namely by the U.S. Bureau of Reclamation, in 1951, Nasberg, in 1951 and Terletskata, in 1954. More recently, Stephens and Neuman (1982 a, b and c) contributed in a more significant way to improvement of the methodology for the interpretation of tests.

The equations developed about 4 decades ago are based on analytical solutions and on electrical analogies of Laplace's equation. Some of these equations assume that the flow in the soil is limited to a zone involved by a streamline, along which the pressure is equal to the atmospheric pressure. In these conditions, it was assumed that the soil inside the confined zone was completely saturated and completely dry on the outside. The set of equations developed in the USBR (1974) allow to determine k_s from data of constant-head tests, both on cased boreholes open only at the bottom (open-end casing tests) and on uncased or partly cased boreholes (open-borehole tests). The equations valid for the various geometries of the boreholes are presented in table 5.

More recently, in the beginning of the 80's, Stephens and Neuman (1982 b and c) studied the accuracy of some of the equations developed in the beginning of the 50's, which assumed that the existence of a free surface was the correct representation of the flow conditions of the soil. To evaluate the accuracy of these approximate equations, the authors used a numerical model of finite elements developed by Neuman and Witherspoon (in Stephens and Neuman, 1982 b). This model allows to simulate the flow conditions by assuming as valid the free surface concept and ensures that all the boundary conditions are fulfilled for the axisymmetric simulations. The results of the numerical simulations and of the analytical solutions showed that there were no significant differences between the two calculation methods. It can be therefore affirmed that in these conditions the analytical solutions of Laplace's equation are approximately correct and lead to reliable results.

Stephens and Neuman studied the accuracy of these same solutions of Laplace's equation, by comparing them with the results obtained from two numerical models which allow to study the flow in the saturated and unsaturated zones. They intended thus to evaluate

the effects of capillarity in the constant-head tests carried out in boreholes. The results obtained by Stephens and Neuman (1982 b and c) showed that the greatest discrepancies between the models that consider the flow in the saturated and unsaturated zones and the classical theory that considers the existence of a free surface occur mainly in the fine soils. In these soils a very high percentage of the total flow occurs in the unsaturated zone of the soil. According to the authors the constant-head tests carried out above the water table cannot be interpreted from solutions of Laplace's equation or from numerical models which will consider the free surface principle. In the coarse soils the flow that occurs in unsaturated conditions is smaller, because the capillary effects are less significant. The results obtained for different types of soils, with the above mentioned calculation methods showed that in the case of coarse sand the differences are not significant and are approximately of the same magnitude for values of h/r up to approximately 50, r being the radius of the test cavity.

On the basis of the results obtained from a set of numerical simulations, Stephen and Neuman (1982 b) redefined the Glover's equation that started to establish a relation between c_u and (h/r), h and the capillary absorption α_1, through the equation:

$$\log c_u = 0.658 \log\left(\frac{h}{r}\right) - 0.238\sqrt{\alpha_1} - 0.398\log h + 1.343 \quad (16)$$

where h and α_1 are m and m^{-1} respectively, being α_1 represented by:

$$\alpha_1 = \frac{\ln 1 - \ln 0.5}{0 - \psi(para\ k_s=0.5)} = -\frac{0.693}{\psi(para\ k_s=0.5)} \quad \psi \le 0$$

TABLE 5 - Summary of equations for calculating saturated hydraulic conductivity

TYPE OF TEST		EQUATION		AUTHOR	COMMENTS
Open-end casing		$k_s = Q_s/5.5rh$	(7)	unknown (USBR)	variable borehole geometry and water table depth cannot be accomodated
Open-hole	Deep water table	$k_s = Q_s/c_s rh$	(8)	Glover (USBR)	applies if L=h;assumes linear increase in flux with depth along borehole
		$k_s = \frac{0.423Q_s}{h^2} \log\left(\frac{2h}{r}\right)$	(9)	Nasberg-Terletskata	applies if L=h;assumes uniform flux along borehole
		$k_s = \frac{Q_s[\sinh^{-1}(L/r) - (L/h)]}{[2\pi(2Lh - L^2)]}$	(10)	Zanger (USBR)	applies if L<h;assumes linear increase in flux with depth along borehole
		$k_s = Q_s/c_s rh$	(11)	Cornwell (USBR)	oribinally intended for use below the water table and in cases in which L<h; pressure head of water, h, is at center of open interval
	Shallow water table	$k_s = \frac{2Q_s}{[(c_s+4)rT_u]}$	(12)	unknown (USBR)	gravel packed borehole; applies if L=h
		$k_s = \frac{2Q_s}{[c_s+4(r/r_e)]r_e(T_u+h-L)}$	(13)	unknown (USBR)	perforated pipe in borehole
		$k_s = \frac{2Q_s}{[c_s+4(r/r_e)]r_e T_u}$	(14)	unknown (USBR)	perforated pipe in borehole; applies if L=h
		$k_s = \frac{2Q_s}{c_s r_e(T_u+h-L)}$	(15)	Zanger (USBR)	perforated pipe in borehole
$c_s = [2\pi(L/r)]/[\ln(L/r)]; \quad c_u = [2\pi(h/r)]/[\sinh^{-1}(h/r)-1]$					
T_u-vertical distance from the water level in the borehole to the water table; r_e-effective borehole radius defined as the ratio of the perforated area to the length of perforated casing					

396

Due to the complexity of the conditions that preside over the execution of the tests some assumptions are generally assumed: (a) the soil is homogeneous, isotropic and incompressible; (b) air is not trapped in the saturated volume during infiltration; and (c) the water content in soil before the test begins it is constant in a volume sufficiently large in the vicinity of the test cavity.

3.2 Experimental programme

The constant-head tests were carried out at Corroios (Seixal), which is located 20 km south of Lisboa. From the point of view of the regional geology the site is located in the Tertiary basin of the river Tagus. At the site, the outcropping formations belong to the Holocene and are constituted of sand dunes and colluvial sands from the underlying Plio-pleistocene series. The thickness of the Holocene superficial deposits is highly variable from point to point, the most frequent values being about 2 to 3 m.

At the site of the study three boreholes with 0.380 m diameter were made, 65 m apart, with depths of 4.50 m (CO1), 4.59 m (CO2) and 4.70 m (CO3). At the time of execution of the tests the water table was at the depth of about 5.30 m. The soil in the three boreholes was a medium to coarse sand. In table 6 the physical characteristics of the soil are shown.

TABLE 6 - Physical characteristics of the soil

HOLE Ref.	SAMPLE Depth (m)	d_{10} (mm)	d_{50} (mm)	C_u	w (%)	γ_t (kN/m³)	γ_d (kN/m³)	n (%)	e (%)
CO1	3.60–4.10	0.13	0.46	4.2	---	---	---	---	---
CO2	3.60–4.20	0.16	0.46	3.4	8.6	19.7	18.2	30	42
CO3	3.60–4.20	0.09	0.39	4.9	7.3	18.7	17.4	34	50

Some suction tests were carried out in order to estimate the water retention capacity of the soil where the punctual tests were going to take place. The results have shown that the soil has a high retention capacity for a small interval of suctions, and a low water content for high suctions. The greatest variation in the water content occurs for values of pF (suction) of about 1.3. This expressed in terms of height of the water column is equivalent to about 20 cm. The greatest water retention capacity obtained for samples from CO1 borehole is due in principle to the presence of metahalloysite, which is absent in CO2 and CO3 according to the results of the mineralogical analysis.

In each borehole a pipe was introduced with the following characteristics: interior diameter $\phi_i=0.158$ m; length of the screen zone L=0.770 m in CO1 and CO3 and 0.730 m in CO2. The position of the intermediate point of the screen zone in relation to the surface of the soil was 3.50 m in CO1, 3.59 m in CO2 and 3.68 m in CO3. The materials and the construction method of the testing cavities of CO1, CO2 and CO3 were identical to those described in section 2.2.

At the time of the tests there was no possibility of using all the required equipment, namely, the neutron probe and the tensiometers. The use of this equipment would have made it possible to evaluate the hydraulic characteristics of the soils in unsaturated conditions. Ignorance of these characteristics led to the adoption of a methodology that would allow to reduce the impor-

tance of the capillarity effects on the value of the flow rate infiltrated from the boreholes. Besides the tests having been made in a coarse soil (medium to coarse sand), the tests were designed according to the Stephens and Neuman's studies (1982 b, c) in order to attenuate the importance of the flow in the unsaturated zone: test cavity in the vicinity of the water table, large diameter of the test cavity, hydraulic head between about 1 and 2 m and a ratio between the hydraulic head and the radius of the cavity of about 10.

Before starting the constant-head test, the soil was soaked in the vicinity of the test cavity. The water was injected into the borehole for 10 minutes, with a constant hydraulic head of 1m. According to the experience of USBR (1974) this leads to a quicker stabilization of the water level in the borehole for a certain rate of flow injected. In the tests the variation of the water level was measured by means of an electric probe at intervals of 5 minutes. The test ended when the variation of the water level in the borehole was equal or less than 0.01 m in three consecutive readings.

During the punctual tests it was possible to follow the evolution of the saturated zone in soil. The volume of the saturated soil was only recorded in CO2, by means of the apparatus for the piezometers shown in figure 4.

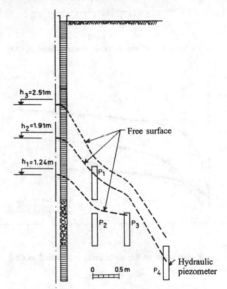

Fig. 4 - Volume of the saturated soil recorded in CO2

For the use of this type of equipment some drawbacks were taken into due account, namely: (a) weak time of response - its value for the test soil, calculated from the expressions deduced by Hvorslev, is about 36 seconds, a value considered satisfactory for the aims in mind; (b) seepages along the pipe - so as to eliminate as much as possible this potential source of error, the borehole above the tip of the piezometer was filled with impermeable soil; and (c) bubbles of air near the tip - to prevent them from happening the previous soaking of the soil was made in the vicinity of the tip by adding water until the piezometric pipe was saturated.

3.3 Results and discussion

Figures 5, 6 and 7 show the variation of the water level in the boreholes versus the time for each of the three flow rates injected. Table 7 gives the results of every test carried out in CO1, CO2 and CO3. Equations (13) and (15) were used, as they can be applied in the geometric conditions of tests, i.e.: water table at low depth in relation to the test cavity, a partly cased borehole and L<h. The same table shows the discrepancies, which are analyzed below, between the values of k_s obtained by the equations (13) and (15) and those from the equation (10) which can only be applied when the water table is at great depth.

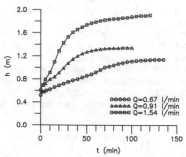

Fig. 5 - Variation of the water level versus time in CO1

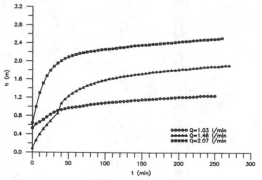

Fig. 6 - Variation of the water level versus time in CO2

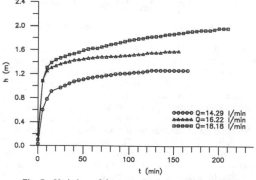

Fig. 7 - Variation of the water level versus time in CO3

The results from the boreholes and hydraulic piezometers allow to draw the following considerations:

(a) Two straight lines may be adjusted to each curve (figures 5, 6 and 7): an initial linear portion corresponding to the quick rise of the water level in the borehole, and a second linear portion for which the water level in the borehole is almost constant along the time. In CO1 the stabilization of the level occurs after about 95, 47 and 35 minutes, for Q_s=0.67, 0.91 and 1.54 l/min, respectively; in CO2 after about 41, 28 and 59 minutes, for Q_s=1.03, 1.46 and 2.07 l/min, respectively; and in CO3 from about 23, 7 and 7 minutes, for Q_s=14.29, 16.22 and 18.18 l/min, respectively.

(b) The curves obtained for each borehole show that the water level in the boreholes had a tendency to stabilize earlier in the tests carried out with the highest flow rates. This trend did not occur as regards the establishment of the quasisteady state. In fact, the quasisteady flow only occurred in the tests carried out in boreholes CO1 and CO3 for the lowest flow rates, i.e, Q_s=0.67 and 14.29 l/min, respectively. These results have shown that a smaller hydraulic head will correspond to a decrease in the quantity of water infiltrated, as well as a reduction of the time necessary to reach a certain percentage of the flow in steady state.

(c) The record of the variation in the hydraulic head along time in CO2 and in the hydraulic piezometers P1 to P4, for the three flow rates injected, have shown that the tendency for stabilization of the water level in the borehole and in the piezometers occurs with a delay of few minutes. The linear portions of the curves corresponding to the stabilization of the hydraulic head, in CO2 and in P1 to P4, are more or less parallel.

The values of the hydraulic conductivity presented in Table 7 allow the following conclusions:

(a) The hydraulic conductivity was calculated in conditions of a quasisteady flow only in CO1 for Q_s=0.67 l/min and in CO3 for Q_s=14.29 l/min. The values of the hydraulic head, recorded in the end of tests, h_f, and those from the ordinates of the straight lines of adjustment to the second portion of the curve, h_{adjust}, are equal. In CO1 for Q_s=0.91 and 1.54 l/min and in CO3 for Q_s=16.22 and 18.18 l/min, the quasisteady flow was not reached. For these flow rates $h_f>h_{adjust}$. In CO2 the quasisteady flow was not reached in any stage of Q_s, in these case being $h_f>h_{adjust}$.

(b) The determination of k_s from equations (13) and (15) have shown an error of about 10 to 35%, for the cases in which the quasisteady flow was not reached. It can be assumed by extrapolation that the errors that may affect the experimental values of k_s in CO2 and in CO1 and CO3, where the quasisteady flow was not reached, are the same order of magnitude.

(c) Comparing the values of k_s obtained in each borehole using equations (13) and (15) it can be verified that as regards Zanger's equation (equation 15), the equation (13) leads to values by defect of about 20%. The same tests interpreted by the equation (10), valid to deep water table, lead to values of k_s about 50 times lower than those obtained with equations (13) and (15). This shows the importance of the exact knowledge of the geometric conditions of testing, in order to apply the most adequate equations.

TABLE 7 - Summary of test results carried out in CO1, CO2 and CO3

HOLE Ref.	TEST Ref.	WATER VOLUME $(x10^{-3}\ m^3)$	TEST DURATION (min)	SATURATED HYDRAULIC CONDUCTIVITY $(x10^{-6}\ m/s)$					
				eq. (13)		eq. (15)		eq. (10)	
				h_f	h_{adjust}	h_f	h_{adjust}	h_f	h_{adjust}
CO1	1	90.5	135	4.30	4.30	5.43	5.43	0.10	0.10
	2	91.0	100	5.16	5.30	6.15	6.40	0.16	0.16
	3	184.8	120	6.52	7.17	·6.62	7.69	0.39	0.35
CO2	1	257.5	250	6.34	7.86	7.73	10.50	0.17	0.13
	2	394.2	270	6.28	7.87	6.32	8.99	0.38	0.29
	3	538.2	260	7.01	8.03	6.10	7.61	0.70	0.60
CO3	1	2357.9	165	89.90	89.90	110.00	110.0	2.43	2.43
	2	2514.1	155	85.50	96.90	95.70	115.0	3.37	2.93
	3	3817.8	210	79.50	96.90	80.30	109.0	4.66	3.73

(d) In CO1 and CO2 the permeability of the soil is of about the same magnitude, reflecting a certain lateral homogeneity of facies. However, the permeability of the soil in CO1 and CO2 is approximately 20 times lower than in CO3, corresponding to a local heterogeneity of the soil. The porosity and the void ratio bigger in CO3 than in CO2, as well as the very structure of the soil are factors that may explain the difference recorded in the value of k_s.

3.4 Comparative analysis with studies found in the bibliography

Stephens et al. (1987) carried out constant-head infiltration tests in boreholes in order to determine the hydraulic conductivity of the soil in the unsaturated zone. The profile of the tested soil consisted of a fine sand of fluvial origin with the following physical characteristics: d_{50}=0.26 mm, C_u=2.4, n=36 to 46% and w=7 to 10%. In this study the authors compare the values of k_s determined, both by equations which take into account the capillary effects and by solutions that assume the existence of a free surface. The results of the tests for each analytical solution showed that the values of k_s are less different than a factor of 2. According to the authors this fact results from the small influence of the capillary effects of the soil, a small percentage of the total flow occurring in the unsaturated zone.

In view of the results obtained by these authors and assuming as likely that the effects of the capillarity also have a reduced importance in the soils tested in Corroios (medium to coarse sands), there seems reasonable to assume that the errors from having calculated k_s without considering the flow in the unsaturated zone, are not relevant.

Apart from the tests made in a soil where the capillary effects are not very significant, the following factors were considered, which according to studies published in the bibliography reduce the influence of the effect of the capillarity of the soil on the flow rate infiltrated from the borehole:

(a) Location of the water table in relation to the test cavity: Stephens and Neuman (1982 b) showed that for the same soil the error that results from disregarding the flow in the unsaturated zone is much lower when the distance of the tests cavity to the water table is very small. According to the results obtained for a silty soil, the use of the Glover's equation (equation 8) overestimates the value of k_s in 60% while the equations (12) and (15, with h=L and r_e=r) overestimate the real value in only 34%.

(b) The influence of h and r: in order to reduce the influence of capillarity, the most favourable conditions of execution of tests in coarse soils occur when (i) the value of r is higher (in the study r=0.19 m); (ii) the hydraulic head is approximately equal to 1.0m (in the study h varied between 1.21 and 2.51m); (iii) the h/r ratio is about 10 (in the study h/r varied between 6 and 13).

4 FINAL CONSIDERATIONS

As regards the type of soils tested, characterized by medium to low permeabilities ($10^{-3} < k < 10^{-7}$ m/s), the following conclusions may be pointed out:

- Tests carried out in the saturated zone (Salvaterra de Magos)

(a) The interpretation of the variable-head tests, by considering Hvorslev's theory led to hydraulic conductivities significantly higher than those obtained from the constant-head tests. The values obtained using the Rat's theory for the interpretation of the variable-head tests were very different from those obtained by the Hvroslev's theory. Those values are closer to the values obtained by the constant-head tests and certainly the most representative of permeability of the soil.

(b) The results of this study are in accordance with the results obtained by Rat & al. (1971) and Rat & Laviron (1974) for soils with a medium to low permeability. These researchers also mention that there is a better agreement between the results of the variable-head tests and those of the constant-head tests, when the former are interpreted for an unsteady state conditions.

(c) The hydraulic conductivity obtained from the constant-head and variable-head tests are the same order of magnitude whenever the first tests are interpreted for steady state conditions and the second tests are interpreted for unsteady state conditions.

(d) When correctly planned and carried out, the constant-head and the variable-head tests allow to obtain reliable results of values of hydraulic conductivity.

- Tests carried out in the unsaturated zone (Corroios)

(a) As regards the present study, the tests were carried out in a medium to coarse sand and under geometric conditions of the borehole and position of the water table which reduce the importance of capillary effects. Thus, and since approximate solutions were used, the error in the values of k_s must have a reduced importance.

(b) The values of k_s determined from tests that stopped before a quasisteady state was reached, are differents from the values of k_s determined in the tests in which there was a stabilization of the water level in the borehole. The error in the value of k_s for the cases in which the quasisteady flow was not reached, is approximately 10 to 35% in CO1; and 4% in CO3 adopting the value of h_{adjust} in the determination of k_s.

(c) The values of k_s obtained from the equations (13), (15) and (10) have shown that the tests may lead to the very differents values if the geometric conditions of execution of the tests are not taken into account.

(d) The borehole infiltration tests performed above the water table in soils where the effects of capillarity are reduced, allowed to obtain reliable results from approximate solutions if these are planned, carried out and interpreted correctly.

REFERENCES

Custódio, E. & Llamas, M.R. 1976. Hidrologia subterránea. Barcelona: Omega

Earth Manual 1974. 2nd ed., United States Bureau of Reclamation. U.S. Government Printing Office, Washington, D.C.

Hvorslev, M.J. 1951. Time lag and soil permeability in groundwater observations. Bulletin N°36, Waterways Experiment Station, Corps of Engineers. Vicksburg, Miss.

Rat, M 1968. Etude théorique des essais Lefranc. Fiche-Programme du Laboratoire Central des Ponts et Chaussées - Departement des Sols. No publish.

Rat, M, Laviron, L. & Rousseau 1971. Mesure de la perméabilité des terrains en place et interprétation de l'essai Lefranc a niveau variable. Fiche-Programme du Laboratoire Central et Laboratoire Régional d'Autun des Ponts et Chaussées - Departement des Sols. No publish.

Rat, M & Laviron, F. 1974. Mesures du coefficient de perméabilité par essais ponctuels. 2nd International Congress of the International Association of Engineering Geology. Vol.6, pp.16.1-16.6. São Paulo.

Roque, A. 1990. Permeability testing in situ in the saturated and unsaturated zones. Thesis for the Degree of Master of Science. New University of Lisboa. (in portuguese)

Stephens, D.B. & Neuman, S.T. 1982a. Vadose zone permeability tests: Summary. Journal of the Hydraulics Division, ASCE. Vol.108, n°HY5, pp.623-639.

Stephens, D.B. & Neuman, S.T. 1982b. Vadose zone permeability tests: Steady state results. Journal of the Hydraulics Division, ASCE. Vol.108, n°HY5, pp.640-659.

Stephens, D.B. & Neuman, S.T. 1982c. Vadose zone permeability tests: Unsteady flow. Journal of the Hydraulics Division, ASCE. Vol.108, n°HY5, pp.660-667.

Stephens, D.B., Lambert, K. & Watson, D. 1987. Regression models for hydraulic conductivity and field test of the borehole permeameter. Water Resources Research. Vol.23, N°12, pp.2207-2214.

Tremblay, M. & Eriksson, L. 1987. Use of piezometers for in situ measurement of permeability. IX ECSMFE, in Dublin - Groundwater effects in geotechnical engineering. Vol.1, pp.99-102. Rotterdam: Balkema.

Soil sampling for geochemical purposes

Echantillonage des sols pour des objectifs géochimiques

J. D. Nieuwenhuis & F.T. J. Poot

Delft Geotechnics, Netherlands

ABSTRACT: In the framework of the storage of radio-active and chemical wastes in The Netherlands geochemically undisturbed samples of the sandy layers bordering the rock salt deposits are needed. For this purpose a ball-valve sampler was developed and built which is capable of confining and sealing the sample during retrieval from a depth of 600 m below surface. The prototype was tested down to 120 m and the sample quality was compared with in situ measurements of electrical conductivity, pH (acidity), ambient pressure and temperature obtained from a pushed-in chemoprobe.

Résumé: Dans le cadre du stockage des déchets nucléaires et chimiques aux Pays Bas il faut avoir des échantillons peu disturbées chimiquement obtenues des couches de sable voisines des diapirs salifères. Pour atteindre ce but un joint à boulet creux tournant dans un cylindre remplis de sol est bâti qui peut prendre et isoler des échantillons quasi-sphériques jusqu'a 600 m de profondeur. Le prototype était déployé à 120 m de profondeur et la qualité des échantillons était comparée au mesurages in situ du conductivité, pH, pression et température fait par une sonde géochimique poussée dans le sous-sol.

1 INTRODUCTION

In the framework of the storage of nuclear waste an elaborate study has been done in The Netherlands regarding storage in rock salt (Wildenborg et al., 1993). The overburden and adjacent rocks of the salt domes consist of varied sedimentary lithologies ranging from quarzites to unlithified sand and clay layers. In view of the potentially high flow and dispersion rates in sand layers coring of these layers to determine their properties was considered relevant. Apart from porosity and permeability the ambient geochemical equilibria determine migration of dissolved and adsorbed or precipitated chemical substances. It is expected that at great depths the equilibria are governed by anearobic, saline conditions which dictate, amongst others, the migration of heavy metals including nuclear isotopes. The least to be prevented in a sample is the escape of gases such as CO_2 and H_2S and the entrance of O_2 during retrieval.

In a general sense it was requested to have a sampler which would retain the original liquid or liquid-gas mixture occurring in the in situ soil at ambient pressure, temperature and pH (acidity).

The Netherlands Ministry of Economic Affairs assigned Delft Geotechnics together with the drilling company H. Haitjema en Zn. BV to conceive, design and build a sampling device capable of retaining in situ geochemical equilibria after retrieval and mountable in a good but traditional counter-flush boring device.

To meet the original requirements not only an air and water tight, very stiff and chemically inert sampler was needed but also a check on the real quality of the sample if compared with in situ conditions. Delft Geotechnics has the disposal of a so-called chemoprobe; a sensor-carrying rod pushed into the soil by hydraulic means (Olie, 1994). The chemoprobe (figure 1) measures electrical conductivity, pH (acidity), redoxpotential, ambient fluid pressure and temperature. In situ measurements of electrical conductivity and pH have been compared with the values obtained from the new sampler. The agreement was satisfactory.

2 PRELIMINARY STUDY OF THE FEASIBILITY OF ALTERNATIVES FOR GEOCHEMICAL SAMPLING

Initially three technical conditions were formulated for taking the development at hand:

- conception of a sampler capable of taking geochemically undisturbed samples from 600 m below ground level

- sampling be combined with a normally drill method for these depths, for example a reversed flush drilling
- testing of a prototype at a depth of at least 100 m. As a preliminary step a study on references in the literature that could well be of interest for the success of the development of a new sampler was made. Also a great deal of research has been done in patents that possibly could give answers to some of the most delicate questions on the subject.

The designwork was focussed subsequently on detailing the aforementioned conditions into specifications:
- the sampling apparatus should be developed for sampling at depths of 50 to 600 m below ground level
- the samples must be geochemically undisturbed. Hence changes in chemical macroparameters such as acidity, redoxpotential, specific electrical conductivity and temperature must be minimal
- the developed sampling system should be based on the wire line principle in order to incorporate the system into a common drill technique
- there are no demands on the geomechanical or on the geohydraulical quality of the samples
- the sampler should penetrate a disturbed zone of 5 cm thickness at the bottom of the borehole before sampling the "undisturbed" subsoil

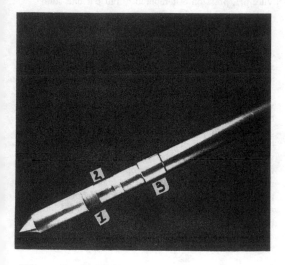

1: in- and outlet filter
2: analyse chamber
3: friction reducer

Figure 1. Chemoprobe

- the sample must be insulated downhole, and the overall pore-water pressure should be unchanged
- a prototype should be capable of sampling at 100 m below ground level.

In figure 2 a table is presented, giving an abstract of the demands and wishes on the new sampler coming forth from the preliminary study.

In figure 3 an estimation of the in situ-stresses is presented.

During brainstorm-sessions 12 possible solutions for the design of the sampler were generated. These ideas can be collected in four sorts of solutions, namely: sampling by cutting, by freezing, vacuum controlled or combined.

Two solutions have been selected as the most successful when compared to the program of demands and wishes. First a sampler with ball-valve and rotating cutting shoe and second a spiral freezing device combined with rotating cutting shoe.

The sampling process comprises two phases, namely the mechanical dissociation of the sample from the surrounding soil, and next locking up the sample or a part of it in such a way that the geochemical structure is kept unchanged.

Phase 1: Borrowing a sample by a rotating cutting shoe does have a few negative effects: the soil structure will be disturbed, resulting in changes in the chemical equilibrium and most certainly mingling with the drill fluid. Non-rotating cutting of the sample only seems possible with a ram. Downhole there is a zone where the soil structure is loosened up by taking away material from above. It is likely that the wider the drill hole is made, the easier the bottom of the hole is penetrated by a ram device. Within the limitation of a hole diameter of maximum 35 cm, because the production of drilling material will be much more when wider holes are used, a field test with a thick-walled tube with cutting shoe driven in the bottom of the drill hole by a ram has given good results in penetration depth. Therefore this method of removing soil material is chosen for the design of the geochemical sampler.

Phase 2: Locking up sampled material can best be done by freezing or with a ball-valve. Freezing is mechanically more complicated than a ball-valve sampler because retaining the in situ pressure would involve the necessity to keep the sample frozen during retrieval. Freezing moreover gives no guarantee that the enclosed gasses will be retained during retrieval because of cracking of the sample by decompression. If a ball-valve sampler works in the specified conditions such a sampler is preferred.

ITEM OF JUDGEMENT		DEMAND/ WISH
Sample macropara- meters	no pressure drop	wish
	no pressure drop > 10% of the in situ-stress	demand
	no loss of sampled material	demand
	no increase of sample volume	wish
	maintenance of the in situ-percentage of moisture	wish
	no exchange of materials with the apparatus itself	demand
Working range	operational at 100 m below ground level	demand
	operational at 600 m below ground level	wish
	up-datable to operations at 600 m below ground level	demand
Sample struc- ture	sampler completely filled with soil	demand
	maximum particle diameter 2 mm	demand
	maximum particle diameter 100 mm	wish
Sampler use	sampling compatible with flush drill technique	demand
	sampler diameter compatible with flush drill technique	demand
	test results comparable with chemoprobe of Delft Geotechnics	wish
	resistant to chemical wear	demand
	usable for more than one sample	demand
	usable for more than one sample by exchanging sampling chamber	wish
Interaction with drill hole	no confinement of drill fluid	wish
	sample area deeper than disturbed zone	wish
Sample dimen- sions	sample length nett 5 cm	demand
	sample length nett 10 cm	wish
Miscellaneous	robust	demand
	simple in use and maintenance	wish
	accurate depth indication	wish

Figure 2. Program of demands and wishes on the geochemical sampler

TYPE OF SAMPLING		TEST SAMPLING		DESIGN SAMPLING	
PARAMETER	UNIT	MAXIMUM	MINIMUM	MAXIMUM	MINIMUM
depth	m below ground level	100	100	600	600
total stress	kN/m^2	2.000	1.875	12.000	10.500
pore pressure	kN/m^2	1.000	1.000	6.000	6.000
mud pressure	kN/m^2	1.050	1.030	6.300	6.180
effective stress	kN/m^2	1.000	875	6.000	4.500

Figure 3. Estimation of in situ-stress before sampling

3 DESCRIPTION OF THE SAMPLING APPARATUS

The developed sampler for geochemical purposes has an outside diameter of 100 mm and a length of circa 1.5 m. The apparatus is used in drill holes with a diameter of 350 mm that are made with a conventional flush drill technique. A ram with a total length of 5 m is used to penetrate the bottom of the drill hole through the cutting shoe of the sampling tube (see figure 4). The 5" drill rod is used for centering.

250 mm made from stainless steel. When placed on the bottom of the drill hole the sampler is initially fully filled with drill fluid. Through penetrating the soil by ramming the sample holder is completely filled with soil, while the drill fluid escapes through the holes in the upper part of the bore. Only the upper part of the column of soil is mixed with drill fluid.

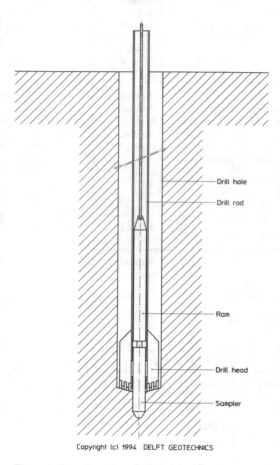

Figure 4. Sampling tube suspended in the drill rod

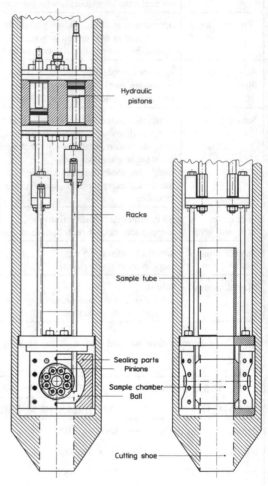

Figure 5. Details of the ball-valve sampler and its driving racks

After reaching the desired penetration depth, a part of the soil that fills the sampling tube is fully locked up. In this particular feature this apparatus differs from other soil samplers.

The main components are the sample holder and the closing mechanism (figure 5).

While penetrating the sample holder forms a cylindrical bore with a diameter of 38 mm and a length of

The sample chamber that partly forms the bore contains after closing the geochemically undisturbed part of the sample. This tubular chamber of 38 mm in diameter and circa 60 mm in length is enclosed in a ball, which is beared between two sealing parts.

404

Closing the sample chamber is done by rotating the ball 90°. The soil in the cylindrical bore is sheared off by cylindrical cutting edges at the upper and lower ends of the ball. Once closed the sample is fully pressure tight locked-up.

Above the closed ball-valve an approximately 10 cm long cylindrical soil sample is left which is partly mixed with drill fluid and can be used for further geological interpretation.

The pressure on the sealing surfaces against the ball is adjustable, depending on the working depth. The sealing part has been made from a special filled soft PTFE that has the capability of storing squeezed-out soil particles without losing its sealing qualities.

The ball is plated with a ceramic finishing, to prevent damage to a minimum.

The chosen combination of counterparting materials stands for a very low and constant friction and is also chemically resistant. The pair of sealing parts is not re-usable.

The rotating of the ball with tubular sample chamber is done by a system of four low friction hydraulic pistons to which racks are connected. The racks turn the the pinions two by two, thus capturing and completely confining geochemically a cylindrical soil sample.

With this compact system the ball is symmetrically loaded, which keeps the loading of the sealing parts minimal.

The hydraulic pressure needed for the displacement of the pinions comes from an accumulator mounted in the sampler.

The great advantage is that no hydraulic hose is needed from above the ground surface. Loss of energy is minimised, the handling of the apparatus is better and the chance of pollution of the environment with hydraulic fluid is minimised.

With a signal through an electric cable from the surface a hydraulic valve is opened through which the pistons are moved with oil pressure. The ball-valve closes in 5 to 10 seconds.

4 PROTOTYPE TEST

A prototype of the sampler for geochemical purposes was built in december 1992 and was successfully used at depths of 90 and 120 m below ground level, at a site managed by a drinking water supply company in the province of Utrecht in the centre of The Netherlands. Between 90 and 120 m of depth fine sand was expected to occur. The sample at 90 m of depth contained moderately coarse sand and the sample at 120 m furnished fine and slightly clayey sand. Sample handling is briefly described in paragraph 5.

The quasi-continuous in situ measurements performed by the chemoprobe (figure 1) were performed at 95 m of depth in between the sampling depths of 90 and 120 m. The chemoprobe records pressure through a piezo- meter and ground water electrical conductivity, pH, redoxpotential and temperature through sensors built in a small water-sampling reservoir. The reservoir is emptied by nitrogen gas pressure (to prevent oxidation) and cleaned with demineralized water.

In the sampling reservoir a second fluidpressure transducer is mounted. This transducer monitors the influx of ground water.

Between measurements while penetrating the nitrogen gas fills the reservoir at slight excess pressure to prevent the influx of ground water.

Two in situ measurements were performed consecutively at the same depth (figure 6).

The results are approximately similar; the temperature difference is probably due to the presence of the instrument.

measure-ment	tem-perature (°C)	pH	redox (mv)	el. conductivity (ms/cm)
1	13.3	7.3	-155	0.9
2	12.4	7.3	-180	0.9

Figure 6. Chemoprobe measurements

5 SAMPLE QUALITY

The pore pressure in the retrieved sample was measured immediately and again before opening the sampler. In both cases the recorded pressure differed from a few percents to 10 percent of the expected value.

The sampler was opened in an anaerobic atmosphere: a hydrogen gas-filled glove bag.

In that atmosphere electrical conductivity, pH (acidity) and Fe^{2+} - ion content were determined.

The sample contained sizable amounts of Fe^{2+} demonstrating that anaerobic conditions were conserved in the sample. No effort was made in this first case to perform the test in a pressurized environment although the sampler design comprises this possibility.

It was considered acceptable to compare the in situ measurements with the measurements in the soil sample. The agreement of electrical conductivity and pH obtained from chemoprobe and sampler was quite acceptable.

6 CONCLUSIONS AND RECOMMENDATIONS

- The prototype of the ball-valve sampler performed according to the specifications of figure 1.
- The electrical cable used to switch on the hydraulic drive of the sampler will be too cumbersome at larger depths.

- Ball-valve sampler and hydraulic drive must be made easily disconnectable in order to take several samples in one boring.
- Pressure compensation in the sampler is needed to cope with temperature changes.
- The chemoprobe with pressure-adapted sensors (the normal working depth is less than 30 m) performed quite acceptably.
- For greater depths all sensors must be in microchip format.
- Penetration of the chemoprobe by ramming is preferred over hydraulic drive for greater depths. In that case the chemoprobe must be strengthened mechanically.

The future of waste storage options in The Netherlands is unclear. Further developments of geochemically undisturbed sampling are therefore uncertain too.

REFERENCES

Kunst D.J.P., Nieuwenhuis J.D., Weststrate F.A. 1993. Development of a coring device for geochemically undisturbed sampling at great depth (in Dutch). Delft Geotechnics report CO-327954/02: 16 p.

Olie J.J. 1994. Application of ISFET-technology in in situ measurements of ground water quality: pH in chemoprobe. Proceedings of the National Symposium on Sensor Technology, Enschede: pp. 25-30.

Wildenborg A.F.B., Geluk M.C., de Groot T.A.M., Remmelts G., Klaver G.T., Obdam A.N.M., Ruizendaal A., Steins P.J.T. 1993. Assessment of storage of radioactive waste in rock salt deposits in The Netherlands. Summary of the results (in Dutch). RGD-Report 30.012/ER: 116 p.